To Staff—

Thanks for your help with sources, and for the many valuable discussions together on various topics.

Best personal Regards.

T Michael Peters

Insects and Human Society

Insects and Human Society

T. Michael Peters
University of Massachusetts

An avi Book
Published by Van Nostrand Reinhold Company
New York

An AVI Book
(AVI is an imprint of Van Nostrand Reinhold Company Inc.)

Library of Congress Catalog Card Number 87-13363

ISBN 0-442-27593-5

Printed in the United States of America

Van Nostrand Reinhold Company Inc.
115 Fifth Avenue
New York, New York 10003

Van Nostrand Reinhold Company Limited
Molly Millars Lane
Wokingham, Berkshire RG11 2PY, England

Van Nostrand Reinhold
480 La Trobe Street
Melbourne, Victoria 3000, Australia

Macmillan of Canada
Division of Canada Publishing Corporation
164 Commander Boulevard
Agincourt, Ontario M1S 3C7, Canada

16 15 14 13 12 11 10 9 8 7 6 5 4 3 2 1

Library of Congress Cataloging-in-Publication Data

Peters, T. Michael.
Insects and human society.
"An AVI book."
Includes bibliographies and index.
1. Insects. 2. Insect pests. 3. Insect pests—Control. I. Title.
QL463.P425 1987 595.7 87-13363
ISBN 0-442-27593-5

Contents

Preface

Entomology is expanding into a new educational role at many colleges and universities—as an elective for students other than majors in the biological sciences. This has enabled a number of departments to develop courses directed at the nonscience major. The general experience is that these entomology courses quickly become popular alternatives to other science electives. There is a reason for this: insects are fascinating animals and we encounter them more frequently than any other kinds of animals in our everyday lives.

As the title suggests, the primary emphasis is on the significance of insects in the lives of humans, both their negative impacts and the positive ones that are so frequently overlooked by other texts. Thus, the nature of our competition with insects for foods and natural materials is explored, as is the extent and severity of insect-borne diseases on human welfare. The importance of insects as pollinators and their use in scientific research are presented as primary examples of the value of insects to humanity.

This book is designed so that students can easily access the material contained within it. I do not believe that a student should be required to ferret through dozens of paragraphs to find out specific important points. Much of the material contained in this book is presented in a systematic way. Insects are compared with other organisms as to their physiological and cellular similarities as well as to the characters unique to insects. Then, the various attributes of insects that contribute to their success as a group are analyzed. This is followed by an evolutionary discussion that puts insects into perspective with the modern vertebrates and with previously successful groups, such as trilobites and dinosaurs. A holistic approach is employed in the chapter on structure and function.

The book also includes chapters on reproduction, on behavior, including communications and coloration, and on interaction with other organisms in the environment, including insect sociality and insectivorous plants. Control of pest insects is approached through the stategy of integrated pest management after the ecological bases of population dynamics have been presented. This is followed by discussions of the various chemical, biological, physical, and cultural tactics currently being employed in IPM.

After a brief treatment of systematics, the diversity of insects is treated through synoptic coverage of each order. In Chapter 11 the major groups of insects and their relatives are presented. Eighteen orders are covered by study sheets, which use illustrations showing mouthparts, wings, type of metamorphosis, general characteristics, life history, and economic impact. These study sheets have been included because most entomology courses require that the students learn a reasonable standard set of characteristics about each major group. It has been my experience that a pictorial presentation of these characteristics is the easiest way to learn them. The major groups of insects also spotlight at least one key economic species, which will help you focus on that group. You will find, however, that you already know most of the important groups of insects. Indeed, some are old friends from childhood that you will remember with pleasure, or in the case of wasps and yellowjackets, pain.

Since many students will make insect collections as a part of their classwork, or because of individual interest, Chapter 12 covers how to collect insects, including equipment and techniques, and a system entomologists use to identify the insects they have collected. The Taxonomic Key to the Orders will allow you to determine the major category in which each of your insects belong, based on the insect's appearance.

Our intention in this book is to focus on the real reason why humans spend so much time and effort studying insects—insects have a tremendous impact on our survival. Some are crucial in the production of food crops, others infest our homes, pets, or even our own bodies. Considerable expense and effort is spent to control insect populations. We therefore hope that the considerable space devoted to explanations of what insects do for us or to us will be of interest and help in understanding these creatures.

The goal of this book is to be a useful and interesting guide to the field of entomology for undergraduates who are not science majors, but who will continue to encounter insects for the rest of their lives, and who will have developed an appreciation of the importance of insects in mankind's survival.

Acknowledgments

A number of people contributed to this work in a wide variety of ways. Many illustrations were provided by colleagues, and some of the most critical topics were the subjects of long discussions. Among those at the University of Massachusetts were Dave Ferro, John Stoffolano, Ron Prokopy, Bill Coli, Joe Elkinton, Ted Sargent, and Don Eaton. I thank Will Tressler for his faith in the project and Claudia O'Connor for her patient attention to detail as she guided the work through the editorial process.

A number of illustrations were created by Kathy Sargent and Nancy Haver from preliminary sketches I provided. It was a pleasure to work with these two fine artists, and I appreciate their ability to execute my ideas, but any factual errors in the artwork is my responsibility alone.

Although in the public domain, I acknowledge the many illustrations I have borrowed from works published by the United States Department of Agriculture and the Center for Disease Control.

The following illustrations are reprinted with permission from the sources listed:

Fig. 3.1 S.M. Muller and A. Campbell. 1954. The relative numbers of living and fossil species of animals. *Syst. Zool.* **3:**168–170. Published by the Society for Systematic Zoology.

Fig. 3.3 R.E. Snodgrass, 1935. "Principles of Insect Morphology." McGraw-Hill Book Co., New York.

Fig. 3.4 *Honeybee:* R.E. Snodgrass. 1975. The anatomy of the honey bee. *In* Dadant and Sons, "The Hive and the Honey Bee." Dadant and Sons, Hamilton IL. *Butterfly, cicada, housefly:* R.E. Snodgrass. 1935. "Principles of Insect Morphology." McGraw-Hill Book Co., New York. *Grasshopper:* R.E. Snodgrass. 1928. "Morphology and Evolution of the Insect Head and Its Appendages," Smithsonian Misc. Col. Vol 81, No. 3. 36A(p. 95), Smithsonian Institution, Washington, D.C., with permission of the Smithsonian Institution Press. *Flea:* Redrawn from Hubbard 1947. *Mosquito:* Redrawn from Ross 1965, after Waldbauer.

Fig. 3.7 *Upper left:* E.H. Forbush and C.H. Fernald. 1896. The Gypsy Moth, Massachusetts State Department of Agriculture. *Upper right:* Michael McManus, U.S. Forest Service, Hamden, CT. *Lower right,* Forest Service, USDA.

Fig. 3.8 Modified from L.A. Dione, Agriculture Canada, Fredericton, N.B.

Fig. 4.34 P.S. Callahan. 1971. "Insects and How They Function." Holiday House, New York, with permission.

Fig. 5.5 Redrawn from W.M. Wheeler. 1910.

Fig. 5.6 P.J. Parrott and B.B. Fulton. 1914. Tree crickets injurious to orchard and garden fruits. N.Y. Agric. Expt. Sta. (Geneva) Bull. No. 388 (also part of Plate 11.I).

Fig. 5.11 W.R. Horsefall. 1941. Biology of the black blister beetle. *Ann. Entomol. Soc. Amer.* **34:**114–126. Published by the Entomological Society of America.

Fig. 5.12 Redrawn from S.D. Beck, 1962. *Biol. Bull.*

Fig. 6.6 Audiospectrographs from R.D. Alexander. 1957. The taxonomy of the field crickets of the eastern United States. *Ann. Entomol. Soc. Amer.* **50:**584–602. Published by the Entomological Society of America.

Fig. 6.7 A.M. Stuart. 1969. Social behavior and communication. *In* "Biology of Termites," Vol. 1, pp. 193–232. K. Krishna and F.M. Weesner, Eds. (q.v.). Academic Press, New York, reproduced with permission.

Fig. 6.10 All photos with permission from Thomas Eisner, Cornell University.

Figs. 6.11B, 6.12, and Fig. 6.13 Redrawn from K. von Frisch. 1967.

Fig. 6.16 Clemson University Extension Service in cooperation with the Federal Extension Service, USDA.

Fig. 6.29 and 6.32 Modified and redrawn from W. Wickler. 1968. "Mimicry in Plants and Animals." McGraw-Hill Book Co., New York.

Figs. 7.5, 7.6, and 7.12 "The Insects of Australia." 1970. Copyright Douglas Frew Waterhouse. Published for CSIRO by Melbourne University Press with permission from P.B. Carne.

Fig. 7.7 USDA.

Fig. 7.8 and Table 7.4 Lawrence R. Goltz, Ed. "ABC and XYZ of Bee Culture," A.I. Root Co., Medina, OH, with permission.

Fig. 7.9 Modified and redrawn from E.O. Wilson. 1971, after Skaife, 1954.

Figs. 7.13 and 7.21 Drawn from photographs by Y. Heslop-Harrison and R.B. Knox. 1971. *Planta.*

Fig. 8.1 Charles H. Southwick. 1976. "Ecology and the Quality of Our Environment," 2nd Edition. Copyright 1976 by PWS Publishers. Used with permission of Willard Grant Press. All rights reserved.

Figs. 8.2, 8.4, 8.5, and 8.6 With permission from the Forest Service, USDA.

Fig. 8.8 With permission from Bertil Kullenberg, University of Uppsala, Uppsala, Sweden.

Figs. 8.11, 8.12, and 8.13 Michael Proctor and Peter Yeo. 1972. "The Pollination of Flowers," Taplinger Publ. Co., Inc. Copyright 1972 by Michael Proctor and Peter Yeo, reproduced with permission.

Fig. 8.17 D. Brncic, P.S. Nair, and M.R. Wheeler. 1971. Cytotaxonomic relationships within the mesophragmata species group of Drosophila. *Stud. Genet* **VI:** 1–16. The University of Texas Publ. No. 7103, with permission.

Fig. 8.18 Modified and redrawn after C.A. Villee. 1977. *Biology.*

Fig. 9.7 Courtesy of John C. Nord, Research Entomologist, Forest Service, USDA.

Figs. 9.6, 9.25, 9.26, and 9.27 Clemson Univ. Ext. Service in Cooperation with Federal Extension Service, USDA.

Figs. 9.9, 9.12A, 9.13A and B, 9.18, 9.19A and B, 9.20A, B, and C, 9.22A, B, and C, 9.23, 9.24, 9.28, 9.29, 9.30, 9.41B, 9.45, 9.46, 9.47, 9.53, 9.54, 9.55, 10.27A and B, 10.29, 10.32, and 11.4 USDA.

Figs. 9.31, 9.37, 9.38, 9.42, 9.43, 9.46A, 9.48 and 9.50 Center for Disease Control, Atlanta, Georgia.

Fig. 9.40 L.L. Pechuman. 1972. The horse flies and deer flies of New York. Search, Agriculture. Vol. 2, No. 5. Cornell Agric. Expt. Station, Ithaca, New York.

Figs. 9.49 and 9.44 Modified from Herm's Medical Entomology 1969.

Fig. 9.51 R.R. Askew. 1971. "Parasitic Insects." American Elsevier, New York, copyright R.R. Askew. Modified from T.W.M. Cameron. 1956. "Parasites and Parasitism." Methuen, New York.

Fig. 9.52 "Craig and Faust's Clinical Parasitology," 8th Edition, 1970, Lea and Febiger, Philadelphia, PA, reproduced with permission.

Fig. 9.56, 9.58 G. Agrios. 1969. "Plant Pathology," 1st Edition. Academic Press, New York, reproduced with permission.

Fig. 10.1 Modified from Engelmann, 1968.

Fig. 10.7 North Carolina Cotton Insect Pest Management. First Annual Report-1972. with permission from Dept. of Agric. Communications. North Carolina State University, Raleigh, NC.

Fig. 10.8 North Carolina Tobacco Pest Management Second Annual Report-1972. With permission from Dept. of Agric. Communications, North Carolina State University, Raleigh, NC.

Fig. 10.9 D.L. Haynes, S.H. Gage, and W. Fulton. 1974. Management of Cereal Leaf Beetle pest ecosystem. *Quaest. Entomol.* **10:**165–176.

Fig. 10.10 Modified from B.A. Croft, 1975.

Fig. 10.13 With permission from Malcolm A. MacLeod for use of his original photo.

Fig. 10.14 With permission from the Ries Memorial Slide Collection, Entomological Society of America.

Fig. 10.20 and 10.24 Courtesy of Jean R. Adams, Research Entomologist, USDA, Beltsville, MD.

Fig. 10.28 W.M. Tingey and E.A. Pillemer. 1976. Leafhopper resistance of dry bean cultivars. *N.Y. Food Life Sci. Quart.* **9**(3),3–5. With permission from N.Y. State College of Agric. and Life Sci., Cornell University.

Fig. 10.34 Courtesy of P. Bhowmik, University of Massachusetts.

Orthoptera plate (Chap. 11) *Images:* M. Hebard. 1934. The Dermaptera and Orthoptera of Illinois. Ill. Nat. History Survey, with permission. *Mouthparts:* R.E. Snodgrass. 1928. "Morphology and evolution of the insect head and its appendages," Smithsonian Misc. Coll., Vol. 81, No. 3 36A(p. 95), Smithsonian Institution, Washington, D.C., with permission of the Smithsonian Institution Press.

Diptera plate (Chapt. 11) *Mosquito mouthparts:* J.F. McAlpine et al. 1981. "Manual of Nearctic Diptera," Vol. I. Monograph No. 27. Agriculture Canada. Also crane fly and black fly. Reproduced by permission of the Minister of Supplies and Services, Canada.

Siphonaptera plate (Chapt. 11) *Flea adult and life history:* C.A. Hubbard. 1947. "Fleas of Western North America." Iowa University Press, Ames, Iowa, with permission.

Coleoptera and Hemiptera plate (Chapt. 11) All images from USDA.

Hymenoptera plate (Chapt. 11) *Sawfly ovipositor:* R.E. Snodgrass. 1935. "Principles of Insect Morphology." McGraw-Hill Book Co., New York. *Honey bee:* R.E. Snodgrass. 1975. The anatomy of the honey bee, *In* "The Hive and the Honey Bee." Dadant and Sons, Hamilton, IL., with permission.

Plecoptera plate (Chapt. 11) *Adult and wings:* J.G. Needham and P.W. Claassen. 1925. "A Monograph of the Plecoptera or Stoneflies of America North of Mexico." Thomas Say Foundation. Entomological Society of America. *Nymph:* P.W. Claassen. 1931. "Plecoptera Nymphs of America." Thomas Say Foundation. Copyright Charles C. Thomas, Publisher, with permission.

Trichoptera plate (Chapt. 11) All images from H.H. Ross. 1944. The Caddis Flies, or Trichoptera, of Illinois. Bull. Illinois Nat. History Survey, with permission.

Odonata plate (Chapt. 11) *Adult:* USDA. *Nymph:* A. Peterson. 1959. "Larvae of Insects," Vol. 1. Copyright A. Peterson. Published with permission from Mrs. Peterson.

Isoptera plate (Chapt. 11) U.S. Nat. Mus. Bull., No. 108, A Revision of the Nearctic Termites, N. Banks and T.E. Snyder. 1920. Smithsonian Institution, Washington, D.C., with permission of the Smithsonian Institution Press.

Neuroptera plate (Chapt. 11) *Antlion and mouthparts:* A. Peterson. 1957. "Larvae of Insects," Part II. With permission from Mrs. Peterson.

Fig. 12.3 and 12.4 E.S. Ross. 1953. "Insects Close Up." University of California Press, with permission.

1

The Significance of Insects in Human Existence

Why study insects? This question asked by the beginning student of entomology is a valid question that has several equally acceptable answers. First, by increasing our knowledge of insects as a significant part of the world around us, we increase the probability of our own continued survival. Second, insects make such ideal model systems for the study of biological concepts that they should be used as subjects for studies in biology. Third, insects should be studied because they exist.

In order to explain adequately the importance of insects to humanity, we must backtrack a little and start by examining the course of human existence. Then, we can superimpose the interactions of insects and humans onto this framework. This finally will lead us to a specific understanding of the insect world as it relates to our own existence.

WHY ANIMAL POPULATIONS FLUCTUATE IN NATURE

If we compare the recent historical development of man, *Homo sapiens*, with that of almost every other kind of animal, great discrepancies immediately become apparent. Animal populations tend to fluctuate over time, but nearly always within certain

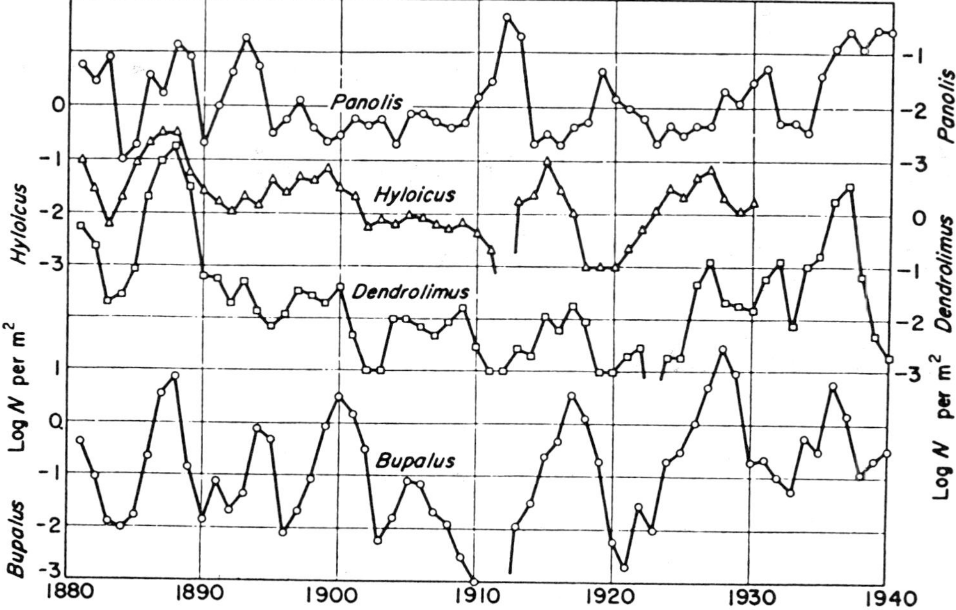

Fig. 1.1 **Successive census counts over 60 generations (one generation per year) of forest moths. Log N of 1 = 10 insects/m², 2 = 100/m². Also note that increasing trends show population crashes after a buildup over several years. Reprinted from G. C. Varley, G. R. Gradwell, and M. P. Hassell. 1974. Insect Population Ecology—An Analytical Approach, Blackwell Scientific Publ., Ltd.**

bounds. Thus, fluctuations in numbers of such insects as the four moth pests of German forests (see Fig. 1.1), while varying from 0.001 caterpillar per square meter to slightly over 10 caterpillars per square meter, did stay within this range for a period of continual observation spanning 60 years. Other animals show the same trends. The oriental migratory locust has been monitored for over 1000 years and fluctuates according to the occurrence of drought conditions. The point is that under natural conditions animal populations fluctuate around some environmental "carrying capacity" (the ability to support a particular weight of organisms per unit of environment).

The factors that are responsible for suppressing populations in nature may also act to cause a population that is increasing to slow its rate of increase until an upper limit carrying capacity is reached and held (Fig. 1.2A). Another kind of population-suppressing mechanism seems to act suddenly at proportionately high population levels, causing a dramatic mortality and a J-shaped population growth pattern (Fig. 1.2B). The factors operating under these growth conditions are said to be "density dependent" (the severity of the factors' action is linked to the density of the animal population). Some factors are "density independent," where severity of action is not linked to density, such as the effects of weather on mortality.

One way of examining the size of a population is to understand that the popula-

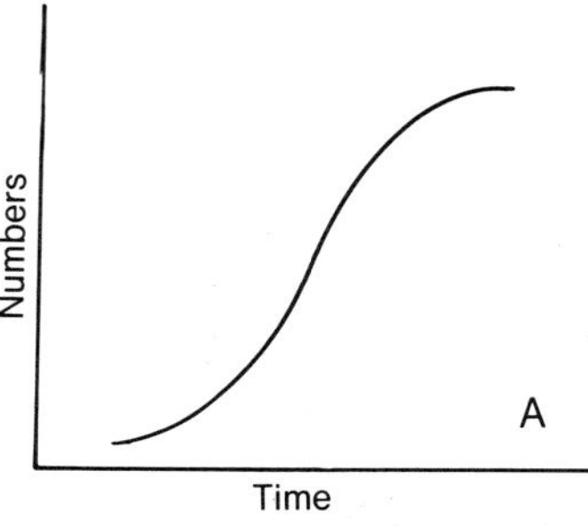

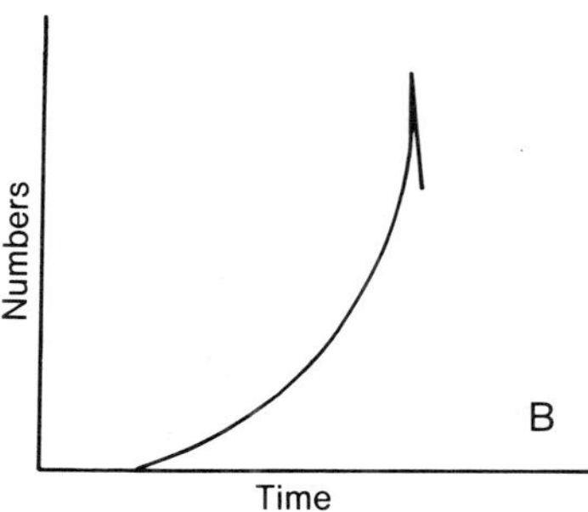

***Fig. 1.2** Basic population growth curves: (A) S-shaped or logistic curve with total numbers approaching and holding at an environment's "carrying capacity." (B) J-shaped curve, indicating a population crash when some factor controls further population growth.*

tion is the result of an interaction between the biotic potential (reproductive capacity) of the species in question and the environmental resistance that the species encounters. The resistance of the environment is most easily visualized by categorizing the environment into units (i.e., a place to live, effects of other organisms, food availability and quality, physical and chemical effects), and then considering the effect each has on reducing population growth. The overall relationship between the animal population and these factors in the environment is represented schematically in Fig. 1.3.

It is easy to see that elimination of a factor that governs density of a population can result in increased population size. For instance, laboratory colonies of organisms increase dramatically when subjected to favorable environmental conditions and adequate supplies of high quality foods, provided that diseases and natural enemies are eliminated (see Fig. 1.4). The ultimate limits to the growth of such a lab colony of animals are (1) how much food and space is supplied, (2) how well diseases are controlled, and (3) how the waste products of the animals (excretory materials and dead bodies) are eliminated.

Figure 1.5 shows the world human population growth curve, which is reminiscent of the initial phases of Fig. 1.2B. Such a growth curve reflects changes in the balance of deaths and births, with more and more humans surviving to the age of reproduction as the year 2000 A.D. is approached. This population growth is largely due to the human ability to ameliorate environmental adversities, thus reducing mortality in the younger segments of the population. Our efforts have been aimed at ensuring an adequate and nutritionally balanced food supply, providing sufficient places to live, eliminating natural enemies and competition, and modifying the physical and chemical components of the environment toward optimal conditions. In short, mankind is essentially approaching population growth curves found in laboratory conditions. Perhaps we should rephrase this by saying that human efforts seem to be aimed at creating optimal laboratory conditions for human population growth on earth. One reason that insects are the subject of scientific study is that the modifications we make to the environment for our own use frequently increase the problems caused by insects.

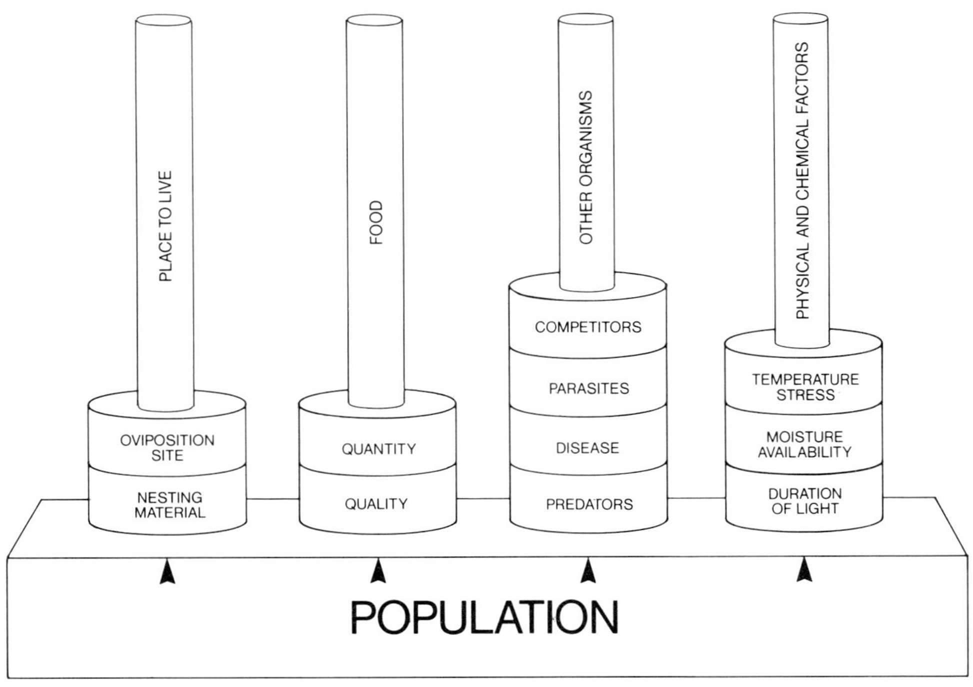

Fig. 1.3 The size of a population is kept in check over time by factors in the environment that curtail expansion by the population. Temporary reduction in the effect of some factors may result in short-term population increases (e.g., summer generations, bountiful food supply). Conversely, sparse populations may result from drought, pesticide application, etc. Some species are governed by population control mechanisms (chemical pollution, territoriality) intrinsic to that species.

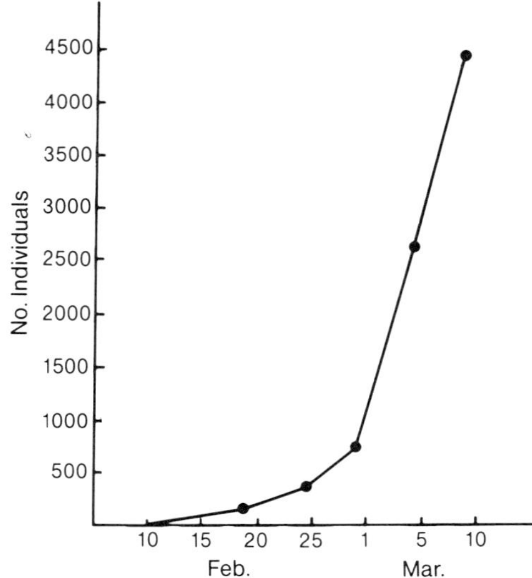

Fig. 1.4 Increase in population of the citrus red mite in the absence of natural enemies, when grown on oranges (a suitable place to live and an adequate food supply) in laboratory conditions. (From Huffaker 1958.)

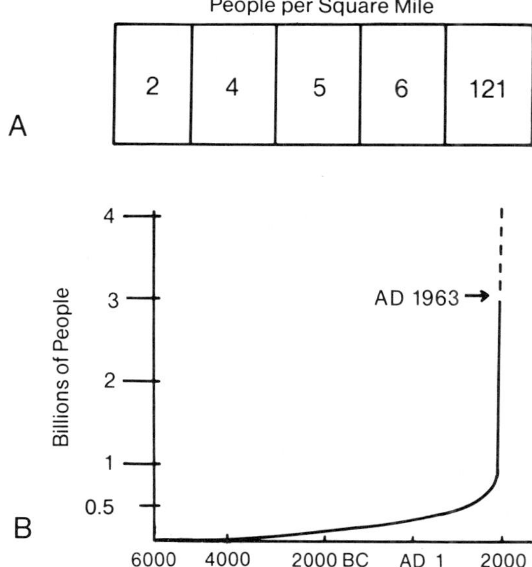

***Fig. 1.5** Growth in the world human population. (A) The increase in density of humans per square mile reflected in the growth curve (B). Modified from UN Science Committee on effects of radiation. United Nations, Suppl. No. 16 (A/5216), 1962.*

HOW HUMAN POPULATION SIZE IS INFLUENCED BY INSECTS

In each of the four groups of environmental factors (Fig. 1.3) that influence human population size, insects can be shown to play a role—often one of great significance. Especially important are the influences that insects have as competitors with humans for environmental resources and the role of the insects in human health and well-being.

Insect–Human Competition

We are in direct competition with insects for foods and for fibers of natural origin. Insect pests damage crops as the crops are growing, after they are harvested or processed, and even the seeds that we reserve to plant the next year. Insects build staggering populations in our stored food supplies in a short time. Insect parasites attack our domestic animals, thereby reducing dairy or meat production significantly, if not killing the livestock outright. Natural fibers of both plant and animal origin such as wool, fur, feathers, hair, hides, felt, and leather are commonly damaged by insects, especially the clothes moth and carpet beetle varieties. Insects are also the major limiting factors in the use of woods. Termites and powderpost beetles, among others, attack not only wood used in construction of dwellings and other structures, but furniture and wooden utensils as well. On the other hand, under conditions when humans and insects are not competing for wooden materials, the insects are valuable agents in speeding up the recycling of organic materials within the ecosystem.

Insects as Natural Enemies of Humans

Generally, natural enemies fall into three basic groups: predators, parasites, and diseases. Human deaths due to insect predation are uncommon, limited to the rare loss of human life to ants in some tropical situations, for example, when unattended infants have been attacked by army ant species. Many more human deaths occur as a result of the stings and bites of some insects, although this cannot be termed predation.

Insects that are parasites of humans are of immense importance. We may harbor permanent populations of lice, including both the body and pubic species. We are also visited by the groups of insects known as intermittent parasites, such as mosquitoes, black flies, tsetses, deer flies, bed bugs, and most flea species. These insects only visit the human body for a blood meal, spending the rest of their lives in other locations.

The third broad category of natural enemy, disease, is where the insects play a major role. The action of insects as transmitters of disease is at least as important as their competition with humans for commodities in influencing human population growth and maintenance. Our control of human disease has been a primary factor in the increase in world population. If we consider that many of the most prevalent and devastating human diseases are transmitted between humans by insects, then we begin to appreciate the insect's importance in our lives. Malaria, yellow fever, filariasis (one form being elephantiasis), typhus in most of its many forms, African sleeping sickness, and the encephalitides are all exclusively transmitted from one human to another by insects. In addition, such diseases as plague and Rocky Mountain spotted fever are in part transmitted by insects or their near relatives.

Increase in Pests Resulting from Human Activities

Humans unwittingly began to create suitable habitats for insect pests when we began to stockpile food reserves, heat our homes, and initiate commerce and agriculture. Early efforts were on a comparatively small scale and therefore had a minor effect on insects. However, in modern society the scope of the environmental changes wrought by mankind is reflected by the magnitude of the pest populations that ensue. When we irrigate dry lands for agriculture, we not only provide food in the new areas for crop pests, but the canals used for carrying water also provide habitat for insect pests with aquatic stages, e.g., mosquitoes. Increased efficiency in intercontinental transportation has allowed hitherto geographically isolated insects to become worldwide pests. Exotic crops also contribute to the increased importance of some pest species. Of greater importance now is the recent trend in agriculture to produce fewer kinds of crops on larger acreages, thereby concentrating the host plant and increasing the devastation by many pests.

INSECTS AS MODEL SYSTEMS IN BIOLOGICAL RESEARCH

In many of its functions, an animal cell performs in fundamentally the same way regardless of the animal. Because of this similarity of cellular function, the scientist studying cells can choose any animal as a source of material that is readily available and inexpensive to raise. Of course, it must still provide the appropriate kinds of responses that enable the scientist to study the particular process under investigation. Since insect reproductive processes, nervous responses, and ecological interactions usually are similar to those of other animals, the insect can be used as the model system for such studies. In genetics, for example, the fruit fly or *Drosophila melanogaster* is often used because

1. The genetic system is fundamentally the same as in other animals.
2. Flies can go from egg, through the immature stages, to adult in as little as 14 days (at 20°C).
3. Experiments can be conducted in pint jar-sized units, each holding a few adults, and ultimately, a few hundred offspring.
4. Polytenic, or giant chromosomes can be dissected easily from the salivary glands and examined without very sophisticated microscopes.
5. The different drosophila genetic types exhibit some rather striking external differences that can be discerned by a beginning scientist.

INSECTS AS OBJECTS OF OUR CURIOSITY

One of the attributes that has contributed to the continued survival of the human species is our intellectual curiosity. We have an insatiable desire to know more about ourselves and our environment. Among the obvious elements in our surroundings are the insects. Since some insects are truly objects of beauty or of bizarre shapes and life styles, various people have become interested in insects and have devoted themselves to investigations of insects. Similar kinds of interest are reflected in our knowledge of many other groups of animals and plants.

SUMMARY

Both density-dependent (e.g., food) and density-independent (e.g., weather) factors act to suppress animal populations below the carrying capacity of the environment. Thus, populations fluctuate over time as food, natural enemies, places to live, and climatic conditions vary.

Human population growth is a result of our increasingly sophisticated methods of producing food, combating disease, and changing the environment to our needs. Insects are our most significant competitors for the natural products we use in feeding, clothing, and providing shelter for ourselves. In addition, most of the human diseases of historical significance are transmitted from human to human by our insect parasites.

Human commerce and modern agricultural practices increase the spread and severity of insect pests.

Because insects are similar in so many ways to other animals, they are used as model systems in many areas of biological research. In addition, they are easily and inexpensively reared.

A final reason why insects have been studied by so many scientists is their incredible variation in form, color, and living habits.

SUGGESTED READINGS

Cloudsley-Thompson, J. L. 1976. "Insects and History." Weidenfeld and Nicolson, London.

Huffaker, C. B. 1958. Experimental studies on predation: Dispersion factors and predator-prey oscillations. *Hilgardia* **27**, 343–383.

Jones, D. P. 1973. Agricultural entomology. *In* "History of Entomology," R. F. Smith, T. E. Mittler, and C. N. Smith (eds.) pp. 307–323. Annual Reviews, Palo Alto, California.

Linsenmaier, W. 1972. "Insects of the World." McGraw-Hill, New York.

2

Insects as Animals

Insects share a number of characteristics with other living organisms on earth. It seems easy to decide whether something like a horse is alive or dead; however, at the microscopic and submicroscopic levels it becomes increasingly difficult to decide what is living or nonliving.

LIVING VERSUS NONLIVING OBJECTS

One characteristic of living beings is their ability to manufacture exceedingly complex chemical molecules, such as proteins. We know of no other ways in which proteins and other *macromolecules* (very large and complex molecules) are constructed except by living organisms. In our most sophisticated laboratories we are only now beginning to be able to construct materials of that complexity. Macromolecules are not only produced by living organisms, they make up the organism itself. Objects such as viruses are, in essence, macromolecules. Since these organisms that are conglomerates of macromolecules do not live forever, they must duplicate themselves in order to insure continuation of their own kind. This "self-replication by macromolecules" is an acceptable definition of "life."

THE TWO MAJOR KINDS OF ORGANISMS

An organism's complex chemical structure requires great amounts of energy to maintain itself. The organism must also replace damaged parts (heal), grow, and develop offspring at various times during its existence. All of these operations require an expenditure of energy, which must be acquired from an external source.

The ultimate source of energy for organisms is radiant energy from the sun. However, the primary distinction between most plants and animals is whether they have the ability to capture the sun's energy directly or not. Green plants have this capability and are called "autotrophs" (organisms that manufacture their own organic chemical foods) whereas animals (and some plants, like mushrooms) are "heterotrophs" (organisms that acquire organic chemicals by eating organisms). Thus, the green plants with their ability to transform sunlight into high energy chemicals (like carbohydrates) by means of "photosynthesis" are the producers in the environment. Such plants produce more than enough high energy materials to continue their own life functions and perpetuate their species. The excess production by plants is used by all consumers as the source of high energy chemicals that enable them to maintain their own life functions. Thus, insects are living heterotrophic organisms, animals that derive the energy necessary for their life operations by consuming high-energy chemicals as foods.

INSECTS AS MULTICELLULAR ANIMALS

The fact that cells are the basic building blocks for all larger organisms was discovered in the 1830s. That they are the basic units of life and are the smallest components of an organism performing all of the functions of life is called the "cell theory." In insects, as in most other multicellular animals, cells are grouped into tissues that have specific functions. For example, muscle tissue consists of cells that are specialized for contractions, nervous tissue is specialized for impulse conduction, and so on. These varieties of cells each have special characteristics of their own, but they typically share some characteristics, too.

A cell from an insect is presented in Fig. 2.1. Note that this cell is "generalized," that is, it is a composite of cellular structures; rarely would an actual cell contain all of these recognizable structures at any one time. In considering the parts of an insect cell, bear in mind that the cells of other animals contain the same basic parts. This is one way of demonstrating that insects are animals.

CELLULAR STRUCTURES IN INSECTS

The *plasma membrane* surrounds the cell and is its outer boundary. In very active locations the membrane may exhibit fingerlike projections called *microvillae* (C in Fig. 2.1) or infoldings (D in Fig. 2.1). Materials external to the membrane may be

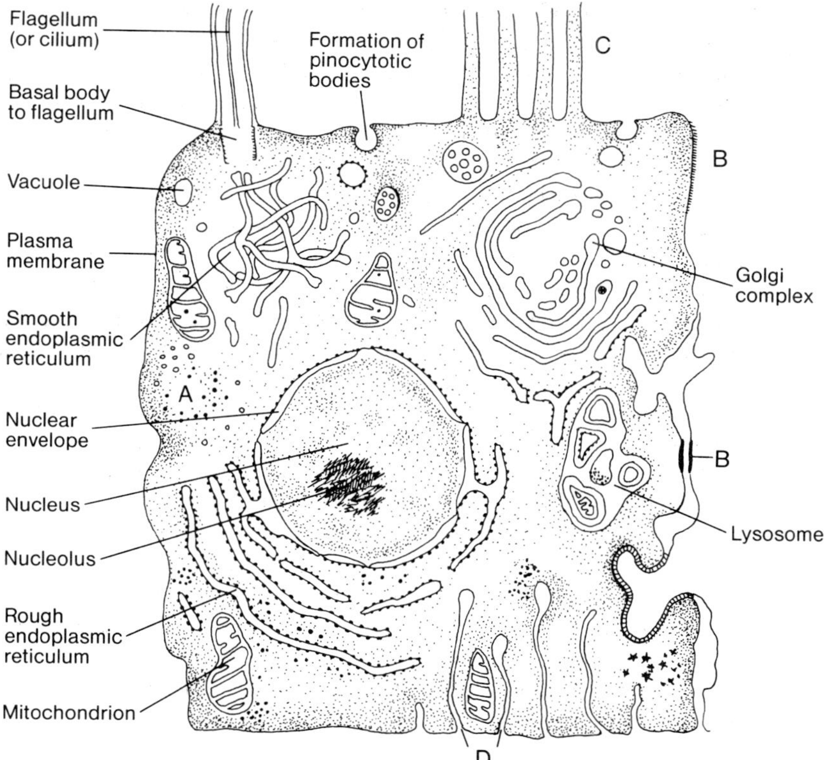

***Fig. 2.1** A generalized insect cell. Since insects are animals, the cell could also be titled "a generalized animal cell," because the components are the same: desmosomes (B), cytoplasmic filaments (A), and membrane convolutions (C and D). The flagellum, though present in many types of cells in other animals, is only present in the sperm cells of insects.*

taken into the cell's interior by "pinocytosis," a pinching off of a membrane-lined inpocketing, with a consequent ingulfing of the adjacent material. This is one way that membrane-lined *vacuoles* (vesicles or pinocytic bodies) are formed. These vacuoles may also store materials or release manufactured materials from the cell by a reverse pinocytosis. Mitochondria occur in varying numbers in cells. The number varies according to the energy requirements of the cell's functions, since mitochondria are the sites at which energy is extracted from foodstuffs, thus becoming available for energy-demanding activities (e.g., muscle contraction).

The *nucleus* is the control center. It contains the information-bearing chromosomes, which may be distinct only during division. One or more *nucleoli* may also be detected when the cell is actively producing proteins, since the nucleolus plays an important role in protein synthesis. In insects and the other advanced organisms, the nucleus is separated from the other cellular components by a variably porous, but double-layered nuclear envelope or membrane. This membrane selectively allows some materials to cross, but halts others. The membrane may be continuous with the complex intracellular membrane network called the *endoplasmic reticulum*. This forms a system of variously shaped membrane-enclosed spaces. If the outer surface of the membrane is lined by ribosomes it is called rough endoplasmic reticulum. If it has no ribosomes attached, it is termed smooth endoplasmic reticulum.

Ribosomes are important in building proteins; thus rough endoplasmic reticulum is the site where many cellular materials of a proteinacious nature are assembled. Smooth endoplasmic reticulum is, at least in part, where some nonproteinacious cellular materials are constructed. The *Golgi complex* is a compound membranous structure. Secretions are packaged and released by the complex and move to different parts of the cell or are secreted through the plasma membrane to the spaces between the cells. *Lysosomes* are membrane-lined vesicles that store digestive enzymes, thus preventing the enzymes from acting on the cell itself.

Flagella or *cilia* are hairlike structures that can be moved by the cell. Although these structures are common in many animals, they are limited to the spermatozoa in insects. A *basal body* is associated with the flagellum and is similar in structure to the *centrioles*—intracellular structures involved in cell division. *Cytoplasmic filaments* of various types are found in cells. Those labeled A in Fig. 2.1 are of a highly ordered type found in muscle cells and are the essential parts of the contraction mechanism of the cell. Similar, but less ordered filaments are present in other types of cells. *Desmosomes* (B in Fig. 2.1) of various types are located in the plasma membrane and may act as points of attachment between cells, or as connecting pores for transfer of materials between cells.

Few major differences exist between animal cells and plant cells. The primary animal cell structures just discussed are also present in plants, but three additional components are also typical of plant cells. These are the *plastids*, *central vacuole*, and *cell wall*. Plastids of the green chloroplast type are the cell structures in which trapping of the sun's radiant energy occurs, whereas clear plastids are storage sites for starch, oils, or proteins. The large central vacuole typical of plant cells is filled with cell sap, basically water along with some dissolved materials. The cell wall is external to the plasm membrane and gives much of the rigidity to the plant because the cell wall is mostly cellulose and pectin.

PHYLUM ARTHROPODA

We have already seen that insects are living, multicellular animals, with the same basic cellular components as other animals. However, insects and their close relatives differ from some other types of animals in several ways. Such differences have been accumulated as the insect–crab–spider type of animal group evolved. A further discussion of the mechanism of evolution will be presented in Chapter 3 after the groups have been introduced.

The Phylum Arthropoda is the large group of animals that contains the insects and their relatives. Because they are descended from a common ancestral form, members of this group share a number of fundamental characteristics, some of which are presented in Fig. 2.2. The exoskeleton of Arthropoda consists primarily of a

PHYLUM ARTHROPODA

(derivation: arthro = jointed, poda = foot)

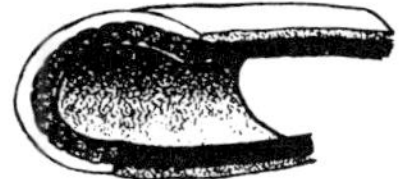

1. EXOSKELETON. Body covered with chitinous (cellulose-like) cuticle, which is usually hardened to form an external skeleton. This creates a problem during growth, because it must be molted (shed) and a larger one regenerated.

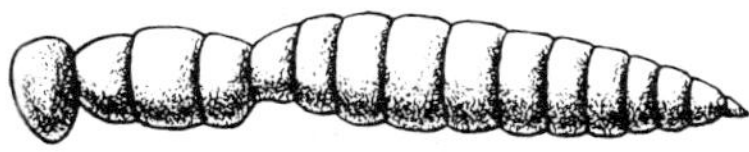

2. METAMERISM. Body composed of numerous segments, although sometimes this condition is concealed. Several segments may be grouped together into a *region,* e.g., the thorax.

3. JOINTED APPENDAGES. Each segment of the body has one pair. These may be lost, or greatly modified for locomotory, feeding, reproductive, or sensory functions.

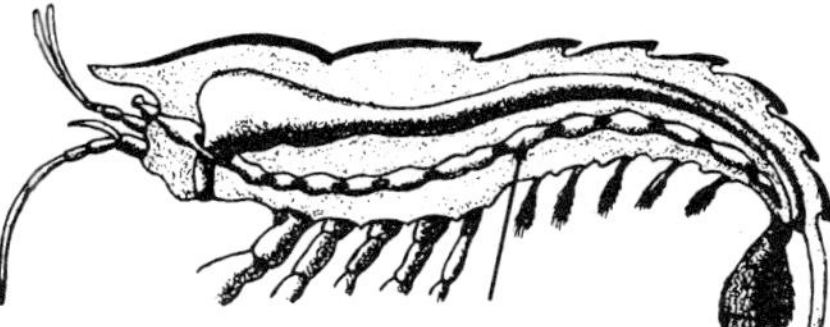

4. DOUBLE VENTRAL NERVE CORD. Primitively with segmentally arranged ganglia (swellings). Only the most anterior portion is above the digestive system.

5. OPEN CIRCULATORY SYSTEM. A dorsal vessel directs the blood forward. Propulsion is by the contractile back part (heart) of the vessel. After leaving the dorsal vessel, the blood percolates back through the body.

6. BILATERAL SYMMETRY. Body can only be divided into two equal sides through one plane (see figure above). Most groups of very active animals are bilaterally symmetrical, e.g., worms, vertebrates, including humans. Compare with radial symmetry of jellyfish (see text).

7. SEXUAL REPRODUCTION. The typical condition, with offspring the result of mating between male and female. Certain groups display other types of reproduction.

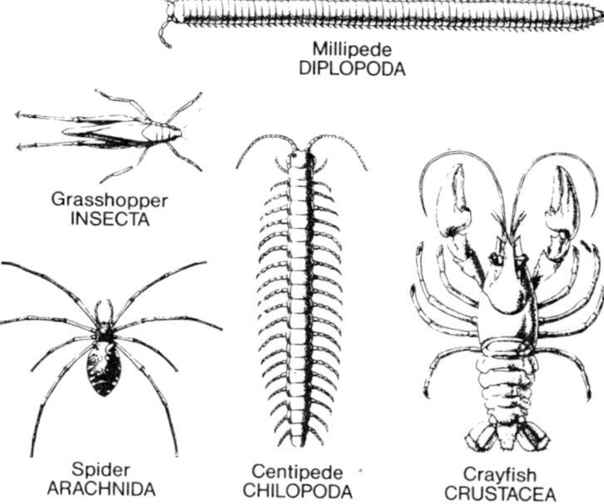

COMMON GROUPS OF ARTHROPODA

Fig. 2.2 Phylum Arthropoda

celluloselike cuticle made up of "chitin," which is tanned (chemically bonded with protein) in insects, or permeated with calcium in millipedes and crustaceans to become hard and either leathery or brittle in some parts, while remaining flexible in others. The differential hardening of the cuticle allows for movement of the body.

The basic arthropod body plan is a bilateral one, consisting of a series of segments, each typically bearing a pair of appendages. The symmetry of some other animals, like jellyfish, is radial, with the body plan having a center and uniformity of structure at all points radiating from that center, like spokes of a wheel. In comparison, all truly active, rapidly moving animals follow the bilateral plan, where there are only two equal halves if the animal were divided through a single plane. In conjunction with bilateral symmetry, an obvious head develops in which many of the senses and the feeding functions are concentrated.

Internally, the arthropod has a central nervous system that is located on the ventral side (underside) of the body, just inside the body wall and below the digestive system. This ventral nervous system consists of a double cord with segmentally arranged masses called ganglia. In comparison, the vertebrate nervous system is dorsal to (on top of) the digestive system and without obvious segmentation. The circulatory system is open, that is, the arthropod's blood is not always contained within blood vessels. There is, however, a dorsal vessel which directs the blood anteriorly (toward the head).

As with most larger animals, Arthropoda are typically sexual, requiring that sex cells from a male and female parent combine to produce an offspring. Several permutations of this method of reproduction exist, however, so that reproduction without mating occurs in some groups. Such "parthenogenesis" may be obligate (the only method employed) or facultative (if no mate is located), depending on the group. Only rarely might a species be found in which both sexes occur in the same individual.

The Phylum Arthropoda is most easily studied if each of the three large subdivisions are discussed individually. Each of the subdivisions are called subphyla (singular, subphylum) and are based on body plan and arrangement of the appendages. The most primitive subphylum is the Trilobitomorpha in which are found only extinct marine arthropods. The Subphylum Chelicerata contains the spiders, mites, ticks, scorpions, and their close relatives. The third subphylum, the Mandibulata, contains not only the insects, but the crabs, lobsters, millipedes, centipedes, and some other related forms. Some minor groups of the two subphyla are given in Table 2.1.

Subphylum Trilobitomorpha

The only class of Arthropoda within the Trilobitomorpha is the Class Trilobita. All of these forms, collectively called trilobites, are now extinct, but at one time they were the most abundant and ecologically significant group of animals on earth (cf. Chapter 3). The trilobite derives its name from the three part appearance of the body. These three parts run lengthwise as can be seen in Fig. 2.3. A single pair of antennae and a pair of compound eyes were also typical of trilobites. These early arthropods had not yet developed specializations of the paired appendages that occurred on each body segment. Therefore, the appendages that were involved in feeding were no different

Table 2.1 The Minor Classes of Arthropoda

Group	Common Name[a]	Environment	Size
Subphylum Chelicerata			
Class Merostomata	Horseshoe crabs (5) Water scorpions	Aquatic (Extinct)	to 60 mm 3 m
Class Pycnogonida	Sea spiders (500)	Aquatic (Marine)	to 5 cm
Subphylum Mandibulata			
Class Pauropoda	No common name (380)	Soil and leaf mold	0.5–2 mm
Class Symphyla	No common name (120)	Soil and leaf mold	2–10 mm

[a]The number of known species is indicated in parentheses

than those elsewhere on the body. Evidently, the mesally (toward the midline) projecting basal lobes of the opposing appendages near the mouth opening were involved in manipulating food and pushing it into the mouth opening. See Fig. 2.4 for a comparison of the feeding appendages of trilobites with those of the other two subphyla. The trilobite appendage also had a branch that appears to have functioned as a gill, exchanging oxygen between the blood and the water. Trilobites ranged in

CLASS TRILOBITA

(derivation: tri = three, lobita = lobes
Common Name: Trilobites
Aproximately 3900 fossil species
have been described

1. No living representatives, known only from fossils.

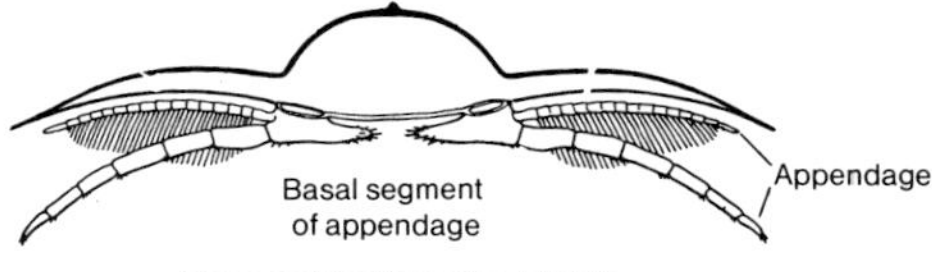

Cross section through a trilobite

2. A pair of appendages occur on all body segments: Although there is some differentiation from front to back, the legs are all practically alike. No specialized appendages occur as are found in more advanced arthropod groups. The appendages are biramous (two-branched).

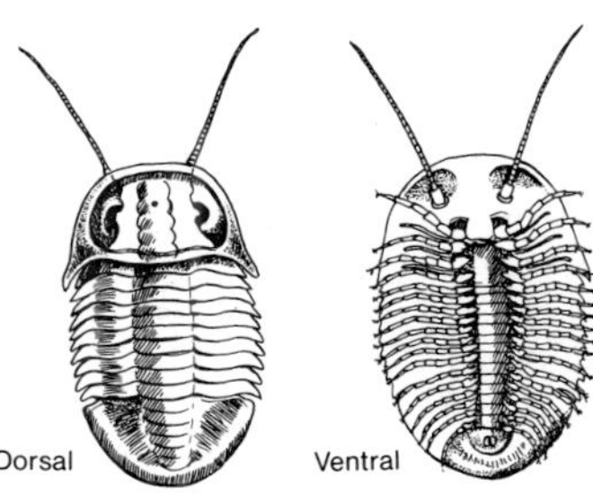

3. Restricted to the marine environment. The feathery leg branches probably functioned as gills.

4. BIOLOGY. Trilobites are the most primitive of the arthropod groups. Although they are now extinct, they were exceedingly numerous in past ages. Their hard, segmented exoskeleton provided protection with flexibility. This was coupled in some groups with an ability to roll up (when disturbed?). They possessed compound eyes and most evidently lived as bottom dwellers, feeding on organic debris in the bottom mud or sand.

Fig. 2.3 Class Trilobita

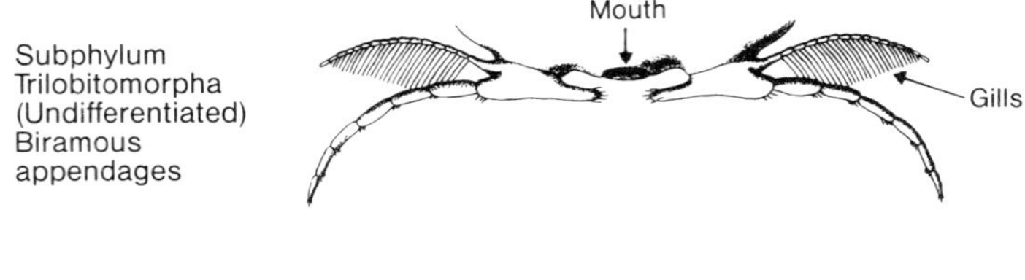

Fig. 2.4 The basic types of feeding appendages among the Phylum Arthropoda. The mandibles and chelicerae did not evolve from the same segment on the ancestral arthropod. The chelicerae each have a movable finger so that they operate as grasping and tearing structures. The mandibles may be highly modified for piercing or cutting in some mandibulate arthropods.

size from a few millimeters to over half a meter. Most were probably feeders on the rich organic bottom muds of the oceans.

Subphylum Mandibulata

The mandibulate arthropods differ from the Trilobitomorpha in possessing paired segmental appendages that have undergone considerable specialization, especially those appendages associated with feeding such as the mandibles (jaws) and the adjacent accessory feeding structures (maxillae) located just behind the mandibles. In addition, the mandibulata typically have antennae and compound eyes. Many familiar animals are encompassed within the subcategories contained by the Subphylum Mandibulata. The most important will be covered by brief discussions and study sheets.

Class Insecta (Fig. 2.5)

Insects occur in many shapes, sizes, and colors, but they commonly share a number of characteristics as adults. Those that are considered diagnostic (used to distinguish insects from all other arthropods) are (1) three body regions—head, thorax, and abdomen, (2) three pairs of legs, one located on each of the three segments that make up the thorax, and (3) the presence of wings on most adult insects. The first two characteristics are almost universally true for adult insects, while the presence of wings is the rule but not without numerous specific exceptions, e.g., fleas, lice,

CLASS INSECTA

(derivation: in = into, sect = cut)
Common Name: Insects
Approximately 750,000 have been described

1. The segments of the body are grouped into three distinct regions: head, thorax, and abdomen.

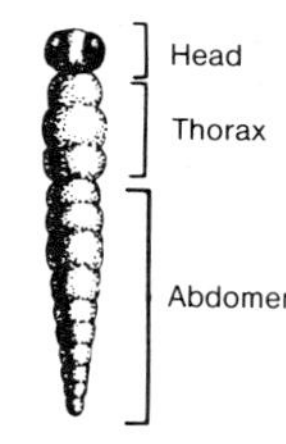

2. Three pairs of legs. One pair is attached to each of the three thoracic segments.

3. Most species have functional wings in the adult stage. A pair of wings is usually attached to the second and third thoracic segments.

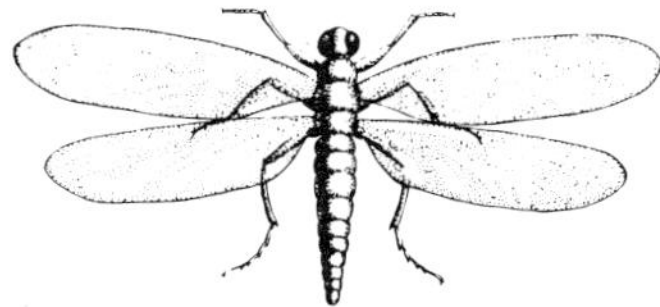

4. One pair of antennae

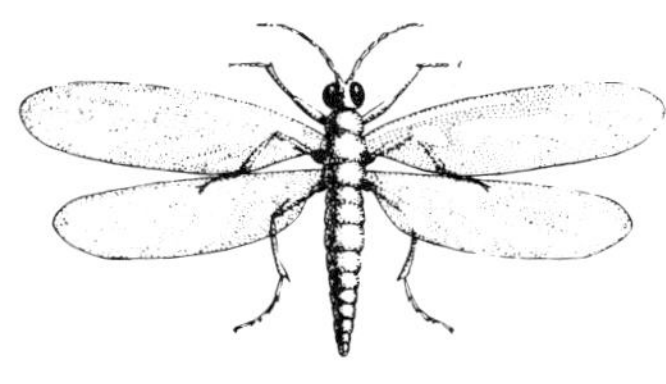

5. Respiratory system consists of a series of interconnected tracheae (tubes) that admit air through pairs of spiracles (openings) on many of the segments.

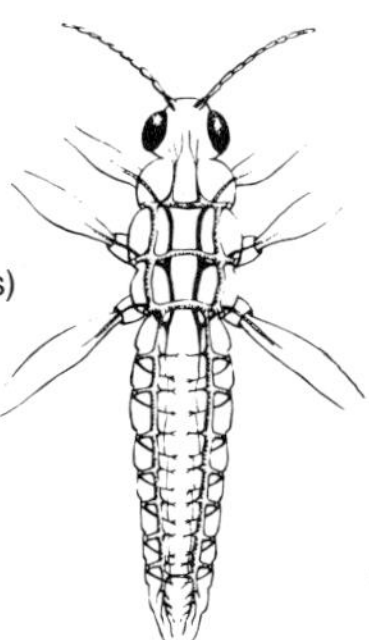

6. BIOLOGY. Insects have successfully invaded terrestrial habitats, but may also be found in fresh water. They are the only invertebrates to have evolved the ability to fly under their own power.

Although most insects reproduce sexually, some are parthenogenetic (reproduce without fertilization). Most lay eggs outside the body. Immatures develop to adult either without change, with moderate changes in form, or with complete changes (egg to larva to pupa to adult). They are typically short-lived, with relatively stereotyped behaviors and small size in comparison to vertebrates. Some have highly evolved social behaviors. Some are of economic importance as pests or disease transmitters. A small percentage are beneficial. Most are neutral, that is, neither harmful nor beneficial.

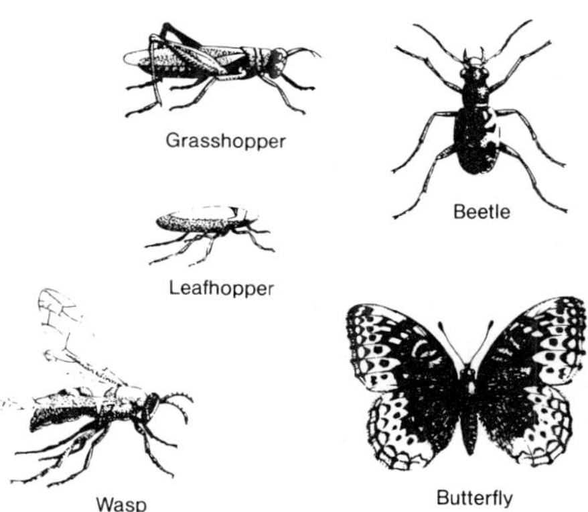

COMMON ADULT INSECTS

Fig. 2.5 Class Insecta

silverfish, some ant forms. The wings, when present, usually consist of two pairs, one pair on the second and another on the third thoracic segment. The wings are only functional for flight in the adult insect. No other invertebrate animal possesses the ability to fly. In fact, flight could rightfully be considered to be diagnostic for insects among all the invertebrates.

Insects possess a single pair of antennae on the head. These antennae function as sense organs for most insects. The insect also possesses a respiratory system of small tubes (tracheae), which branch throughout the body, bringing air from the environment to its site of use. Air gets into the tracheae through openings called "spiracles," often associated with valves, on the sides of the body. The reproductive system of the insect opens at the posterior end of the body. The major groups are presented in Table 2.2.

Table 2.2 The Orders of Insects

Order	Representative common name	Mouthparts (Adult)	Wings	Metamorphosis
Diplura[a]	—	Chewing	None	None
Protura[a]	—	Sucking	None	None
Collembola[a]	Springtail	Chewing	None	None
Thysanura	Silverfish	Chewing	None	None
Ephemeroptera	Mayfly	Reduced, nonfunctional	2 pair; net-veined; nonfolding	Gradual
Odonata	Dragonfly	Chewing	2 pair; net-veined; nonfolding	Gradual
Orthoptera	Cricket	Chewing	2 pair; forewing thickened; hindwing fanlike	Gradual
Dictyoptera	Mantis	Chewing	2 pair; forewing thickened; hindwing fanlike	Gradual
Phasmida	Walking sticks	Chewing	2 pair, when present; forewing narrow, thickened; hindwing fanlike	Gradual
Grylloblattodea	Rock crawler	Chewing	None	Gradual
Dermaptera	Earwig	Chewing	Forewing short & thick; hindwing membranous, folded beneath	Gradual
Embioptera	Webspinner	Chewing	2 pair, when present; weak veins; membranous	Gradual
Isoptera	Termite	Chewing	2 pair; when present; membranous; size, shape, & venation similar	Gradual
Zoraptera	—	Chewing	2 pair, when present; membranous; deciduous, long	Gradual
Plecoptera	Stonefly	Chewing	2 pair, at rest folded flat over body; hindwings wider and pleated	Gradual
Psocoptera	Booklouse	Chewing	2 pair; venation reduced; membranous	Gradual
Mallophaga	Chewing louse	Chewing	None	Reduced gradual
Anoplura	Sucking louse	Piercing–sucking	None	Reduced gradual
Thysanoptera	Thrips	Rasping–sucking	2 pair, when present; narrow; fringed with long hairs	Advanced gradual

(continued)

Table 2.2 *(Continued)*

Order	Representative common name	Mouthparts (Adult)	Wings	Metamorphosis
Hemiptera	Bugs	Piercing–sucking	2 pair, when present; forewing thickened at base or uniformly textured	Gradual
Coleoptera	Beetles	Chewing	Forewing thick, veinless; hindwing membranous	Complete
Hymenoptera	Ant, wasp	Chewing	2 pair, when present; membranous	Complete
Megaloptera	Dobsonfly	Chewing	2 pair, membranous; net-veined	Complete
Neuroptera	Lacewing	Chewing	2 pair; membranous; net-veined	Complete
Raphidiodea	Snakefly	Chewing	2 pair; membranous; numerous veins	Complete
Mecoptera	Scorpionfly	Chewing	2 pair; long; numerous veins; membranous	Complete
Diptera	Mosquito, fly	Variable	1 pair; membranous; second pair reduced	Complete
Siphonaptera	Flea	Piercing–sucking	None	Complete
Trichoptera	Caddisfly	Reduced chewing	2 pair; membranous; densely covered with short hairs	Complete
Lepidoptera	Butterfly, moth	Siphoning	2 pair; membranous; covered with scales	Complete

[a]Considered to be non-insects by some scientists.

Class Crustacea (Fig. 2.6)

This is the only large group of arthropods that is primarily aquatic. Their method of respiration is by means of gills, even in the few forms that have invaded land, thus limiting land crustacea to microhabitats of extremely high relative humidities. Without high relative humidities much of the crustacean's body moisture is lost, and it dries up. Another characteristic of crustacea that is linked to their usual aquatic life is a free-floating planktonic larval stage. It is this larval stage that is responsible for the wide distribution of many crustacean species, because the planktonic larva drifts along in the ocean currents and settles out only after it is large enough to withstand the force of the current. The high density of the crustacean's environment with its consequent bouyancy has freed these arthropods from the limitations on size that acts on terrestrial forms. The result is that many crustaceans reach sizes greater than those of any other arthropod group. It is significant that the largest known fossil arthropods were also aquatic forms (the fossil eurypterid arachnids known as water scorpions).

Many of the larger crustaceans are consumed by humans. Most of the species such as lobster, shrimp, and crab are considered to be delicacies. Other species are the main dietary components of some whale species. Crustaceans are the ecological equivalents of insects in the marine environment.

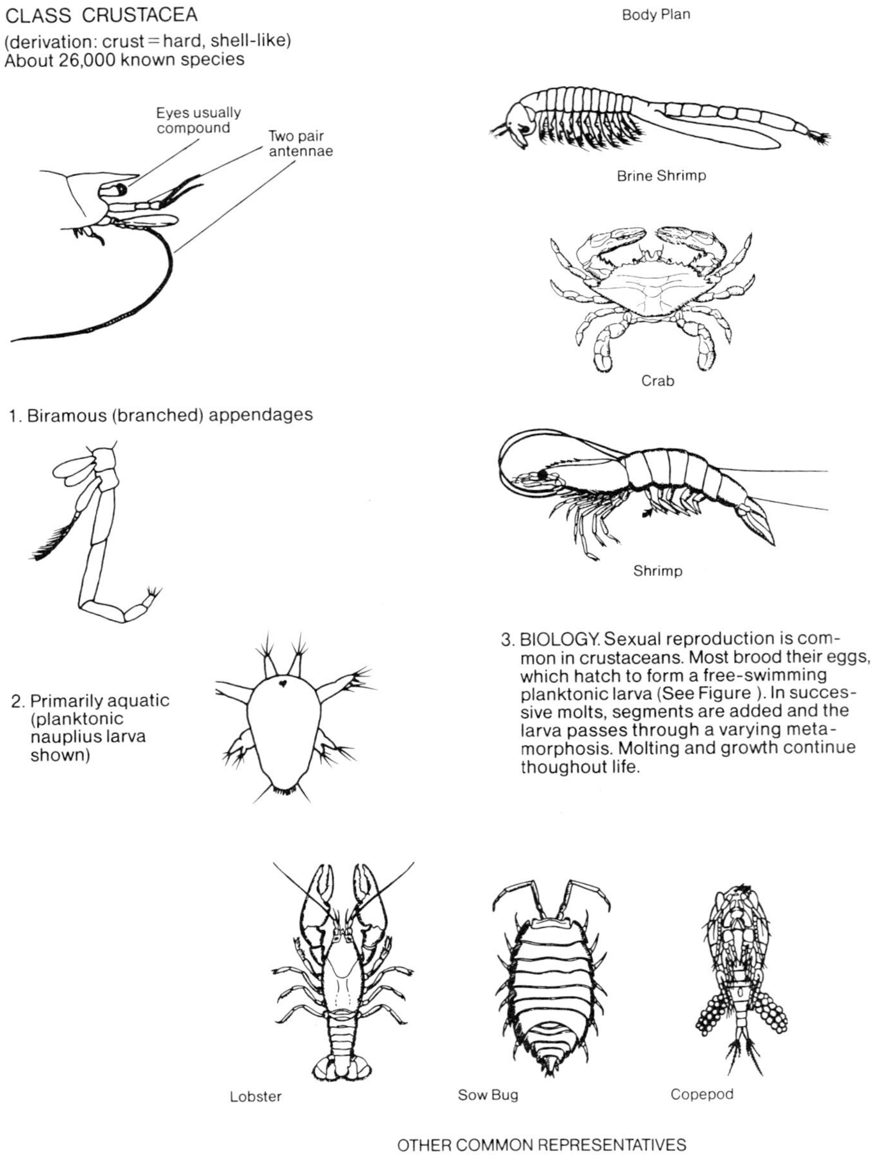

CLASS CRUSTACEA
(derivation: crust = hard, shell-like)
About 26,000 known species

Body Plan

Eyes usually compound

Two pair antennae

Brine Shrimp

Crab

1. Biramous (branched) appendages

Shrimp

2. Primarily aquatic (planktonic nauplius larva shown)

3. BIOLOGY. Sexual reproduction is common in crustaceans. Most brood their eggs, which hatch to form a free-swimming planktonic larva (See Figure). In successive molts, segments are added and the larva passes through a varying metamorphosis. Molting and growth continue thoughout life.

Lobster

Sow Bug

Copepod

OTHER COMMON REPRESENTATIVES

Fig. 2.6 Class Crustacea

Class Diplopoda (Fig. 2.7)

Millipedes are common in most areas of the world, but are seldom noticed because of their secretive habits. They are more common in areas of high humidity, such as leaf litter or in the soil itself. Most feed on decaying organic matter and are slow moving arthropods that may coil up when disturbed. Others can secrete repugnant or poi-

CLASS DIPLOPODA

(derivation: diplo = two, poda = feet)
Common Name: Millipedes
Approximately 8000 known species.

1. Most millipedes have two pairs of legs per obvious body segment. This results from the fusion, during embryonic development, of pairs of true segments. Leg bases of each pair attach closely together along mid-ventral line.

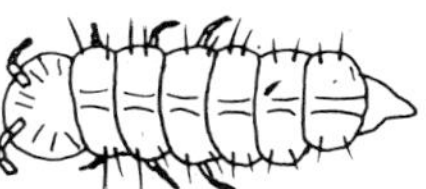

2. Newly hatched millipedes possess only three pairs of legs; additional legs and body segments are added during molts.

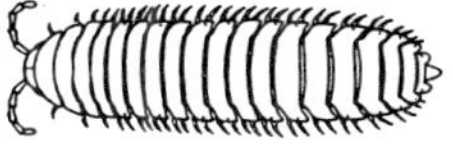

3. Reproductive openings on the anterior part of the body, between the second and third pairs of legs.

4. BIOLOGY. Millipedes usually live in moist conditions with high organic matter, e.g., rotting wood, leaf litter, or soil. They feed primarily on decaying organic materials, although a few attack tender parts of growing plants, and some are predacious. Millipedes are usually slow-moving and may curl up when disturbed. Some tropical forms eject defensive chemicals from the body surface which repel attackers. The many legs enable the slow-moving millipede to exert considerable push-force, allowing it to work its way through the soil.

Fig. 2.7 Class Diplopoda

sonous materials when disturbed. One tropical species has been reported to spray out chemical irritants over a considerable distance when disturbed.

Segments of the millipede's body have fused in pairs (see Fig. 2.7) so that each apparent body segment bears two pairs of legs. Internally there is also a doubling of structures within each apparent segment.

Most millipedes are not of economic significance although some feed on tender foliage and may cause damage in greenhouse crops.

Some scientists group millipedes and the Class Chilopoda into the Class Myriapoda (many legs).

Class Chilopoda (Fig. 2.8)

The predacious centipedes often live in the same soil–leaf litter environment where millipedes are found. However, because of the centipede's typical rapid movements in response to disturbance, they attract more attention. Their rapid movements, coupled with their supposedly "poisonous bite" causes most people to react in fear at the sudden appearance of a centipede. Centipedes in fact do possess a mechanism (see Fig. 2.8) for injecting poisons into their prey and, although the small animals that are eaten by centipedes are killed by chemicals injected through the poison claws, the chemicals do not cause fatalities in humans.

Most centipedes are restricted to humid environments because they lack the

CLASS CHILOPODA

(derivation: chilo = lip, poda = foot)
Common Name: Centipedes
Approximately 3000 known species

1. One pair of legs per segment. Legs attached along sides of body, bases of each pair of legs widely separated.

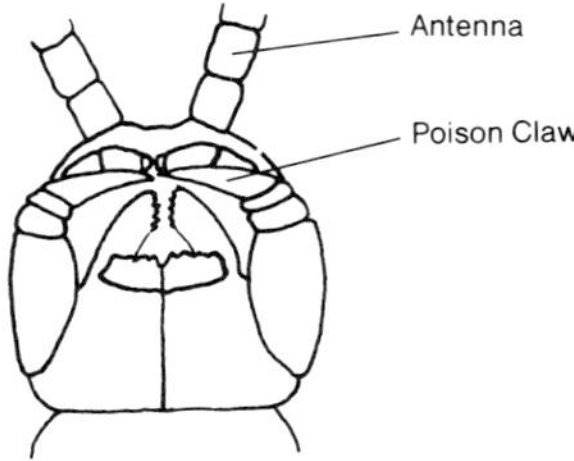

2. Appendages of first body segment modified into poison claws and associated closely with head on ventral surface. Contain poison glands.

3. Predacious. Most feed on small arthropods.

4. BIOLOGY. Live in soil and humus, beneath rocks, logs, and bark. Many require rather high humidity for survival. Some centipedes guard the masses of eggs they deposit until after the young have hatched and dispersed. Others deposit eggs singly. Of the four major groups of centipedes, two have the full complement of legs when they hatch. The other two groups add segments as they molt. Centipede poisons usually are not lethal to humans, with reactions somewhat similar to yellow jacket stings.

Fig. 2.8 Class Chilopoda

waxy coating of the exoskeleton that so effectively prevents water loss in insects. Their cuticle also lacks the sturdiness that the deposition of calcium salts gives to the millipede exoskeleton.

Subphylum Chelicerata

The organisms in this group possess a feeding mechanism that differs fundamentally from the jaws of the mandibulate arthropods just discussed. The feeding appendages are a pair of chelicerae, located just in front of the mouth. Each chelicera has a forked tip and one of the two terminal fingers can be moved against the other fixed finger, thus making an effective grasping structure (Fig. 2.3). The second pair of appendages are known as pedipalps and function as sensory or grasping structures. They may also be involved in reproduction as in the male spiders where they function as copulatory organs.

The chelicerate's body typically is divided into two main regions—a cephalothorax with sensory, feeding, and locomotory functions, and an abdomen. However, variation in body plan ranges from that of a scorpion in which the cephalothorax and anterior abdomen are fused, with a much reduced and taillike postabdomen, to the single body region of ticks and mites, which lacks obvious segmentation.

Included within the Subphylum Chelicerata are the arachnids and two smaller classes. Information on the lesser known chelicerates is presented in Table 2.1.

Class Arachnida

The arachnids are among the largest and most familiar of all arthropod groups. Members of this class include the spiders, ticks, mites, and scorpions (Fig. 2.9). Also included are the less familiar sun spiders, vinegaroons or whip scorpions, false scorpions and harvestmen (daddy longlegs).

CLASS ARACHNIDA

(derivation: arachne = spider)
Approximately 57,000 known species

1. No antennae

2. Most with two (rarely one) body regions. First is cephalothorax, second is abdomen. Rarely wormlike.

Cephalothorax

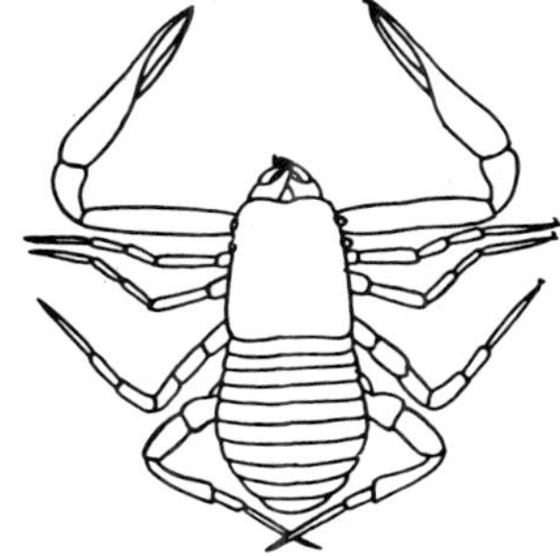

False scorpion

3. Feeding appendages are chelicerae. Most of this group are carnivorous and hold their prey with chelicerae while pouring enzyme-rich salivary juices over the torn victim. Digestion is partly external, followed by ingestion of the resulting juices.

4. Most are terrestrial.

5. BIOLOGY. Primitive scorpion forms may have been the first terrestrial arthropods. Evolution within the group has been along essentially carnivorous lines, with development of poison glands common. Many also have evolved the ability to produce silk. Most have book lungs or tracheae for respiration. Book lungs are leaflike processes in an internal chamber. Blood circulates into the processes and picks up oxygen. Some are of economic importance. The animal parasites are significant as transmitters of some devastating diseases, e.g., Rocky Mountain Spotted Fever, some forms of typhus, and others. The plant-feeding mites have proved to be very difficult to control on crops. Predatory arachnids are common and in some cases exert reductions in population sizes of pest insects.

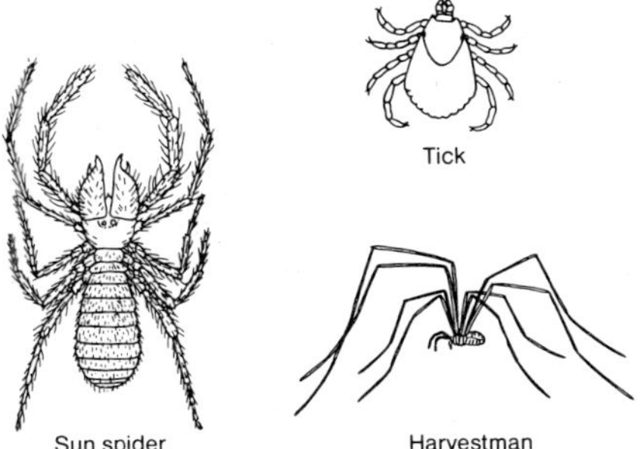

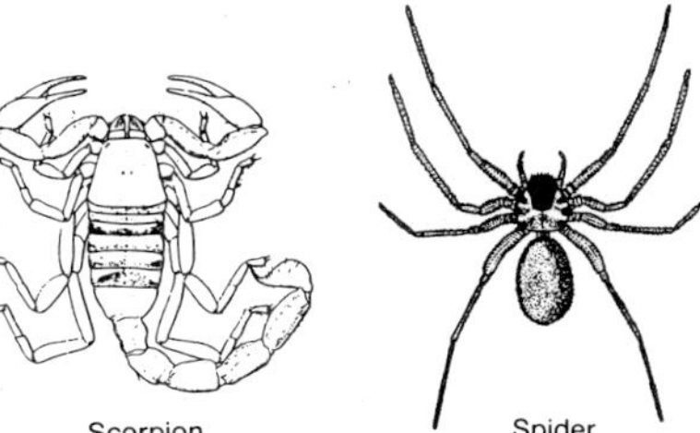

COMMON FORMS

Fig. 2.9 Class Arachnida

With few exceptions, the arachnida is a terrestrial group. It is also generally a carnivorous group with many predatory types (scorpions, spiders) and some that are effective parasites (ticks, some mites) of other animals. One group among the mites has evolved as plant parasites where they feed on cell sap much like the animal-parasitic mites feed on host tissue fluids.

The ability to produce silk evidently has evolved independently in several arachnid groups, i.e., plant-feeding mites, the spiders, and the false scorpions. The uses of silk vary from transport (ballooning by spiderlings and mites) and capture (spider webs) to contruction of nests (false scorpions and spiders). Some species, e.g., the bola spider of South America, capture prey by flinging a strand of silk at the potential victim and ensnaring it.

The typical reaction of most humans to arachnids is one of fear. Such fears may be justified by the occurrence of poison glands in many arachnid groups. While most arachnids will not "bite" and are relatively harmless if they do, several scorpions and spiders can inject venoms that may be fatal. The ticks are also loathed by humans and, although their feeding may not be detected, especially on children, the real danger is in their potential as transmitters of several human and animal diseases, e.g., Texas cattle fever and more recently, Lyme disease.

THE ARTHROPODAN GROUPS—ARE THEY RELATED?

The classification of arthropods just presented is the traditional view, based on the assumption that the arthropods are monophyletic, that is, they evolved from a single ancestral form. This assumption has been accepted by most entomologists because of the many complex structural and functional similarities shared by most of the major groups. Examples are the chitinous cuticle, compound eye, Malpighian tubule, tracheal respiration, and the mandible. However, recent research in the areas of functional morphology and embryology has caused the monophyletic ancestry of arthropods to be questioned.

The polyphyletic theory of arthropod evolution suggests that several groups of organisms independently evolved the "arthropod" characteristics of chitinous cuticle, compound eye, and mandible through convergent evolution (development of similar characteristics by two or more independent lineages due to adaptation to similar ecological status). Thus, each of these lines of evolving organisms became "arthropodized" as they reached similar levels of evolution. The grouping and categories of organisms according to the Tiegs–Manton polyphyletic theory is presented in Fig. 2.10.

The polyphyletic theory is currently being evaluated by many scientists and has been accepted as the current classification in most of the recent texts. However, there is considerable controversy on both the interpretation of the evidence presented and the degree to which convergent evolution could have acted on these organisms. The conservative approach presented throughout this text follows that of Boudreaux (1979). He reasons that it is easier to assume the many complex characters found in

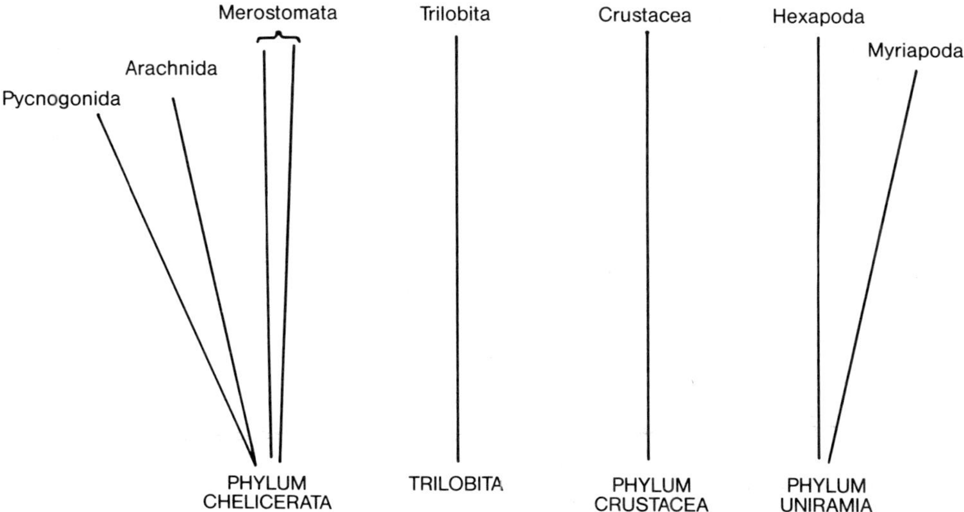

***Fig. 2.10** The arthropods are divided into four groups, each having independently evolved arthropodan characteristics. Three of the four are given phylum rank, but the status of the trilobites is uncertain. Several of the less-common arthropodan groups have been left out to avoid confusion. (Modified from Manton 1977.)*

the Arthropoda are shared by descendents from a single lineage rather than having arisen a number of times in independently evolving lines, that is, through convergence.

SUMMARY

Insects, like other living organisms, have macromolecular systems and reproduce themselves. As heterotrophs, insects must consume their source of chemical energy directly or indirectly from the autotrophs, green plants that capture solar energy through photosynthesis.

Although insects have many specialized cells, the basic cell organelles (nucleus, mitochondria, Golgi complex, endoplasmic reticula, and others) are the same as those in other multicellular animals and function in the same way. Plant cells have these same organelles, as well as a cell wall, plastids, and central vacuole.

The Phylum Arthropoda has a chitinous cuticle and segmented body, usually with one pair of appendages on each body segment. These bilaterally segmented animals have a ganglionated double ventral nerve cord, an open circulatory system, and usually reproduce sexually, but parthenogenesis is common in some groups.

The Trilobitomorpha is a primitive subphylum of marine forms with unspecialized feeding appendages. Trilobites, with their heavily armored, segmented bodies were once the dominant oceanic life forms, but are now extinct.

The Subphylum Mandibulata feeds with opposable mandibles. The modern forms are the most abundant animals in the world. Among the Mandibulata the Class Insecta (Hexapoda) is unique as the only invertebrate group that can actively fly. They are primarily terrestrial, with the typical adult form possessing three pairs of legs, three body regions, one pair of antennae, tracheal respiration, and one or two pairs of thoracic wings. Some variations in development, wings, and mouthparts of the insect Orders are presented.

The Class Crustacea is an important group of Mandibulata, restricted by their gills to aquatic or very moist environments. They have two pairs of antennae and often disperse as larvae in water currents.

The Classes Diplopoda (millipedes) and Chilopoda (centipedes) are myriapods, that is, they have an elongate body with many legs. The carnivorous centipedes have one pair of legs per segment and possess a "poison claw." The millipedes feed on decaying organic matter and have two pairs of legs per apparent body segment.

Members of the Subphylum Chelicerata are without antennae and feed with chelicerae rather than mandibles. The most common class, Arachnida, contains the spiders, ticks, mites, and scorpions. Most Arachnida are carnivorous and produce silk.

New evidence suggests that evolution of the Arthropoda is polyphyletic and that the process of "arthropodization" among the three arthropod phyla is an extreme example of convergent evolution rather than the result of evolution from common ancestry.

SUGGESTED READINGS

Anderson, D. T. 1973. "Embryology and Phylogeny in Annelids and Arthropods." Pergamon Press, Oxford, England.

Barnes, R. D. 1963. "Invertebrate Zoology." W. B. Saunders Co., Philadelphia, Pennsylvania.

Boudreaux, H. B. 1979. "Arthropod Phylogeny with Special Reference to Insects." John Wiley & Sons, New York.

Kristensen, N. P. 1981. Phylogeny of insect orders. *Ann. Rev. Entomol.* **26,** 135–158.

Manton, S. M. 1977. "The Arthropoda—Habits, Functional Morphology, and Evolution." Clarendon Press, Oxford, England.

Savory, T. 1977. "Arachnida," 2nd ed. Academic Press, New York.

Sharov, A. G. 1966. "Basic Arthropodan Stock with Special Reference to Insects." Pergamon Press, Oxford, England.

Wootton, R. J. 1981. Paleozoic insects. *Ann. Rev. Entomol.* **26,** 319–344.

3

Why Insects Are Successful

THE COMPONENTS OF SUCCESS

Insects are successful animals. Yet an animal's success must be measured against the success of other animals, and therefore becomes a matter of comparison among groups. In comparing groups it is easy to fall into the trap of selecting criteria that are biased toward a particular group. One criterion that contains a bias but still seems to be suitable for measuring success is whether or not an animal group is still living or has become extinct. Any animal that is alive today has to be considered successful to some extent, because an estimated 50 to 100 times as many animals have become extinct as those that are still alive.

If the basis of comparison of living animal groups is the relative numbers of species, then insects are the most successful group of animals on earth today. In fact, the preponderance of insects among the species of known animals is overwhelming. As illustrated in Fig. 3.1 almost 75% of all species of animals are insects. Admittedly, the numbers of individuals among the microscopic one-celled organisms and other groups are probably even greater than insects.

There are two obvious difficulties with the preceding statement that insects are the most successful group of animals. First, a rather convincing argument can be made that man is the most successful animal. Second, other animals have rivaled the success of insects, if a different time frame is used. Man is certainly the single most

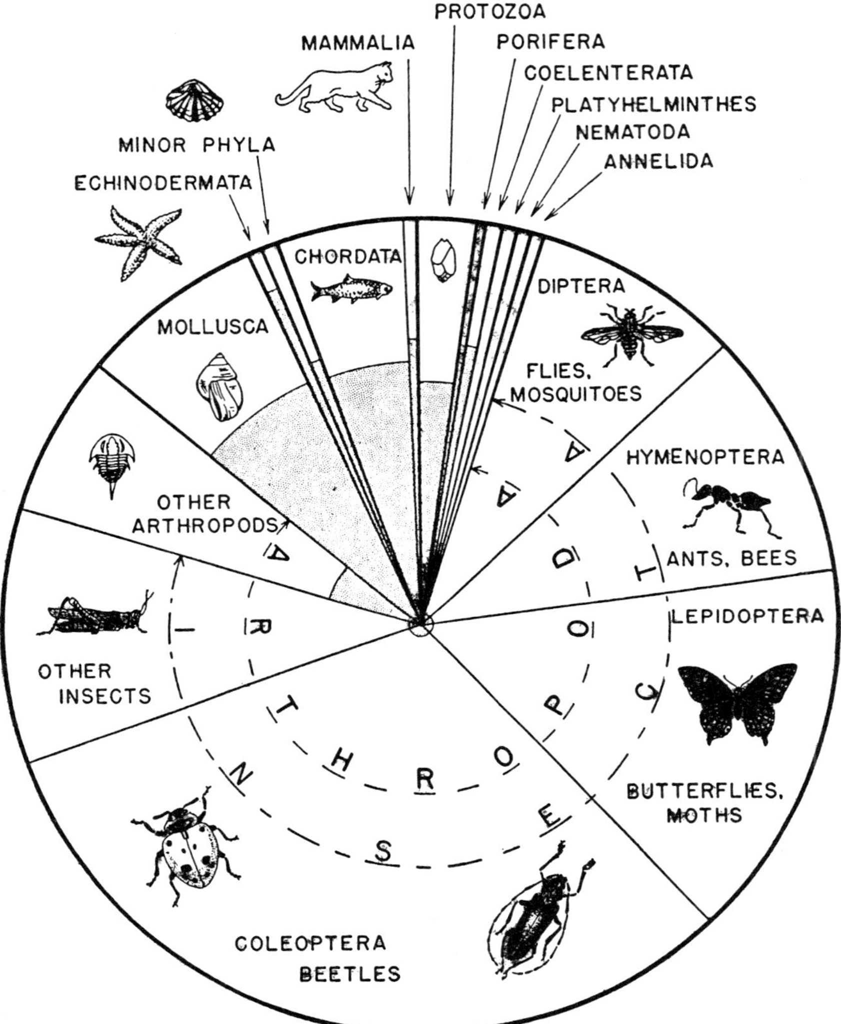

Fig. 3.1 The relative abundance of species among the living animal groups of the world as represented by relative areas of a circle. Note that all of the animal species other than the arthropods are represented within the upper left-hand area.

significant species the world has ever known if we consider the amount of environmental modification any animal species has caused. Also, if we consider the increase in population numbers and spread throughout the world by any single animal species, man would probably be at the top of the list. Even though humans may be atypical as representatives of the mammals, we anthropocentrically consider mammals, including the rodents and hoofed grazers as well as man, to be one of the dominant groups. Although there are not very many species of mammals alive today, mammals exert a controlling influence within most terrestrial and even aquatic areas of the world by virtue of their activities.

If we change the time frame, our argument about the relative success of insects as compared to other animal groups brings several new groups into the picture. Among these are the trilobites and the dinosaurs. The only reason trilobites and dinosaurs were not mentioned before is that they are now extinct, and therefore are not thought of as successful animals. However, both of these groups were also dominant for many millions of years, in fact, a great deal longer than the mammals have been a dominant life form.

It would be instructive to examine why some of the dominant groups of the past were so successful. Trilobites became abundant not long after the brachiopods, in the seas about 600 million years ago. These are the first two groups of animals that are found abundantly as fossils among the layers of rock laid down at that time. One obvious reason why we find so many fossils of these two groups is also why they were so abundant when they lived: they both have hard external skeletons.

Brachiopods may still be found living in the ocean. They are somewhat like clams in appearance with two shells that fit tightly together when they are closed (Fig. 3.2). Brachiopods had a real advantage over their soft-bodied contemporaries because they could close their shells when danger approached. The only flaw in their system of physical protection was that their hard exoskeleton was only protective when it was closed. Trilobites became much more abundant than brachiopods and the trilobite exoskeleton was superior because it was hard, thereby enclosing the animal in armor, yet it allowed maximum movement because of its jointed nature. The undersides of trilobites were relatively soft, and they probably reacted to danger by curling up like an armadillo. However, the trilobite caught unaware still had considerable protection on the entire upper surface of the body. Another aspect of the exoskeleton may have been an even more important factor in the ascendance of the trilobites to dominance—the paired, segmented appendage. Evolution of the paired appendages required a more advanced nervous system for coordination, but provided both precision and power in movement. Dinosaurs and their early reptilian relatives became tremendously abundant and dominated undisputedly for 140 million years. The secret to reptilian success was the kind of egg they produced. Because of the nature of the egg, they were the first large animals that did not have to reproduce in the water.

The reptile egg, typical of the dinosaurs and also of currently living reptiles, was enclosed by a protective capsule with enough nutrient for the developing embryo as well as a supply of water. Between the embryo and the egg shell was a membrane that, along with the shell, allowed the passage of oxygen into the interior of the egg for the newly forming young to use. Although the early dinosaurs probably lived most of the time in the water they crawled out to lay their eggs on dry land, much like turtles do today. By depositing their eggs on land, the dinosaurs escaped all of the predation on eggs and newly hatched young by aquatic predators. Once reptiles were freed from their dependence on water to reproduce, they radiated into many different forms of terrestrial life unimpeded, since the only other terrestrial forms were small organisms like the insects.

It is easy to see that trilobites and dinosaurs reached dominance because of their novel solutions to biological problems. They evolved more efficient ways of doing things and used resources not available to other animals. The trilobites had their segmented, yet armored, exoskeletons and the dinosaurs their freedom from the aquatic environment.

The fossil record shows that dinosaurs disappeared in a geologically brief time, but one that still stretched over millions of years. They were dominant for 140 million years, then "suddenly" disappeared. Of equal importance is the subsequent disappearance of groups such as the large, nautilus-like ammonoid mollusks which evolved into a multitude of forms very quickly once the basic body and life style had evolved (Fig. 3.2). Such explosive evolution is known as "adaptive radiation" or divergence from a single phyletic line into a series of different forms with different niches.

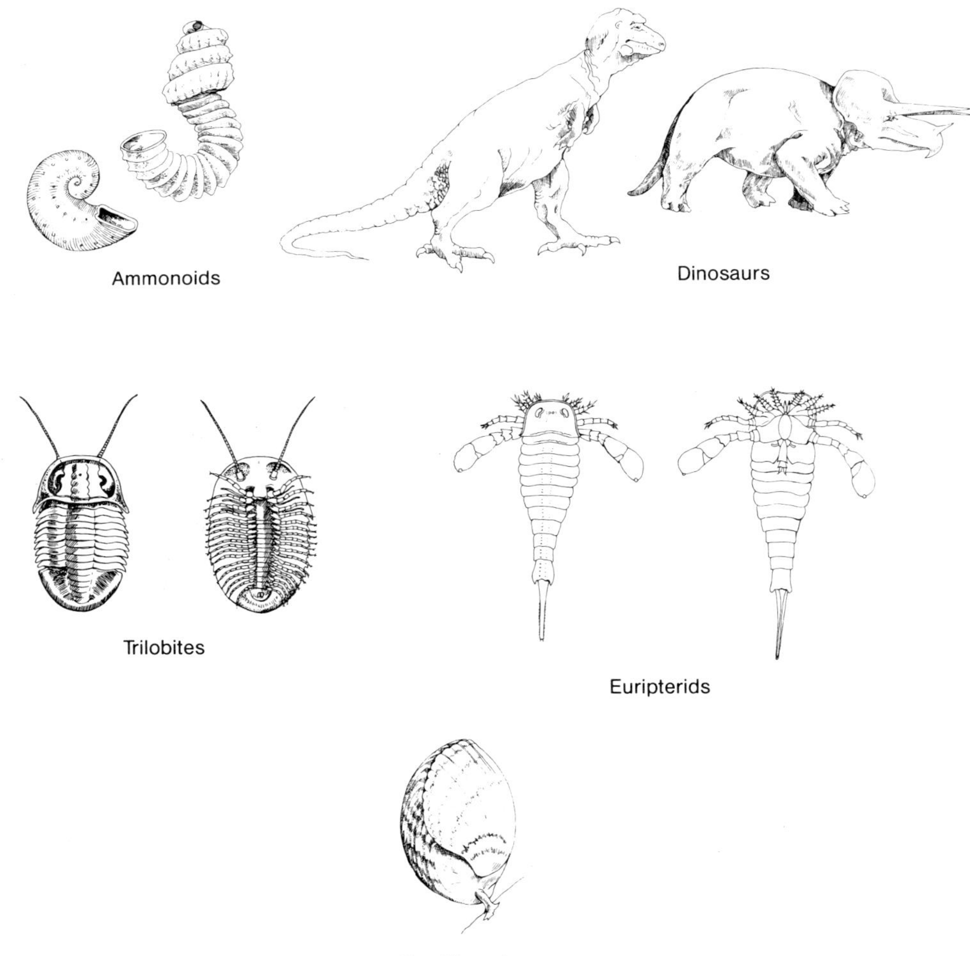

***Fig. 3.2** Some once-dominant animal groups. Only the brachiopods are still represented by a few living forms. It is unclear why Ammonoids and Eurypterids became so very abundant in ancient seas. Redrawn from various sources.*

Why did these life forms disappear? Although we really do not know, their disappearance is thought to be the result of the evolution of even better ways of living by other animals that subsequently outcompeted the trilobites, dinosaurs, and ammonoids in a changing environment.

Instinct Versus Intelligence

On the criteria of dominance and continuing survival, both insects and mammals have to be ranked among the most successful. However, a comparison reveals that each of these dominant groups has been evolving along extreme poles of evolution. The mammals have continued to evolve toward low reproductive rate with high survival through parental care, extended individual life, and training of offspring. In short, the intelligence route. On the other hand, insects have evolved toward stereotyped behavior, short lives, small size, and massive reproduction, which offsets high mortality—an evolutionary path we might call the instinctive route. Neither of these routes is mutually exclusive. Man certainly is instinctive to a degree, and there is considerable evidence of intelligence in insects. Also, neither of these extremes is "best"; both have been effective survival techniques for the respective groups.

INSECT ADAPTATIONS AND THEIR ROLE IN THE SUCCESS OF INSECTS

The success of insects, as a group, is due to a whole series of characteristics that have evolved within this class, rather than to a single key factor. How these characteristics evolved in the insects will be discussed later in the chapter after the characteristics have been presented.

Characteristics

ATTRIBUTES THAT CONTRIBUTE TO THE SUCCESS OF INSECTS

1. Exoskeleton
2. Jointed appendages
3. Wings
4. Size
5. Metamorphosis
6. Ability to escape adverse conditions
7. Methods of reproduction
8. Short generation time
9. Specialization in life styles
10. Methods of solving the water problem

Exoskeleton

Insects, like the trilobites before them, enjoy the benefits of a segmented exoskeleton. Although some immature stages of insects are soft-bodied, most insects have a relatively rigid "suit of armor" composed of overlapping sclerotized (hardened) plates with membranous areas between the plates, which ensure flexibility. The aquatic ancestors of insects did not need much of a support system for their bodies since they were buoyed up by the water. They did, however, evolve this skeletal system, which provides an excellent surface for muscle attachment as well as rigidity, and consequently, protection. The early arthropods were bottom dwellers, probably living much like lobsters. The exoskeleton as the arthropodan solution to life on the ocean bottom was not only efficient for use in their marine life style, but set up the ideal system, with minimal modification, for their subsequent move to land. As is discussed later, the exoskeleton is also supremely suited for inhibiting water loss to the environment.

Jointed Appendages

The ancestral arthropod was probably a wormlike animal with a pair of jointed appendages on each segment of the body. The pattern of evolution in the arthropod's paired appendages is shown in Fig. 3.3. This pattern, along with the evolution of the body regions, is undoubtedly a secret in arthropodan success. Initially, the appendages were all similar to each other and not highly specialized. In other words, here was this large number of general tools with the potential for being modified through evolution into specialized tools for the life processes of the evolving insect. The appendages near the mouth opening of the insect ancestor evolved into feeding tools; those in the middle of the body became more and more responsible for locomotion; and those at the posterior tip of the body, adjacent to the reproductive openings, became specialized in the male for facilitating sperm passage into the female's body or in the female for depositing the completed and fertilized eggs into a suitable and selected location. Several sets of appendages have been lost, notably those on the anterior segments of the abdomen.

Another aspect of the evolution of the jointed appendages is illustrated in Fig. 3.3. Concurrent with the evolution of specialized tools in the arthropod body plan is a regionalization of body segments into recognizable units of head, thorax, and abdomen. This regionalization is known as "tagmosis," with the resulting units identifiable not only structurally, but functionally, too. The head is the sensory center, with the allied responsibilities of coordination and sensory inputs necessary to determine food location, direct the feeding process, and detect conditions external to the body. The thorax has evolved into the locomotory center, bearing the three pairs of legs so characteristic of adult insects. Locomotion is concentrated centrally in the insect body plan in order to balance the body. If locomotion were concentrated either to the front or rear, the appendages would have to be extremely long to lift the entire body off the substrate. Note that the wings that evolved in insects are also thoracic appendages, but not derived from the paired segmental appendages of the arthropod ancestor.

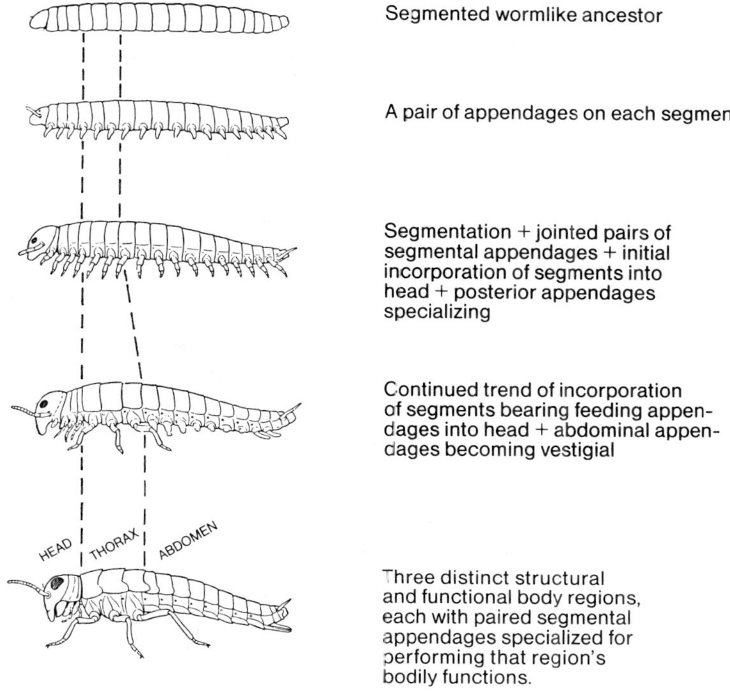

Fig. 3.3 Theoretical steps in the evolution of the modern-day insect body plan from a presumed wormlike ancestor of the arthropods.

Feeding Appendages of the Head. As the appendages adjacent to the mouth opening became more involved with the feeding process, they evolved into the chewing mouthparts, an integral feeding unit that could efficiently detect and tear apart organic material and push it into the mouth cavity. The head of the modern insect has evolved so that it is now composed of the original head bearing antennae and eyes, and a number of body segments, bearing the appendages that have become adapted for feeding, but that were originally appendages just like those that have evolved into the legs and reproductive appendages of the insect.

The basic type of chewing mouthparts typical of grasshoppers, roaches, caterpillars, beetles, and many other insects, is suitable for solid foods but not for exploiting foods in liquid form. From this basic chewing type of mouthparts have evolved a whole spectrum of feeding mechanisms. Some of the numerous special types of feeding mechanisms are illustrated in Fig. 3.4. However, the overall appearance of these different feeding mechanisms is so dissimilar that relationships between them are not easily detected. The relationships are more easily seen if some of the individual parts are shown separately with the intermediate steps indicated where they are known or can be demonstrated with reasonable assurance of accuracy (Fig. 3.5).

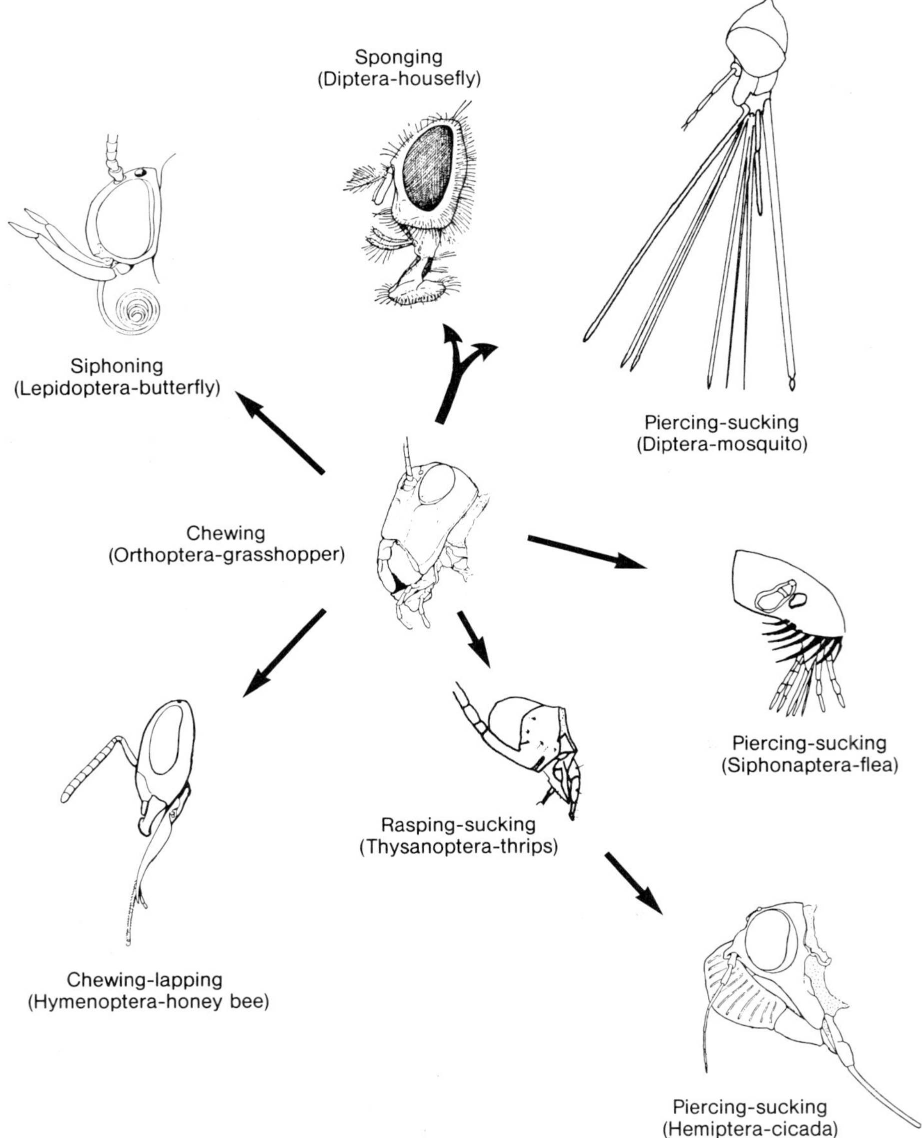

***Fig. 3.4** Each of the mouthpart types in insects evolved from the chewing type of mouthparts like those still possessed by the grasshopper.*

Wings

The majority of insect species have wings (Fig. 3.1). These wings give insects a decided advantage over most other terrestrial animals. The advantage is at least as significant to insect success as was the freedom from the aquatic environment that

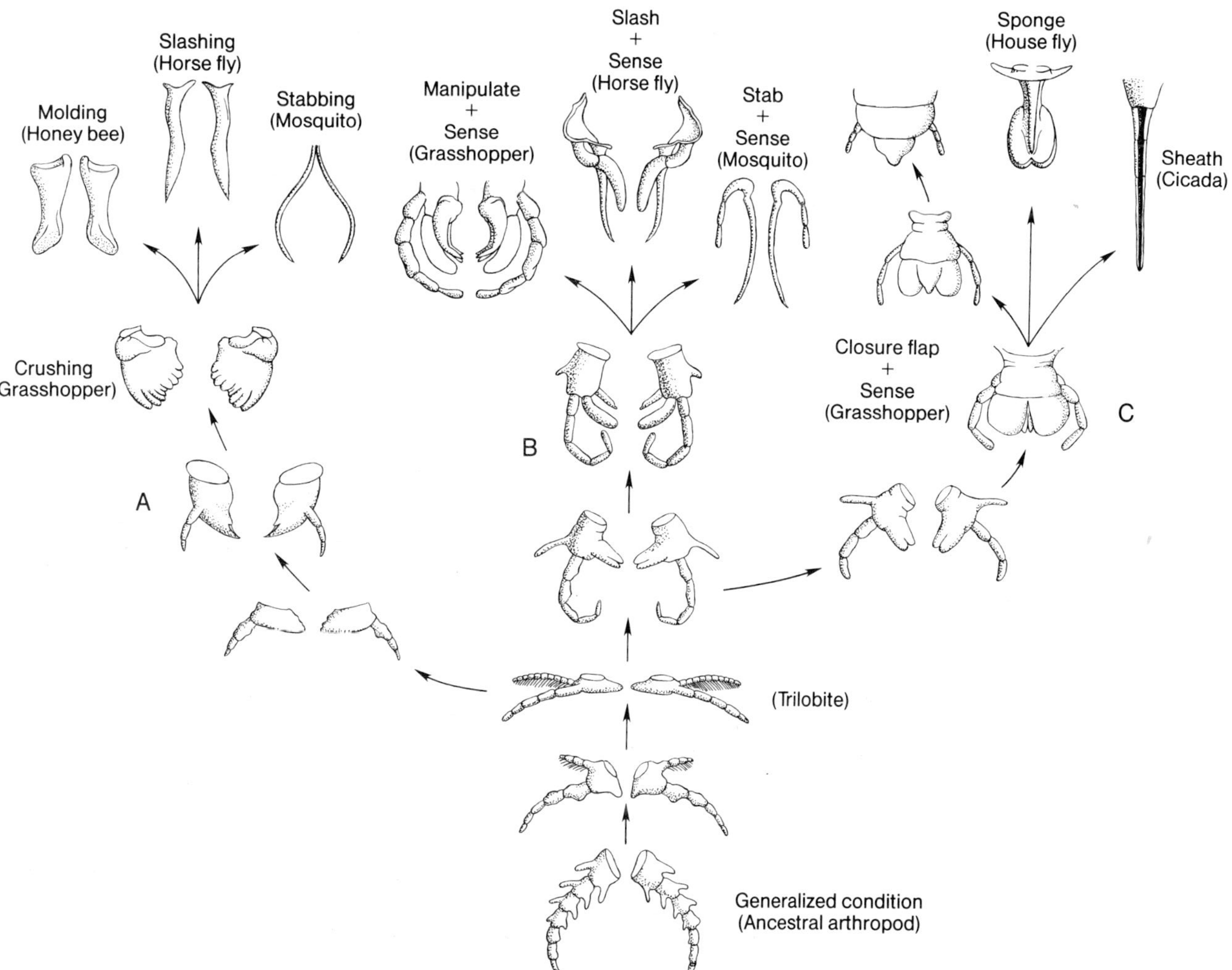

***Fig. 3.5** Evolution of arthropod appendages into the mouthparts of insects, resulting in the widely varying types of mouthparts possessed by modern-day insects. (A) Evolution of the pair of mandibles, (B) the pair of maxillae, (C) the fused labium. Note that all types are not illustrated and some steps are hypothetical, but that similarities of function, position, and embryological formation reinforce the fact that these markedly different structures were all originally locomotory appendages of the arthropodan ancestor.*

the dinosaurs gained with their egg. Active flight gave insects two marked advantages over wingless terrestrial animals: (1) They could escape into the aerial environment where wingless predators could not follow them, and (2) they could cover very great distances, either through active directional flight or through a planktonic drifting wherever the wind currents took them. The freedom in movement by flight has several consequences. It allows insects that may exist as sparse populations to come together for mating. It also allows insects to cover large areas in search of available, but widely scattered, sources of food.

Size

Insects have several distinct advantages which small size confers on them, but also some specific problems due to size. One advantage of being small is the relative ease with which a small animal can hide. Most insects are small enough that they can crawl under leaf litter or low growing plants and be lost to view. In fact, many insects that occur on plants effectively disappear upon disturbance simply by dropping off the plant. Another more significant advantage has already been mentioned in the section on wings, that is, the ability to seek out widely scattered food sources. In combination with small size, this has allowed insects to specialize in the use of nutritious materials in the environment that exist in such small quantities that larger animals could not exploit them and small animals without wings could not cover the distances between the sources of food quickly enough to exist in the low numbers that such sources could maintain.

The disadvantages of small size are tied to the physical rule relating the volume and surface of a body. As the size of a body is increased, the surface area of that body increases as the square while the volume increases as the cube. Thus, the larger a given body is, the lower the ratio between the volume and surface. Conversely, the smaller the animal is, the greater is the discrepancy between the volume it contains and its surface (Fig. 3.6). The surface–volume relationship affects the lives of insects in a number of ways. As will be discussed later in this chapter, the water problem of insects as small-bodied animals is probably the greatest factor in insect evolution. This is easily explained by the fact that the amount of water the insect body can contain is proportional to its volume, while the rate of evaporation from its body is a function of its surface. Thus, the smaller the insect body is, the less water it can hold and the more easily the water can be evaporated. This tendency is only offset by the waterproofing of the insect cuticle.

The surface–volume relationship is exploited by insects in their methods of dispersal. Many wingless insects are carried aloft on rising currents of warm air because they are light in weight and have relatively large body surfaces so that friction with the air (a surface phenomenon) allows for a very slow rate of fall (see Fig. 3.7). The method of dispersal by the dandelion flower with its delicate parachute is essentially the same. This surface–volume relationship explains why very small droplets of liquid, e.g. fog droplets, remain airborne for so long even though the liquid may be relatively heavy. Man exploits this same principle by using sprays of chemicals in fine

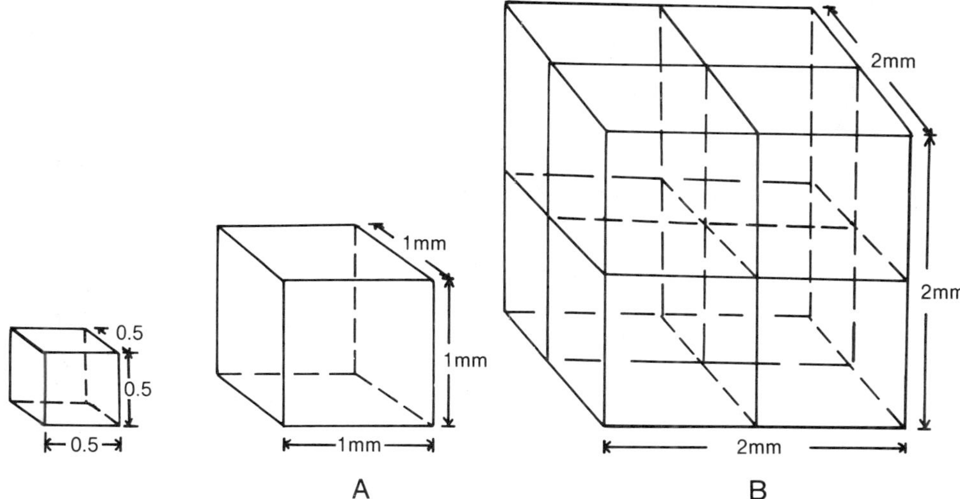

Fig. 3.6 Surface to volume relationships are demonstrated by these three cubes. Cube B results if the linear measurements of cube A are doubled. The surface area of cube A is 1 mm × 1 mm × 6 sides = 6 mm² and the volume is 1 mm × 1 mm × 1 mm = 1 mm³. The surface to volume ratio is 6:1. The surface area of cube B is 2 mm × 2 mm × 6 sides = 24 mm² and the volume is 2 mm × 2 mm × 2 mm = 8 mm³. The surface to volume ratio is 24:8 or 3:1.

droplets so that they will penetrate throughout a warehouse or tree before settling out.

The strength of insect muscles is in great part a function of the surface–volume relationship. In absolute units, the muscles of vertebrates and insects are of approximately equal strength. However, strength increases directly with its cross-sectional area while weight (volume) increases with the cube of its linear measurements. Therefore, if a flea were the same size as a man, it would be only slightly better at jumping than man.

Metamorphosis

Metamorphosis or change in form was originally viewed as a literal change in body structure. However, of more importance are the consequent changes in behavior and habitat that occur with these changes in form. Insects are divided into three basic groups, those without metamorphosis, those with simple metamorphosis, and those with complete metamorphosis. The first two groups, as presented in the previous chapter, show essentially the same feeding habits, location in the environment, and appearance, regardless of the stage being considered. The functions of the stages change only slightly, that is, the adult insect with simple metamorphosis can fly. The real advantage in metamorphosis, then, becomes more apparent as insects with complete metamorphosis are considered. The advantage is in specialization, for it

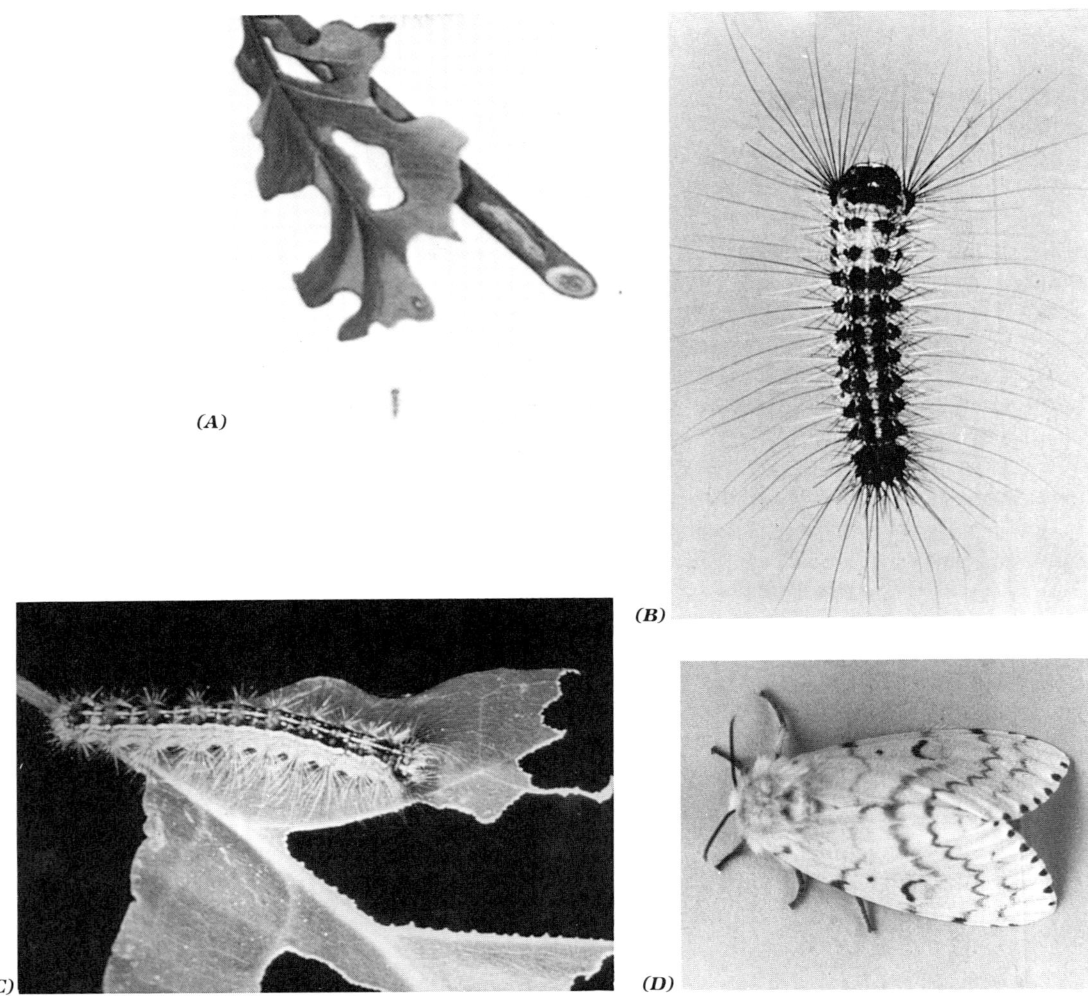

Fig. 3.7 After hatching, the minute gypsy moth caterpillar rests for 18 hours, then climbs up the structure on which the eggs were laid. It then disperses by spinning down on a piece of silk (A), which breaks, releasing it along with its silk. The newly hatched, parachuting caterpillar is very hairy (B) compared with the mature caterpillar (C), and, combined with the silk, presents a large surface area on which friction with the air acts to produce a very slow rate of fall. These caterpillars have been known to travel 25 miles. (D) The adult gypsy moth female has wings, but cannot fly.

confers competitive advantages when each stage can evolve toward its own special role in the life of the insect. Insects with complete metamorphosis show such specialization in each stage. A part of the specialization is in the different habitats each stage occupies. The advantage in occupying more than one habitat is that most habitats are temporary to some degree. If the insect is restricted to a single habitat, when the habitat disappears, so does the insect. However, if a different stage of that insect lives in another environment, the chance that both habitats will disappear concurrently is much less, even if the habitats are relatively unstable. For example, large organisms that are restricted to life in freshwater are to be found only in permanent bodies of water, whereas insects like mosquitoes easily reinvade temporary pools of water because although the larva and pupa are aquatic, the adult stage lives an aerial existence.

Escape from Adverse Conditions

Insects, as small "cold-blooded" terrestrial animals, encounter two specific types of adverse climatological conditions: drought and cold. Especially in the temperate regions of the world, most insects have evolved mechanisms that allow them to escape the adverse conditions of winter or periods of drought. The mechanisms allow for escape in time or in space, that is, the insect essentially "turns-off" metabolically, migrates away from the inimicable site, or both.

Escape in time, the condition called *diapause*, is a kind of "suspended animation" in which the insect's activity is minimal. In some cases the insect may appear dead, with barely detectable respiration. If active, it may only respond sluggishly to stimuli. The diapause condition is initiated by environmental cues, most commonly by changes in the duration of light during each 24 hour period. Hours of light per day is the environmental cue to the approach of winter because it changes in a regular and constant manner, not erratically like temperature fluctuations.

Escape in space or migration has been shown to be an integral part of the life history of many insects. For insects with more than one generation per year the migratory stimulus may elicit a response easily in spring or fall, but may have to be much more intense in the long day period of summer. Again, duration of light is a common environmental cue, but changes in food quality or population density may also act as cues. Many insects with one generation per year have been found to undergo a migration soon after they reach the adult stage and their wings become functional. Such an event aids in the spread of the species.

Diversity in Reproductive Types

Reproduction is the process by which organisms give rise to offspring. It consists fundamentally of the segregation of a portion of the parents' body and the subsequent growth and differentiation of this segregated portion into a new individual. Reproduction is required because individuals are lost to the population through destruction by physical (e.g., weather) or biotic factors such as predation, parasites, and disease. Because the bodies of individual animals also undergo aging processes and the end result is death, individuals must also be replaced to maintain continuity of numbers from generation to generation. Senescence is the sum total of the alterations in structure and function in an individual that accumulate over time that tend to decrease the individual's capacity to survive and ultimately result in its death.

The basic sexual process in reproduction is the contribution by two parental insects to the formation of a new individual. Such contributions are possible without doubling the information contained within the cell at each generation because only half of the information is contributed by each parent. Most insects, like the majority of animals, are diploid with respect to the genetic material carried within their cells. "Diploidy" is the condition of carrying paired chromosomes within the cells. A part of the reproductive process occurs in both the male and female reproductive system when the diploid condition is reduced to haploid in the resulting sex cells (sperm and egg). "Haploidy" is the condition in which a cell has only one set of chromosomes. Thus, when the haploid sperm from the male and the haploid egg from the female fuse, the diploid chromosome number of the newly formed offspring is the same as its parents.

The advantage of this process is at least twofold. First, the organism may only express part of the information it carries within its cells. For instance, it may have the same color eyes as its male parent, which differ from the color of its female parent. Yet, its own progeny may have eye color like its maternal parent. This is retention of genetic information. The second advantage is the mixing of parental genetic material in the offspring. Although each pair of chromosomes is separated into the forming sex cells so that one of each pair goes into a different cell, which one of the pair goes where is random. Therefore, although the number of chromosomes is the same in each sex cell, the combination of chromosomes is extremely variable.

Sexual reproduction in insects, therefore, contributes to their success by introducing individuals of different genetic composition into the environment through the recombination of genetic material that occurs in the sexual process. The introduction of variability is further facilitated by the diploid condition found in most insects, which permits the carrying of information that may not be expressed but is not eliminated. Sexual reproduction is the basic type of reproduction in the insect world, but it is not the only type found among the insects. In fact, many species are known that do not undergo sexual reproduction.

Although sexual reproduction is a successful mechanism in producing variable genetic combinations, there is a drawback. Material located on the same chromosome is reasonably well linked together, but the combinations of chromosomes are not maintained in the offspring. Therefore, even if the adult had a particularly advantageous combination of chromosomes, the combination would be lost in the next generation. There are some exceptions to the loss of combinations, but with sexual reproduction the mechanism so effective in recombining parental chromosomes is equally effective in splitting them up again. Some insects have evolved away from sexual reproduction and are locked into a type of reproduction that does produce exact parental replicas at each generation. This type is uniparental; offspring are produced by females that have not been fertilized by a male. Such reproduction is termed "parthenogenesis."

The advantages of parthenogenetic reproduction are that exact replicas of advantageous chromosomal combinations are retained from generation to generation. Also, very sparse populations can maintain themselves because there is no requirement that a female that is reproductively mature must locate or be located by a male prior to producing offspring. Many species of insects exist in which only females have been found and which have been shown to have obligate parthenogenesis, that is, parthenogenesis is the only type of reproduction existing in that particular species. This is common among the weevils (snout beetles) and some stick insects.

Parthenogenesis occurs more commonly among insects as a facultative condition. Facultative parthenogenesis allows for reproduction sexually if a male is present or parthenogenetically if a male is not. Such a system optimizes the female's reproductive capacity without sacrificing the recombination value inherent in sexual reproduction. This last point emphasizes the inherent problem with reliance on parthenogenetic reproduction and is probably the reason why more insects do not exist with obligate parthenogenesis. Once locked into parthenogenesis, the only way now recognized in which variation can occur is when mutations within the genetic material alter the chromosomes. Since the vast majority of mutations are deleterious, only small, short-lived organisms can successfully exploit reproduction without genetic recombination.

The advantages of both parthenogenetic and sexual reproduction have been combined in some insects in such a way that the disadvantages of each kind of reproduction have been minimized. The combination occurs in "metagenesis," which is defined as the alternation of generations. With respect to reproduction, the alternation is between sexual reproduction and parthenogenesis. However, insects may not only alternate between the two types of reproduction irregularly, but may alternate in other ways. Figure 3.8 shows an extreme case of metagenesis in aphids in which alternation occurs not only between sexual and parthenogenetic reproduction, but between winged and wingless adult body forms, between production of young as eggs or live births, and between different kinds of host plants. Variability is maintained in the sexual production of overwintering eggs, but parthenogenetic production by isolated females during the growing season eliminates the search for mates in the population as they spread out from overwintering sites.

Short Generation Time

Two principal advantages accrue to organisms with rapid turnover of generations. The first is that large numbers of generations, especially in conjunction with sexual reproduction, present a great deal of genetic variability to the environment. Thus, the possibility of an advantageous genetic combination being produced is increased. The second advantage is that ephemeral food sources become available to animals that can complete their life history in a short period. Foods available for very brief periods like carrion, pollen, and mushroom fruiting bodies, are easily exploited by animals that are highly mobile, small, and have short generation times. Such a description fits the insects.

Specialization in Life Style

Although a number of insect species can feed on a wide variety of foods and may be found in a number of habitats, the insects as a group are specialists in the type of life requisites needed. In fact, the route toward specialization in type of food or habitat

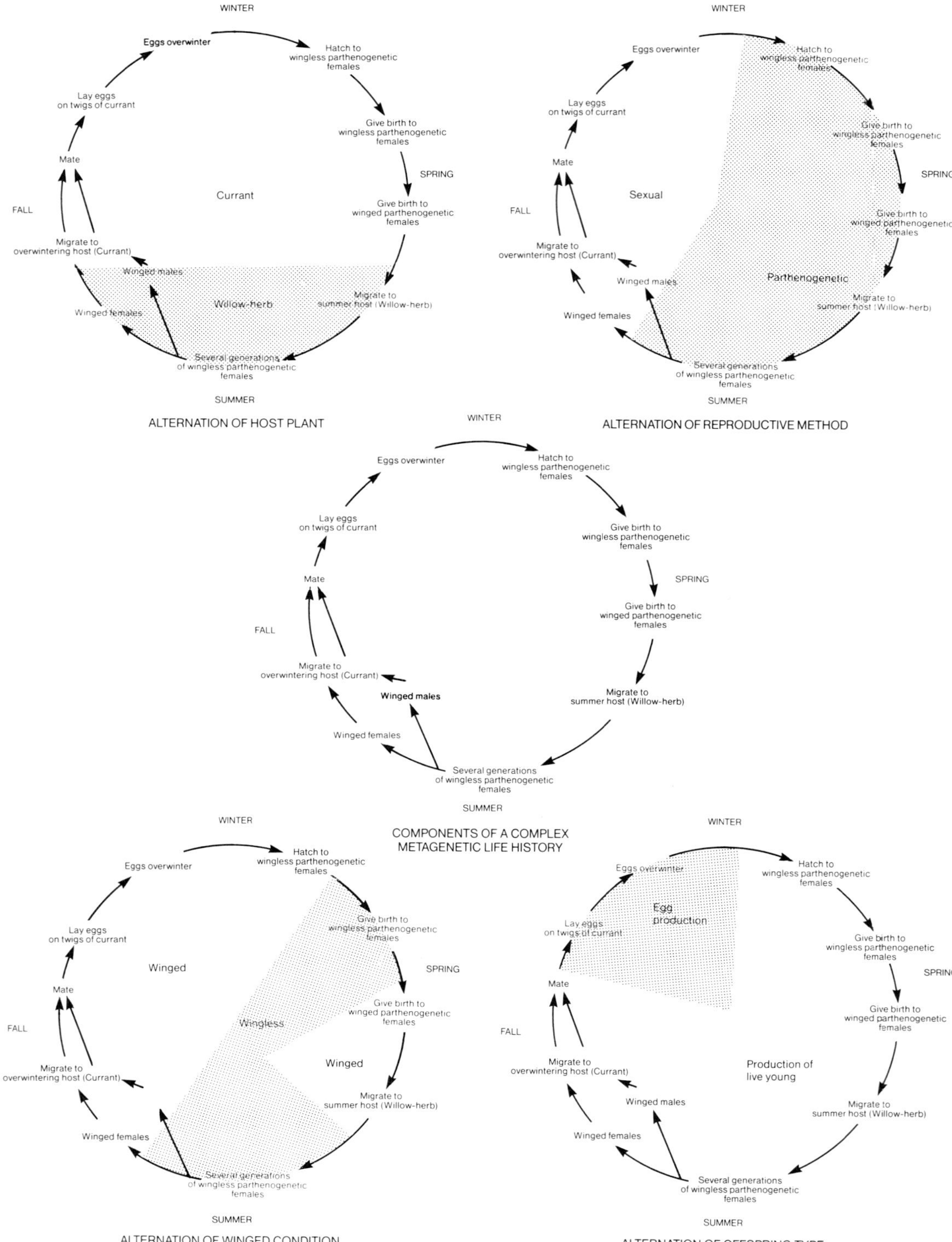

Fig. 3.8 ***In the life history of the variable currant aphid, the generations alternate in four distinct ways: alternation of host plant, alternation of reproductive method, alternation of winged condition, and alternation of offspring type.***

that the insect requires for life is probably the primary reason why there are so many different species in the insect world. Some of the factors that contribute to the insect's evolution toward increasing specialization are its small size, high mobility, and rapid generation time.

There are some distinct advantages to specialization. Of particular importance is that an insect specialized on a single type of plant will have sense receptors that allow it to discriminate between its host plant and all other types of plants. The advantage in host discrimination is that the insect can find its food supply, even when the host plant occurs sparsely in the environment. Therefore, the insect is assured of adequate nutrition once the host plant is located. Insects that are general feeders have been known to attack plants that are not nutritionally adequate, and consequently will not support production of vigorous offspring. An obvious disadvantage in becoming specialized into a very restricted life style, i.e., dependent on a single kind of plant, is that existence depends on the continued survival of that plant. If some other factor, e.g., a plant disease, comes along and eliminates the host plant, then the insects dependent on that plant also become extinct.

For the most part, insects have gone beyond specialization in a life that is restricted to a single host. They have become specialized in attacking particular hosts or parts of hosts at particular periods of the host's existence. Figure 3.9 illustrates the division of typical plant and animal life forms into habitats that have become occupied by individual insect species. Such a list for apple trees would yield over 500 insect species, some restricted to apples, others that attack apples among the other species utilizable by them.

Methods of Solving the Water Problem

As animals evolved away from a marine existence to a terrestrial life, water became of paramount importance because cells of the insect's body require an aqueous environment just like those of marine forms. The water problem of insects is twofold: acquiring sufficient water for life processes and conserving the water that has been acquired. Acquisition of water is not a problem for insects that feed on such materials as leaves, fruits, plant juices, animal flesh, or blood. The water level of the food is sufficiently high to supply the insect's needs. As everyone knows, insects are frequently encountered in containers of dried foods, e.g., flour, raisins, cereals, and pet foods. Control of such pests would be greatly facilitated if they had to leave their dry environments to secure water. Insects, however, have several mechanisms which allow them to live continuously in dry environments, securing enough water to maintain themselves and to reproduce. Since foods are seldom absolutely dry, some moisture can be extracted from them. Several stored-product pests are capable of using this residual moisture. Some are also able to extract moisture from water vapor in the air, especially when the air is humid. However, the majority of stored-product pests utilize "metabolic" water, which is the end result, with CO_2, of metabolism.

$$C_6H_{12}O_6 + 6O_2 \rightarrow 6CO_2 + 6H_2O + \text{energy}$$

In fact, at low humidities, such insects consume increased quantities of food to increase the amount of water available as a result.

Fig. 3.9 Insects have specialized with regard to habitat. Arrows indicate (A) the variety of microhabitats available to insects that specialize on a single tree species and (B) microhabitat division between the two species of human lice (head and body louse strains of **Pediculus humanus** *L. and the public louse)*

Conservation of water by insects is behavioral as well as structural and physiological. The primary structural mechanism is the deposition of one or more layers of wax on the cuticle by cellular secretion. The wax layers are effective barriers to water as witnessed by the use of this technique in preserving moisture in cheeses. If the insect's wax layers are disrupted by high temperature or scratched by abrasives, the insect soon dehydrates and dies.

Another aspect of water conservation by insects is exemplified by the excretory system. Ammonia is the form of the nitrogenous wastes released by aquatic animals. Ammonia, however, is very toxic and must be flushed out of the body by large volumes of water. Such a system is feasible for aquatic animals, but would unduly stress a terrestrial form by dehydrating the tissues. Therefore, terrestrial forms have evolved whereby wastes are metabolized to urea, which is less toxic than ammonia, and consequently, less wasteful of the body's water. This is the product excreted by man. A system that is the ultimate in nitrogenous waste production has evolved in some insects, reptiles, and birds. The end product is uric acid, which is almost insoluble in water and relatively nontoxic. Uric acid and fecal matter are both produced in a dry form; the insect's body has extracted almost all moisture.

In order to protect sex cells from desiccation and death, fertilization takes place internally; the male's sperm are introduced into the female's body. The typical insect egg is then expelled from the female's body with a covering similar to the insect's body wall for protection. The egg may also have elaborate structures that allow for entry of moisture into the egg, but greatly retard water loss out of the egg to the environment. In respiration, gaseous exchange takes place across a moist membrane. This system mandates some water loss, but the insect minimizes such loss by regulating ventilating movements and also by regulating the types of activity that could result in fatal levels of water loss. Insects that live in very dry environments usually regulate essential behaviors around the least hazardous part of the year, season, or day. For instance, dryland *Drosophila* fruit flies emerge from their pupal cases early in the morning. This behavior is triggered by light levels, but has developed because those that emerged at dawn had a better chance to survive. Because dawn is usually the coldest part of the 24 hour temperature pattern in the desert, it is consequently the time when relative humidity of the air is highest. Therefore, dawn-triggered emergence allows for spreading of the wings and hardening of the adult's cuticle (the period of maximum vulnerability) during the wettest part of the day.

Insects have undergone evolution in many ways in their struggle to survive in the terrestrial environment. Modifications of all systems have allowed insects to withstand fatal water loss through their relatively large body surface. In addition, water loss is minimized in the essential activities of life: respiration, reproduction, digestion, and excretion. In combination with their ability to extract moisture from their foods and the air, the development by some insects of the ability to use metabolic water has truly allowed them to conquer the terrestrial environment.

THE MECHANICS OF EVOLUTION—HOW INSECTS CHANGE

The material presented throughout this book is based on evolution and the mechanics of the evolutionary process. "Evolution" is the principle of progressive change. The theory of evolution is not the only theory that has been forwarded in explanation of the natural world, e.g., special creation is commonly offered as an alternative by opponents of evolutionary theory. Evolution however, is the theory that most adequately explains not only the diversity in our natural surroundings, but the processes we detect in our changing world. With evolutionary theory it is possible to follow through from the initial descriptive level of understanding our world to an understanding of its operational mechanics, that is, how it works. More importantly, evolutionary theory explains how our natural world got to be the way it is and how it may change in the future. Another advantage of this theory is that understanding evolutionary mechanics allows for a high degree of predictability when it is applied to a new situation, e.g., a new species or a different environment.

The mechanics operating within evolution that give it the characteristic progression of changes are presented in Fig. 3.10. The units shown are "The Survival Test" and "Reproduction," with survival of the fittest as a result of the action of the survival test on the various genetic types within the large numbers of young. Variation and overproduction are characteristics of organisms that enhance the evolutionary process.

Variation

The various genetic combinations produced by insects are a result of diploidy and mutation, plus the recombination and mixing of genetic material by parents if reproduction is sexual.

Overproduction

All animals produce more offspring than could exist as adults on the available environmental resources (food, place to live). Most of the offspring are eliminated by natural enemies, lack of food or nesting sites, and environmental extremes prior to reaching the reproductive stage. The degree of failure in survival of offspring is reflected in the reproductive capacity of the insect, since those that produce fewer offspring than are necessary for continuance of the species become extinct.

The Survival Test

The result of the insect's interaction with the selective forces of the environment is the survival test.

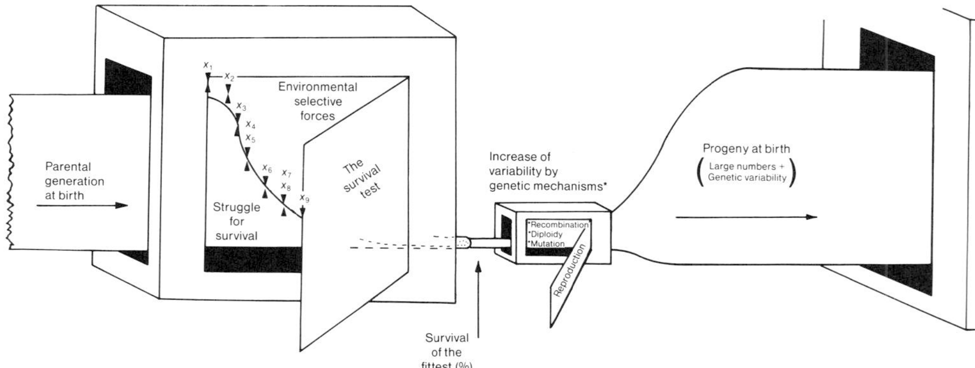

***Fig. 3.10** The evolutionary process of natural selection in sexual animals. Succeeding generations cope with the various selective forces ($x_1 \ldots x_n$) of the environment, yielding greater survival of those individuals that are most fit. Greater inputs by the more fit parental stock permit continued selection of fitness in each generation to the selective forces acting during its existence.*

Struggle for Survival

All organisms struggle to obtain the necessities of life within the limitations of the environment. Since the individuals within a generation are of different genetic combinations, some will have combinations that are more advantageous than others, that is, different genetic combinations make some more fit than others. Thus, they are more capable of coping with selective forces than others.

Selective Forces

The action of environmental stresses on the individuals of the population include such factors as competition for nesting sites or overwintering locations as well as competition for food. The ability to survive exposure to parasites, diseases, and predators is important, as is that of surviving physical excesses of the physical environment, e.g., drought or cold temperature. The forces are selective only if some individuals can survive exposure while others perish.

Survival of the Fittest

The result of the action of selective forces on the genetic variability of the population is survival of the fittest. However, this is a statistical phenomenon, not an absolute. Those individuals that are more fit than others tend to survive and contribute their genetic material to the next generation. Although some of the less fit individuals may survive, the tendency is for fewer of the less fit to survive. Thus, each generation has a higher percentage of survival of the most fit genetic material and combinations.

Natural Selection

The result of the evolutionary process is the natural selection of animals capable of coping with the environment to some degree. The ability to cope (fitness) is improved generation by generation as the biological units respond to the various selective forces. These events demonstrate that evolution occurs today and that it sometimes is rapid enough to permit us to measure changes in populations. The effectiveness of the system is dramatically illustrated when new selective forces are introduced. An example of this is the principle of preadaptation in insects, such as the case of pesticide resistance, which will be discussed after the following section.

Preadaptation

As with all animals, insects exposed to conditions to which they are not adapted die as a result of such exposure. They seldom become able to tolerate new conditions as a result of repeated exposure. If this is so, how then can animals survive exposure to conditions with which they have never been confronted? The answer is that the animal is preadapted to the new conditions. It possesses some characteristic that may have been unimportant in its original environment, but is the key to its survival in the new environment. An example of this preadaptation is the evolution of terrestrial animals from marine forms. Aside from the problem of maintaining water within its body, the most severe problem to overcome when moving to land is both support of the body and a way of moving over the land. Air is so much less dense than water that an animal's best chance at solving this problem is by supporting itself on the hard substrate and moving by some means on that substrate. In essence, bottom-dwelling marine forms were already suited (preadapted) for locomotion in a terrestrial world. The arthropods with a previously evolved exoskeleton and paired appendages had the support and locomotion problems solved prior to the invasion of land, and the exoskeleton solved much of the water loss problem as well. Snails and their bottom-dwelling relatives have been less successful in invading land because of their method of locomotion. A layer of mucus is formed by glands on the bottom of the muscular foot, and the snail glides along the mucus layer by muscular contractions. This plan of locomotion with its expenditure of materials on land is inefficient.

The lobe-finned fishes that gave rise to terrestrial vertebrates were swimmers that were preadapted to a clumsy, but moderately effective form of terrestrial locomotion because of their lobe-fin appendages. Undoubtedly, some forms occurring in shallow drying pools faced the same locomotion problems that the modern lung fish do and crept out of their dwindling pools on their fins, preadapted for survival under these conditions. Thus, the three groups that have most successfully invaded land were all preadapted to one of the major difficulties of terrestrial existence because of their types of locomotion.

Resistance to Pesticides—Evolution in Progress. When DDT was first introduced about the time of World War II, it caused a major change in insect control procedures, against insects affecting both public health and agricultural production. DDT was a spectacular new tool with which to fight insects because of two significant characteristics: only minute amounts were needed to kill, and its toxic activity persisted for a very long period. These two characteristics in combination meant that effective insect control could be accomplished at very low cost.

DDT and some related compounds were seen as the solution to many of the major insect problems of mankind. Pests that could not be controlled economically prior to DDT could now be suppressed. Pests like the housefly, a common pest in dairy barns, could be greatly reduced by spraying the walls of the barn with DDT. The material deposited would continue to kill flies for months before more had to be applied. Unfortunately, in a very short period it was reported that DDT was losing its effectiveness against some insect pests. A closer look revealed that the loss of effectiveness was linked to the constancy of exposure to the DDT and the proportion of the population in any given area that was exposed. Two solutions to the loss of effectiveness seemed obvious. The amount of DDT used could be increased, or some other related compound could be used instead of DDT. The former was the most common response to the problem, but before long even the increased concentrations of DDT became ineffective.

Examination of recent history reveals that most populations of the housefly have evolved to tolerate DDT in their environment. When DDT was first used, it became a new selective force in the housefly environment. Therefore, the genetic combinations most prevalent among the housefly populations being exposed to DDT were no longer the fittest types. They were susceptible to DDT and died when a lethal concentration was contacted. How then did the population become tolerant to DDT? A survey of unexposed populations revealed that a small percentage of flies had the ability to make an enzyme which detoxified DDT by breaking it down. These flies were not as well adapted to conditions in their pre-DDT environment as the majority of the population was, so they were few in number, being continually outdone by their more fit (better adapted) relatives. When DDT was introduced into the environment the flies that were preadapted to DDT, but otherwise poorly fit, became more abundant because of their ability to cope with DDT.

As the use of DDT increased, a greater proportion of the housefly populations came under the new selection pressure. Only the flies that had avoided contact with lethal levels of the chemical and the DDT-tolerant flies survived to reproduce. A high percentage of the offspring of tolerant flies have tolerance too, so that repeated exposure simply wiped out the susceptible flies and allowed the tolerant ones to increase. Since tolerance is not an all-or-none situation, an increase in concentration of DDT eliminated the less tolerant and sped up the process of selection. Populations of flies now exist that can tolerate over 100 times the concentrations that previously decimated housefly populations. They have truly developed a resistance through natural selection from forms that had been preadapted to a new selective force.

The DDT-resistant populations have improved their fitness in the environment to the point where they are about as fit in a DDT-free environment as when DDT is present. In other words, they have now reached the level of fitness that the housefly populations had prior to the introduction of DDT. Such fine tuning of fitness to an environment with reasonable continuity is termed "postadaptation."

SUMMARY

Insects are among the most successful groups of animals. There are more kinds of insects than all other animals combined. Along with the vertebrates, insects are ecologically dominant, exerting a major and controlling influence in most terrestrial and freshwater environments.

Organisms now extinct have also been ecological dominants. Examples are the trilobites and dinosaurs. Both evolved characteristics that gave them an advantage over other groups. Trilobites were the first organisms with a segmented exoskeleton that provided continuous protection. Dinosaurs were the first vertebrates to escape dependence on the aquatic environment for embryonic development because of their unique egg. Both groups probably disappeared because other life forms evolved that made more efficient use of the environment, thereby outcompeting them.

Mammals have evolved around low reproductive rates, extended individual life span with parental care and training of offspring, and with considerable intelligence. Insects have evolved small size, massive reproduction, and stereotyped behavior.

The following characteristics have contributed to the success of insects.

Exoskeleton. The "suit of armor" provides support and protection from both physical injury and water loss.

Jointed appendages. The numerous paired appendages of the insect ancestor have evolved to perform three special functions: feeding (head), locomotion (thorax), and reproduction (abdomen).

Wings. Most insects have wings, but they function only in the adult stage. Wings permit insects to escape natural enemies and cover great distances seeking food, mates, or egg laying sites.

Size. The small size of insects is an advantage in concealment and the ability to use and specialize on minute or widely scattered food sources, especially if the insect is winged. Small size, because of the surface-volume rule, is a disadvantage because of the potential for increased evaporation of water. This rule also aids in insect dispersal and in their apparent strength.

Metamorphosis. Whereas insects with direct or simple development are found in a single habitat, those with complete metamorphosis often exist in different habitats depending on their stage of development. Thus, temporary habitat destruction is less catastrophic to insects with complete metamorphosis. Each stage in complete metamorphosis is highly specialized to exploit its own environment.

Escape from adverse conditions. Many insects either migrate or enter diapause prior to the onset of annual changes in weather. A common cue is photoperiod, the number of hours of light per day, because of its regularly changing nature.

Diversity in reproductive types. Most insects reproduce sexually, benefiting because of the variety produced as a result of random assortment of chromosomes when sex cells are formed and of recombination. Some species produce parthenogenetically, without mating. This limits variability among offspring. The advantages of both reproductive types are combined in metagenesis, the alternation of generations.

Short generation time. The benefits of the short life typical of most insects are the use of ephemeral food supplies and the increased evolutionary rates resulting from the high number of generations per year.

Specialization in life style. Although some insects are generalists, most have become highly specialized in food or place in which they live. Such specialization optimizes location and use of requisites.

Methods of solving the water problem. Acquisition and conservation of water are two prime problems of terrestrial animals, including insects. Insects have minimized water loss by excreting uric acid rather than ammonia or urea. Their waxy coating, internal fertilization, and use of metabolic water are aids in conserving and acquiring water.

Evolution, the principle of progressive change, accounts for the diversity of organisms and adequately explains the way in which such diversity has arisen. Selective forces act on the various genetic combinations in the offspring so abundantly produced. The more fit tend to survive and pass on their genes.

Organisms survive new selective forces if they are preadapted. In this way, preadapted house flies survived exposure to DDT that killed most of the fly population. Survivors produced offspring with a high percentage of DDT tolerance. Development of highly tolerant populations over a few years shows the rapidity with which evolution acts.

SUGGESTED READINGS

Bronowski, J. 1973. "The Ascent of Man." Little, Brown, and Co., Boston, Massachusetts.

Darwin, C. 1859. "On the Origin of Species by Means of Natural Selection, or the Preservation of Favored Races in the Struggle for Life." John Murray, London.

Dobzhansky, T., Ayala, F., Valentine, J. W., and Stebbens, G. L. 1977. "Evolution." Freeman and Co., San Francisco, California.

Futuyma, D. 1979. "Evolutionary Biology." Sinauer Assoc, Sunderland, Massachusetts.

Mayr, E. 1963. "Animal Species and Evolution." Belknap Press (Harvard Univ. Press), Cambridge, Massachusetts.

4

Insect Structure and Function

The fact that insects are so much smaller than humans often leads us to the erroneous conclusion that their construction and life functions are much less complex than our own. They are not. The systems that function to keep the insect alive are the same as those found in human bodies, the primary differences being one of magnitude, not function. The insect has the same environmental problems to solve as other animals do. These environmental problems, acting as evolutionary selection pressures, have resulted in animals with organ systems that often function the same as ours do or, in some cases, by an acceptable, or even superior, alternative method.

The basic arrangement of the systems within an insect's body is presented in Fig. 4.1. The figure includes only portions of each system, some (e.g., digestive) represented almost in their entirety, and others (e.g., musculature) showing only a small fraction of the total system. The life functions that each system performs for the insect are covered in Table 4.1.

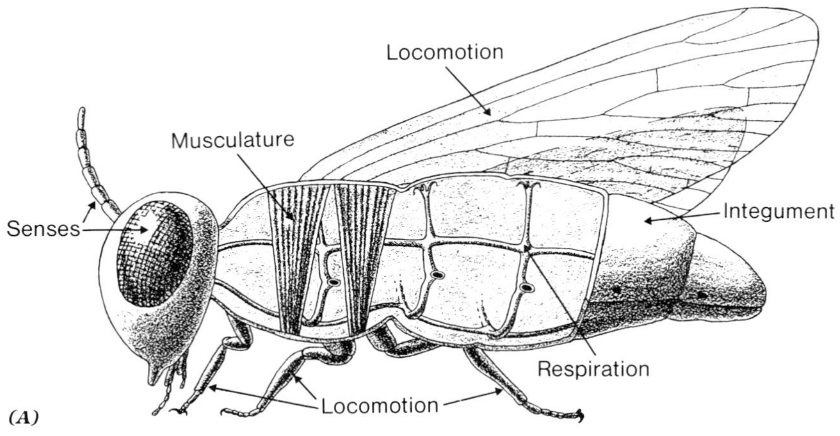

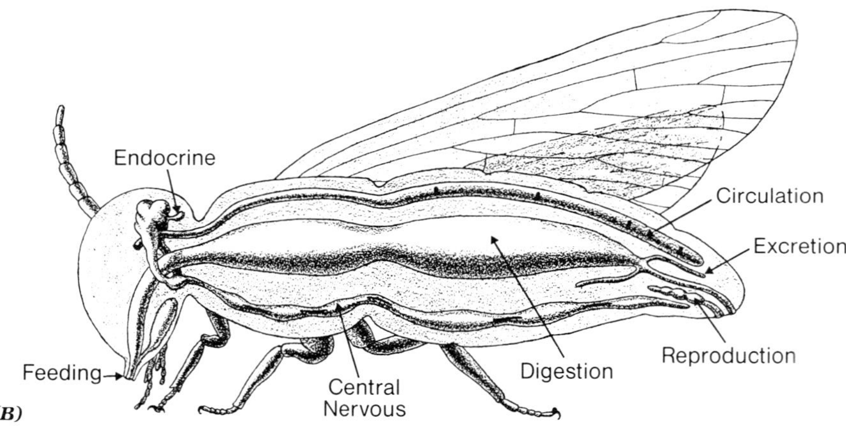

***Fig. 4.1** The major systems of the insect body: (A) The sensory, muscle, locomotion, respiration and integument; (B) endocrine, digestive, circulatory, reproduction, and nervous systems.*

***Table 4.1** Basic Life Functions Performed by the Systems of the Insect Body*

System	Function
Digestive	Breakdown and absorption of high energy chemicals (food) and elimination of indigestible materials
Excretory	Removal of metabolic wastes from the blood
Respiratory	Acquisition of oxygen and its transport within the body
Circulatory	Transport and storage within the body of foods, metabolic wastes, other products
Integument	Protection and physical support
Musculature	Locomotion and movement of body parts
Nervous	Rapid coordination, conduction and integration of information
Endocrine	Coordination of long-term activities such as growth and development
Reproductive	Production of offspring
Sensory	Perception of external and internal environment and its changes

INTEGUMENT

The "integument" or body wall of insects is the first system to be discussed because an understanding of its characteristics and functions will provide insights into many facets of the other systems. The integument is the interfacing structure between the environment and the insect. In fact, it is more than this, because it protects the insect from physical harm as well as from many chemicals, it determines the body form, and it protects from water loss. It is also an integral part of the sense organs, which act to monitor the external environment. Finally, it acts as a reserve of recyclable materials for the insect.

The integument has three basic components: epidermis, cuticle, and basement membrane. The cells that compose the epidermis, the outermost living layer of the insect's body, are the key to the entire integument because they secrete the cuticle. The basement membrane is a unifying layer of material deposited onto the epidermal cells by other cells within the body.

The complete cuticle consists of several layers, each one unique, that combine to give the cuticle its many characteristics. One of the cuticle's features is its relative inelasticity, so the entire cuticle must be destroyed and a larger one formed periodically in order for the insect to increase in size. This entire process is termed "molting," and will be described after the components of the integument have been discussed.

As indicated in Fig. 4.2, the primary layers of the cuticle are epicuticle, exocuticle, and endocuticle. The innermost layer "endocuticle" is made up of minute fibers that when first secreted are not aligned, but are later arranged into sheets. Therefore, the fibers closest to the cell do not show the orderly arrangement that is apparent in endocuticle that is further away and therefore older. Between periods of molting, the secretion of endocuticle, while a continuous process, forms layers differently during the light and dark periods each day. The "exocuticle" is the zone of cuticle where protein plasticized by a tanning process with quinones makes the exocuticle practically invulnerable to attack by chemicals as well as making it very hard. The "epicuticle" is a proteinaceous coating with a layer of waxes on its surface. It is this

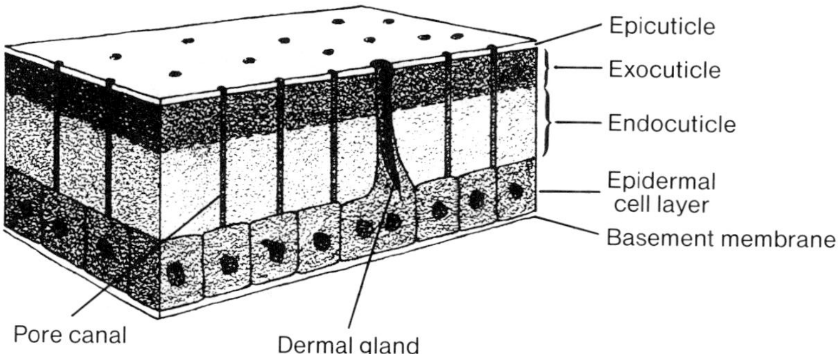

***Fig. 4.2** The structure of the body wall, or integument. The exceedingly thin wax layer lies external to the epicuticle and is vital as the barrier to water loss. The cement layer on top of the wax constitutes the outermost layer.*

layer, more specifically the waxy part, that provides the essential waterproofing to the entire cuticle. External to the epicuticle may be a cement layer secreted by the dermal glands through channels in the cuticle.

The chemical composition of cuticle varies among the different layers, with the most common constituent being chitin (25–50% of the dry weight of the cuticle). Chitin is a polysaccharide similar in structure to cellulose. It has a chainlike series of subunits, and the adjacent chains themselves are joined together to form microfibers. Chitin is associated with various proteins, thus forming a complex glycoprotein. Among the protein components are arthropodins (water-soluble forms), resilin, and sclerotin. Resilin is like rubber in that it can store energy as it is distorted, then release the energy as it returns to normal. Such proteins are found in some joints of the body, e.g., the legs of fleas. Arthropodin is the protein that undergoes a tanning process, or "sclerotization," thereby forming sclerotin. The sclerotin is what gives some parts of the insect integument its rigidity. Since sclerotization only occurs in some areas, commonly forming plates, the more flexible areas between the plates provide for movement. This is especially advantageous if telescoping of structures results in overlapping sclerotized plates (sclerites), with flexibility maintained by the protected membranous portions (see Fig. 4.3).

Molting

The relatively inelastic cuticle forms the exoskeleton that encases the rest of the body. It must be molted at intervals during growth and development.

The process of molting involves destruction of the old cuticle, formation of a new cuticle, and shedding the remnants of the old cuticle. The epidermal cells become active and divide, thereby increasing in number. Then the old cuticle, in a process

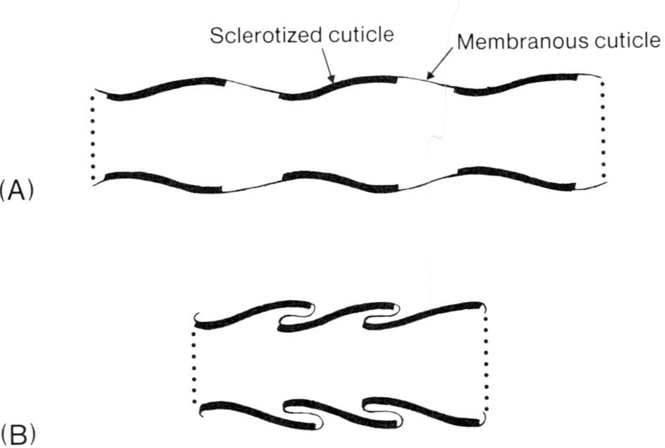

Fig. 4.3 Differential sclerotization: (A) distended abdominal segmentation, demonstrating the sclerotized and membranous portions of the cuticle, (B) the same segments telescoped in to increase protection while maintaining flexibility.

called "apolysis," separates from the epidermal cells and the cells begin to secrete molting fluid into the forming space. The molting fluid digests the endocuticular part of the old cuticle. The digested proteins and chitin molecules are absorbed by the epidermal cells and may be reused, leaving only the exocuticle and epicuticle intact in the old cuticle. This last process is viewed by many scientists as conservation of resources by the insect.

The first part of the new cuticle to be deposited by the epidermis is the epicuticle. Significantly, this part of cuticle is not subject to attack by the enzymes within the molting fluid. Undifferentiated cuticle is laid down between the new epicuticle and the epidermal cells. Just before "ecdysis" (shedding the old cuticle), the waxy layer is deposited on top of the epicuticle of the newly formed cuticle, thereby waterproofing it. The waxy deposition is made through "pore canals," which are very minute openings that are left in the cuticle as it is deposited.

When the molting insect exerts pressure on the old cuticle, the cuticle splits. This happens where the exocuticle was thinnest. The insect withdraws from the old cuticle and swells out the new cuticle, including the wings, by swallowing air or water. At this time the arthropodins are converted into sclerotins by quinones secreted from the epidermal cells, and the cuticle hardens.

The process of molting is controlled by the endocrine system, both as to when the process occurs and what the form of the new cuticle will be. This will be discussed later in the chapter.

DIGESTIVE SYSTEM AND DIGESTION

The primary function of the digestive system is to extract utilizable chemicals from the food that has been taken into the body. In primitive insects, the digestive system consists of a relatively simple tube through the body. The decaying organic matter that is the food of these primitive insects has already been physically broken down by other organisms (bacteria and fungi) and may also have been reduced to less complex chemical forms by these same organisms. This simple digestive system, however, limits the insect to specific habitats (e.g., soil humus) not allowing the insect to exploit the many other potential foods present without some modification. Therefore, the digestive system has evolved, whereby different portions of the system have become specialized to perform tasks in addition to the chemical breakdown of food.

The basic digestive system of a chewing insect is presented in Fig. 4.4, where the parts of the system and associated structures are linked to the functions each performs. Note that the mouthparts are not a part of the digestive system, but do function to chop the food up just finely enough to allow it to pass through the esophagus and store it in the crop. The salivary glands lubricate the mouthparts.

Storage in the crop, coupled with the additional grinding which the gizzard (proventriculus) provides, and the valve into the stomach (midgut) allows the system to function continuously, without the insects having to feed continuously in order to use the system. The system can also be smaller than if it had to accommodate large bursts of intense activity.

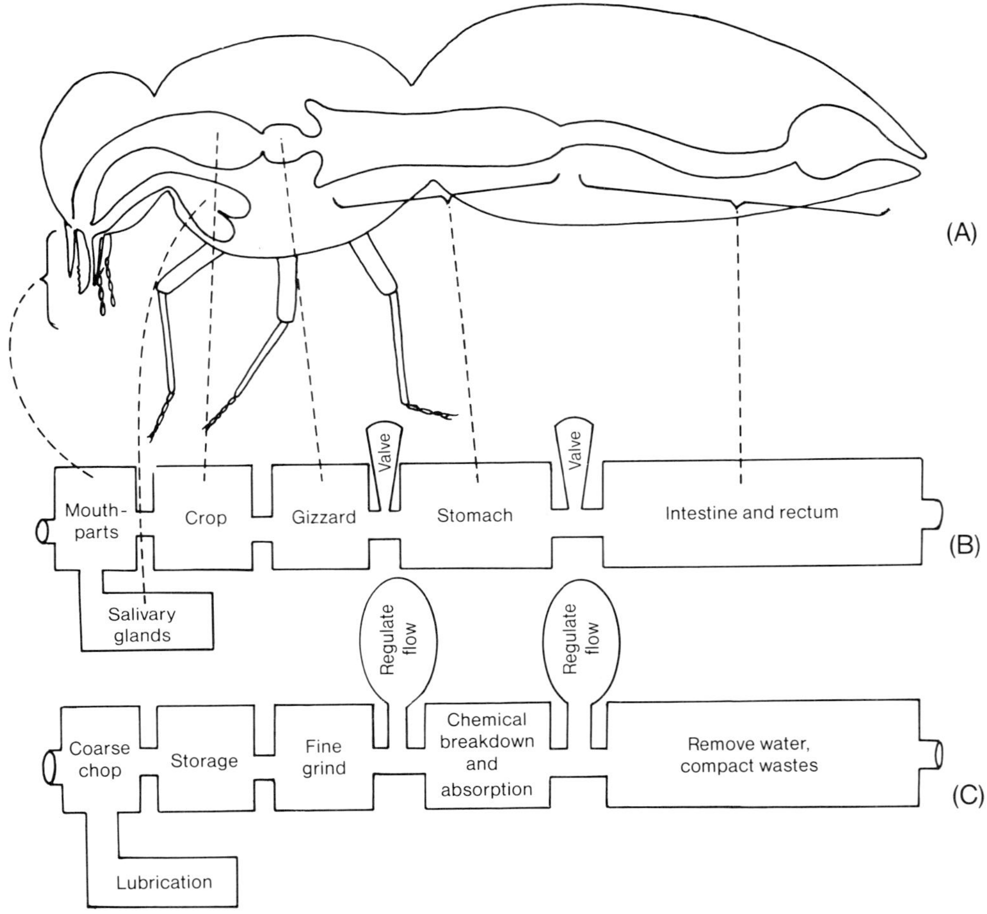

Fig. 4.4 ***The digestive system and associated parts as they appear in a chewing insect (A) names (B) of each part, and the functions (C) of each specialized section.***

All of the foregut (esophagus, crop, gizzard) has a cuticular lining that is shed at the molt. Within the gizzard the lining thickens into the grinding teeth. The hindgut (intestine and rectum) is also lined with cuticle. This occurs because the cells of the fore- and hindgut are derived from the same layer of cells in the embryo that forms the body wall of the insect (see Fig. 4.5). The midgut is formed from a different layer of embryonic tissue, and therefore does not have a cuticular lining, but it does have a peritrophic membrane that encloses the food and protects the stomach walls.

The primary function of the hindgut is to allow indigestible materials to pass out of the digestive system. Additional functions have evolved in which the hindgut is able to withdraw all possible water from the wastes before it compacts and expels them. Often other materials of value to the insect such as potassium and sodium are absorbed in the hindgut.

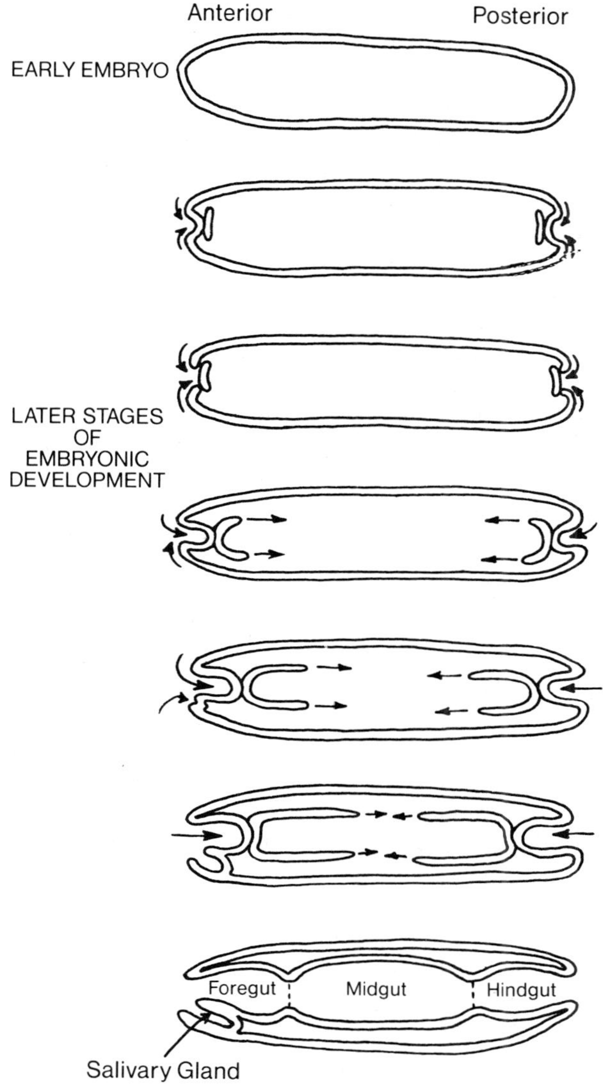

Fig. 4.5 Formation of the rudimentary digestive system in an insect embryo. Note that the foregut, hindgut, and salivary gland are formed from the outer layer of the embryo. This is the layer that characteristically produces a cuticle.

Some insect digestive systems have adapted to a completely liquid diet. The peritrophic membrane, which protects stomach cells in the chewing insect is not present in liquid feeders like aphids. However, aphids and other plant sap feeders using very dilute food solutions, have the problem of too much water. In many of these insects a "filter chamber" or similar arrangement has evolved that actively removes water from the ingested food and expels it from the anus. The chamber is a complex area of contact between the foregut and hindgut. Water is shunted from the foregut directly to the hindgut through the gut walls rather than passing down through the stomach. This concentrates the nutrients passing into the stomach and eliminates the necessity for excessive enzyme production. Blood feeders are able to eliminate the liquid part of the blood meal selectively, since most of the nutrients in the blood are contained within the blood cells.

Digestion of food in most insects occurs when enzymes are released from the stomach cells into the gut cavity (stomach). The "digestive enzymes" (proteins that break down food) act in the gut cavity by digesting (breaking down complex food substances into smaller components) it, then these smaller components are taken back into the gut cells and released into the blood. Some insects, such as the predatory suctorial groups, pump powerful enzymes that start the digestion process into the bodies of their victims and then withdraw the liquefied body contents after this "extraintestinal digestion" occurs.

Digestive enzymes are quite specific as to which foods they attack. For instance, carbohydrates must be broken down into the simplest forms (monosaccharides) before they can be absorbed. If the enzymes that break down the more complex di- and polysaccharides are absent in the insect then such materials as cellulose from plant cell walls cannot be digested. In fact, most plant-feeding insects do not produce cellulase (the cellulose-cleaving enzyme) and, therefore, the insects either utilize only the contents of the cell of their food plant, or the cellulose is attacked by microorganisms that may reside in the insect's gut. Carbohydrates in general serve as the most common source of energy used by insects.

Proteins must be broken down into their basic units (amino acids) before they can be used. Digestion involves a series of steps that cleave proteins into their component parts. There are many kinds of protein-attacking enzymes; they are called proteases. Initially, proteins are broken down to peptones and peptides; subsequently, other proteases complete the breakdown to amino acids.

Fats (lipids) taken in by insects are split by enzymes called lipases, into fatty acids and glycerol.

The complex molecules that comprise structural components in animals are difficult materials to digest. In fact only bird lice, carpet beetles, and clothes moth larvae have the capability of digesting vertebrate keratin (the horny material in nails, hoofs, feathers, and hair) and no predatory or scavenger insect can digest arthropod sclerotin. The difficulty lies in the chemical bonding which forms these long chain molecules.

Activity of enzymes in the digestive system is affected in two ways: by enzyme release in discrete regions of the gut and by changes in gut pH (acidity–alkalinity). Thus, an inactive enzyme in a low pH region becomes activated when it reaches a region with higher pH. Activity is also determined by changes in the reduction oxidation characteristics of each region of the gut.

EXCRETION

The function of the excretory system is to remove metabolic wastes from the blood (hemolymph). These nitrogenous wastes are the result of the metabolism of such materials as proteins by the body. If these wastes accumulate to any degree in the blood, cells throughout the body are poisoned, and the animal succumbs.

The primary organs of excretion are named Malpighian tubules. They remove the metabolic wastes from the blood and are thus comparable to the kidney in vertebrates. The number of Malpighian tubules may vary from two in some scale insects up to 250 in other forms. The tubules are blind-ended, elongate projections opening into the digestive system (see Fig. 4.6).

The nitrogenous wastes that result from protein metabolism by the insect may be excreted in a number of forms, most notably ammonia, urea, and uric acid. In its pure form, uric acid is a solid; however, as it moves into and through the Malpighian tubule cells it is in a water-soluble form associated with potassium. Once within the lumen of the Malpighian tubule, it is moved toward the opening into the gut and thus passes through a gradient of acidity, where the uric acid solids crystallize out, and potassium and sodium ions are reabsorbed along with the water.

Malpighian tubules may perform functions other than excretion. For instance,

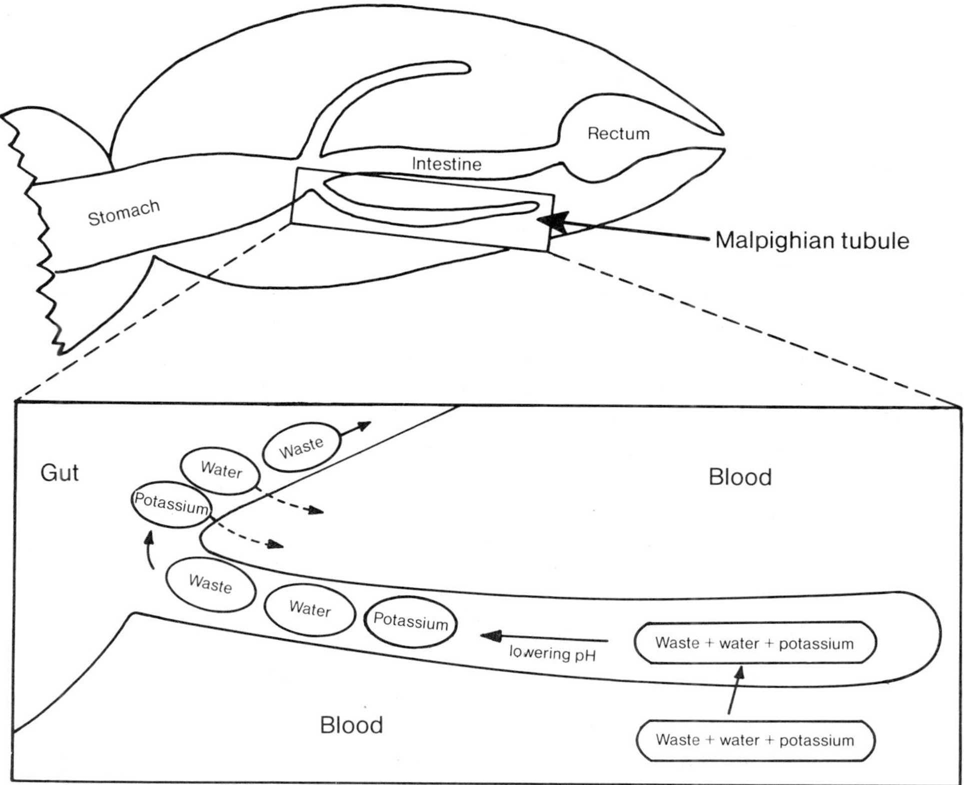

Fig. 4.6 The excretory system of insects consists of the Malpighian tubules that "purify" the blood by removing nitrogenous wastes. Once in the hollow tubule's cavity, the materials are physically emptied into the gut where water and other valuable materials are reabsorbed back into the blood.

they are the sites of silk production in some neuropterans and are the luminous organs of some "glow worm" types of fly larvae. Some of these additional functions are possible because wastes in the form of uric acid are so low in toxicity that they can be stored in specialized "urate" cells instead of being excreted.

TRANSPORT

In order for a cell to function, it must be supplied with food and oxygen. We have already discussed how the digestive system modifies the high energy molecules into utilizable form. These materials are then transported from the digestive system throughout the body by the circulatory system. The oxygen that all cells require in order to release energy from the food is supplied to the cell directly by means of the respiratory system. Thus, we have the two principal transport systems, the respiratory (moving oxygen and most of the carbon dioxide) and circulatory (moving foods, wastes, and other materials) systems. We will not discuss the respiratory system as it deals basically with the problem of supplying oxygen to the cells. Internal respiration, the exchange and utilization of gases within the cell, will be covered in the discussion of energy supplies and utilization later in this chapter.

The Tracheal Repiratory System

Some small, slow moving, thin skinned insects that live in moist environments have no discernible respiratory system. They are able to survive because the large surface-to-volume ratio of their body permits oxygen to pass in through the body wall. Such movement of oxygen is by "diffusion" (the spread of molecules from regions of high concentration to those of lower concentration, tending to equilibrate over time). This movement is only sufficient if the membrane (the body wall) is moist, because oxygen will not diffuse readily through a dry membrane. These membranes are subject to water loss, therefore, insects require a system that minimizes water loss from the respiratory membranes, while it admits greater supplies of oxygen, which is necessary if the insects are going to evolve into larger sizes, be more active, or possess thicker cuticle. A system, that accomplishes this, called the "tracheal respiratory system," has evolved in insects and several other animal groups.

In most insects the system is an elaborate interconnected labyrinth of tubes called "tracheae" (singular, trachea) (Fig. 2.5), which admit atmospheric air into their lumens by means of external openings called "spiracles." The spiracles are often of a complex form, with valvular arrangements, which open into large longitudinal trunks with multiple, radiating branches. The very smallest branches, called "tracheoles," ultimately penetrate or lie next to the cells of almost all of the body's tissues (see Fig. 2.5). The entire tracheal system is lined with a cuticle similar to that found in the integument. The cuticle tends to be deposited in spirally arranged thickenings (taenidia) that provide for maximum flexibility without collapsing.

Diffusion is an adequate mechanism for supplying oxygen if the oxygen demand is low. However, insects that are large or move quickly use oxygen at a rate greater than can be replaced by diffusion alone. These insects have evolved the ability to physically exchange the air in the trachea by "ventilation." In order for ventilation to be efficient the insect must be able to regulate the spiracular valves. Then if muscles apply positive pressure to the body while the posterior spiracles are closed, the tracheal air is expelled through the thoracic spiracles. Closed thoracic and open abdominal spiracles during abdominal muscle relaxation allow new air to enter the system. In this way the system can be flushed of exhausted air with efficiency. Insects that can ventilate their trachea in this way do so only when it is necessary because tracheal ventilation is associated with heavy water loss. In some insects, weakly fortified expansions of the trachea form air sacs that function in many different ways, such as in increasing the ventilatory capabilities of the system, in controlling heat loss during maximum exertion, and in lightening the body.

Carbon dioxide (CO_2) escapes from the body by means of the tracheal respiratory system. However, since the tissues are 36 times more permeable to CO_2 than to oxygen, its escape through the body wall is less a problem in respiration than the acquisition of oxygen. CO_2 is a problem when its removal causes a relaxation of the water conservation mechanisms, as during some of the immobile stages.

Modifications for Aquatic Life

A tracheal respiratory system equipped with spiracular valves and capable of active ventilation is easily adaptable to aquatic life. All that is needed for the insect is to make frequent trips to the surface in order to replenish the tracheal air. Some insects that spend their lives in water have such systems. Others have additional refinements such as extensible snorkels that can be pushed through the water's surface (see Fig. 4.7).

Many aquatic insects carry bubbles of air with them when they submerge. Some of them capture the air beneath their wings; others carry a fine layer of air over much of their body surface. These bubbles on the body surface are held by finely packed hairs that are coated with wax, and therefore repel water. Thus, the insect is like an immersed piece of velvet, carrying air on its surface from which it can remove oxygen. If the water has sufficient dissolved oxygen in it, the used oxygen is replaced by dissolved oxygen from the water, and the insect can continue to use the bubble as a physical gill. This system provides sufficient oxygen to enable the insect to stay submerged for up to 13 times longer than if it relied only on the oxygen within the original bubble.

Some groups of aquatic insects are able to survive by extracting oxygen dissolved in the water. Most species are able to do this because one or more areas of their bodies have many fine branches of the trachea packed closely together and are covered by very thin cuticle. These respiratory surfaces are often extensions of the body and take the form of leaflike or filamentous "tracheal gills" (see Fig. 4.7), where oxygen diffuses from the water into the closed tracheal system. In dragonfly nymphs the gills are within the rectum.

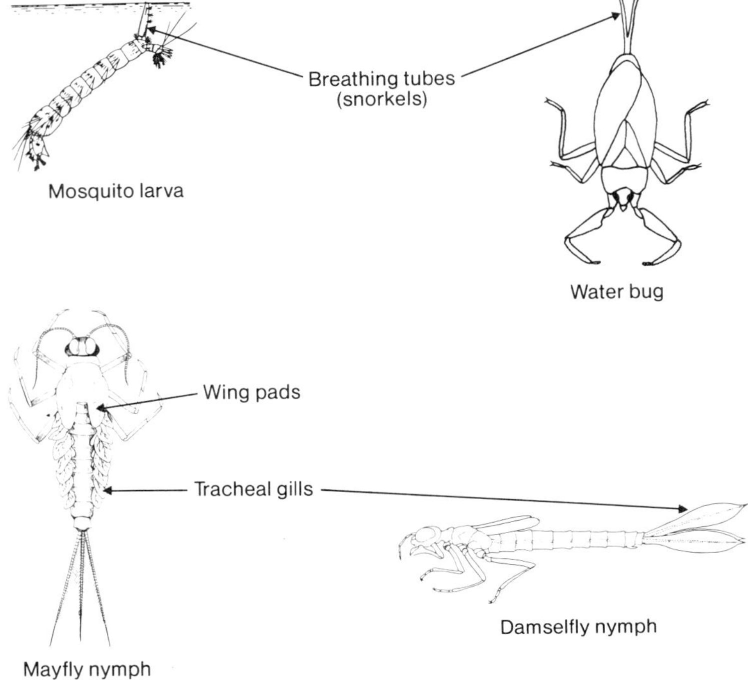

Fig. 4.7 Some respiratory modifications of aquatic insects.

Circulatory System

Transport of materials other than gases within the body is accomplished by the circulatory system. Nutrients are moved from the digestive system or from storage organs (fat body) to sites of utilization. The waste products from cellular activities are moved to the Malpighian tubules. Hormones and other secreted materials also move by means of the circulatory system. In addition, the system serves in a defensive capacity by countering the activities of foreign organisms (infectious disease organisms and parasites). This last activity is also the means by which dead or dying cells of the body are removed.

The system consists of the "hemolymph" (blood) and those structures and functions associated with circulation of the hemolymph (see Fig. 4.8). There are both functional and structural differences between the system in insects and that in vertebrates. Functionally, the blood in vertebrates (e.g., humans) is vitally involved in respiration since oxygen is carried from the lungs by red blood cells to sites of use. This function is carried out directly by the many-branched tubular repiratory system in insects, so that the circulatory system is not involved in oxygen transport. Structurally, the blood in a vertebrate body is always contained within blood vessels (arteries, capillaries, or veins), thus forming a closed circulatory system. In insects, as

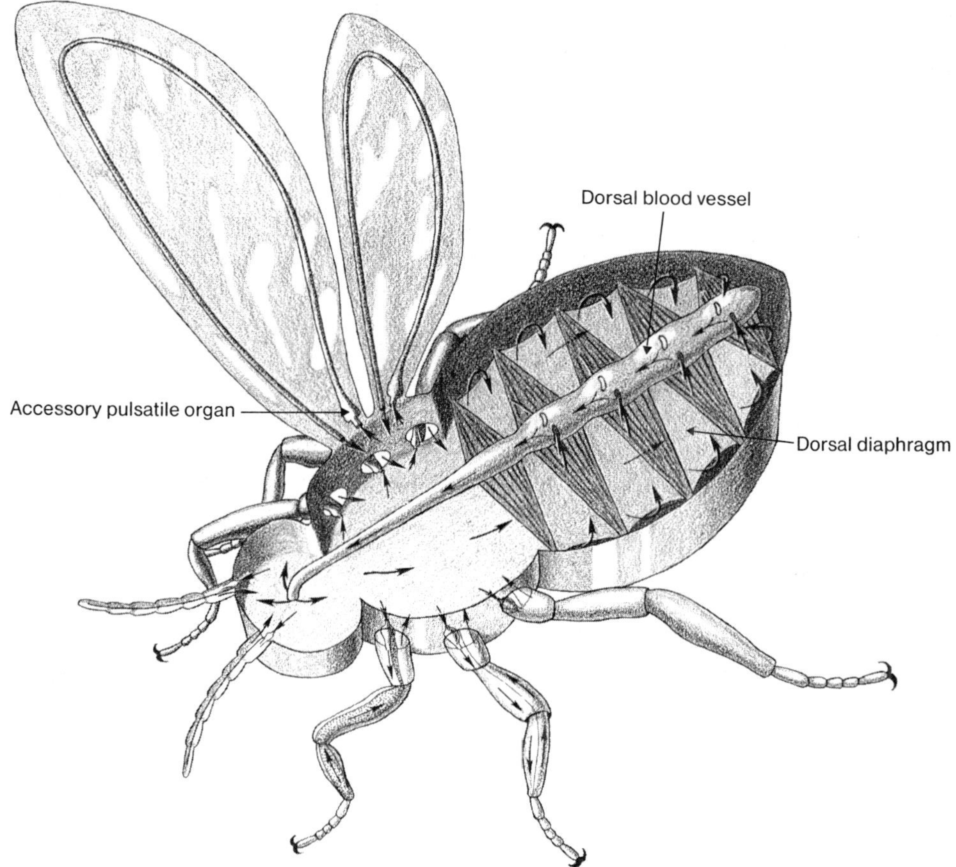

Fig. 4.8 Structural elements of the circulatory system. The key structure is the dorsal blood vessel, which is a contractile heart with segmentally arranged valves in the posterior portion. Immediately beneath the dorsal blood vessel is the dorsal diaphragm around which blood percolates as it flows back through the body cavity. Legs, wings, and antennae often have accessory pulsatile organs and diaphragms, which aid blood flow within them.

in all Arthropoda, the system is an open one, that is, only a small part of the pathway through which the blood travels is within a blood vessel, most of the time the blood is not inside vessels, but bathing all of the internal organs as it moves.

The only blood vessel present in most insect groups lies just inside the integument and runs lengthwise along the midline of the insect's back. The posterior part of this "dorsal blood vessel" is the heart, the circulatory pump of the system. It functions, as do our own hearts, by alternate contraction and relaxation. However, the "heartbeat" is not as consistent and rhythmic in insects as it is in vertebrates. Since the heart is elongate, a wave of contraction passes from the posterior forward, squeezing the blood ahead of it. Whereas the necessity for continuous oxygen supplies to nerve centers is tied to continuous blood flow in humans, in fact, in insects the heart may stop contracting for relatively long periods, or the wave of contractions may be reversed.

The blood usually leaves the dorsal vessel at its point of termination in or near the head. The path of the blood back through the body is directed by "diaphragms" (sheets of tissue), which divide the body into ventral, middle, and dorsal compartments. Since each diaphragm is not "water-tight," the blood percolates from the lower compartments into the upper one and eventually back through a valve into the heart. Much of the flow is caused by general body movements as well as by contractions which create differences in pressure near the entrances and exits of the dorsal blood vessel. Flow is also enhanced by the very distance between the blood's entrance into the dorsal vessel and its point of exit. In a much shorter vessel, eddies would form around the heart itself and considerable stagnation would occur in areas some distance from it.

The system as described so far would be sufficient to handle the blood flow within the body provided there were no appendages opening off of the main body. However, the muscles and other tissues within the legs, wings, mouthparts, and antennae also require the blood and its transport functions. "Accessory pulsatile organs" located at the bases of these appendages provide this blood flow. These organs function as hearts and pump blood into the appendage. Since the appendages often have a diaphragm oriented along their length, efficient blood flow is accomplished.

Hemolymph

The blood of insects as in humans is composed of a portion containing blood cells called "hemocytes" and a noncellular or "plasma" portion. These two distinct components of blood perform rather different functions. The blood cells are actively involved in isolating foreign objects or dead cells from the blood. They accomplish this by flowing around an object and engulfing it, a process called "phagocytosis." Some insects have hemocytes capable of encapsulating foreign objects (e.g., eggs of parasites) that are in the blood-filled body cavity (haemocoel). During the encapsulation process, hemocytes first adhere to the surface of the foreign object, then changes in the layer of hemocytes eventually wall off the object with a relatively inert melanin pigment. This defense reaction has been shown by several species of mosquitoes against nematode parasites. Blood cells are also involved in the formation of the basement membrane of the integument, in the coagulation and eventual healing of wounds, and in the secretion of materials, especially those associated with the molting process.

Microscopic examination of blood cells reveals a bewildering array of cell types. This is so because cells change in size, shape, and textural characteristics as they mature from the juvenile, or "prohemocyte," type; only three or four fundamentally different types exist. Blood cells are produced by division and maturation of some of the prohemocytes.

The noncellular or plasma fraction of blood is the medium through which the blood cells, nutrients, hormones, and waste products are transported. This plasma is primarily water, with some inorganic materials (e.g., sodium, chlorides, phosphates, and carbonates) and organic molecules (especially amino acids). Only a few insects have pigments in their blood, and those, such as bloodworms, do not have their hemoglobin within the blood cells but in the liquid portion. In addition to its just-mentioned functions, plasma serves as the primary reservoir of water in the insect's body.

MUSCULATURE

The most abundant type of tissue in the insect body is muscle. It fills much of the thorax and head, operating the organs of locomotion and feeding. Muscle also occurs within the legs and antennae. These muscles are referred to generally as "skeletal" muscles, that is, they attach to the exoskeleton and are involved in both movement (walking, swimming, flying, feeding) and in maintaining a given posture (see Figs. 4.9, 4.10, and 4.11). Other muscles occurring in sheets or layers surrounding various organs (e.g., gut and reproductive ducts) are the "visceral" muscles. These muscles move materials within organs, such as the food within the gut, and seldom are attached to the exoskeleton.

The contractile capability, which all animal cells possess, has become highly evolved in muscle cells. In fact, the key characteristic of muscles is their contractility. This is accomplished by telescoping intracellular fibers at the cost of considerable energy. The energy is released from high energy phosphate molecules called adenosine triphosphate (ATP). The heavy demand for oxygen is evidenced by the mass of tracheae that cover and infiltrate muscles such as those used for flight. Since most muscles require a nerve impulse to contract, the muscles also have numerous nerve endings. In fact, some muscles have two sets of nerve endings. One results in a powerful twitch of uniform intensity and short duration. The second set has a graded type of response in which rapid successive impulses produce progressively stronger

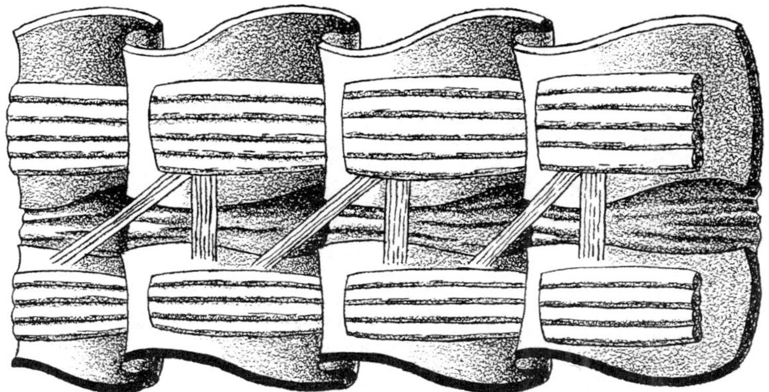

Fig. 4.9 The basic skeletal muscles in the abdominal segments. The muscles shown provide depression and telescoping of the segments, movements involved in respiratory ventilation.

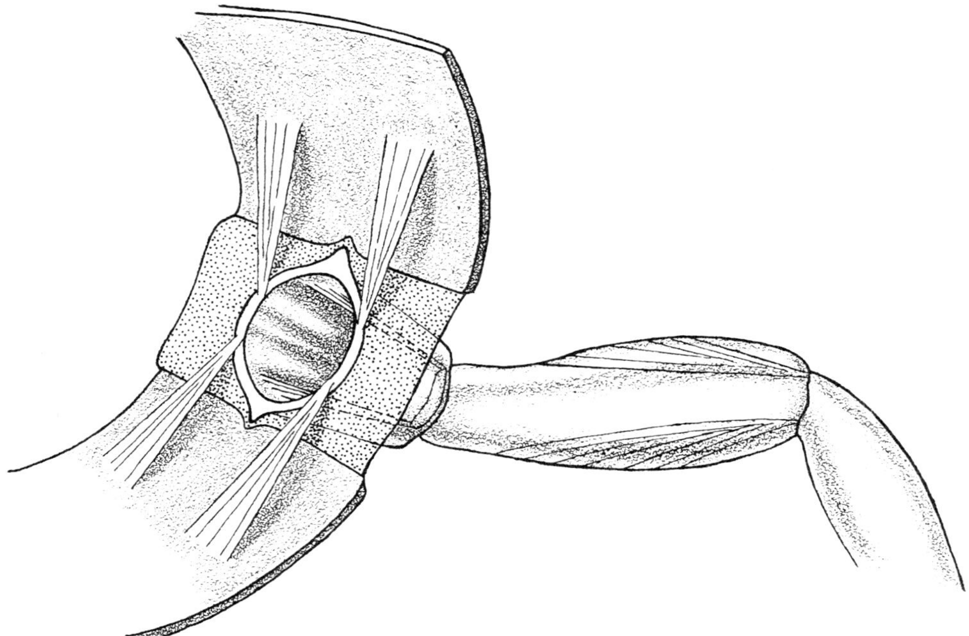

***Fig. 4.10** Diagrammatic sketch of the simplest sets of opposing muscles necessary to provide a walking function. Note that each muscle has an opposing one that can return the appendage to its original position upon contraction.*

contractions. Thus, a single muscle, for example, one in a grasshopper's jumping hind leg, can provide a powerful contraction resulting in a jump when one set of nerve endings is fired, or can maintain position or exhibit slow movements (as in walking) when the other set operates. Although a number of different types of muscle have been described in insects based on structural rearrangements of their components the most important differences are functional. The flight muscles of insects with relatively slow wing beats (e.g., dragonflies and butterflies) are "nonresonating," that is, they require a single nerve impulse for every contraction, usually reaching a rate of about 25 per second. Smaller insects with very-high-speed muscle contractions have "resonating" muscles. This type contracts more than once for every nerve impulse so that the stimuli for repeated contractions come from the muscle itself. This is set up when the pull from one muscle in the insect's flight mechanism increases the length and tension of the opposing muscle, causing it to contract, which in turn causes another contraction of the opponent as it is stretched. Flight muscles in insects with 200–1000 wing beats per second can only function as resonating muscles because more frequent nerve impulses would cause one prolonged contraction rather than many short successive contractions and relaxations.

A last important point in discussing the insect muscle system is an evaluation of the relative strength of insect muscles. It has been stated that if a flea were the size of a human, the flea could easily jump over a multi-storied building. Other insects, like

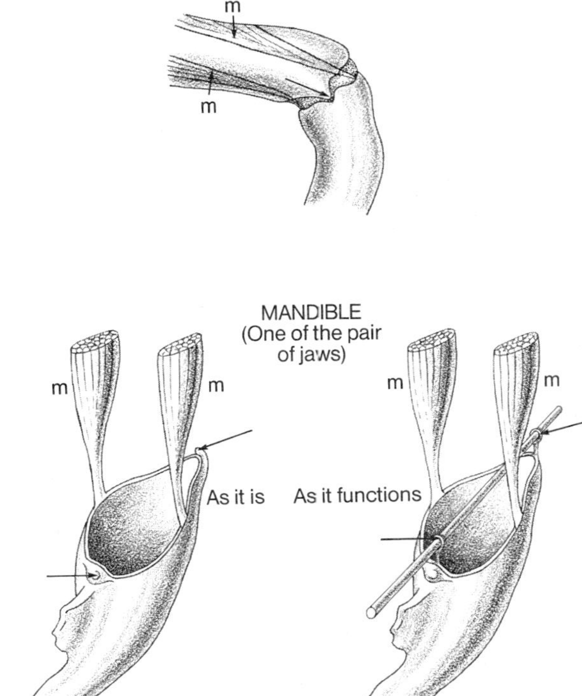

Fig. 4.11 Articulations (movable points of contact between two hard structures) at joints between segments. Opposing muscles (m) are indicated to show how movement is controlled. The arrows indicate the points of articulation.

ants, would supposedly be able to lift huge weights, if they were man-sized. These generalizations are made from observations of the distances real fleas jump in relation to their size, or the amount of weight a living ant is able to carry. However, insect muscles are about as powerful in their contractile force as are those of humans. The difference lies in the absolute size of each organism, because the power of the muscle is proportional to the area of its cross-section, whereas the weight is proportional to the volume. Therefore, as absolute size is increased the weight of the body increases at a faster rate than the strength.

Storage and Utilization of Energy

As mentioned in the preceding paragraphs, the energy required by the muscle as it contracts is supplied by high energy phosphate molecules (ATP), which ultimately are derived from the food taken in by the individual. Foods such as carbohydrates are either stored by the insect or utilized as a source of energy. The "fat body" is a storage organ that allows energy to be accumulated by the insect so that it is available for later use (e.g., in movements such as flight, in rebuilding the body when the insect changes form, or in living through harsh conditions). The fat body is a mass of cells scattered in many parts of the body, but loosely tied together. These cells fill up with glycogen (a storage form of carbohydrate) or lipids. Lipids are better forms of energy storage because the droplets can be stored directly in the cells, whereas glycogen requires 2.3 units of water for each unit of glycogen to remain in solution. This means that the more energy rich (because of its higher caloric value) fat may be up to eight times as efficient as glycogen on a weight-to-weight basis. The energy from glycogen or lipid molecules is released in a chain of chemical reactions within the cell in which the energy is used. The series of steps, called a "metabolic pathway," allows the familiar reactions

$$\text{Carbohydrate} + \text{Oxygen} \rightarrow \text{Carbon dioxide} + \text{Water} + \text{Energy}$$

If this reaction were allowed to occur without control of the rate, most of the energy would be released in the form of heat and light rather than in a useful form. The cellular mechanism both controls the rate and subdivides the reaction into a great many small steps, thus enabling it to maximize conversion of the energy into usable (high energy phosphate, or ATP) forms. Each step is controlled by an enzyme, which converts one existing molecule (substrate) into another, which (in turn) becomes the substrate for the next enzyme to attack. This sequence happens because each enzyme acts very specifically on one substrate.

Figure 4.12 represents the biochemical pathways whereby energy is released from glucose, a simple sugar, and made available for muscle contraction or other work. Glycogen (storage carbohydrate) and tehalose (insect blood sugar) are converted into glucose before being metabolized. Other energy forms, like lipids, are converted into intermediate molecules that enter the Krebs cycle part way through its sequence.

Locomotion

Insects, as a group, employ many different types of locomotion, including various kinds of walking, running, jumping, swimming, burrowing, crawling, and flying. Of these many types, walking is the most common and flying the most unique.

In the adult, six legs are employed in a sequence of movements that result in walking. The classic form of walking is illustrated in Fig. 4.13. The three points of contact (the front and hind legs of one side and the middle leg of the other) form a stable tripod that supports the body, while the other tripod (the rest of the legs) are moved into position. Cinematographic studies have revealed, however, that many patterns are followed, depending on which species is being considered and the speed with which it moves. Cinematography has been helpful in making direct observations because it can record rapidly occurring events, such as an insect's walk. Then the film is rerun, but at much slower speeds. Thus the variety of movements can be analyzed

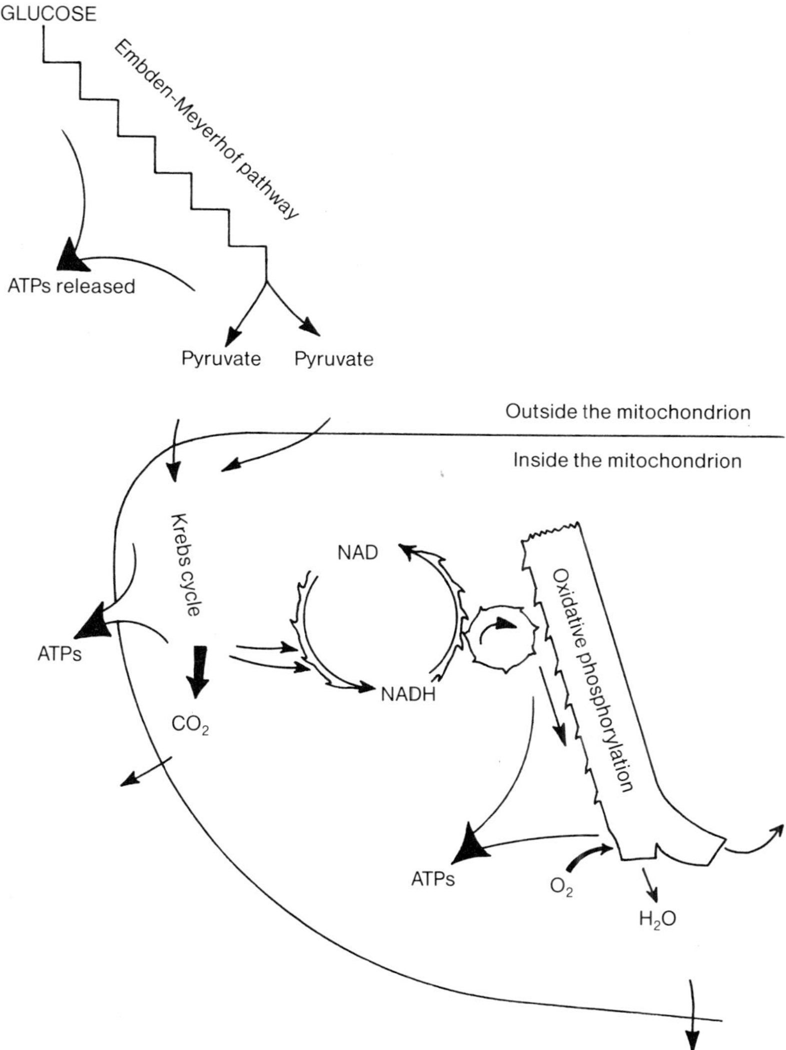

Fig. 4.12 ***The basic sequence of pathways in the cellular release of energy from foodstuffs. Here the energy release involves approximately 140 reactions (many are repeated) in the efficient extraction of energy within a single molecule of sugar. Energy could be released in a single step (i.e., by combustion) but most of it would be in unusable forms (i.e., heat and light energy). Yield of energy from a single molecule of glucose is about 72% efficient, resulting in 38 molecules of ATP.***

and differences in gait can be noted. From these studies, it is apparent that a running cockroach is not moving in the same way as a cockroach that is walking very quickly.

The legs of fleas, some beetles, some leafhoppers, and the grasshoppers are used in jumping. Although somewhat different mechanisms provide the thrust in each case, the effect of vaulting into the air is the same. Several of these insects fly after

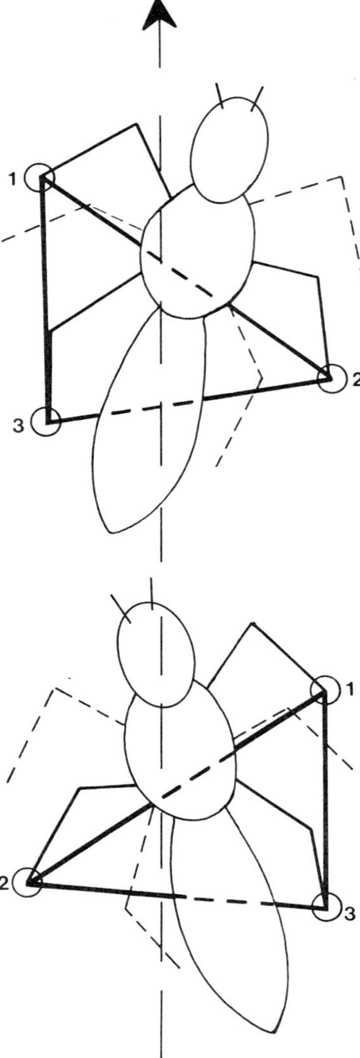

Fig. 4.13 Use of the insect's six legs to form a pair of tripods, which are placed alternately as the insect walks.

initiating the jump, but the wingless fleas and the immature stages of grasshoppers and leafhoppers fall back to earth after a jump. It is usually easy to distinguish which of a particular insect's legs are those that provide the power for the jump, since they are usually larger than the other legs. It is worth noting that these insects only jump when alarmed and that walking is a normal method of locomotion. In this way, they are similar to toads, where hopping is an escape mechanism only employed in stressful conditions.

Some insects have other mechanisms that they employ in jumping. Insects in the Order Collembola are known as springtails because of the elongate terminal structure they usually possess. Prior to the jump, the structure is bent forward beneath the body and clasped by a gripping organ. When it is released, either its muscles or the elastic cuticle (depending on the type of springtail) straighten out the "tail," thereby propelling the insect into the air. Click beetles throw themselves into the air when the first thoracic segment is suddenly straightened out from a previously lowered position. As it is released by a catch mechanism an audible click is produced, hence the name.

Insect larvae, such as caterpillars, maggots, and grubs often move by crawling. Such movements may or may not involve the use of legs, but are basically an alternating contraction and extension that results in telescoping body movements and thereby locomotion. The insect may grab the surface by using its mouth hooks, as in maggots, or by grasping with the segmented thoracic legs. Some larvae have fleshy abdominal legs that grip the surface and these "prolegs" may be tipped with a variable number of hooks that aid in grasping.

Most caterpillars (larvae of moths and butterflies) have prolegs on four or five abdominal segments, so that telescopic crawling is possible, but some measuring or inch worms have prolegs only on the back of the abdomen. This arrangement of legs results in a type of crawling as illustrated in Fig. 4.14.

Aquatic Insects

Locomotion by aquatic insects in or on the water may take the form of walking, swimming, crawling, or even flying. A number of insects live on the surface of water. They are small enough to be buoyed up by the water's surface tension, and use their legs to walk across the surface, usually moving the two legs of a segment in unison. Surface-walking forms do not have claws at the very tip of each leg, as do other insects. The claws are located a little back from the tip of the leg and the tip is covered with tightly packed hairs, giving it a velvety appearance. The many hairs are wax coated and, collectively, keep the leg from breaking through the surface tension of the water. These insects, like the water-strider, lose all ability to move on water if the surface tension is broken.

Many insects walk or crawl about on the bottom of lakes, ponds, and streams. Their mode of locomotion is similar to that used on land, except that in moving water they must contend with currents that could sweep them downstream. Most of these forms are heavier than water so that they do not bob to the surface if they lose contact with the substrate.

Buoyancy in the true swimming insects ranges from positive in many aquatic bugs and adult water beetles, to negative in many midge and primitive fly larvae. Most of those with positive buoyance respire atmospheric gases from bubbles or snorkels, whereas negative buoyance is often linked to insects with gills. The swimming insects usually have legs that are long and flattened or have a fringe of hairs that result in an oarlike structure. The two legs of a segment are used by most swimmers in pairs, rather than alternately as in walking.

Several aquatic insects swim without using legs at all. Dragonfly nymphs forcibly

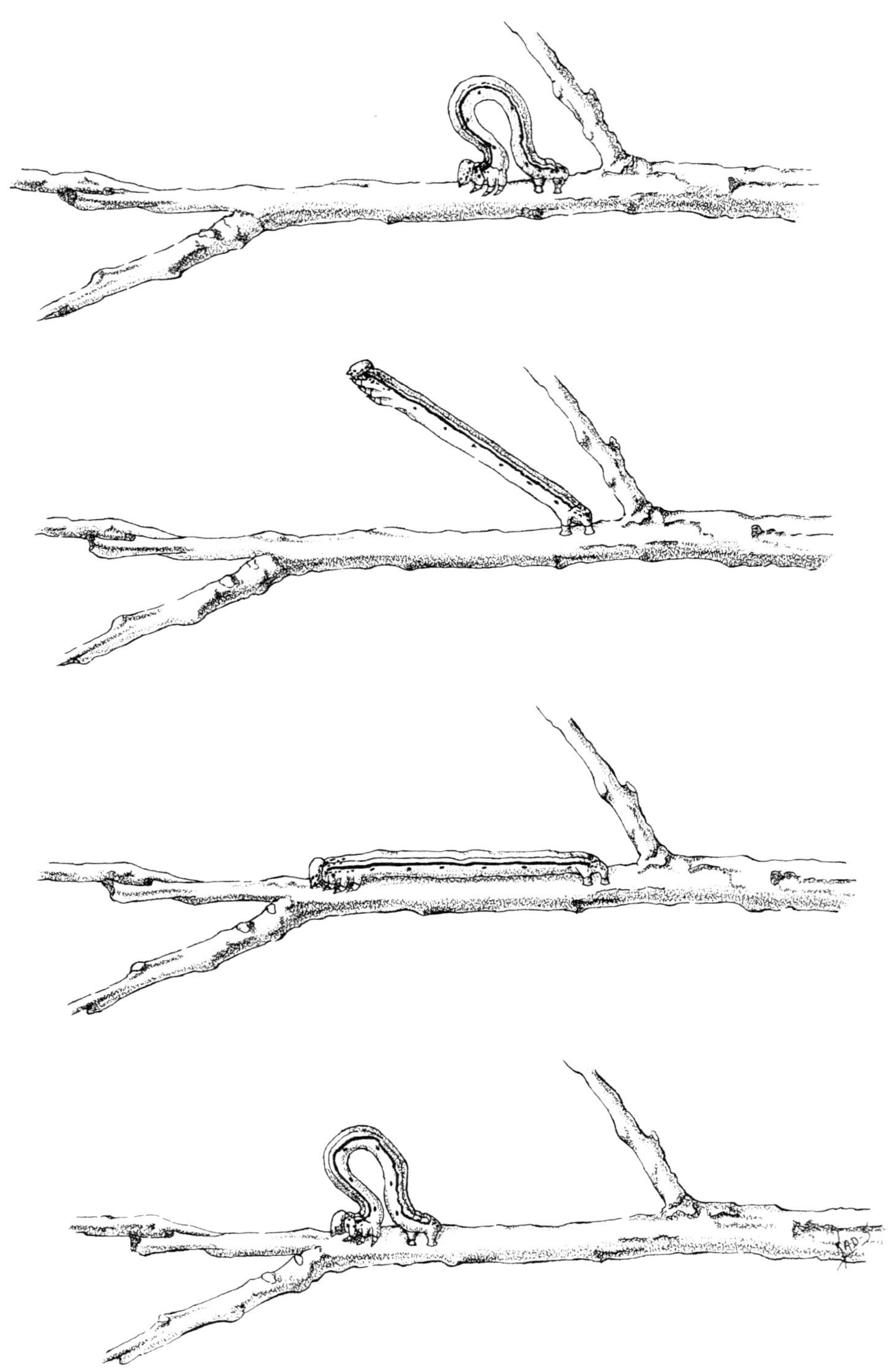

Fig. 4.14 Locomotion by measuring worms (also known as span worms, inch worms, and loopers). Note the forward progress from the starting point indicated by the twig.

expel the water that bathes their rectal gills through the anus, thus effecting a jerky, pulse-jet that propels them through the water. Many larval insects swim by undulations of the whole body. Such swimming is frequently aided by a tuft of hairs or other structure that effectively acts as a flattened fin or tail, as in the larval and pupal mosquitoes. A few species of adult insects propel themselves under water by rowing with their wings. This type of propulsion is limited to adults, since the wings are not functional except in the adult stage.

Flight

Among all invertebrates, insects are the only forms that have conquered the air by evolving active flight as a method of locomotion. Other invertebrates such as spiders and mites may undergo passive transport by ballooning in the air on strands of silk, but only the insects can move through the air with relative freedom from moderate wind currents. The evolution of wings not only simplified long-distance travel for food and mating, but also helped in avoiding natural enemies. The fact that the majority of the insect groups possess wings is tribute to the significance that active flight has had in the evolution of insects.

The existing groups of primitive insects (Subclass Apterygota = without wings) are relatively few in number and restricted to a basically detritus-feeder lifestyle. Yet from these modest evolutionary beginnings the rest of the insects (Subclass Pterygota) evolved. They all have wings or had wings, which have been lost. Scientists conclude, therefore, that wings have evolved only once in the insects.

It is interesting that the evolution of wings in insects differs so radically from the evolution of wings and flight in the vertebrates (bats and birds). The primary difference is that the flying vertebrates use greatly modified appendages that originally were walking legs, whereas insect wings evolved from structures having nothing to do with the early segmental appendages of arthropods. These structures are called "paranotal lobes" and are lateral extensions of the body wall (Fig. 4.15).

A difficulty in establishing that paranotal lobes are the forerunners of insect wings is that the "missing link" is still missing. No one yet has found a fossil insect with large paranotal lobes, but without wings. Fossil insects have been found that are either wingless or have fully formed wings. Some winged fossils, however, also have paranotal lobes on the first thoracic segment (Fig. 4.16). Such findings lead us to assume that the forerunners of winged insects must have looked like that in Fig. 4.17A and that they must have had some advantage at this evolutionary stage over types without paranotal lobes (Fig. 4.17B) in order to have survived.

Paranotal lobes could have provided an advantage in several different ways. The most widely accepted theory is that the lobes allowed the insect to glide, if it jumped from an elevated area like a tree or rock. Gliding would be an effective way to escape from a natural enemy. Paranotal lobes might also have functioned to cover legs and other parts, thereby camouflaging the insect's outline from predators, while increasing its apparent size. This would explain the prominence of lobes on the leg-bearing body segments and may have been a step preceding the glider function.

Several changes are required to make an insect that can glide into one that can

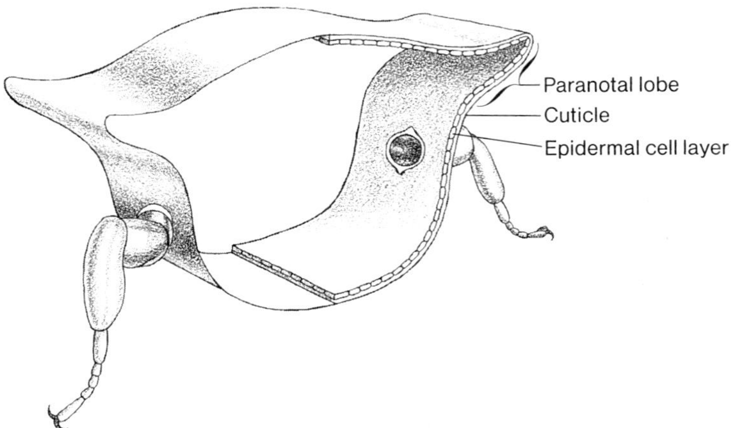

Fig. 4.15 Cross-section through a thoracic segment showing that paranotal lobes are integral parts of the body wall, with epidermal cells covered by the cuticle they have secreted.

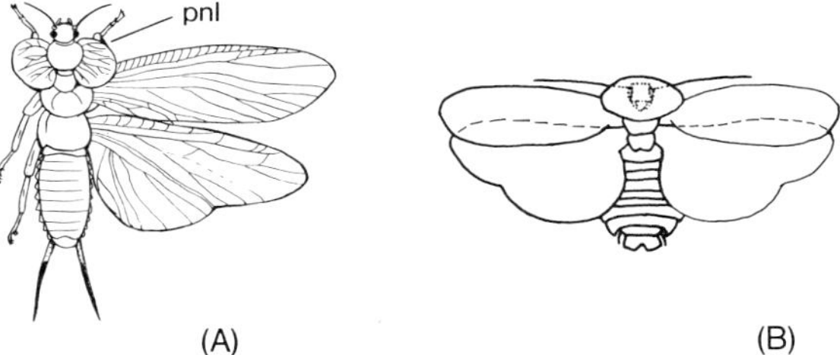

Fig. 4.16 (A) A fossil insect with paranotal lobes (pnl) on the first thoracic segment and fully formed wings on the second and third. (B) A common family of tropical roaches showing that some living insect groups also have expansions of the notum.

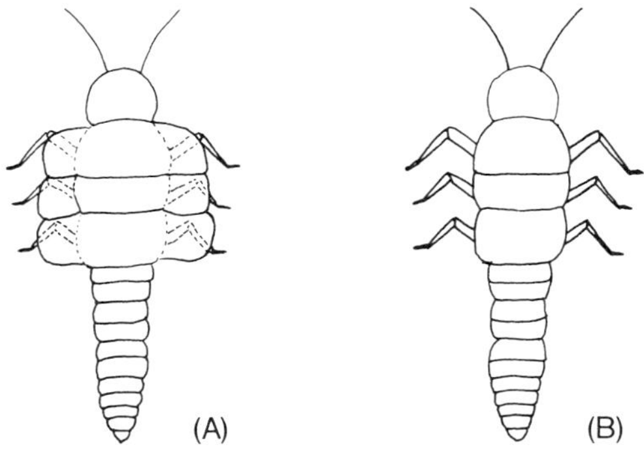

Fig. 4.17 A comparison of the paranotal lobe stage (A) of wing evolution with a lobeless form (B). Some selective advantage must have been operating in this stage of evolution.

fly. Among these changes are hinging at the wing base, strengthening of the lobe as it gets progressively longer and thinner, developing a mechanism to move the wings, and strengthening of the body wall adjacent to the wing area.

The musculature for moving the wings of insects was present in the ancestral animal, even before it became an insect. In fact, the muscles that provide much of the power for the wings' strokes are not attached directly to the wing base, but to various regions of the thorax. These muscles are the "indirect flight muscles" (see Fig. 4.18). The vertical muscles pull the top of the segment down, thereby raising the wings. The longitudinal muscles lower the wing as they contract and draw the front and back of the segment together, thus popping the top of the segment up. These powerful muscular movements of the thoracic elements require that the unit also have considerable rigidity. This is provided by an internal skeletal brace called the furca and the muscles attached to it. The furca has become highly evolved in winged insects, but is rarely present in insects whose ancestors never had wings.

As the paranotal lobes lengthened and became thinner, the wing veins evolved. Originally these veins were arranged in a fluted pattern, which maintained the wing's rigidity. Early flight wings might have looked like those in Fig. 4.19a. This fluting

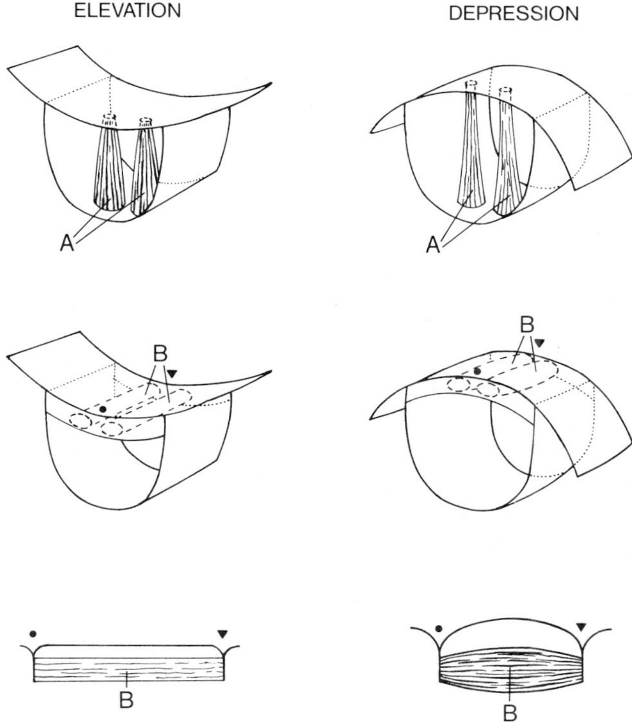

***Fig. 4.18** The indirect flight muscles move the wings up and down by raising and lowering the top of the thoracic segment. On the upstroke A contracts and B relaxes. On the downstroke B contracts and A relaxes. B is a band of longitudinal muscles that runs from the front (●) to the back (▼) of the segment.*

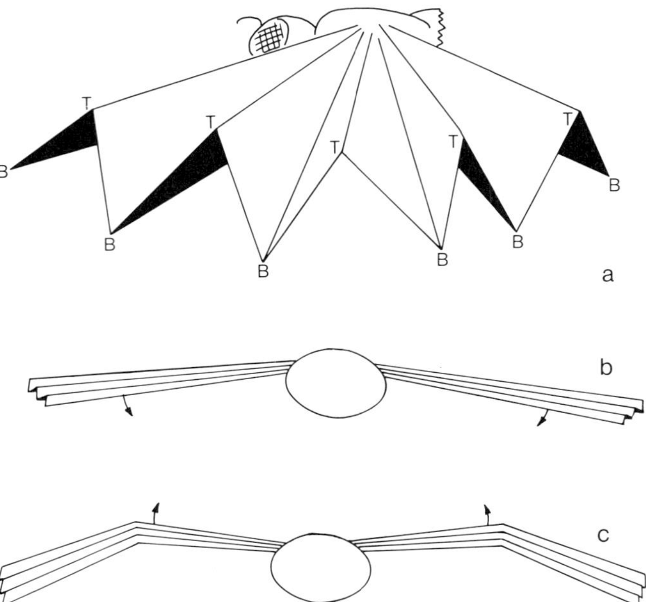

***Fig. 4.19** The fluted wing and flapping flight of primitive insects. (a) The convex (T, top) and concave (B, bottom) veins are labeled in the single wing that is illustrated. (b) The wings retain their rigidity on the downstroke, while the areas of weakness in the concave veins allow the wing to bend on the upstroke (c). These wing movements yield vertical patterns and are typical of the mayflies (Order Ephemeroptera) which have this type of flight mechanism.*

(corrugation) is used today to strenthen cardboard boxes and sheet metal roofing. However, although rigid corrugations would allow for lift on the downstroke, it would be counteracted by the upstroke, unless some method were evolved to allow the air to slip by. This is accomplished by weak points in the veins at the bottoms of the corrugations. The top veins (convex) maintain rigidity on the downstroke, but the weak areas along the concave veins (at the bottoms of the corrugations) let the wing bend on the upstroke (see Fig. 4.19b,c).

Although it is evident that all of the early winged insects used flapping flight (as do most mayflies still), most modern-day flying insects have a type of wing movement that differs from the original. Instead of fluting, flapping, and vertical pathways, the action is more a sculling type of flight with emphasis on control of the angle of the wing as it moves up and down. Change of wing angle on the downstroke is accomplished by "direct flight muscles" connecting the thorax wall to the wingbase. On the upstroke, the wing angle change is passive. Change is facilitated by a thoracic arrangement whereby the wing sits on a single point of hard cuticle, surrounded by membrane (see Fig. 4.20). In addition, most of the rigid wing veins in modern fliers occur in the leading edge of the wing so that this, in combination with the thoracic contact point, angles the wing forward and backward as it moves up and down. In fact, the wing moves more in a figure 8 pattern with the axis slanted from the vertical; a change in the axis of the figure 8 controls whether the insect moves forward, hovers, or moves backward.

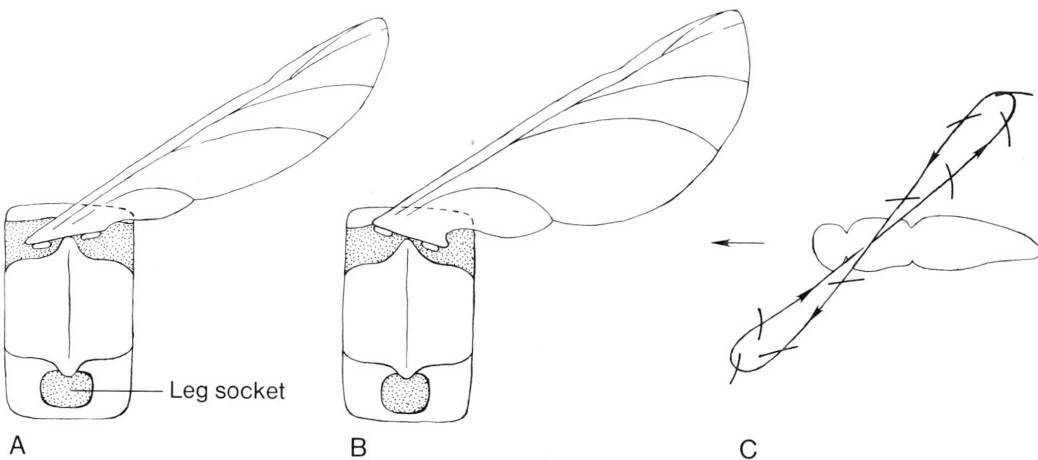

Fig. 4.20 The sculling flight of insects. (A,B) The wing resting on the single hard point of cuticle and otherwise surrounded by membrane. The wing is angled down in (A) and back in (B) by direct flight muscles which attach at the wing bases. (C) The figure 8 pathway with the angle of the wing given at several positions.

The discussion of flight thus far has involved what a wing does as it is moved. However, most insects have two pairs of wings, so that it is necessary to consider how the two pairs move with respect to each other. The most common situation is one that is followed by most insects, wherein the front and hind wings move in unison. This pattern of movement is easily accomplished by a variety of wing-coupling techniques that have evolved in insects, such as simple overlapping or attachment by a lobe, hooks, or spines. The most effective way to achieve a single flight surface is to have only a single pair of wings remaining, as in the Diptera. The second pair of wings in Diptera has been reduced to a pair of clubs (halters) that have a sensory function.

Wings have enabled insects to use the aerial environment better than any other invertebrate groups. However, when the primitive insect was not actually in flight, the presence of those large, fragile structures must have been a hindrance. It would be nearly impossible to run around on the ground or burrow beneath the surface with the wings sticking out. Even today the two most primitive fliers, the Ephemeroptera (mayflies) and Odonata (damselflies and dragonflies), cannot fold their wings and are limited by this inability, but in different ways. Odonata capture their insect prey on the wings, but cannot chase down insects on vegetation covered ground. Ephemeroptera do not feed in their brief (1–3 day) adult existence, so that their wings are used solely for dispersal and mating. The rest of the insect groups are derived from ancestral types that evolved a wing-folding mechanism. In this way, the wings could provide maximum mobility, but not otherwise limit the insect when it was not in flight.

Not only do the modern wing-folding insects get their wings out of the way when at rest, but several additional problems have been solved in various groups through wing modifications. In the beetles the forewings have thickened into wing covers that protect the delicate flight wings folded up under them. The same kind of protection is afforded by the much shortened forewings of earwigs, but is complicated by the intricate folding mechanism of the flight wings. The short forewings allow the insect to have much more body flexibility and to facilitate its secretive nature. Wings of moths and butterflies are covered with scales, as well as being variously colored. The forewings of the true bugs are also thickened but only in the basal half; the terminal half and the hindwings are membranous.

COORDINATION

Many activities occur simultaneously within the insect's body, including growth, movement, respiration, digestion, and so on. These activities must be coordinated so as to optimize the insect's probability for survival. The problems in coordination are of two distinct types. Long-term coordination of activities over an extended period leads to the maturation of the individual and completion of its contribution to the next generation. This long-term coordination is the function of the endocrine system. Short-term coordination is responsible for getting the insect out of danger, for finding a food supply, or for other activities that require a rapid series of coordinated responses. This is the function of the nervous system.

Endocrine System

The endocrine system controls activities within the insect's body by means of chemical messengers. The chemicals, termed "hormones," are usually secreted directly into the blood from discrete glands where the hormones are produced. These endocrine glands are located in various parts of the body, and the hormones may, through the circulatory system, affect many separate tissues.

As is indicated in Table 4.2, hormones determine many body functions and

Table 4.2 Effects of the Endocrine System in Insects

Effect	Cause	Insect
Diapause	Lack of activation hormone from brain	*Hyalophora* moth pupae
Polymorphism	Altered balance between developmental hormones produces solitary or gregarious forms	Locusts
Color	Altered balance between developmental hormones produces differently colored forms	Locusts, Walking sticks
Metabolism	Corpora allata increase the basal metabolism	Colorado potato beetle
Water balance	Secretions from central nervous system cause excretion of excess water	*Rhodnius* (blood-sucking bugs)
Daily activity rhythms	Secretions from central nervous system and glands synchronize activity to light–dark intervals each day	Roach, Ground beetle
Development	Glandular secretions control changes in form of body	*Rhodnius*
Molting	Glandular secretions control initiation of the molt	*Rhodnius*, Silkworm

behavior patterns, but the most important is the maturation of the insect. All of growth and development after the insect hatches from the egg is orchestrated by hormones. As with the other hormones, the developmental hormones are produced by both secretory cells of the nervous system and by discrete glands that have no direct link with nervous tissue.

The classical endocrine control of development is presented in Fig. 4.21. In this two-part operation epidermal cells (along with many others) are given two chemical commands. The first command is, "molt." The second command tells the cells what to form at the molt, that is, "what to molt into." The molt command is initiated by the brain, which manufactures "ecdysiotropic (ecdysio = molt, tropin = target) hormone" and releases it from the corpora cardiaca, a pair of storage organs. However, ecdysiotropic hormone does not act directly on the primary end target (the epidermal cells) but on the ecdysial gland. The ecdysial gland secretes ecdysone, the molting hormone, which stimulates the epidermal cells to molt.

The second command starts in the corpora allata, which are a pair of glands located behind the brain and which secrete juvenile hormone. This hormone causes production of juvenile structures at the molt, in addition to being responsible for allowing the individual to mature to an adult. Passing through the three stages of development is accomplished by changes in concentration of hormone. High concentrations of juvenile hormone yield the larval form, low concentrations result in the pupa, and absence yields the adult form.

One may wonder how hormones can influence epidermal cells to form different structures at different times. To comprehend this subject, one must first understand how cells carry and utilize information, involving the chromosomes within the nucleus of every cell. As mentioned in chapter 2, chromosomes are the information-bearing materials, with the information carried in genes. Genes occur in specific numbers along the chromosome and are responsible for individual structures or functions (like eye color or production of a specific enzyme) or they may control more than one function. The genes consist of deoxyribonucleic acid (DNA). Information within the DNA is used by the cell in all of its metabolic activity. Included in this activity is construction by the cell of complex structures, the most complex ones being proteins.

The structure of DNA is essentially like a ladder, with long uniform sides that consist of regularly alternating phosphate and deoxyribose (sugar) molecules. The DNA macromolecule is twisted into a helix. Since the long sides of the DNA ladder are made up only of phosphate and deoxyribose molecules, they cannot be the message-bearing portion of the molecule; this function must be contained within the "rungs." The rungs contain four different molecules: adenine, cytosine, guanine, and thymine, although the four do not occur necessarily with equal frequency.

Analysis of a piece of DNA reveals a structure as indicated in Fig. 4.22. While the sequence of adenine, cytosine, guanine, and thymine is extremely varied along any one side of the ladder, whenever cytosine occurs, its complementary molecule (the other half of the rung) is guanine, and adenine is always linked to thymine. These links are by relatively weak hydrogen bonds, so that when the DNA molecule is split during cell division, each of the two sides of the ladder, with their ½ rungs of adenine, cytosine, guanine, or thymine, can be reconstituted as two complete DNA molecules, one for each of the two daughter cells being formed. This DNA duplication (termed

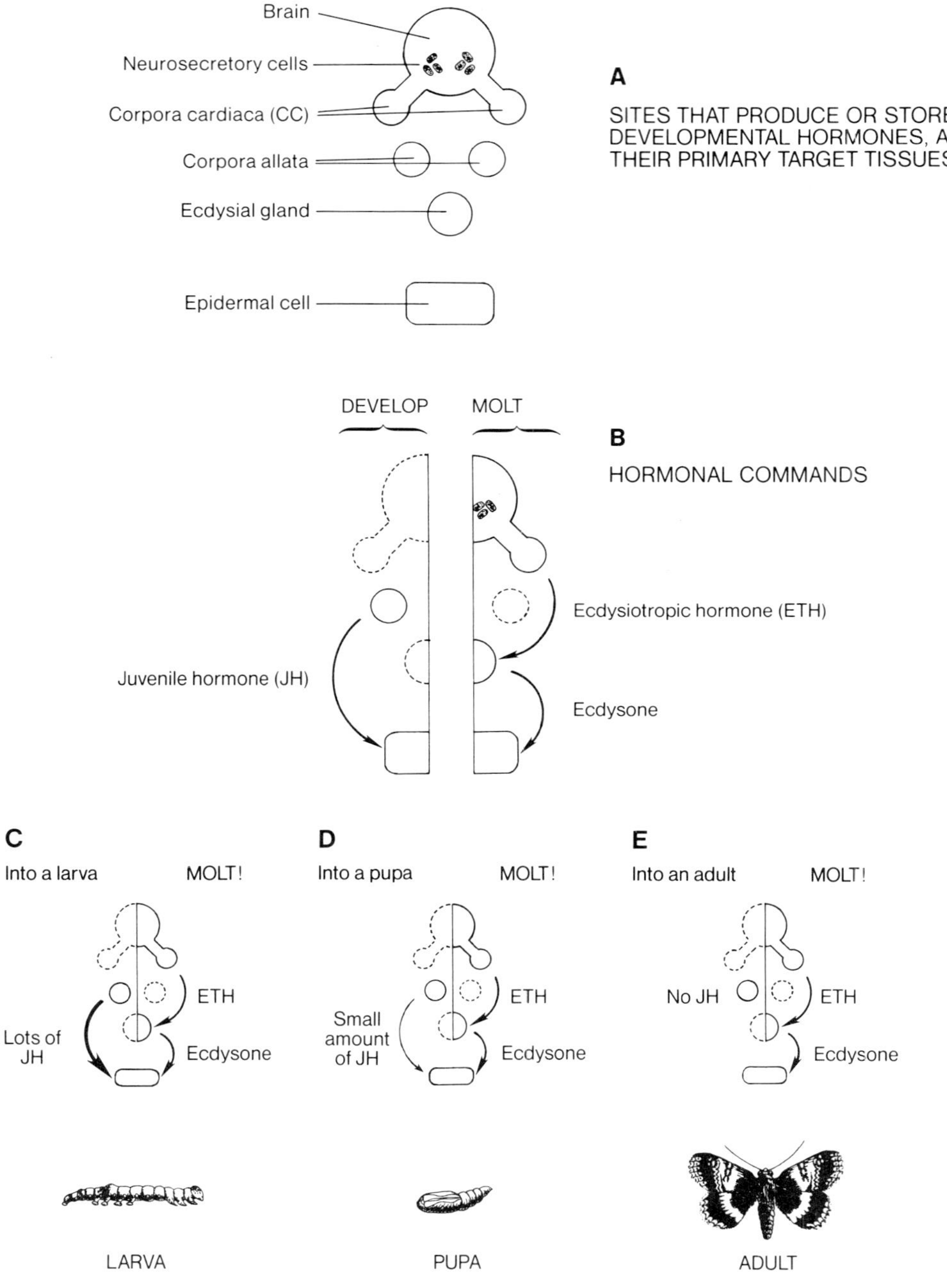

Fig. 4.21 ***Hormonal control of development. (A) The glands that produce or store hormones and their target tissues. (B) The two separate hormonal controls for molting and development within the general system in a split diagram. Note that "develop!" is a one-step command, but that "molt!" is a two-step command. (C,D,E) The combined activity of the molting command and that for development in split diagrams as it changes in larval–larval, larval–pupal and pupal–adult molts by means of varying amounts of juvenile hormone.***

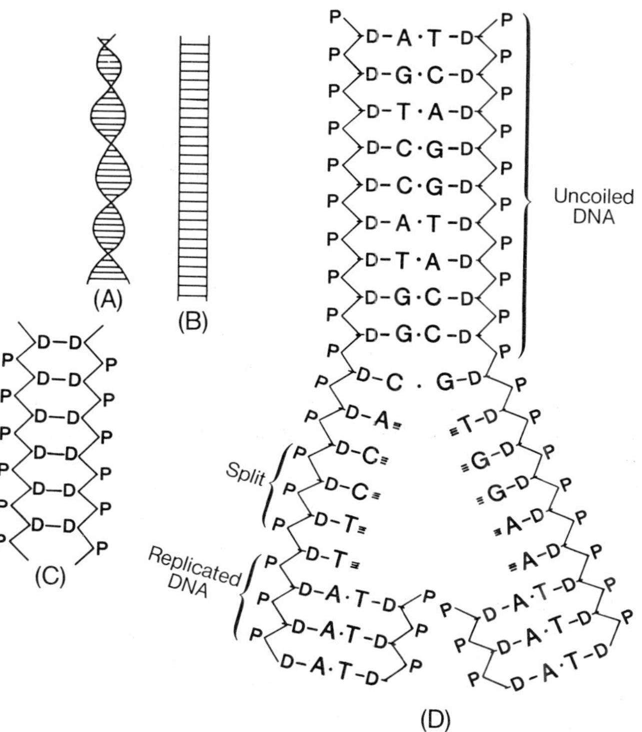

***Fig. 4.22** (A) Deoxyribonucleic acid (DNA) is usually helical (twisted) in form. (B) The essentially ladderlike arrangement if a piece is untwisted. The sides of the ladder are chains of regularly alternating D (deoxyribose sugar) and P (phosphate) molecules, not capable of containing the hereditary information (C). (D) Where the message lies. It is in the sequence of molecules that make up the rungs. Note that the rungs consist of paired molecules, with adenine (A) always matched up to thymine (T) and guanine (G) always with cytosine (C). When the molecule is split as in cell division, each half of the DNA can reconstruct the original molecule exactly as it existed in the parent cell.*

replication) allows each daughter cell resulting from cell division to possess the exact DNA information that the parent cell possessed. It is the precise sequences of adenine, cytosine, guanine, and thymine in DNA that direct the cellular functions and control such aspects as synthesizing complex proteins.

Proteins are assembled by ribosomes, which occur outside the nucleus (see Chapter 2). These ribosomes construct the proteins according to the message they receive from the DNA in the nucleus. The message is carried by another complex molecule, ribonucleic acid (RNA).

Messenger RNA has a construction similar to half of a DNA molecule, except that several molecules are substituted for those in DNA (Fig. 4.23). The sequence of the bases (A, U, G, and C) controls the sequence of amino acids that will compose the protein. However, because there are more essential amino acids than RNA bases, no direct base-to-amino-acid relationship exists. Even every combination of the four RNA

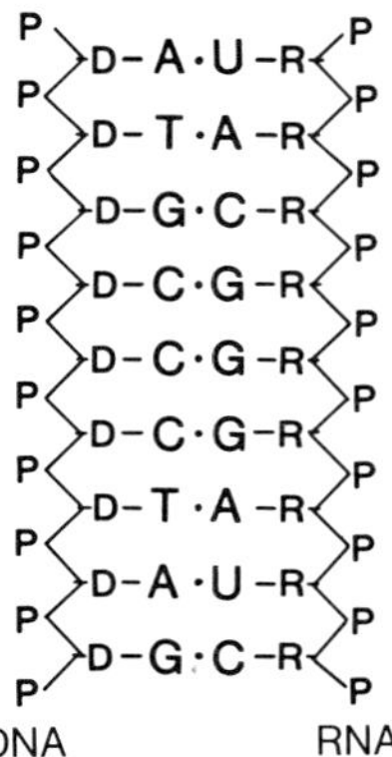

Fig. 4.23 Half of a DNA strand and its complementary RNA. RNA differs from DNA by being single-stranded, by having ribose sugar rather than deoxyribose sugar in the regular alternation on the long side, and by having uracil rather than thymine as one of its four bases. Thus, when DNA manufactures RNA within the nucleus, the RNA is actually an exact complement of that portion of the DNA on which it was made.

base pairs could not code for all the amino acids, but RNA bases in groups of three have been shown to provide the minimal number of combinations to code for all the amino acids. Experimental evidence for these three-base sequences suggests that the ribosome "reads" messenger RNA in triplets called codons. The codons are sites at which the ribosome lines up the appropriate amino acid to be added to the protein under construction.

A final point to consider about the endocrine system is the manner in which hormones act to coordinate developmental processes, that is, by orchestrating the sequence or appearance of specific materials (e.g., particular proteins). Evidence for the direct activity of developmental hormones on DNA within the cell's nucleus is based on changes in giant chromosomes. These giant chromosomes are present in the tissues of several midges and fruit flies. The giant chromosomes are visible under a microscope in the salivary glands. Particular areas along the chromosome may be swollen at different times. The swellings or "chromosome puffs" occur in areas where DNA is actively producing a specific RNA to be used in directing construction of a protein by the ribosomes. For example, a high concentration (titer) of ecdysone (the molting hormone) in the blood causes chromosomal puffs that control synthesis of salivary proteins in *Chironomus* midges, that are used in construction of the silken cocoon in which the larva lives.

Nervous System

As mentioned earlier, the nervous system is responsible for rapid coordination of bodily functions while the endocrine system coordinates long-term activities. However, the nervous and endocrine systems are so closely allied that it may be difficult to determine where endocrines leave off and nerves take over. For instance, the neurosecretory cells of the brain obviously originate as nerve cells, but function as hormone producers, thereby acting as part of the endocrine system.

The only way that the nervous system can control and coordinate the many functions of the body is to reach into all of the parts. Thus, the nervous system is analogous to the electrical wiring in a house; it conducts signals to all areas contacted by that "wiring" because some parts of the system permeate almost every structure. This section, then, describes the nerves that radiate throughout the body, but will not attempt to explain what signals are being relayed, since that is determined by the sensory system, which is covered next.

The nervous system is arranged segmentally, with origins tied to control of the segmental appendages that were found in the arthropod's wormlike ancestor. Concentrations of nerves called *ganglia* that control movement of each appendage developed initially, followed by a pattern of nerve connections between the ganglia that provided coordination within the pairs of nerve centers in each segment and between the centers of adjacent segments (see Fig. 4.24). These segmentally arranged pairs of

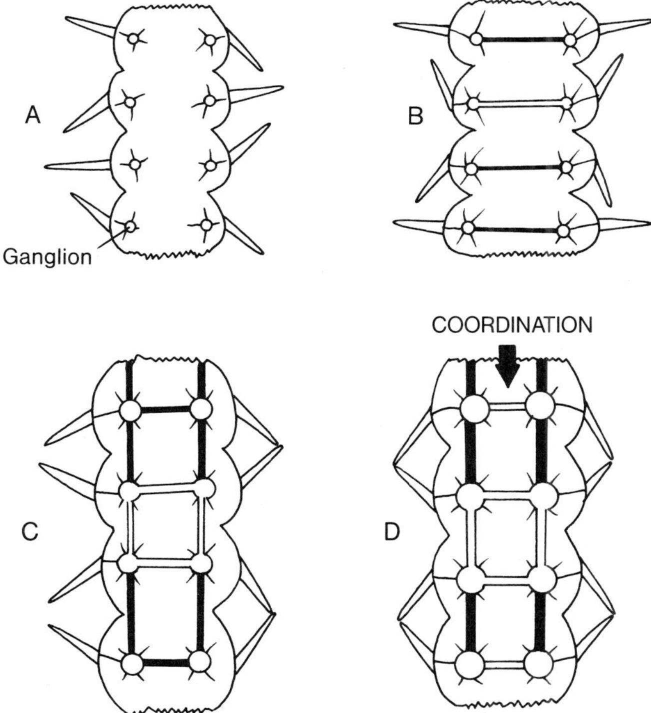

***Fig. 4.24** The evolution of coordination among the nerve centers (ganglia). (A) Each appendage has its own ganglion, but lacks communication with other nerve centers. (B) The two ganglia within a segment are linked by a connective, providing control for simultaneous movement of both appendages of the segment. (C) Not only are the ganglia of each segment linked by a connective, but they are linked by a double nerve cord with ganglia in adjacent segments. (D) A coordination center (the "brain") exerts some control over the segmental ganglia. It is probable that the changes illustrated in (B) and (C) evolved simultaneously.*

ganglia have undergone modification as the insect evolved distinct body regions (Fig. 4.25). A side view of the primary ganglia (or central nervous system) shows the relationship between the body regions and the primary functions controlled by the various ganglia (Fig. 4.26). The supraesophageal ganglion controls the eyes and antennae, as well as the front lip. The subesophageal ganglion innervates (supplies the nerves to) the mouthparts. The legs and wings are controlled by the thoracic ganglia and the abdominal activities are controlled by the abdominal ganglia. In addition to the central nervous system, several smaller nerve complexes innervate parts of the digestive, respiratory, and reproductive organs.

The cells that compose the "wiring" of the insect nervous system are called "neurons." They are elongate, are often covered with a sheath, and have one to several projections (see Fig. 4.27). The nerve impulse may cross the synapse (junction) between the axon of one cell and the next cell's dendrite in several ways, but most commonly by a chemical transmitter known as "acetylcholine." This chemical is released by the

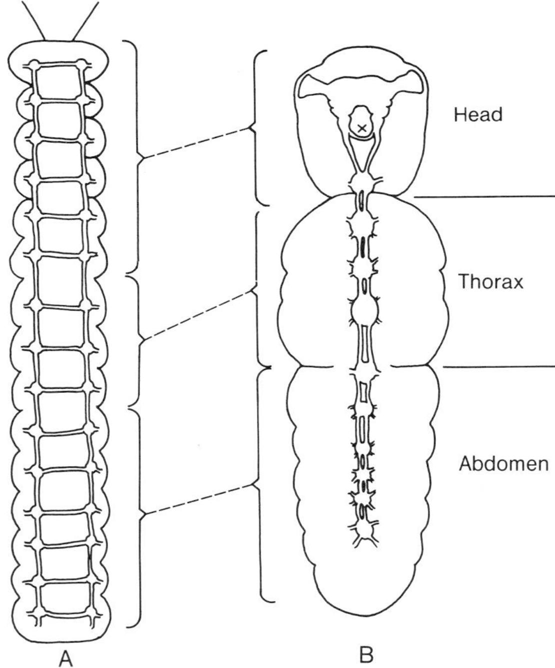

***Fig. 4.25** The segmental arrangement of the central nervous system. (A) This essentially ladderlike system has undergone fusion, reduction, and centralization as the insect evolved. (B) The first three pairs of ganglia in the head can be distinguished, but the next three pairs are fused into a unified ganglion. The X indicates where the digestive system passes between the cross connectives of the central nervous system. The cross connective still loops around behind the digestive system even though the nerve center for the third body segment has become fused into the anterior part of the central nervous system. This dramatically illustrates the complex makeup of the insect head, in which six or seven segments are incorporated into the single structure.*

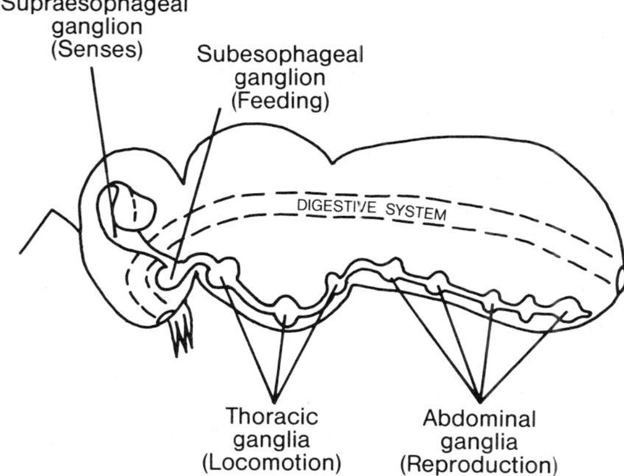

Fig. 4.26 ***Centers of control within the nervous system of the insect. Note that the portion of the digestive system that passes through the anterior part of the nervous system is the esophagus. The sensory center, or brain, is above the esophagus and is termed the supraesophageal ganglion. The center that controls the mouthparts is the first ganglion below the esophagus. It is called the subesophageal ganglion.***

terminal branches of the axon when the nerve impulse reaches them. The chemical passes across the synapse and initiates the wave in the dendrite branches of the next cell. Acetylcholine is then destroyed by the enzyme "cholinesterase." This destruction is critical because if the acetylcholine is not destroyed, it continues to cause the neuron to fire.

Organophosphate insecticides are cholinesterase inhibitors, that is, they interfere with destruction of acetylcholine and thus cause the neurons to fire continually. A tetanic paralysis results because all of the muscles are in a contracted state (similar to extreme cases of human tetanus, or lockjaw). It is important to note that these organophosphate insecticides may affect other organisms, including humans, similarly, since most animals have the same cholinergic type of synaptic junction in some or all of their nervous system.

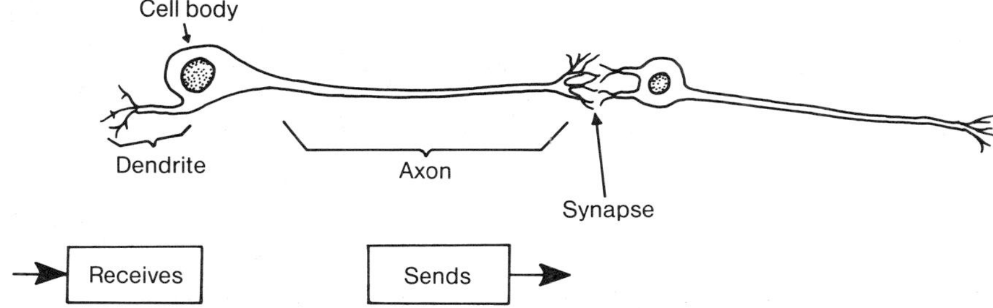

Fig. 4.27 ***The neuron (nerve cell). The nerve impulse, or wave of excitation, can enter the neuron only at the dendrite and can leave the neuron only at the axon. A small but distinct gap between the terminal branches of a cell's axon and the dendrites of the next cell is termed the "synapse."***

As indicated in Fig. 4.27, the structure of the neurons makes the neuron capable of conducting nerve impulses one way only. So the nervous system must be a conglomerate of neurons having different roles. "Afferent neurons" carry signals away from sense organs to the central nervous system, "association neurons" relay these signals to one or more "efferent neurons," which effect a change, as in contraction of a muscle. Some insects such as the American cockroach have giant axons, which provide greatly accelerated conduction and thereby a rapid alarm system.

Sensory System

In order to survive in an environment where food, shelter, and mates have to be distinguished from natural enemies and hostile conditions, insects must receive external signals. Reaction to these signals is directly related to the variety and intensity of environmental inputs. The sensory system is the one responsible for monitoring the environment, and it does so by means of a variety of "sensilla" and complex sense organs. Sensilla have evolved into organs that detect environmental stimuli in many forms—light, heat, chemical, and mechanical. The locations of common sense organs possessed by many insects are illustrated in Fig. 4.28. It is easy to be misled by Fig. 4.28 into believing that insects are poorly equipped to detect changes in their environment. In actuality, the density and variety of sense organs with which an insect monitors its environment staggers the imagination. Table 4.3 lists the common sensory receptors of insects. Some of these, especially the hair, peg, plate, and pit organs can occur almost anywhere on the body and may number in the hundreds or even thousands.

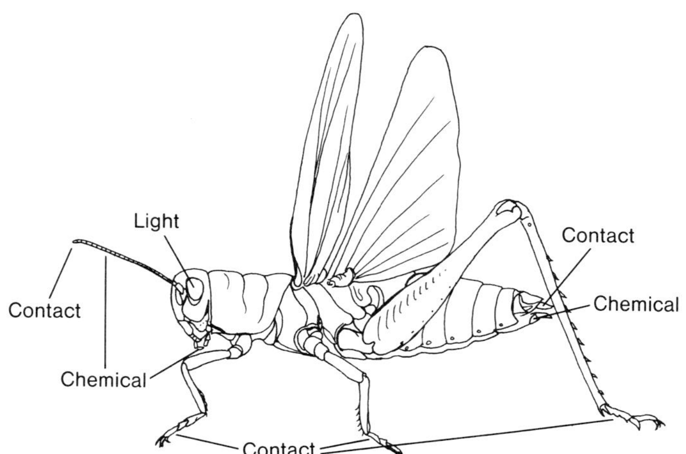

***Fig. 4.28** Common locations of receptors for external stimuli. Sound may be received by a specific sensillum, like the pair of tympanic membranes on a grasshopper's abdomen, or by hairs scattered over the body. Heat is received by the antennae or specific organs located elsewhere, or may be detected by a generalized sensitivity over the whole body surface.*

Table 4.3 Classes of Sensory Receptors in Insects

Receptor Type	Sense	Receptor
Mechanoreceptor	Gravity and Pressure	Hairs, Hair pads
	Sense of Movement	Hair plates, Halteres
	Sense of Current	Antennae, Hairs
	Touch	Tactile hairs
	Vibration of substrate	Subgenual organs
	Hearing	Tympanum, Hairs
Photoreceptor	Vision	Compound eyes
	Light intensity	Ocelli
Chemoreceptor	Olfaction (smell) } Gustation (taste) }	Pegs, Hairs, Plates, Pits

Mechanoreceptors

Mechanical stimuli, such as contact with a solid structure, are important cues to an insect about changes in its environment. In fact, a variety of information is available to the insect, provided it is equipped to detect changes in position of its cuticle, the basis of mechanoreception.

The basic mechanoreceptor is a hair that is jointed at the base and has a nerve cell attached to it (see Fig. 4.29a). This hairlike sense organ, or "hair sensillum" is thought to be the type of mechanoreceptor from which a variety of receptors evolved (Fig. 4.29b–e). These modifications changed the sensitivity or restricted the degree and type of mechanical stimulus that could be detected.

The hair sensillum type of receptor responds to distortion by generating a nerve impulse in one of two distinct patterns. The pattern may be a "tonic response," in which the impulse is generated throughout the time that the sensillum is distorted, or a "phasic response," which only occurs when the hair is bent or straightened out. These responses are illustrated in Fig. 4.30.

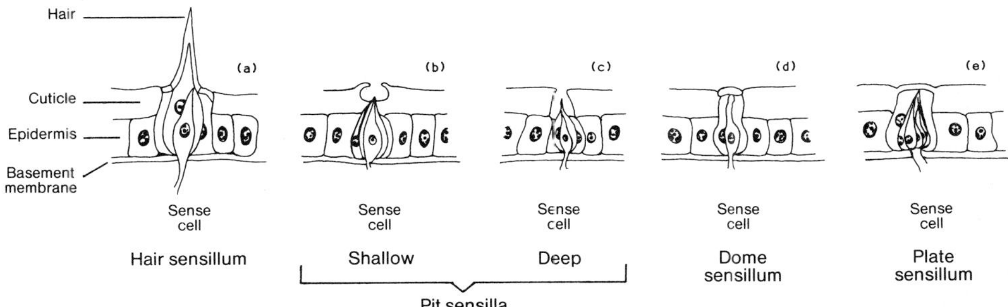

Fig. 4.29 The hair sensillum and related types of sense organs. If distorted, the articulated hair triggers a nerve impulse into the nervous system. The other modifications provide limits on the intensity or direction of distortion required to stimulate the sense cell.

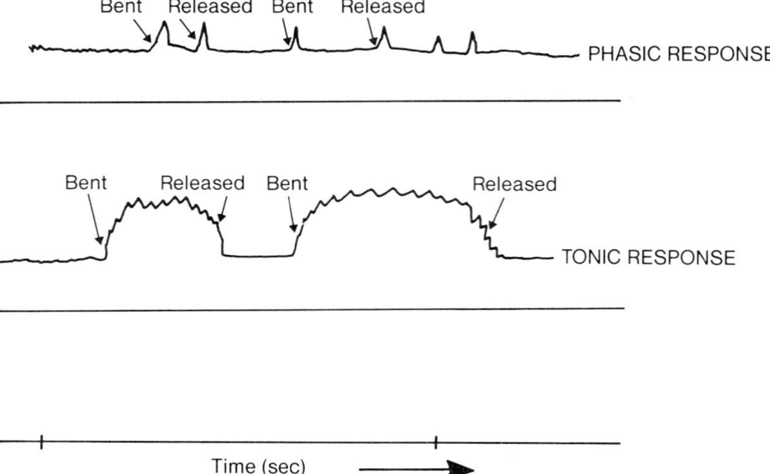

***Fig. 4.30** Electronic recordings of the different responses shown by various types of hair sensilla to distortion (as when the substrate is contacted). The phasic type responds only when the sensillum is bent or released. The tonic response is continuous throughout distortion of the sensillum.*

Most of the hairs on an insect's body are not for insulation, as in a mammal. Instead, each functions as an individual sense organ. It seems logical, therefore, that these sense organs are not just randomly distributed over the body, but occur in specific and highly uniform patterns for each species. Hair sensilla, or modifications of them, located at the ends of the legs or other structures give information about the substrate. The same types of hairs on other parts of the body provide information about body or appendage position and about gravity. These hairs are often located in dense clusters called "hair plates" or "hair pads" positioned between segments or joints. In ants, hair plates at the segmental lines in the abdominal constriction ("waspwaist") just behind the thorax are used to determine gravity. The pendulous abdomen in a relaxed state hangs down and the degree of surface contact made by the hair plates with the other segments indicates the vertical position. Plates between the head and thorax in other insects (dragonflies) function during flight to correct for changes in position, whereas the hair plates in the neck region of the praying mantis relay positional changes in its remarkably mobile head so that the "strike" by the grasping forelegs is at the same angle to the body axis as is the head.

The second antennal segment of insects contains a series of sense cells (collectively called Johnson's organ) that attach between the wall of the segment and the membrane where the next segment joins it. Any change in the position of the terminal antennal segments with respect to the second segment distorts these sense cells and is relayed into the central nervous system. In a flying honey bee, distortion of the antennal tips is picked up by Johnson's organs, which indicates velocity of the air stream. This in turn causes the insect to adjust the wing beat amplitude. In male mosquitoes the antennae function as sound receptors. The female's wing noise differs from the noise produced by their own wings, thus, the male locates his mate by sound.

Insects in some groups are able to hear because they possess tympanic organs—a thin taut membrane, which picks up the air vibrations—and sensilla that attach to the membrane and feed into the nervous system. The number of sensilla range from two in some cutworm moths to many hundreds in cicadas. They are sensitive to intensity and patterns of sound, but not to pitch. Tympanic organs usually are found in pairs and thus provide a directional aspect to sound reception. The organs may be located on the front legs (crickets and katydids), abdomen (cicadas, grasshoppers, and some moths), or thorax (other groups of moths). A moth with only two sensilla per membrane can function effectively enough to elude a bat after detecting its cry.

Photoreception

Many insects possess receptors that perceive light energy. There are several different types of receptors, but all utilize the mechanism whereby a pigment that is sensitive to light and absorbs it, and then transduces (changes) it into a nerve impulse. Insects that live in habitats where sunlight does not penetrate (subterranean forms, cave dwellers, etc.) usually have no discrete photoreceptors, but insects that are exposed to light have one or more types of photoreceptors. Even insects without photoreceptors usually have a generalized sensitivity to light by the body surface.

The discrete photoreceptors of insects are located in the head. The adult insect is usually equipped both with ocelli (simple eyes) and compound eyes. Two or three ocelli, if present, lie near the compound eyes. The compound eyes are paired, but as the name implies, each compound eye is made up of a number of separate visual units. Ocelli and compound eyes are found on both adult insects and on immature forms of insects that develop without metamorphosis or with only gradual metamorphosis. Insects that have a larval immature stage have a third type of eye that is only found in the larval stage. These eyes are called stemmata and may function somewhat similarly to the compound eyes of the adult.

Compound Eyes. The compound eye consists of separate visual units called ommatidia. In insects that are fast moving, especially those that are aerial predators, the number of ommatidia in each compound eye can reach over 10,000. In slow-moving wingless insects there may be only a few of these units in each eye. Among the ants, where some individuals are winged and others wingless, the winged forms may have 400 or so units and the wingless ones only a dozen.

The ommatidium is a cone-shaped, multicellular visual structure. The base of the cone is a lens (the surface layers) and the deeper portions consist of a rodlike receptor apparatus (the rhabdom) surrounded by pigment cells. Some diurnal insects have "apposition-type" ommatidia, arranged so that the pigment cells block off any light that might enter from adjacent cells. In many nocturnal insects, with "superposition-type" eyes, the ommatidia show marked migration of the pigment in the cells around the receptor in response to level of illumination. Thus, the pigment tends to cover most of the receptor rod in bright light, but retreats deeper into the ommatidium in darkness, exposing the receptor rod to more light passing in through the lens of the unit as well as light angling in obliquely through neighboring cells. This superposition eye therefore tends to gather more of the available light in its dark-adapted (pigment near the base of the receptor rod) state than when it is light-adapted (pigment dispersed along the receptor rod), but the "image" is probably not as sharp. This occurs because the light-adapted eye allows only the light passing directly through the lens and being focused by the lens to fall on the receptor rod, much like the apposition-type ommatidium.

The insect's compound eye differs in many ways from the human eye. The insect eye is not movable and cannot be focused. Since there is no eyelid, the eye cannot be shut. In general, insects are nearsighted and have poor visual acuity (the ability to separate objects close together). However, scientists working on insect vision have no way of knowing exactly what image is transmitted from the compound eye to the central nervous system. The generally accepted idea is that insects see images as mosaics. According to this concept each ommatidium receives visual information from the portion of the image within the field of view of its lens. Thus, each ommatidium's "piece" contributes to the mosaic image (see Fig. 4.31). Whether the mosaic is comparable to looking through a handful of straws with fragments of complex images in each straw or whether each ommatidium contributes to an image as do the dots that make up the photographs in this book is a matter of conjecture. The latter seems more likely because of the restricted number of receptors in each ommatidium.

Many insects can see colors. This was proved by Nobel Laureate Karl von Frisch when he showed that honey bees could discriminate between blue and the same intensity of gray in food location trials. The range of colors that insects see is often as wide as our own, but also somewhat different. For instance, the honey bee cannot see into the red end of the color spectrum as humans do, but they see much further into the ultraviolet end (see Fig. 4.32).

Because of the mosaic arrangement of ommatidia, insects are particularly efficient at detecting motion. One explanation for this sensitivity is that the moving object could be easily detected as it swept across the ommatidial field as compared to the background that remained stationary. Another unusual characteristic of insect vision is the ability to discriminate forms. Whereas the insect usually cannot discriminate between forms such as circles and squares, it can easily discern between hollow and solid circles. The edges of objects appear to have particular significance in insect discrimination.

Other Light Receptors. Ocelli appear to serve as light intensity indicators. Image formation seems to be physically impossible given the lens–receptor systems evident in those species that have been studied. However, in behavioral studies, an increased sensitivity to light is detectable in those specimens with intact ocelli.

The stemmata of larval insects probably also serve as light sensitive organs, which relay little information into the nervous system. However, with as few as six stemmata on each side of its head, a caterpillar can distinguish between short and tall objects (e.g., trees).

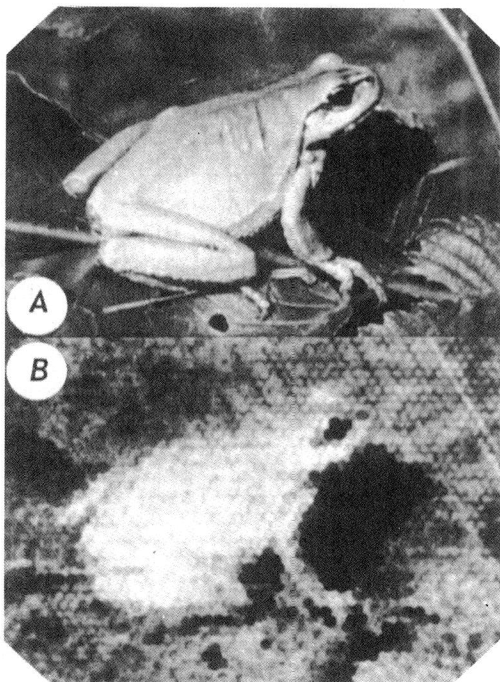

Fig. 4.31 A frog as viewed by a human (A) and a fly (B). The insect's mosaic view is a result of visual inputs from each unit of the fly's compound eye. (Reprinted with permission from G. A. Masokhin-Porshnayakor, 1969. **Insect Vision.** *Plenum Publishing Corp. New York.)*

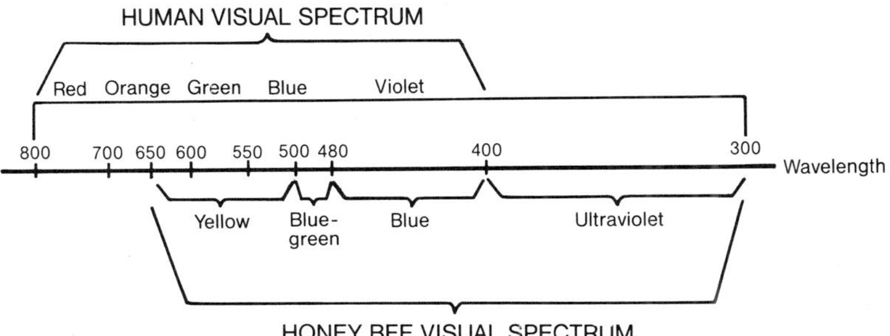

Fig. 4.32 The differences in color vision between human and honey bee. Honey bees cannot see true reds, but see ultraviolet as a distinct color even though it is undetectable by humans. Humans can distinguish between about 60 hues, whereas honey bees see only four.

Chemoreception

Insects live in a more chemically oriented world than humans do. No doubt the high visual acuity of the eye has enabled humans to survive despite poor abilities at detecting tastes or smell. Conversely, the limitations of the insect visual system, along with its small body size, has mandated development of great sensitivity to chemicals in the environment. These chemicals may be detected in direct contact in relatively high concentrations of material via the sense of taste or from a distant source in a very dilute mix with the air or water via the sense of smell. These senses are critical to the survival of an insect because they often are directly related to food selection and reproduction, including mate location and egg-laying site location. Regardless of how the insect uses chemical information, a primary aspect to be considered is how the chemicals are perceived by the sensors involved.

The current theories on most sensory stimulation involves the principle whereby energy somewhere in the electromagnetic spectrum is received by the sensor, which is then modified. Thus, in the eyes a pigment is changed when photons of light energy contact it. Similarly, when the mechanical energy of sound waves deflects the tympanic membrane of a grasshopper, the wave's energy is transduced by the tympanic mechanism into a pattern of nerve impulses. However, the situation with respect to the way chemical reception operates is not as firmly established. It may involve a sensor's sensitivity to the specific energy a chemical releases when exposed to external radiation (that is, sunlight) or to the specific form of the chemical's structure. The theory based on molecular form states that a molecule will be detected if it fits into a receptor site on the surface of the sensillum. Such receptor sites are so specifically formed that closely related (and therefore closely similarly shaped) molecules may not fit well enough to initiate a signal.

One interesting chemical reception hypothesis theorizes on the specific form of the minute receptors on the insect's body. Dr. Philip Callahan proposed that the receptors are, in effect, miniature antennalike structures whose specific form determines precisely which chemical is perceived and the distance from the source of the chemical. In this hypothesis it is the specific infrared emissions given off by the particular molecule (for example, a sex attractant) some time after it has been exposed to an energy source (like light) that is detected. The minute antennae pick up these emissions if their antennal shape "resonates" with the emission. Therefore, the variety and specificity of sensor shapes, which have been demonstrated in scanning electron micrographs can be explained adequately for the first time (see Fig. 4.33).

The chemoreceptors are located primarily on the extremities of the insect's body. The antennae, mouthparts, and reproductive structures all bear sensilla. Even the tarsus, the last segment of the leg, may have chemoreceptors. This explains why a house fly extends its mouthparts when it lands on certain materials, for example, cake frosting. It tastes the substrate as it walks around. Receptors of this type are called contact chemoreceptors and are responsible for the sense of taste (gustation).

The most common contact chemoreceptors are hairlike, and not only have a nerve cell that can detect bending, but several others that detect different chemicals such as water, sugar, and salt. These hair sensilla are located in a fringe around the

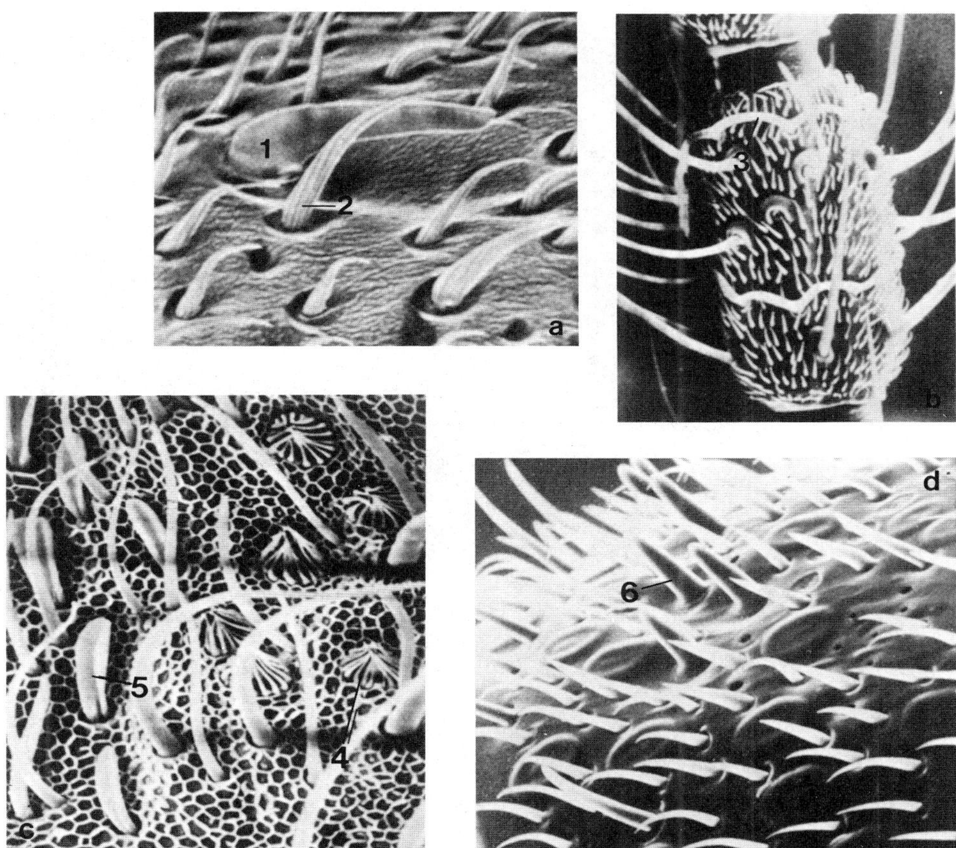

Fig. 4.33 Sensilla on the antennae of insects. These scanning electron micrographs show the antenna's surface and a few of the many types of sensilla located on them. It has been hypothesized that the specific shape, size, and sculpturing of each sensillum allows for perception of specific molecules of significance in the insect's life. (a) Sickle-shaped (1) and spirally sculptured (2) sensilla of an ant. (b) Loop sensilla (3) on gall midge. (c) "Picket-fence" (4) and "folded-shoehorn" (5) of pyralid moth. (d) Several blunt-straight (6) sensilla in a mass of pointed-straight and pointed-curved sensilla on a honey bee. Note that the stimuli for these sensilla have not been identified. (Reprinted with permission from P. S. Callahan, Insect antennae with special reference to the mechanism of scent detection and the evolution of sensilla. **Int. J. Ins. Morph. Embryol. 4,** *381–430 (1975). Pergamon Press, Ltd.)*

periphery of the sponging lobe in advanced Diptera (see Fig. 4.34). Placement of these hairs is so uniform among individuals of a species that scientists working on them have been able to map their locations and then record the variations in their responses. Similar hairlike sensilla are located on the mouthparts and elsewhere on the insect's body.

***Fig. 4.34** The blow fly's extensible sponging mouthparts, as viewed from below. The "fringe" consists of hairlike mechano- and chemoreceptors. (Reprinted with permission from P. S. Callahan, 1971.* **Insects and How They Function.** *Holiday House, N.Y. © P. S. Callahan. The USDA, Ms. Thelma Carlysle, and Dr. Philip S. Callahan are credited for this photo.)*

SUMMARY

Although insects are small, they are about as complex structurally and functionally as larger animals. Their organ systems accomplish the same life processes as those of humans.

The integument consists of the cuticular exoskeleton, its underlying epidermal cells, and a basement membrane. In the molting process the epidermal cells digest parts of the old cuticle for reuse in the construction of a new larger cuticle of either essentially the same form or a different form.

The digestive system stores food, digests it into absorbable form through both physical and enzymatic action, absorbs the digestible materials primarily in the midgut, and then expels undigested materials. Once the indigestible wastes have passed into the hindgut, water and other reusable materials (potassium, sodium) are resorbed from the contents before the wastes are expelled.

Most solid food eaters have a protective peritrophic membrane produced in the midgut. Liquid feeders often have a filter chamber that removes excess water, thereby concentrating the nutrients. Although many insects possess enzymes that attack proteins (proteases), fats (lipases), and those for carbohydrates, few can digest cellulose, keratin, or sclerotized chitin.

Malpighian tubules excrete metabolic wastes (urates) from the blood into the gut, functioning as the insect's "kidneys." Water and other reusable materials are resorbed.

Oxygen is supplied to the body through the tubular tracheal system. Spiracles, often with valves, admit oxygen, either through diffusion or ventilation, but the more permeable CO_2 escapes from the body through the trachea and body wall. Some aquatic insects take in surface air through snorkels or by trapping bubbles on the body surface. Others have gills that remove O_2 dissolved in the water.

Hemolymph is circulated by the pumping of the heart and by general body movements. Flow is directed by diaphragms. Since O_2 transport is not a function of the blood, heart beat is less regular and rhythmic than in vertebrates. The blood functions to transport nutrients and wastes, to isolate or remove foreign or harmful materials (encapsulation and phagocytosis), and as a reservoir for water and nutrients.

Muscles constitute a major portion of the body. Visceral muscles encase organs. Skeletal muscles act in movement and in position maintenance. Skeletal articulations are usually associated with powerful functions (chewing, flight, walking). Rapidly moving flight muscles are resonating (multiple contractions per impulse), but most muscles are nonresonating (one contraction per impulse).

Energy is usually stored as carbohydrate (glycogen) or fats. Its conversion into a usable form, ATP, involves a complex sequence of enzymatically controlled steps.

Most six-legged insects employ the alternate triangle method for walking. Grasshoppers, fleas, springtails, and click beetles each use different methods for their violent propulsive movements. Most larval forms crawl, employing telescopic movements of the body. Most aquatic insects swim, either by leg movements or by body undulations.

Insects are the only invertebrates that actively fly. Their wings evolved from paranotal lobes and may have functioned initially as gliding surfaces. Indirect flight muscles move the wings by distorting the thoracic exoskeleton. Primitive fliers employed fluted wings and flapping movements. Direct flight muscles attach to the wing base and alter the angle of the wing on a pivot point throughout its figure-eight pathway in the sculling flight of more advanced insects. A single flight surface is achieved by wing coupling or atrophy of one of the pairs of wings. Wing veins provide rigidity. Beetles, flies, and most other advanced insects flex the wing along the body when not flying.

Hormones produced by the endocrine system coordinate maturation through control of growth and development, as seen in the changes that occur as a larva becomes an adult form. Ecdysiotropic hormone from the brain stimulates release of ecdysone, which causes a molt to occur. Change in form is controlled by the amount of juvenile hormone secreted by the corpora allata. Action of the hormones determines which part of the cell's repertoire of information (DNA in the nucleus) is employed.

Rapid coordination of body functions is performed by the nervous system. The "wiring" consists of segmentally arranged nerve ganglia that control functions within each segment. The ganglia are connected by a double ventral nerve cord. Functional centers for coordination—senses, feeding, and locomotion—exist as regional ganglia fuse. Most nerve impulse transmission employs acetylcholine at the synapse, a system easily disrupted by anticholinesterases like the organophosphate insecticides.

The insect employs a variety of senses to monitor its environment. Common mechanoreceptors include those sensilla that detect distortion, movement, vibration, position, and sound. Impulses from such receptors may be tonic or phasic. Tympanic organs employ thin membranes that pick up sound vibrations and nerve cells that transduce the vibrations into nerve impulses.

Most insects possess a complex of photoreceptors including both ocelli and compound eyes. The compound eye is not movable, cannot be focused, and has no eyelid. In addition, acuity is poor. However, motion and outlines of objects are easily distinguished by the mosaic of ommatidia.

Chemical reception is highly developed. Chemoreceptors are used by many insects to locate mates, food, and oviposition sites.

SUGGESTED READINGS

Chapman, R. F. 1969. "The Insects, Structure and Function." American Elsevier Publ. Co., New York.

Gillott, C. 1980. "Entomology." Plenum Press, New York.

Harbach, R. E., and Knight, K. L. 1980. "Taxonomist's Glossary of Mosquito Anatomy." Plexus, Marlton, New Jersey.

Matsuda, R. 1965. Morphology and Evolution of the Insect Head. Mem. No. 4. American Entomology Institute, Ann Arbor, Michigan.

Matsuda, R. 1970. Morphology and Evolution of the Insect Thorax. Mem. No. 76. Entomological Society of Canada.

Matsuda, R. 1976. "Morphology and Evolution of the Insect Abdomen." Pergamon Press, Oxford, U.K.

Snodgrass, R. E. 1935. "Principles of Insect Morphology." McGraw-Hill, New York.

Tuxen, S. L. (ed.) 1970. "Taxonomist's Glossary of Genitalia in Insects," 2nd ed. rev. Ejnar Munksgaard, Copenhagen.

Wigglesworth, V. B. 1974. "Insect Physiology." Willmer Bros. Ltd., Birkenhead, U.K.

5

Perpetuation of the Species

The insect is mortal; therefore it must reproduce. In essence, this function of life is an effort to perpetuate the genetic material within the organism even though the individual itself does not survive. The various methods of reproduction (parthenogenetic, sexual, and metagenetic) have been introduced in Chapter 3, as was the haploid–diploid chromosome condition and the advantages of internal fertilization.

FUNCTIONAL MORPHOLOGY OF THE REPRODUCTIVE SYSTEM

Because most species of insects are sexual, that is, the male and female parents both contribute genetic material to the offspring, the reproductive system is primarily a production line for sex cells. In the more primitive groups of insects the male sex cells (the sperm) are protected by being imbedded in spermatophores. These may be extruded by the male onto the substrate and then actively picked up by the female. More advanced groups rely primarily on internal fertilization with or without spermatophores. This necessitates that the reproductive assembly also includes copulatory organs, as well as storage capacities for the sex cells prior to copulation in both sexes, and after copulation in the female. Thus the system has taken on a number of tasks which enhance the survival of the species: production, protection, nurture (of the stored sex cells), storage, and copulation. Another function, which will be covered later, is that of actually bringing the sex cells together. A generalized reproductive system is illustrated in Fig. 5.1, along with the units of the system as they exist in each sex.

Male

The male's sex cell production lines (testicular follicles) are enveloped in a pair of structures called the gonads or "testes." The germ cells in the tips of the follicles give rise to sperm through a series of cell divisions, including one that halves the number of chromosomes within each sperm. The sperm from each testis are then passed down the "vas deferens" (duct) to the "seminal vesicle" where they are stored until copulation. At copulation they usually are passed out of the male's body and introduced directly into the female's body by means of the aedeagus (penis) without exposure to the air. The accessory glands supply a seminal fluid that may act as a medium to carry and protect the sperm.

Female

The female's equivalent of the male's testes are called "ovaries." However, this pair of structures appears more complex in the female because each sex cell production line

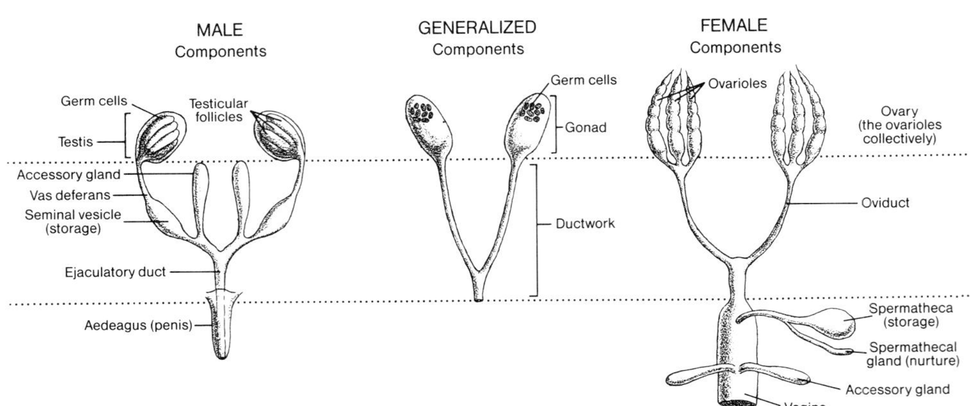

***Fig. 5.1** The reproductive system. (Middle) The basic components of a generalized reproductive system are shown. (Right) The female system, (left) male system. Note that the individual sex cell production lines (ovarioles) in the female are separate, but those of the male (testicular follicles) are encased in a compound gonadal unit, the testis. The system in the female is more complex because it is responsible not only for storage of the eggs it produces as well as storage and maintenance of the sperm it receives from the male at copulation, but for bringing the sperm and ova together. The male stores sperm in its seminal vesicles.*

(ovariole) is separate from the others; the ovarioles of each side are tied together only by the oviduct they share. The number of ovarioles varies from one to more than 2000. The individual oviducts join together for a short distance in a common oviduct then lead into the "vagina" (the female's copulatory structure). Several other structures also open into the vagina. Among them is the "spermatheca" (sperm storage vessel) and its associated spermathecal gland, which nurtures the stored sperm. Accessory glands attached to the vagina frequently are the source of adhesives, which are used to glue the fertilized eggs to some object at oviposition.

Formation of the mature sex cell (unfertilized egg or ovum) in the female is somewhat different than the comparable sperm formation in the male. Differences include the time at which the number of chromosomes in the egg's nucleus is halved (it does so only after sperm penetrate the egg) and the formation of a shell (*chorion*) around the egg.

Deposition of the chorion around the egg prior to entry of the sperm presents a problem to sperm penetration. This problem is solved partly by the presence of minute holes called micropyles in the chorion through which the sperm gain entry to the egg's interior. Another aspect of this problem is that of conserving the number of sperm that the female has received at copulation. Such conservation is critical since many insects mate only once, or at least during one short period. Thus, the female has only a limited supply of sperm with which to fertilize her eggs. She must release only a few to each egg during oviposition.

Most insect eggs fail to develop if not fertilized. Others, like the honey bee, produce males from unfertilized eggs and females from fertilized eggs. Therefore, if the queen honey bee runs out of sperm she produces only males causing her colony to replace her while there are still some young developing female larvae that can be transformed into queens. This last process will be discussed later. The significance here is that the sperm the queen acquired during her few mating flights (over a period of 3–4 days) must be released from the spermatheca parsimoniously or else she soon runs out of sperm and is replaced. Since she may produce up to 200,000 offspring in her life, the sperm release mechanism within her reproductive system must be very efficient.

EGG

The egg starts its development as a small single cell in the terminal part of the ovariole. The carbohydrate, fatty, and protein components of the yolk are contributed either by the cells that make up the walls of the ovariole or by specialized nurse cells.

After the yolk nutrients have been supplied to the egg, its chorion is secreted by the cells of the ovariole wall. The chorion is functionally similar to cuticle, that is, it is made up of several layers, including a tanned protein and an oily layer. It functions to protect the egg from physical damage and water loss after the egg has been deposited, while allowing exchange of oxygen and carbon dioxide. Gaseous exchange is facilitated while water loss is minimized by surface sculpturing of the chorion and by special structures that admit air. All of the basic structures of the insect egg are

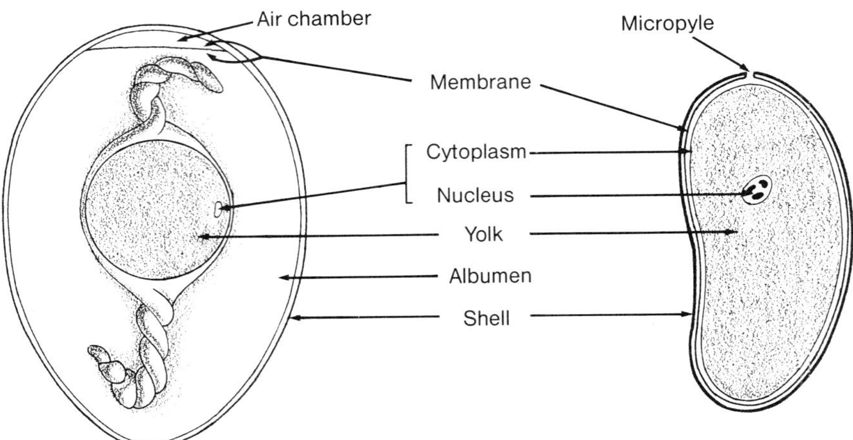

Fig. 5.2 A comparison of eggs: chicken versus insect. Basic structures, some of which are shared by the two types, are indicated. Note that a micropyle for sperm entry is not found in the chicken egg since fertilization occurs prior to shell secretion, a major difference between insect and bird.

shown in Fig. 5.2 in comparison to a larger and more familiar egg, that of the chicken.

EMBRYOLOGY

The following account is a synopsis of the major embryonic events that occur in most insects. The details of the process for various groups may differ, but all share the process described below.

After the sperm enters the egg through the micropyle and penetrates the yolk, its haploid nucleus fuses with the haploid egg nucleus forming the diploid, fertilized egg with its double set of chromosomes. The nucleus begins to divide and continues dividing, migrating to positions exterior to the yolk. Membranes are laid down so that the yolk is surrounded by a layer of cells. Those cells at the bottom, destined to form the embryo itself, thicken and form more than one layer.

The embryo at this point is not segmented, nor are there any traces of appendages (see Fig. 5.3). Then, a segmentation wave starts at the front and progresses backward resulting in a segmented embryo. Before the more posterior segments are entirely formed, each of the segments in the front begins to develop a pair of limb buds. Even the abdominal segments develop limb buds, although the limbs do not continue on to become limbs of the postembryonic form.

As is illustrated in Fig. 5.3, the limb buds are destined to form a variety of

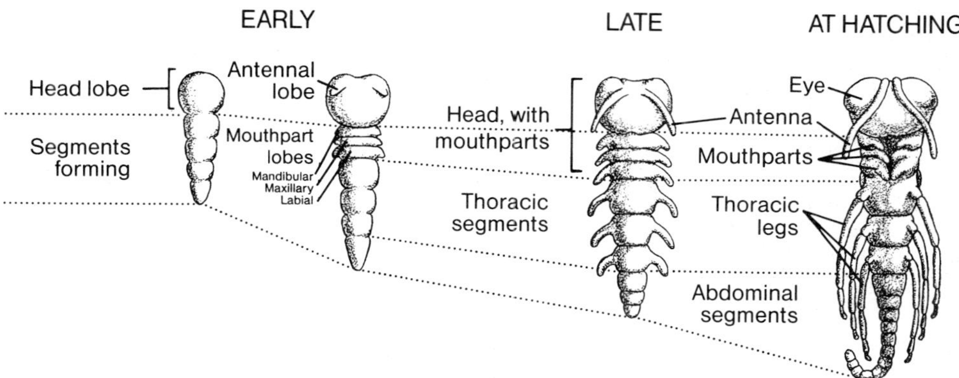

***Fig. 5.3** Formation of segments and appendages in a generalized insect embryo. Note that the definitive insect head consists of the embryo's head, plus the first few appendage-bearing segments of the body, with the appendages forming the mouthparts.*

appendages of the postembryonic form. Those segments nearest the front fuse with the primitive head and the appendages become antenna and mouthparts (mandibles, maxillae, and labium, in that order). Thus the head is a compound structure consisting of the primitive head and several of the immediately posterior segments fused with it. The exact number of segments incorporated into the head is a subject of great controversy among embryologists.

The next three segments become the thorax, and their appendages become the legs of the insect. The rest of the segments make up the abdomen, which bears no appendages in the mature insect, at least on the anterior segments. At the posterior end, however, appendages are retained and are modified in the mature insect as copulatory and egg-laying appendages. The limb buds that appear on the embryonic abdominal segments usually are resorbed and disappear, or sometimes take on functions that are of significance only in the embryonic stage, such as aiding in escape from the egg.

During the sequence of events just described, the digestive system is forming (see Fig. 4.5) as are the other systems. It is worth noting that the entire embryonic sequence can occur in as little as 24 hours. Thus, the "immobile" egg is in reality a writhing bundle of tissue undergoing significant changes.

POSTEMBRYONIC DEVELOPMENT

The initiation of postembryonic development occurs after the insect hatches from its egg. However, not all insects are "oviparous" (laying an egg, which develops outside the female parent's body). In some species, the fertilized egg is retained within the female's reproductive system, where it completes its embryonic development. In this "ovoviviparous" condition, the immature hatches from the egg at or soon after the egg is deposited. A very few insects are "viviparous" in which the embryo hatches from the egg in an early embryonic state while still in the female's body. It completes much or all of its growth and preadult development within the female's reproductive system, which supplies nutrients to it from glands located within the system.

Some eggs hatch as soon as embryonic development is complete. Others require specific stimuli to initiate hatching, such as reduced environmental oxygen levels by some species of mosquitoes. Many hatching young escape by means of a sharp spine that pierces the chorion, or by having part of the egg dissolve. Some eggs have an "operculum" (lid) that the hatching insect pushes off. Several insects molt the cuticle that they formed while embryos. These "skins" often are left behind with the egg as the insect escapes.

All insects share some common characteristics of development, regardless of the specific type of development they undergo. For instance, the insect must periodically molt if it is to increase very greatly in weight and size, because it is encased in a relatively inelastic exoskeleton. In molting, a process explained in Chapter 4, the insect sheds its exoskeleton and secretes a new one. Prior to the tanning (plasticizing) of the new cuticle, the insect swallows air or water, thus puffing itself up. It maintains this expanded state until the tanning is complete, thereby creating some room inside the cuticle in which it can grow before it has to molt again. A plot of weight gain for an insect would show a rather smooth increase as the insect goes through its growth period. The pattern of changes in size undergone by the same insect, however, would show a much greater influence of molting on growth, since structures like the legs only increase in length at a molt (see Fig. 5.4). Because changes in form of external structures or their appearance usually only occur at the molt, the insect may look entirely different after a molt.

"Instar" is the term used by entomologists for the insect between molts. Thus the first active form is the first instar. After it goes through a molt it is a second instar and so on until it becomes an adult. The time the insect spends between molts is termed the "stadium;" the remnant of the exoskeleton discarded at the molt is termed the "exuvium." The number of instars varies according to the species and sometimes also according to the sex of the individual, or the previous environmental conditions to

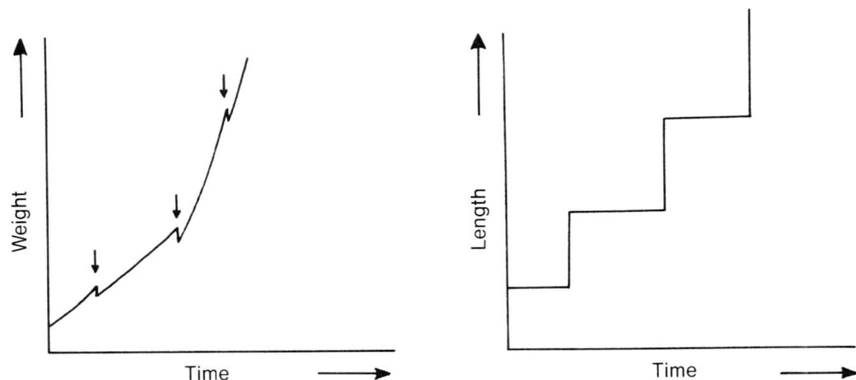

Fig. 5.4 Change in body weight and in appendage length of a typical insect undergoing growth. The vertical arrows indicate the points at which molting occurs.

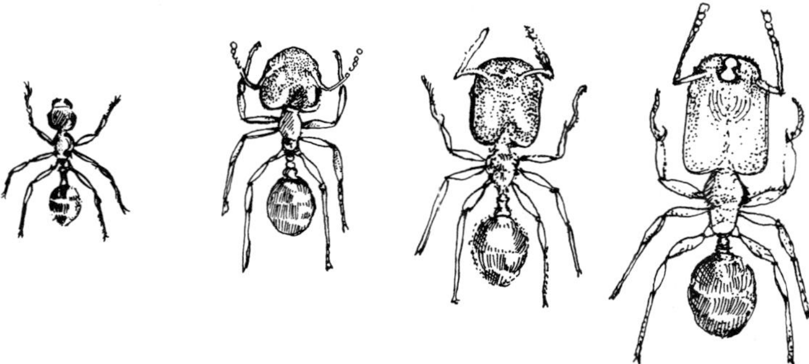

Fig. 5.5 Allometric growth. These ants, all of the same species and in the comparable, nonmolting adult stage, collectively demonstrate the results of allometric growth. The ants to the right are larger but have disproportionately larger heads than the ant at the far left. Thus, the head grows at a faster rate than the rest of the body in this species.

which it has been exposed. For example, female *Trogoderma glabrum* carpet beetles typically have one more instar than do the males. The number of instars may also differ if populations are put under stress, as with short food supplies. Not only will these starved larvae survive for a long time, sometimes over a year, but they will continue to molt at fairly regular intervals. These individuals get smaller and smaller at each molt during starvation. Thus, an insect that has six instars in the larval stage under optimal conditions may have as many as eight additional molts under stress conditions.

The ultimate body size reached by an individual is controlled in part by its genetic makeup and also by the amount and quality of food available. However, experiments with some species have shown that optimally fed individuals may not be just generally larger versions of the species as compared to the same species on suboptimal rations. Careful measurement of these insects reveals that the proportions of the body are different. That is, the different body parts may grow at different rates. This phenomenon is termed "allometric growth" or "heterogonic growth." Figure 5.5 demonstrates the result of allometric growth in ants.

Although allometric growth may be a term that is unfamiliar, the phenomenon is not restricted to insects. In fact, humans show allometric growth in development. Whereas an adult has no trouble grasping hands over the top of his head, young children cannot do so. Children's heads are larger in relation to limb length than are adults', but since the head does not grow as rapidly as the arms, the previously impossible task is easily accomplished after a few years.

Types of Postembryonic Development

Insects follow three basic pathways whereby the newly hatched young reaches adulthood. The stages of development are given in Table 5.1. As with most schemes established to explain groups of organisms, numerous slight exceptions and a few

Table 5.1 Functional and Structural Synopsis of the Life Stages of Insects

Stage	Structure	Function
Egg	Inside a shell; contains own food supply	To complete embryonic development without external support
Nymph	Looks like miniature of its adult; external wing pads, if present, grow larger at successive molts	The growth stage; eats same food and lives in same habitat as adult; lives like the adult
Larva[a]	Wormlike; no resemblance to its adult; wing development, if any, internal; mouthparts usually chewing, without compound eyes	Energy gathering and storage; growth stage; lives in different place than its adult; eats different food than adult
Pupa	Quiescent; nonfeeding; may be in cocoon or protective cell; wing pads, if present, directed down	Conversion stage from a larva to an adult; does not live where the adult does
Adult	Only form with functional wings; reproductive system is functional; nonmolting	Dispersal and reproductive stage

[a]Note that insects have either egg–nymph–adult stages or egg–larva–pupa–adult stages.

significant ones exist in this series of developmental types. These exceptions will be explained in Chapter 11.

Direct or ametabolous development (without metamorphosis) is the type found in very primitive groups of insects such as the Order Thysanura (silverfish) and the Order Collembola (springtails). These insects live in the same habitat throughout their life so that all stages are found in the same locality. Also, the active stages all feed on the same types of food. The immature silverfish is called a nymph, and a number of nymphal instars intervene between hatching and the point at which sexual maturity is reached. As is illustrated in Fig. 5.6, few obvious differences exist between the first instar silverfish nymph and the sexually mature silverfish. These insects, like all other ametabolous forms, continue to molt throughout life. This differs from the more advanced insects who stop molting once they are sexually mature.

Insects with "hemimetabolous development" undergo gradual metamorphosis. The basic difference between this type and direct development is in the formation of wings, which most hemimetabolous adults possess. Postembryonic development of hemimetabolous insects starts with the nymph emerging from the egg. It passes through several nymphal instars until it finally molts into an adult. After the first or second instar, structures called wing pads make their appearance on the outside of the nymph's thorax. They increase in size at each molt (see Fig. 5.6), but do not reach full development or become functional until the adult stage. The only exception to this pattern is the mayfly (Order Ephemeroptera). These insects leave the water, in which the growth stage occurred, as winged forms and then molt again before becoming adults. Thus, mayflies are the only winged insects that molt after they are able to fly.

Most hemimetabolous insects live in the same habitat in both the immature and adult stages. They also typically eat the same food. However, sometimes the term "hemimetabolous" is used to describe the metamorphosis of three groups whose nymphs and adults live in different habitats. Ephemeroptera (mayflies), Odonata

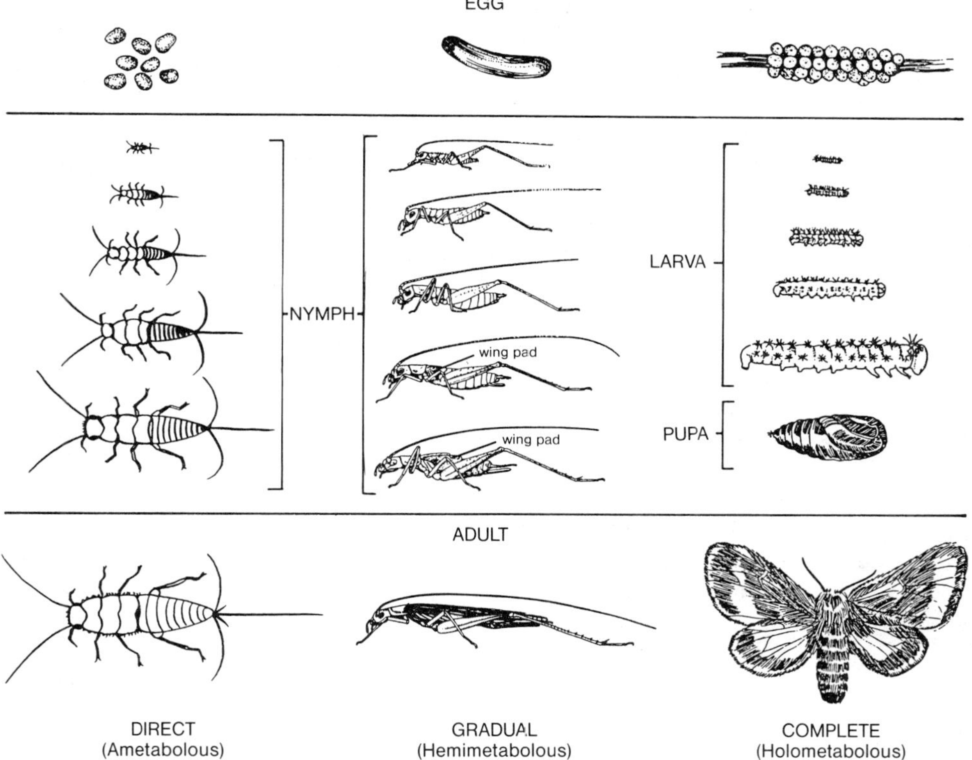

Fig. 5.6 Insect development, the stages and the changes in appearance. Note the progressive development of the external wing pads in gradual metamorphosis, a feature lacking in ametabolous forms (which never develop wings), and in the growth stage (larva) of the holometabolous forms. In the latter, large wing pads appear only in the pupa before they become the fully formed wings of the adult.

(dragonflies and damselflies), and Plecoptera (stoneflies) have aquatic nymphs that usually possess gills, and look less like their adults than do other insects with simple development. The adult lives in a different habitat (aerial-terrestrial) than the nymph (often called a naiad). The term "paurometabolous" is used to describe the insects in which the nymph and adult share the same habitat.

The most advanced of the basic types of development is known as complete metamorphosis, or "holometabolous development." Holometabolous insects come out of the egg as a wormlike larva. This growth stage is comparable to the nymphal stage of gradual metamorphosis (see Fig. 5.6). After passing through several larval instars, the insect molts into a pupa. The pupa is usually a quiescent (relatively immobile), nonfeeding form. It is the pupa that first shows any external evidence of wing formation because it has wing pads, though they are often glued tightly to the body. The adult emerges from the pupal exuvium (skin).

Holometabolous insects tend to live in one habitat in the immature stages and in a different one as an adult. For instance, the house fly is a terrestrial-aerial form in the adult stage, whereas the egg, larva, and pupa are found in or very near semiliquid decaying organic matter, like garbage. The Japanese beetle spends it egg, larval, and pupal stages underground, but the adult is a highly mobile flier in a terrestrial-aerial habitat.

The appearance of the immature holometabolous insect is very different from the adult. In fact, most larval forms look completely unrelated to their ultimate adult form. Many adult holometabolous insects are nonfeeding, but those adults that do feed eat foods very different from the foods of the larvae.

The wing pads of a holometabolous insect do not become visible until the pupal stage. When they appear they are large and in an advanced stage of development. In fact, they have been developing throughout much of the larval stage, but development has been internal. This allows the wing tissues to continue developing between molts without the restrictions inherent in producing and molting a cuticular cover like the wing pads of hemimetabolous nymphs. Internalization also provides maximum protection for the developing wing. Then at the larval–pupal molt the wing pad is everted, rather like the finger of a glove that starts out directed internally and then is popped out. Figure 5.7 demonstrates the differences between internal and external wing development. There is such a fundamental difference in development that the hemimetabolous insects are sometimes called the Exopterygota (external wing development), and the holometabolous forms are called the Endopterygota (internally developing wings).

The difference between the development of exopterygotes and endopterygotes is more than just whether the wing buds are internal or external. Most exopterygote immatures are so much like their adult forms in structure that about all that is necessary to make the adult is to "tack on" some functional wings to the nymph. This is not true for the endopterygotes, where the larvae bear little resemblance to the adult. The pupal period becomes a conversion stage in which the larval insect is reconstructed into a radically different adult form. With some insects, like the true flies, most of the larval structures are histolyzed (destroyed) and a histogenesis (building) of new adult structures occurs. The adult structures originate from groups of embryonic tissues (imaginal discs) located in various parts of the larva's body. These imaginal discs may have a prolonged developmental period throughout the larval stage, and then an intense burst of activity to form specific adult tissues. For example, a third instar *Drosophila* fruit fly maggot has a pair of eye-antennal imaginal discs in its anterior end. Each will form the parts of one of the adult fly's antenna, part of the head capsule epidermis, and several types of mechanoreceptors, in addition to part of one of the compound eyes. Other discs are destined to form most of the structures in the adult's body.

Adult exopterygote insects are rather obviously the result of a single body plan to which are added the adult structures and functions. Such a relationship is not as apparent in the endopterygote insect. However, it is generally accepted that the larva is a highly specialized nymph, which has evolved its own set of characteristics that permit survival under the selection pressures peculiar to the larva's mode of existence. In fact, each life stage evolves in response to its own set of problems. Since the problems of the adult differ from those of the larva, each form evolves and therefore

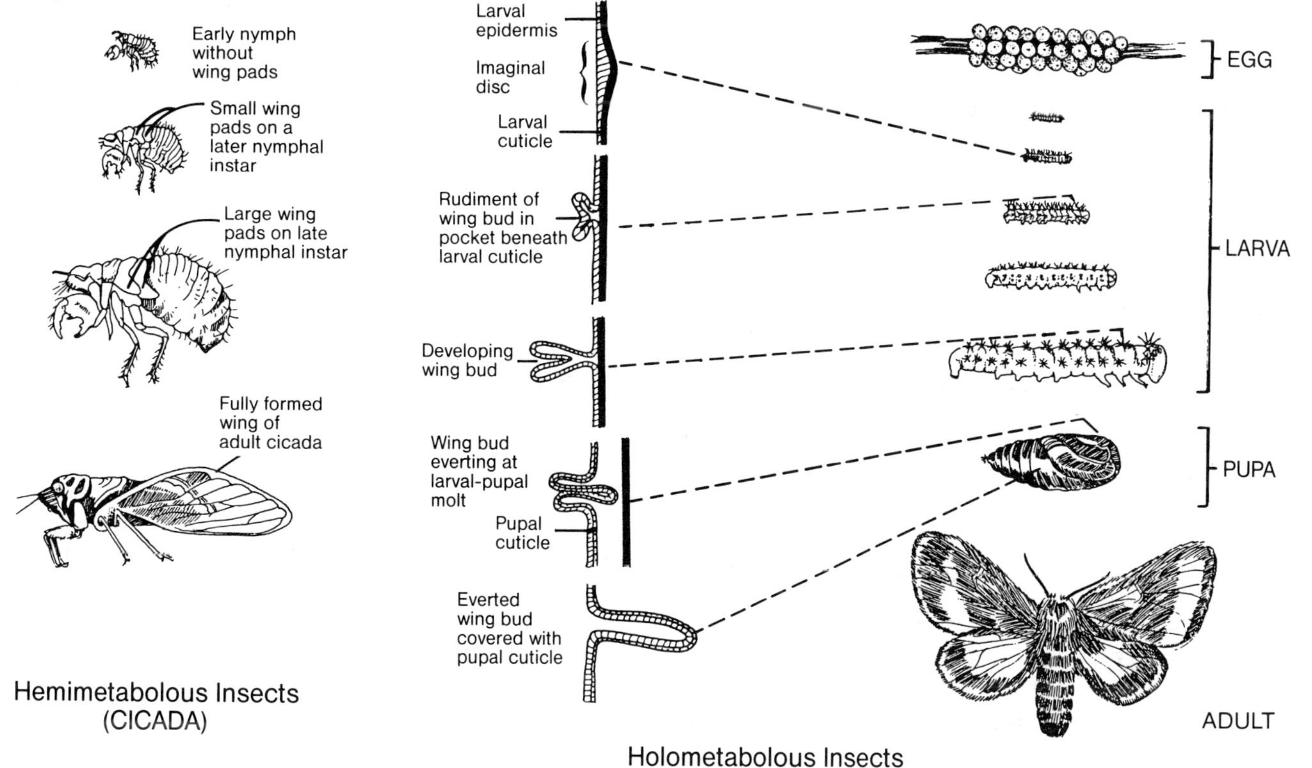

Fig. 5.7 ***A comparison of wing development in hemimetabolous and holometabolous insects. The hemimetabolous cicada has wing pads that develop externally, growing larger at each molt. In the holometabolous moth, the wings develop internally in pockets beneath the larval cuticle. These wings grow continuously as masses of undifferentiated cells (imaginal discs), only everting and forming a cuticle at the larval–pupal molt.***

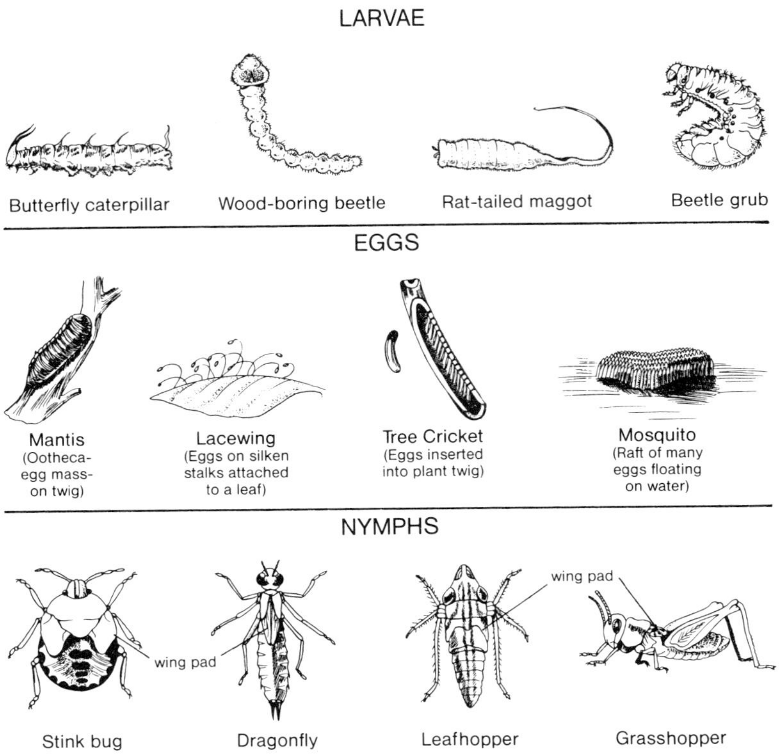

LARVAE
Butterfly caterpillar
Wood-boring beetle
Rat-tailed maggot
Beetle grub
EGGS
Mantis
(Ootheca-
egg mass-
on twig)
Lacewing
(Eggs on silken
stalks attached
to a leaf)
Tree Cricket
(Eggs inserted
into plant twig)
Mosquito
(Raft of many
eggs floating
on water)
NYMPHS
wing pad
wing pad
Stink bug
Dragonfly
Leafhopper
Grasshopper

***Fig. 5.8** Some common examples of the various immature stages of insects.*

Wing pads
Beetle
Moth
Fly
Exarate
Obtect
Coarctate

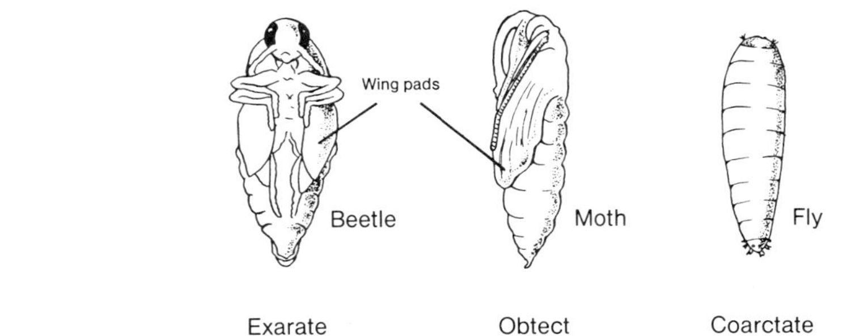

***Fig. 5.9** The basic types of pupae found in holometabolous insects. The exarate type has free appendages, whereas the obtect pupa has the appendages glued tightly to its surface. The coarctate pupa consists of the larval skin that has hardened around the exarate pupa inside. The hardened larval skin is referred to as the puparium.*

diverges from the other. Thus, the number of structures in the larva's body that can be carried over into the adult become fewer and fewer. This requires progressively more extensive remodeling of the insect during its pupal stage. Coupled with the differences in organs and structures is the need to evaginate the internally developing wings in order to provide space for development of the huge muscles used in flight and for maximal growth of the wings.

Unusual Types of Development

In a group as diverse as the insects, there are a number of bizarre aspects of development to be found. Among these are polyembryony, pedogenesis, polymorphism, the subimaginal stage, and hypermetamorphosis.

Polyembryony

The eggs laid by most insects produce single individuals. However, several species have been reported to undergo a type of multiple twinning called *polyembryony*. In these insects the very early embryonic divisions, which in most insects give a two-cell stage, then a four-cell stage, eight-cell, 16-, and so on, yields instead two separate embryos, then four embryos, eight embryos, and so on. The final result of one egg may be as many as 1000 offspring (see Fig. 5.10).

Those insects reported to have polyembryonic development are primarily insect parasites of other insects, especially some of the parasitic wasps that attack lepidopterous caterpillers. Polyembryonic forms produce relatively few eggs, but the host that is attacked is usually entirely consumed because of the large number of parasitic larvae that result from polyembryony. This large number facilitates survival of the parasitic species by maximizing the production of offspring once the food source (the host insect) is located.

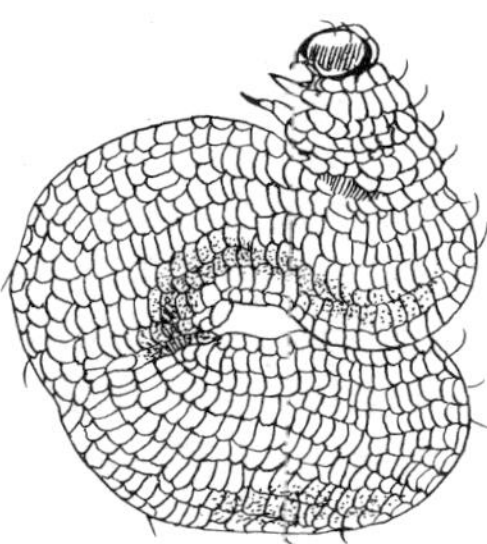

***Fig. 5.10** Numerous larvae of the polyembryonic and parasitic wasp* **Litomastix** *completely fill the interior of the host caterpillar's body after consuming its soft parts. All of these larvae may have originated from a single egg.*

Pedogenesis

In a few species of gall midges (Order Diptera) and also in a beetle, the immature forms are capable of producing offspring. Such reproduction by structurally immature forms is called "pedogenesis." In these insects the ovaries mature while the individual is still in the larval stage. The eggs develop parthenogenetically; the offspring, while still in the larval stage, may also produce young or they may develop normally and not produce young until they become adults.

Aphids also exhibit pedogenesis in that embryos within the parthenogenetic female already have embryos developing within their own bodies before they are born. Such telescoping of development is rare in other insects.

Polymorphism

Although literally translated polymorphism means "many forms," use of the term by entomologists is restricted to several specific conditions. Such restrictions are necessary since almost all species are literally polymorphic with respect to changes that occur with age, or to those differences between the sexes. The sum of the "obvious differences" between males and females of the same species is called "sexual dimorphism," meaning two forms. Thus, the definition of polymorphism is restricted to cases within a single species in which the following hold true.

1. Individuals of the same sex occur in two or more recognizably different forms. This may be true for both sexes.
2. The different forms (morphs) occur regularly. This may include those that occur only at particular seasons.
3. The less common forms make up a significant proportion of the population, at least 5–10%. Of course, this would not be true for social insects in which a single insect (e.g., the queen) may be the sole representative of her morph (called a caste) in a colony of 60,000–75,000 honey bees.

At least three kinds of polymorphism have been identified thus far. Control of production is a very complex interaction between a species and its environment. For instance, in the "social polymorphism" exhibited by honey bees, the differentiation between the queen and the rest of the workers, which are also females and result from fertilized eggs, is based on nutrition. However, the drone caste (the males) is produced from eggs that are not fertilized, and nutrition plays no part. See Chapter 7 for more on social polymorphism.

"Phase polymorphism" is common in many species of Orthoptera (see Chapter 11) and is also known to occur in moths. The production of different morphs seems to be controlled by hormones, but the triggering of changes in hormone levels may be due to the characteristics of the food or to stimulation from other individuals, whether visual, chemical, or physical in nature. In fact, the alternation of generations in aphids (Chapter 3) produces a type of phase polymorphism based, in part, on seasonal progression, but often triggered by food, daylength, or physical contact.

A "mimetic polymorphism" exists in some butterfly species. In these cases the insect gains a survival advantage by mimicking (looking and acting like) a species (the model) that has a method of protecting itself. Such protection through mimicry has limits because the protected species has to be numerous enough to provide frequent encounters with the potential enemy. Thus, the mimic's population is limited by the model population. In those insects that exhibit mimetic polymorphism the various morphs are mimics of different protected species. See Chapter 6 for an elaboration of this topic.

Subimaginal Stage

The adult (imago) is a nonmolting, reproductive form. It is also the only stage in which the wings are functional, if wings are present. Thus, the adult stage (imaginal stage) is the dispersal and reproductive stage for most winged insects. As discussed previously, the Order Ephemeroptera (mayflies) is a notable exception. It has a flying form which undergoes a molt prior to reproducing. This molting form with functional wings is called the "subimago" (subadult).

Some early authors have equated the subimaginal stage in Ephemeroptera with the pupal stage of holometabolous insects. However, it is now generally accepted that the subimago is really just the first adult instar, although the subimago and imago of a mayfly species often look somewhat different. The wings of the more drably colored subimago tend to be somewhat opague. The same insect after its subimaginal to imaginal molt is often more brightly colored and the wings are often transparent, or partly so.

Hypermetamorphosis

A few groups of holometabolous insects exhibit some specialization within the larval or growth stage, that is, succeeding larval instars differ in appearance from one or more of the previous instars. In the parasitic Stylopoid beetles the hypermetamorphic development is as described by Pierce (1964):

> Free-living first larvae emerge from a dorsal canal behind the head of the larviform parent female, which (with a single exception) is permanently within the host. These larvae enter the bodies of new host insects. They lose legs and eyes, and reduce the mouthparts to utter simplicity, for they receive nourishment by direct absorption from the host's organs, without devouring anything. They develop through a series of stages, in which sex begins to differentiate. The females, on maturing, stick their heads out between the abdominal segments of the host, and await fertilization. The male larvae form a puparium with detachable head cap, or cephalotheca; they also protrude their heads from the host's abdomen. In these puparia true pupae are formed. When mature, the male pushes off the cephalotheca and emerges to fly about until it locates a female head in the abdomen of another host.

The change in appearance in larval form is evident in the blister beetle larvae shown in Fig. 5.11. The first instar larva hatches out from an egg mass buried in the soil. It must emerge from the soil and find a grasshopper egg pod or die. This first instar larva is known as a "triungulin" (crawler), and it can cover considerable

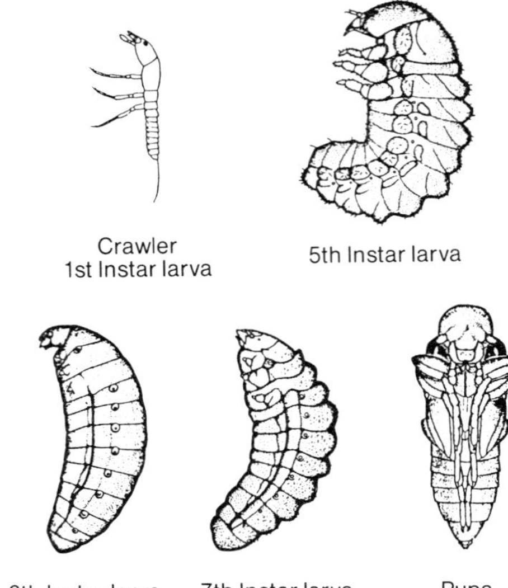

Fig. 5.11 ***Hypermetamorphosis in the larval stage of a blister beetle. Note the relative size of legs in the 1st instar (crawler) larva and in the 6th instar. The 6th instar is quiescent, nonfeeding, and buried deeply in the soil where it may overwinter. The 7th instar is nonfeeding, but active. It burrows up almost to the soil's surface where it transforms into the nonfeeding pupa, which in turn transforms into the adult blister beetle. Note that the various drawings are not drawn to the same scale.***

distance on its long legs. Once it locates a grasshopper egg pod and starts to feed on it, there is a progressive reduction in leg size in the succeeding instars.

DIAPAUSE

Insects in the temperate zones are limited in their development from generation to generation by the annual drop in environmental temperatures, often accompanied by cessation of plant growth, and, in more extreme areas, the accumulation of precipitation in the form of snow. Many insects cope with such seasonal harshness of their environment by migrating out of the temporarily unsuitable area or by cessation of activity through the condition known as "diapause."

Diapause is not just the drop in metabolic rate that occurs in a "cold-blooded" insect in low environmental temperatures resulting in cold torpor (sluggishness). This rate change is controlled by the immediate conditions. Diapause, on the other hand, is a physiological state that may superficially resemble the inactivity common to cold torpor except that it is triggered by much earlier events. Insects entering diapause accumulate energy by undergoing fat body hypertrophy (excessive growth) and by ceasing development of the reproductive system. The condition is a physiologically resistant state that has evolved to maximize survival during adversity. Diapause has been reported to occur in each stage in the life history of insects. Some species enter diapause as embryos in the egg, others as larvae, as pupae, as nymphs, or as adults. However, a particular species usually only enters diapause in one particular stage rather than in several.

Some species always enter diapause at a given point in their life history. These insects are "univoltine" (one generation per year). This obligatory diapause provides synchrony of the insect with its food supply as well as with others of its own kind, so that the probability of mating is increased. However, not all species exhibit diapause in every generation. Some, like the face fly (*Musca autumnalis*) have continuous generations during the milder months and only as winter approaches do any of the flies enter diapause. This is facultative diapause in a multivoltine (more than one generation per year) species.

Diapause is induced by environmental cues to approaching adverse conditions. Among the cues that have been identified are changes in temperature, food, crowding, moisture levels, and photoperiod. Of these cues, photoperiod (the environmental rhythm of daylight and darkness) is the most reliable indicator of seasonal progression. The insect that enters diapause based on changes in photoperiod is more likely to be in the physiologically resistant state prior to arrival of adverse conditions than insects using other less reliable cues.

In order for an insect to be able to use photoperiod to detect the approach of adverse conditions it requires some kind of internal clock. We do not yet understand

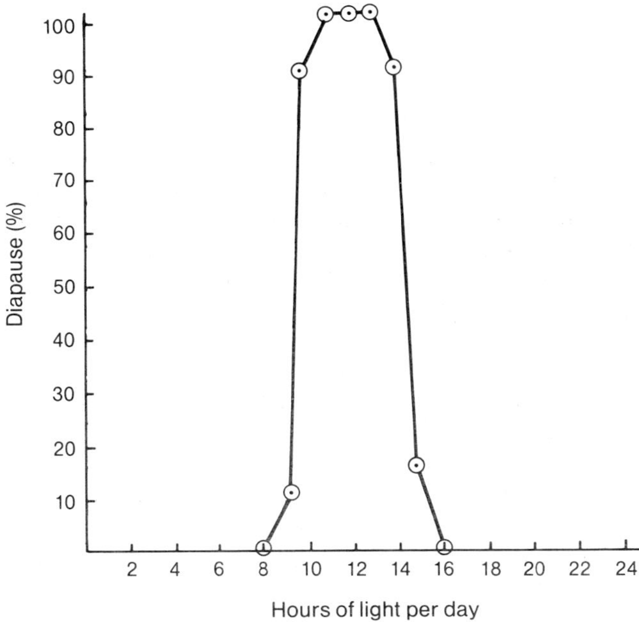

***Fig. 5.12** The response of European corn borer larval populations exposed to different photoperiods. The change in response (numbers in diapause) when populations are exposed to 8, 9, and 10 hours of light demonstrates the accuracy with which the insect reads relatively small changes in the duration of light.*

the mechanism of this internal clock, but we do know that it works. For instance, insects can be induced to enter diapause by changing the amount of light to which they are exposed. As is evident in Fig. 5.12, the European corn borer responds to slight differences in the amount of light and dark to which it is exposed each day. Other experiments have shown that the dark part of the 24 hour cycle is the most important phase in controlling entry into diapause. Diapause induction in other species has been shown to be caused by a hormone.

The duration of diapause carries the insect through hazardous times. For instance, if brief exposures to warm temperatures broke diapause, insects would come out of diapause during midwinter thaws and be caught by the return to usual winter temperatures. Experiments have shown, however, that most insects in diapause must remain in that condition for a certain period before activity will resume. For example, many overwintering moth pupae will not produce adults unless the pupae are exposed to cold for long periods or to alternating cold and warm conditions. It is evident that some internal functions must be completed under cold temperature before diapause can be broken.

The insect in diapause usually has a different hormone balance than does the active insect. In fact, changes in hormone balance influenced by photoperiod have been shown in several cases to be the mechanism whereby diapause is induced. However, regardless of the exact mechanism that controls diapause in each insect, its significance as a physiological timing mechanism for escape from adversity cannot be overemphasized.

SUMMARY

The male and female reproductive systems are production lines for the sex cells that are formed from the germ cells contained within the gonads. The completed sex cells pass out through ducts, which store and nurture the completed cells. Associated with the reproductive system are organs of copulation that facilitate internal fertilization: a process that minimizes desiccation. The fertilized egg, complete with chorion, is deposited by oviparous females to complete development outside her body. Ovoviviparous and viviparous females retain the egg until embryonic development is complete.

The developing embryo undergoes a segmentation wave and limb buds are formed on each segment; some are destined to be modified into mouthparts, legs, or reproductive structures.

Postembryonic development involves growth and molting, with three basic patterns: direct development of primitive insects, gradual metamorphosis—usually associated with externally developing wings in the nymphal stage, and complete metamorphosis. In this last type the growth stage is the wormlike larva, with wings developing internally. Next is the quiescent pupal stage, in which larval tissues are histolyzed and rebuilt, through histogenesis. The pupa is characterized by external wing pads. The final stage, the adult, is usually winged. It emerges to disperse and produce the next generation.

Major exceptions to the general patterns of development are polyembryony—multiple twinning, pedogenesis—parthenogenetic reproduction by immatures, and polymorphism—different forms within a species due to sex, social structure, mimicry, or environmental conditions. The mayfly's molting, flying subadult stage and hypermetamorphosis (specialization within the larval stage) are also major exceptions.

Annual environmental extremes of temperature and moisture present major hazards to insect survival. Although some species migrate to less harsh areas, many survive such adversity by entering diapause. This physiologically resistant state is obligatory in some species, facultative in others, and may occur in any stage of development, but it is usually restricted to one stage in a given species. Diapause is often triggered by changes in daylength, but other environmental cues such as temperature, food, etc., may also induce it.

SUGGESTED READINGS

Anderson, D. T. 1973. "Embryology and Phylogeny in Annelids and Arthropods." Pergamon Press, New York.

Chapman, R. F. 1969. "The Insects, Structure and Function." Elsevier, New York.

Davey, K. G. 1965. "Reproduction in the Insects." Freeman and Co., San Francisco, California.

Engelmann, F. 1970. "The Physiology of Insect Reproduction." Pergamon Press, Oxford, England.

Hinton, H. E. 1970. Insect eggshell. *Sci. Amer.* **223,** 84–91.

Hinton, H. E. 1979. "Biology of Insect Eggs." 3 vols. Pergamon Press, New York.

Lees, A. D. 1955. "The Physiology of Diapause in Arthropods." Cambridge University Press, Cambridge, England.

Tauber, M. J., and Tauber, C. A. 1976. Insect seasonality: diapause maintenance, termination, and postdiapause development. *Ann. Rev. Entomol.* **21,** 81–107.

6

Insect Behavior

THE TYPES OF BEHAVIOR—INNATE VERSUS LEARNED

Behavior is what an animal does and how it does it. In other words, behavior is how an animal acts and reacts in its environment. Such actions are subject to the types of senses possessed by the insect and the range of sensitivity within which the senses operate. Other limitations to the insect's activity are imposed by the relatively low number and consequent reduced complexity of the neuronal pathways within the insect's nervous system. Thus, most behavior exhibited by insects is *innate*, that is, there is an inherited disposition for a particular behavior. At the other extreme is *learned behavior*, in which behavioral response is altered as a result of experience.

The difficulty in describing insect activity is compounded for the ethologist (someone making a scientific study of behavior) by the human tendency to impose anthropomorphic interpretation into another organism's behavior. Anthropomorphism is the reading of human motivations, characteristics, and behaviors into nonhumans. In addition, descriptive behavioral studies of insects are difficult because of the rapidity with which behavioral sequences may occur. For instance, of the seven separate behavioral elements in courtship of the house fly, four are concluded in $\frac{3}{64}$ of a second each or less, while the rest of the elements complete the mating sequence in a total of under 4.5 seconds. Such elements have been overlooked in simple observational studies and are only disclosed when a different time frame is employed in analysis. In rapid behavioral sequences such as insect courtship studies, high speed photography of the events subsequently can be analyzed frame by frame to reveal the behavioral components.

HOW WE DESCRIBE INSECT BEHAVIOR

Behavior has been described from two different approaches. The *mechanistic* approach describes behavior as the actual physical elements of body movement that make up the behavior. For instance, a mechanistic description of behavior might read as follows: "upon tarsal contact with a 10% sugar solution, the fly's proboscis was extended. The cibarial pump then begins contractions." The second type of behavioral description is termed *consequential* because it describes the result of the behavior. A consequential description of the behavior just presented would read: "The fly drinks when a 10% sugar solution contacts its tarsal receptors."

Neither of the above types of behavioral description is necessarily the right one. Each has its own inherent problems. With a strictly mechanistic description such complex and lengthy behavior as nest building would involve a list of so many repetitive appendage movements that the description would be confusing. On the other hand, a consequential description tends to ignore the many separate components that enter into a behavior. This loses the details that often provide the clues to our understanding of the behavior. Without some analysis of the details of behavior we often slip into anthropomorphic interpretations that imply moods, drives, or appetites that may not exist in the insect.

Much of the nerve activity of insects is based on the *reflex arc*, a physiological link between a sensor through a part of the central nervous system (via an association neuron) and including a motor neuron (see Fig. 6.1). With this type of "wiring" a simple stimulus of a single sensillum tends to cause a specific response. The response may be a movement if the motor neuron leads to a muscle, or result in some other response depending on the termination of the neuron. Such reflex arcs are straightforward and logically lead to a quite simplistic view of insect behavior as consisting of a series of push-button stimulus-response arcs which allow the insect to function.

Early work in animal behavior was based on just such a stimulus-response theory, proposed by Jacques Loeb. The difficulty with this theory is that the animal does not always respond to a given stimulus in the same way. Responses differ as the animal ages, at different seasons, or times of day. Thus, the current view is that the organism responds in one of several ways available from its repertoire of behaviors, depending on the conditions to which it has previously been exposed. Certain ac-

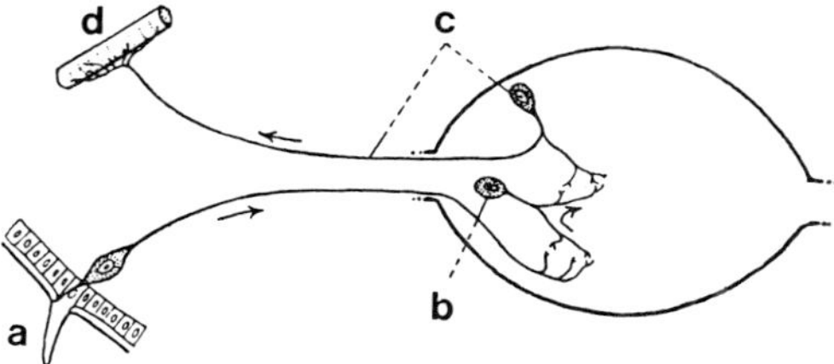

***Fig. 6.1** Elements of a simple reflex arc. A stimulated sensillum (a) initiates a nerve impulse in the sensory neuron (nerve cell). The impulse passes down the sensory nerve into a ganglion of the central nervous system, where it passes across a synapse into an association neuron (b). The association neuron passes along the impulse to a motor neuron (c) across another synapse. The impulse in the motor neuron causes contraction of the muscle (d) to which it is attached. Reprinted with permission from V. B. Wigglesworth,* **The Physiology of Insects.** *Chapman/Hall Publications.*

tivities "prime" the system for other activities, while concurrently inhibiting other behaviors. The stereotyped pattern of behavioral acts, or *fixed action patterns* are performed by the animal when triggered by the "behavioral releaser" (appropriate stimulus).

Flaws in the Loebian stimulus-response theory (called the theory of tropisms) caused subsequent workers to investigate further the subtleties of behavior. From this research more complex explanations of behavior have evolved which are based in part on the role played by the central nervous system as a primer or inhibitor of behavioral events.

Fixed action patterns are often made up of simpler behavioral components. As scientists have observed, a particular behavioral component, just by being emitted, tends to do two things: (1) its immediate repetition is inhibited, and (2) its satisfactory completion primes the system for a subsequent behavioral pattern. Although most workers in insect behavior ignore this principle of *successive induction* as proposed by C. S. Sherrington many years ago, the current trends in analysis of behavioral patterns in fact are studies of the variations in sequences of successively induced behaviors.

In a stereotyped behavior (fixed action pattern) such as mating, the elements involved and their sequence may be identified (Fig. 6.2). The various elements for two closely related species are recorded for each sex as the events occur. Such a sequence between two individuals is termed a "reciprocal behavior chain." If the chain is interrupted, or if the sequence is modified, the behavior does not run to completion. In this case, the result would halt the mating behavior before copulation. Thus, reciprocal behavior chains for two closely related species need be only slightly different in order to eliminate mating between different species. Since interspecific mating is biologically disadvantageous, the modification in behavior chains would be selected for. The behavior is consummated if the actions by each individual induce responses by the other individual that are in the correct sequence. Interestingly, only slight modifications of sequence can allow the same components of a behavior to be used by many different species.

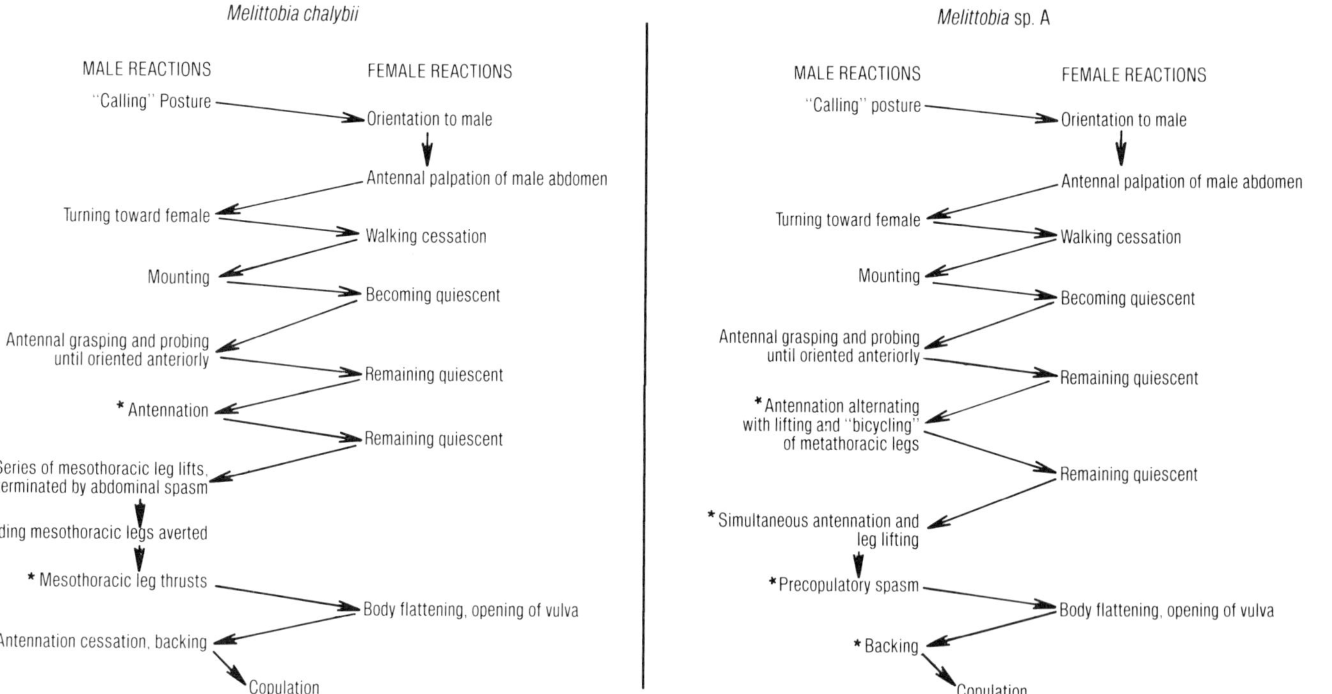

Fig. 6.2 ***A comparison of reciprocal behavior chains in closely related species. The courtship of the two parasitic wasp species differs only in the components marked with an asterisk. Note that no indication of time involved in each of these steps is indicated. The duration of each step is as important in the chain as is the sequence in some species. Reprinted with permission from D. E. Evans and R. W. Matthews. 1976. Comparative courtship behavior in two species of the parasitic wasp Melittobia.*** **Animal Behaviour.** ***24:46–51. Published by Bailliere Tindall, London.***

ORIENTATION

Loeb's early work on stimulus-response theory was too simplistic to provide adequate explanations for all insect behavior. It did, however, result in explanations for the behavioral mechanisms involved in insect orientation to a stimulus. Although the types of orientation have been classified in a number of ways, the classification proposed by Fraenkel and Gunn (1961) is best. According to this classification body orientations are of two basic types. *Primary orientation* refers to a static body position in space, such as maintenance of horizontal or vertical position of the long body axis to the source of the stimulus. Activity in response to an external stimulus is termed *secondary orientation*. One example of secondary orientation is movement away from a light source.

Primary orientations do not necessarily involve locomotion, but orient the body at a fixed angle to the stimulus. Because the fixed angle is usually 90° these primary orientations are also known as *transverse orientations*. If locomotion is involved, at least it is not toward or away from the stimulus.

Primary Orientations

The *dorsal light reaction* orients the insect so that the source of light is above the insect's back. This reaction is especially critical in aquatic animals, and many aquatic insects as well as fish use this orientation to maintain body position. It is interesting that water boatmen maintain their upright position in the open water by exhibiting a classic dorsal light reaction whereas the closely related backswimmers maintain an inverted position by using a ventral light reaction. The dorsal (or ventral) light reaction requires that the insect possess a pair of intensity-sensitive light receptors.

The *light compass reaction* is a primary orientation in which the insect moves at an angle to the source of the light. Such angular movement is possible in insects that have multifaceted compound eyes where the point source of light can be maintained in a particular position on the eye by orientation of the body. Such a system permits insects to move in a straight line, provided the light source is very distant (e.g., the moon). This system is easily disrupted in insects when a light source such as a streetlight is turned on. In this case, the insect must continually reorient its body to the light source in order to maintain the source on the appropriate part of the mosaic image it receives through its compound eyes. The insect travels in spirals trying to orient itself to the light source, trapped by its stereotyped behavior (see Fig. 6.3).

Ventral earth reaction, another primary orientation mechanism (previously explained in Chapter 4), is based on detection of gravity using hair plates. These masses of sensory hairs, located at joints, are especially common in the thoracic–abdominal area of ants. They sense their position with respect to the vertical by position of the pendulous abdomen as it hangs down. Different hairs respond depending on position of the abdomen and whether or not they are being distorted.

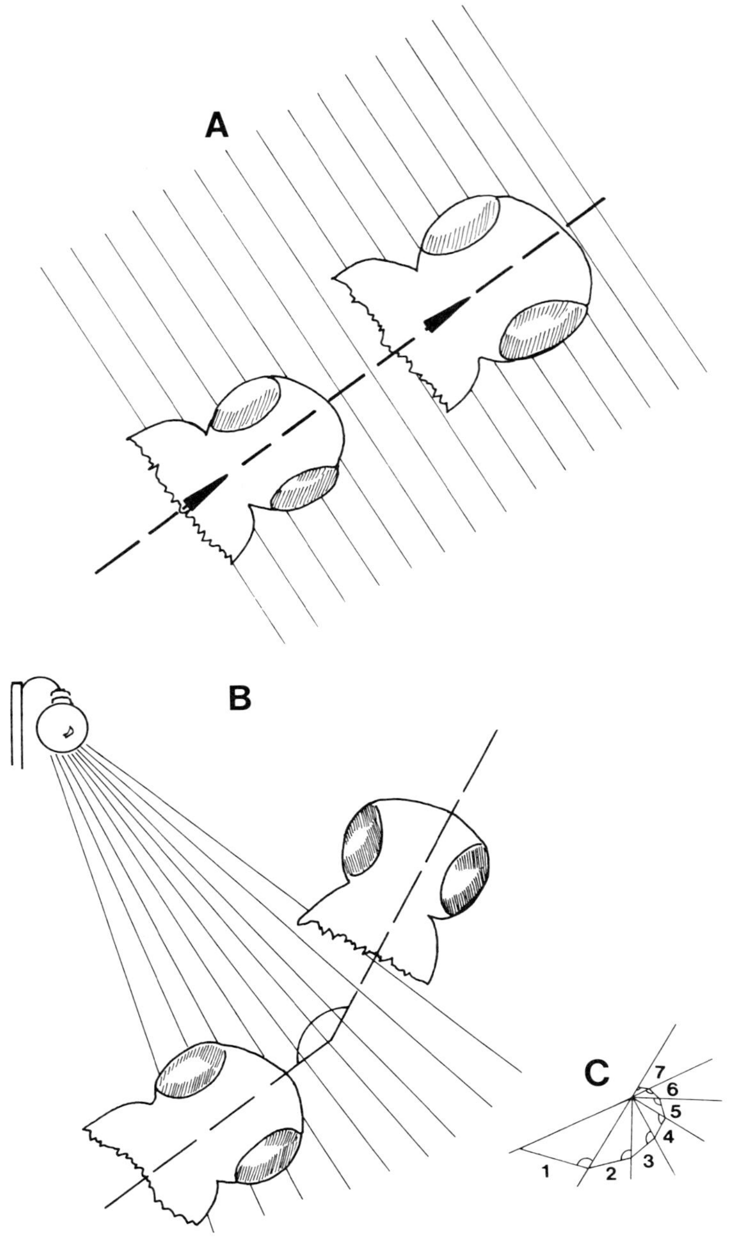

Fig. 6.3 The light compass reaction. (A) The insect is able to continue in a straight line for a considerable distance by maintaining position of the sun (or moon or stars) on a particular portion of the compound eye since these light sources are so far away (indicated by the parallel light rays). (B) The insect must adjust its direction of flight constantly in order to keep the light emanating from a light source relatively close to the moving insect (like a streetlight) on the same portion of the compound eye. (C) The spiral an insect follows as it repeatedly (here shown in seven steps) attempts to correct its orientation to a nearby light source.

Secondary Orientations

Activity orientations to stimuli involve one of two responses. *Kinesis* is an undirected response to a stimulus, that is, the animal responds, but without movement toward or away from the source of the stimulus. *Taxis* is a directed response in which the stimulated insect moves toward or away from the source of the stimulus. The kind of response exhibited by the insect varies with the type of stimulus, the kinds of receptors possessed by the insect, and the physiological state of the insect. Reversals in orientation (+ to − or − to +) are common in insects at different developmental stages.

Kinesis

Kineses are the simplest kinds of secondary orientation since they require only a single intensity-sensitive receptor in the insect. The single receptor is capable of controlling the kinesis provided there is a gradient of intensity in the stimulus (e.g., humid conditions at one side of a cage ranging to dry at the other). A common example in many insects is called orthokinesis, in which the *rate of activity is changed with the intensity of the stimulus.* This kinesis operating within a humidity gradient tends to keep the insect in the area of optimum humidity because the rate of movement is fastest in the most stressful situations and slower in more optimal conditions. Individuals may aggregate as a result of common orthokinetic responses to a stimulus. Orthokinesis also adjusts the rate of activity in many insects to the amount of light. Diurnal (active during the day) insects require a minimum intensity of light, a threshold, before they will start moving, and their rate increases with increasing light intensity.

Klinokinesis is another type of kinesis exhibited by many insects in a gradient of stimulus intensity. It involves rate of turning during locomotion and depends on the level of stimulation received. An insect tends to turn as it proceeds through areas of high stimulation, effectively holding itself in favorable areas. Insects pass through locations without stimulation with few turns. The importance of such a reaction to survival of an individual has been demonstrated in laboratory studies of prey-searching ability with predatory insects. Ladybug larvae thoroughly search flat environments with uniform light distribution when no prey are present by patterns of turns as they crawl (Fig. 6.4A). The same predatory species concentrates its searching in areas where prey are contacted because the rate of turns per unit time are greater once prey has been found (Fig. 6.4B). Since these predators perceive prey only when the prey is contacted, the klinokinetic behavior tends to keep the predator in areas of high prey density, such as aggregations of aphids, scales, or mites.

Taxis

In the directional responses typifying a taxis, the long axis of the body is oriented in a line with the source of the stimulus and the locomotion is either toward (positive) or

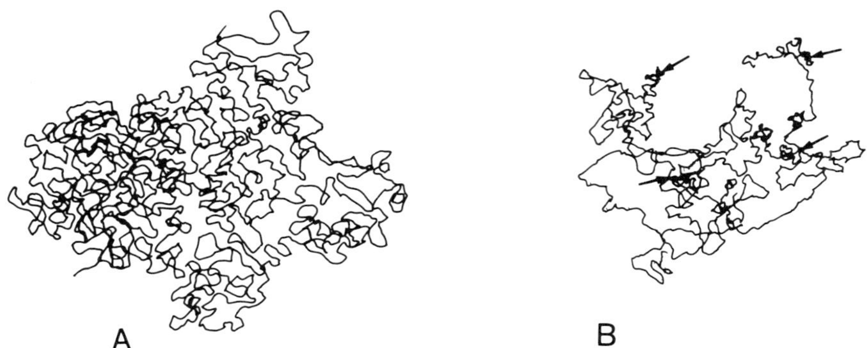

Fig. 6.4 Klinokinetic responses to stimuli. A record of the movements of ladybug (Stethorus) larvae in 1 hour in a flat, uniformly illuminated environment. The searching pattern differs between (A) an environment lacking any prey, and (B) where prey (citrus red mite) exists at a density of one mite per square inch. (B) An encounter with prey increases the amount of turning in that area, keeping the predator in the area so that prey aggregations are eliminated. Several of the prey encounters are indicated with arrows.

away from (negative) the source. The method of expressing a tactic response may differ according to the number of receptors possessed by the insect that are sensitive to the stimulus. With two or more receptors, simultaneous comparisons of intensity from each receptor allow the insect to orient directly by turning to the more (or less for negative responses) stimulated side. This is called *tropotaxis*. Thus no deviations from the straight line path are required as they are in a typical klinotaxis.

The insect's klinotactic use of a single intensity receptor involves successive comparisons of the receptor response as it is waved in the environment. Such successive comparisons of intensity allows an insect like a fly maggot to move away from a light source even though it possesses a single light sensitive structure. The structure is in the pointed (head) end of the body and, as the head is waved laterally, the insect's own body comes between the light source and the sense organ (see Fig. 6.5).

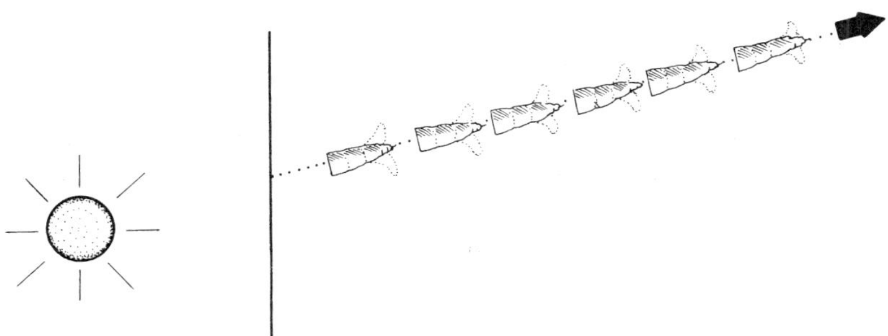

Fig. 6.5 Klinotactic progress involves successive comparisons of stimulus intensity. The fly maggot's light receptor is in the pointed (head) end of the body. At each movement forward the head is swung from side to side, thereby comparing intensities. The lowest intensity received at each comparison occurs when the posterior part of the body is between the receptor and the light source.

The various types of taxes are named according to the type of stimulus and the reaction invoked. Thus, a phototaxis is a tactic response to light and it may be either negative or positive. Rheotaxis and anemotaxis are both responses to current, the former in water and the latter in air. Geotaxis is a response to gravity, and thigmotax-is to contact. Although a knowledge of these various taxes and kineses does not, in itself, explain the totality of insect behavior, it is useful in clarifying many of the mechanisms involved.

LEARNING AND MEMORY

As previously stated, insects are animals that exhibit primarily stereotypic behavior. They do, however, also show abilities to learn and to remember. This is demonstrated by the solitary wasps, when they fly off from their individually constructed nesting burrows in the ground to seek food. That they can find their way back to their own particular burrows is in itself a demonstration of memory.

Learning is usually defined as a relatively permanent change in behavior which occurs as a result of practice. The relative permanence and the requirement of practice are aspects of learning that differ from behavioral changes that occur because of maturity. According to the above definition, insects can learn such tasks as running a maze. In experiments demonstrating this, cockroaches learn to run a maze in relatively bright light reaching an opaque "home cup." They learn the fastest if the cup has previously been occupied and thus has the appropriate "roach odor," and if it is not transparent to light. The stimulus in this experiment is the light, since cockroaches are nocturnal and usually avoid exposure to light.

In experiments with grain beetles, both larvae and adults can learn to turn left or right consistently in simple T-mazes. The learning was retained over periods of several days to several weeks and occurred even after metamorphosis. In other words, adult beetles retained memory of the maze they had learned as larvae.

Either overt physical activity or nervous system activity is thought to interfere with retention of memory. A corollary of this interference theory is that inactivity after learning will extend memory of that learning. Experiments with cockroaches prove this to be true. Roaches trained in a simple task (as in avoiding a shock) retained the training for an extended period if they were cooled down (with consequent lowered metabolism) immediately afterwards. The same results were obtained when the roach was restrained in paper. Thus inactivity, whether thermally or physically induced, extends memory retention. Conversely, forced activity, as with a treadmill, shortened retention.

THE RHYTHMS OF LIFE

Ecological rhythms exist in the environments of almost all insects. These rhythms are usually linked to light in some way, whether they be day, month, or seasonal in duration, and are often termed collectively as photoperiodism. Since many rhythms are on a roughly 24-hour basis, they are known as *circadian* (*circa*= about, *dia* = day) rhythms, and insects exhibiting them may be referred to as *nocturnal*, *diurnal*, or *crespuscular*, depending on whether the activity is greatest during night, day, or at dawn–dusk periods, respectively.

Insects respond to ecological rhythms by adjusting their life processes into *biological rhythms*. Such adjustments in process rhythms can be evolutionarily advantageous when they synchronize the biology of the organism to the rhythmicity that occurs in the environment.

In photoperiodic terms, insects may be long-day, short-day, or neutral. Long-day insects actively develop during the portion of the year with the most hours of light in 24 hours, that is, the summer. Short-day insects may enter diapause during the summer and only actively develop during the short daylight periods common to spring and fall. However, regardless of the particular activity pattern followed, it is evident that insects can discern the passage of time and that they have some type of "physiological clock." Phase setting of the physiological clock requires an alternation of light and dark periods and is referred to as *entrainment*. Thus, insects that lack rhythmicity in constant light show diurnal periodicity even when darkness of very short duration is experienced. In this case the period of darkness has acted as an external time cue which sets the phase.

Determining what the target organ for detection of photoperiodism is has been the subject of many experiments. Because the compound eye is light sensitive, it was a likely prospect as the target organ. However, roaches with the optic nerves cut still responded to photoperiod. Interestingly, a mutant strain of *Drosophila* fruit fly without compound eyes also responded. Surgical removal of ocelli proved that they were not the target organs.

In at least some moths the target organ has been identified as the brain itself. In cecropia moth larvae, removal of the brain destroys the clock-controlled timing of the pupal molt. If the brain is implanted into another "brainless" cecropia larva, this "loose-brained" larvae's molting time is set. Therefore, not only is it the brain that is the target organ, but lack of nerve connection between the implanted brain and the rest of the central nervous system provides evidence for hormonal control of molting by the clock.

Theories of Photoperiodic Time Measurement

The early evidence suggested to Bunning in 1936 that the photoperiodic clock was an endogenous (internal) rhythm of sensitivity to light which was set by dawn. A 24-hour period consists of two 12-hour periods, the first half of the cycle requiring light and the second half cycle requiring darkness. If the coincidence of light duration extends into the darkness-requiring phase then long-day effects are exhibited, if not (that is, the light only occurs in the light-requiring phase) then short-day effects occur. This has been observed in Lepidoptera, Hymenoptera, and Diptera.

It is also evident that some insects, like the aphids, use an hourglass or interval timer type of physiological mechanism to measure time. That is, if a given duration of darkness was required to elicit a response, once that period was reached, the response was elicited. If an endogenous rhythm had been operating, the next dark-requiring phase would have elicited a recurrence of the response.

Subsequent work on rhythms has disclosed several permutations of the two basic photoperiodic schemes. Multiple, rather than single timing mechanisms have been reported. Also, cases where the phase-setting stimulus of the light–dark cycle is dusk rather than dawn have been identified. As might be expected from these complications in the system, there is also evidence that some insects have photoperiodic rhythms that have both types of mechanism operating simultaneously and with both light- and dark-sensitive states.

FEEDING AND OVIPOSITIONAL BEHAVIOR

Both feeding and oviposition involve complex behavior chains that result, in the first case, in locating food, in the other, a site for egg laying. The behavioral sequences involve locomotion to the site, some kind of recognition, feeding or oviposition, and termination of the behavior.

Locomotion to food in many insects involves flight. The winged female aphid leaves its host plant and is attracted upward by the ultraviolet waves of the sky and repelled by the longer waves from earth. It continues to be attracted up into the sky for about 3 hours, when a reversal in response occurs. The aphid begins to settle back to earth because it is now attracted to yellows (reflected by plants) and repelled by ultraviolet light. Young adult female apple maggot flies (*Rhagoletis pomonella*) are also attracted to yellows reflected by foilage, which is where the aphid fecal honey dew that the fly consumes is found. However, as the ovaries of the fly mature, the female becomes more attracted to darker colors, like red, which is the color of the fruit she deposits her eggs in.

Insect parasites such as parasitic wasps and flies are attracted initially to the location where the host lives (host habitat selection), then to the host, which is finally accepted or rejected. Recent work suggests that after the adult parasite gets into the right host habitat, it may use chemical cues to locate the host. Those flies that parasitize larval insects are attracted to fecal material of the host insects, and one species has been shown to oviposit in response to a protein found in the host's feces. Wasp parasites of moth eggs search more intensively in those areas that have been walked over by moths. In this case, it turns out that the few minute wing scales that are dropped by the moth while it is on the leaf act as arrestants. They keep the parasite in the area after it has contacted the scales, which act chemically in arresting the insect. Even the particulate nature of the scales adds to their efficacy. By arresting the movement of the parasites, the presence of scales increases the amount of searching by the parasite in the area where the moth has recently been, rather like a bloodhound following the scent of a recent trail.

The insect maintains quality control over its nutrition by reacting to the chemical and physical nature of foods. Common nutrients, like sucrose (cane sugar) are an effective stimulus to feeding in most insects. However, many insects exhibit feeding behavior only in response to stimuli that are not in themselves nutrients. For example, the secondary plant chemical sinigrin is not a nutrient for the cabbage aphid, but it does cause continued feeding when present in material probed by the aphid. The reaction is so strong that aphids can be induced to feed on nonhost plants if the plant is allowed to absorb sinigrin through its stem. Sinigrin is typical of the nonnutrient feeding stimuli that are called "token stimuli."

It is not surprising that many nutrients stimulate feeding, but rather that so many of the stimulants to feeding are not nutrients. The specialized feeding habits of the majority of insects is a result of rejecting foods that possess deterrents or that lack nutrients, as well as acceptance of one or more token and feeding stimuli.

Clues to some of the additional controls to feeding are given by modifying the insect's ability to perceive its food. Normal tobacco hornworms will feed only on plants in the family Solanaceae, such as tomato and tobacco. Maxillectomy (surgical removal of the maxilla with its sense receptors) removed the inhibition to feeding on nonhosts. With the maxilla or its sensory structures removed, the insect will feed on plants of other families, e.g., dandelion (Family Compositae) and complete development. That they do not normally do so is tribute to the advantage that specialization gives to the insect that can recognize its own particular host(s) thereby assuring a nutritionally adequate diet (see also Chapter 3). However, it is the central nervous system that is usually the ultimate site for discrimination of diets, since sensilla are not usually very specific in what they respond to; that is, each sensillum may respond to several stimuli, but with maximum responses by different sensilla to different chemicals.

Many blood-feeding insects, such as the mosquitoes and horse flies, also feed on nectar and sometimes on water. The series of behavioral steps leading to a blood meal, therefore, must differ somewhat from those resulting in a sugar meal. Mosquitoes locate their blood source by cueing in to a variety of stimuli including gradients of temperature, moisture, and CO_2, all typically produced by a living warm-blooded animal. They also react to lactic acid, a compound deposited on the body surface in the sweat of vertebrates. Dark-colored objects, especially if moving, are very attractive to mosquitoes, as they are to horse flies (see Table 6.1).

After locating the host and landing, the mosquito probes in response to heat stimuli, in combination with CO_2 and moisture. The meal is terminated by stretch receptors in the abdomen when it is expanded. Readiness to seek and secure a blood meal is determined by the female's age and her state of egg development. Very young females refuse blood meals as do those that have already engorged on blood and are in the process of developing eggs. In many other blood-feeders, the prime signal that induces probing is temperature differential. For example, fleas are attracted to warmth, as can be demonstrated by dragging a hot water bottle covered with skin through their habitat.

Perhaps the best total picture of feeding is in *Phormia regina*, the black blow fly. It begins feeding when taste receptors on the feet are stimulated. The proboscis, or feeding apparatus, is extended and contacts the substrate with food on it. As the insect feeds, sensory hairs and papillae continuously monitor the function. When the sense organs in contact with the food become adapted to the meal's chemical stimuli, the insect stops feeding. In some other insects, however, nerves control the cessation

Table 6.1 Basic Stages in Blood Meal Acquisition and Stimuli Controlling Them in Winged Insects[a]

Stage	Facilitated or initiated by	Stimuli orienting to	Terminated or inhibited by
Rest	Cold, darkness, bright light, engorgement, exhaustion, tarsal contact	Gravity, wind optomotor, odor	Mechanical, visible movement, no tarsal contact
Flight and searching	Moderate wind, mechanical, visible movement, no tarsal contact	Wind, odor, optomotor	Cold, bright light, tarsal contact, exhaustion
Settling on host	Exhaustion, odor, convection current	Convection current, temperature, gradient, gravity	Surface texture and temperature, repellent odor
Probing and biting	Hunger, surface temperature, sugars, ATP	Sometimes gravity	Abdominal tension; all thresholds elevated.
Take-off	Nourishment, abdominal tension, voiding of fluid	Darkness?	Excessive engorgement?

[a]Modified from B. Hocking. (1971) *Annu. Rev. Entomol.* **16,** 1–26.

of feeding, because when the nerves that lead to receptors in the gut are severed, the insect continues to feed until it bursts. These receptors in the gut are sensitive to stretching and inhibit feeding when the gut is distended.

COMMUNICATION

Communication between two organisms involves three distinct and required aspects. First, one organism must provide a stimulus for another organism. Then, the second organism's ability to respond is altered by the stimulus from the first. Last, the reaction of the stimulated organism either benefits itself, or the other organism, or both of them.

The stimulus from the sender is a coded message, that is, the stimulus is a signal to the receiving organism, which relays something about the sender. The code takes one of several forms available, depending on the sense organs of the receiving organism. These signals may be visual, acoustic, tactile, chemical, or a combination of them. They may be effective only between members of the same species, as with mating signals. The communication may also be interspecific such as a defensive reaction by an individual under attack. In fact, communication is not always a process involving activity; it often is passive in nature. For example, a harmless fly mimic of a stinging wasp may escape from attack by potential predators simply because its color pattern inhibited the attack response in an experienced predator. In this case, not only is the communication interspecific and passive, but the communication is false because the mimic is creating a "false impression," which will be discussed later in the section on Mimicry.

Communication between members of a solitary species most often occurs at those brief intervals when it is required, or at least advantageous, for completion of the life cycle, such as at mating, egg laying, or aggregation. These social relationships involved in communication even among so-called solitary forms reach their zenith in the true social insects whose communication systems have evolved a high level of sophistication. Such sophistication is required to control and coordinate the various activities being performed, e.g., reproduction, nest construction, care of the young and queen, food exchange, foraging, and defense.

Acoustic Communication

Among the several groups of insects that use acoustic communication are the crickets, katydids, and grasshoppers of the Order Orthoptera. Communication by sound is advantageous because the caller need not be large nor leave its cover and become exposed to view, as it might if using visual communication. The sounds may be an "incidental" aspect of the insect's activity, as the whine of the mosquito's wings in flight, but it is more commonly produced by special sound producing organs, for example, stridulatory structures or tymbals. Stridulation involves rubbing one rough part of the body against another, often with an identifiable single scraper on one part and an elongate roughened or ridged file (strigel) on the other. In many insects the file is attached to an expanded membranous area that magnifies the sound of the stridulation. The tymbals of cicadas, many other Homoptera, stinkbugs, and some moths are membranes that are caused to vibrate by the muscles attached to them. The sound from the tymbal is a click produced in the same way as the top of a tin can when it is pressed in and out. The vibrations are usually amplified by association with one or more air-filled chambers, and the intensity of the sound produced is surprisingly loud. Recent tests conducted with predatory grasshopper mice showed that they erred more often and took longer to capture normal (noisy) cicadas than those cicadas that had been silenced prior to the test.

Analysis of the sounds produced by a single species of cricket discloses a number of different ways in which sounds are used. Various species have been recorded using sounds as

- Calling signals for aggregation and pair-formation
- Aggressive signals to establish territory and in combat
- Courtship signals
- Copulatory signals
- Alarm or disturbance signals to repel predators and/or warn others

If the calling sounds of several cricket species are recorded and visual graphs of the sounds made (audiospectrographs) several distinguishing characteristics become apparent. Each pulse (homogeneous sound) is produced by the wings closing once, with the individual pulse resulting from the series of teeth hitting the file (see Fig. 6.6). The differences between calling songs are sufficient in themselves to allow each cricket to recognize mates of the same species. However, the song differences are reinforced by differences in seasonal development, geographic distribution, habitat selection, and in preferred daily activity patterns.

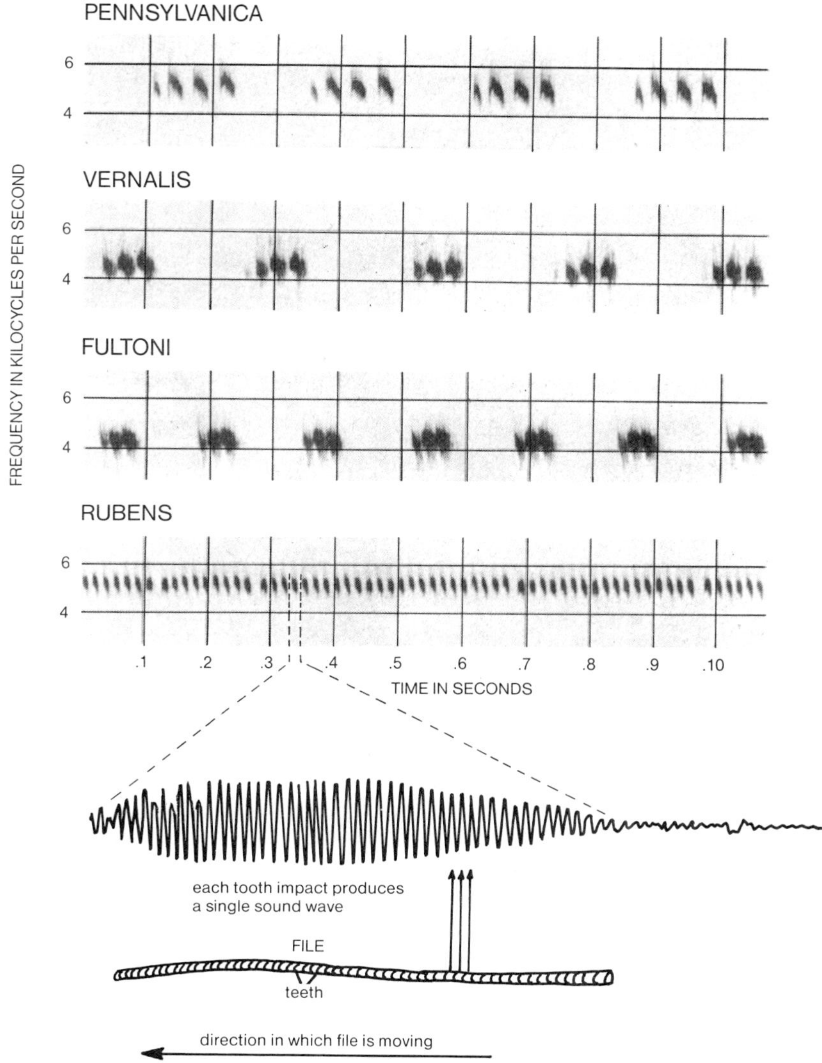

***Fig. 6.6** (A) Audiospectrographs of the calling songs of several* Acheta *species show differences in the character, rate, and grouping of sound pulses when recorded against time at the same temperature. Although the pulse rate of the call changes with temperature, the selective apparatus of the female changes at the same rate so that the variability in the male's song is matched by that of the female's responses. (B) A visual recording of a single pulse from an* Oecanthus *tree cricket, showing how each tooth of the file is responsible for producing a single sound wave.*

Chemical Communication

Although chemical signals between individuals may be the primary mode of communication in all the living world, it has not been studied much until the last few years. Since humans rely so heavily on our own visual and acoustic senses and so little on olfaction, we have tended to emphasize visual and acoustic communication research, while minimizing work on chemical signals. Fortunately, the advances in chemistry, notably those in microanalytical techniques, now allow us to identify the minute quantities of chemicals released by signaling insects. As a result, we are beginning to grasp the scope and complexity of chemical communications existing around us.

The advantages of a chemical mode of communication are several. (1) As with the use of acoustic signals, the sender need not be exposed while it is signaling, that is, it could be hidden from sight in the vegetation but still releasing a chemical message. (2) The chemical medium seems particularly appropriate for animals like insects with such poor visual acuity, because long-distance communication is almost always easier with chemicals than either visual or acoustic means. (3) Not all chemical signals are released into the medium and dispelled by wind or water currents. Some are contact chemicals that stay on the substrate until the actively searching receiver contacts the substrate where the chemical message is located. (4) There is a huge variety of molecules available for use. Thus, the potential for great variability and specificity exists.

The problems with chemical communication involve control of the signal. In most cases, signal duration is not precise because of the lag (seconds or minutes usually) between release of the chemical by the sender and its movement through air or water until it reaches the receiver. Likewise, when the sender stops signaling, it takes a while for the chemical to dissipate. Also, not too much information can be relayed by the use of a specific chemical because the primary variables are concentration, site of communication, and the stage of development of sender and receiver. A tenfold increase in concentration can change a chemical from one that attracts other ants to one that causes an alarm reaction.

The chemicals that are used in communication have been termed broadly as "semiochemicals" (semio = signal), and they may be of a "releaser" type, that is, stimulating an immediate behavioral response or "primer" type, which cause long-term changes in the receiver. The earliest semiochemicals known are active only between two members of the same species. These are called *pheromones*, and are defined as chemicals that are released by one individual and ellicit a response in another individual of the same species.

A classic example of a pheromone is the "queen substance" of the honey bee, *Apis mellifera* L. Queen substance is a complex molecule (9-oxodec-*trans*-2-enoic acid) produced in the queen's mandibular salivary gland. This molecule, released continuously by the queen in the hive, inhibits ovary development in all of the worker (female) bees. As long as a queen (the source of queen substance) is present, the ovaries of workers remain undeveloped; but if the queen is removed, some workers will eventually develop eggs and produce them. However, since the laying workers have not been mated, all of the eggs are unfertilized, which (in the order containing the honey bee, Hymenoptera) results in male offspring.

Queen substance acts as an *aggregation pheromone* (causing aggregation of individuals) when it is released outside of the hive by a virgin queen on her mating flight. In this case the aggregation chemical functions as a sex attractant to the drones (males), a type of response that appears to be very widespread among insects. The classic type of sex attractant is released by one sex, usually the female, as a signal that the individual is physiologically receptive to mating. Social insects, especially ants and termites, use aggregation pheromones to mark trails. A foraging worker ant, upon locating a food source too large to be carried home, returns to the nest marking the return route with chemicals continuously released from the abdomen. Other workers, stimulated by the pheromone, follow the trail to the food and return. As long as the recruited workers find food they mark the trail back to the nest, but when the supply of food is exhausted, those that return without food no longer mark the trail. Termites discovering a break in the wall of the colony will run toward the interior of the nest, marking a trail that attracts others to the site for repair (see Fig. 6.7). Other aggregation pheromones have been reported which (1) attract gravid female mosquitoes to a site where other mosquitoes have laid eggs, (2) attract bed bugs to suitable resting sites, and (3) keep honey bee swarms together.

A high level of complexity in chemical communication using aggregation pheromones is reached in the bark beetles. The insects attack a specific tree initially as a result of chemicals released by the tree. However, after several beetles have arrived and started their attack, they release pheromones that attract other individuals of both sexes to the tree. Individuals of the same sex begin to burrow into the tree to make their own reproduction chambers, but members of the opposite sex enter burrows of the early arrivals to mate with them. In one bark beetle (the western pine beetle *Dendroctonus brevicomis*), the aggregation pheromone consists of three chemicals, one from the male, one from the female, and one from the tree. Overpopulation of the host tree is prevented in several ways. A heavily attacked tree has such a concentration of pheromone around it that approaching bark beetles land at adjacent trees before reaching the primary source. In another species (*Dendroctonus pseudotsugae*), a mechanism has been proposed based on the masking effect of a pheromone, whereby the attraction of the aggregation pheromone is eliminated once the

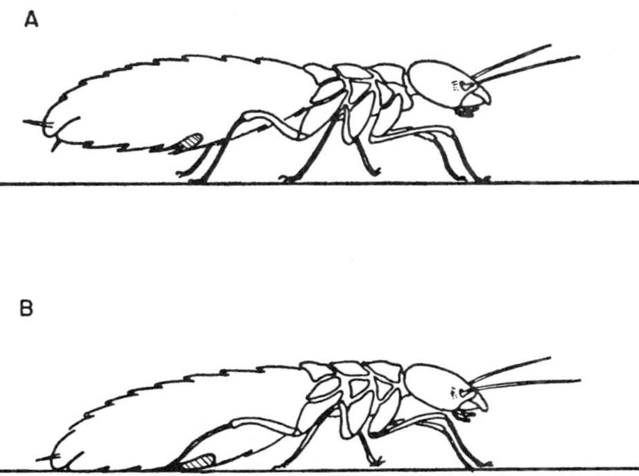

*Fig. 6.7 (A) A termite nymph (**Zootermopsis nevadensis**) resting, without the odor gland (cross-hatched part of the abdomen) touching the substrate. (B) The alarmed termite drags the gland along the substrate, thereby leaving a trail to be followed by other workers to the site of nest damage.*

individual releasing the aggregation pheromone has been contacted by an individual of the other species.

The so-called "aphrodisiac" pheromones of butterflies are released by the male over the antennae of the female after she has been pursued in flight. The chemicals of the pheromone inhibit the female from further flight, and the male continues its precopulatory behavior with the "grounded" female. In butterflies, the aphrodisiac pheromone is produced by scent pencils at the tip of the abdomen (see Fig. 6.8). In roaches a type of aphrodisiac pheromone is produced by a gland on the male's back and when the female mounts the male to feed on the gland exposed by his spread wings she is in the appropriate position for copulation.

One of the sex-related pheromones produced by some female ground beetles might be referred to as the antiaphrodisiac pheromone. The bombadier beetles, a type of predatory ground beetle, are known to produce chemical deterrents when they are disturbed. These materials are used mostly against other species, causing a brief disorientation, with recovery in a few seconds. The chemical produced by female bombadier beetles of the genus *Pterostichus* and its use presents a different story.

Pterostichus females in a mature state of reproductive physiology will mate with a male a few minutes after the original encounter. However, an unreceptive female runs from males after contact and, if pursued, sprays the male with a discharge from the tip of her abdomen. The chemical signal to the male to stop pursuing is rather harsh, but effective. The male halts his progress immediately, and attempts to clean his mouthparts and antennae. Within 10 seconds his movements become wildly erratic and he soon ends up on his back. His legs become stiff and all movement then ceases. After 1–3 hours in a deathlike coma he starts to recover, and in about 30 minutes recovery is apparently complete.

Pheromones that act as antiaggregation signals have been reported for several insect groups. These *dispersion pheromones*, when applied to the substrate by a female after she has laid an egg, causes other ovipositing females to move away. The result of this behavior is to guarantee an adequate food supply for the developing young. Dispersion pheromones are known for several insect parasites and fruit flies, which have relatively immobile larvae that are confined to the immediate habitat into which the eggs are deposited.

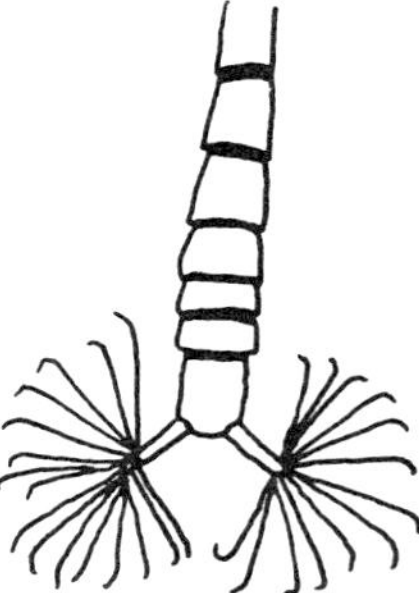

***Fig. 6.8** Scent pencils at the tip of a butterfly abdomen. These modified tubular extensions are everted by the butterfly to disperse the aphrodisiac pheromone.*

The alarm chemical of aphids is one of the few pheromones known in which the same chemical is produced by a number of species and in which so many different species in the same group react to the pheromone's presence. This pheromone is also one of the few alarm pheromones that have been identified among the "nonsocial" insects. However, many aphids occur in aggregations ranging from several individuals up to many thousands. The chemical, *trans*-β-farnasene, is released from the aphid's cornicles when the aphid is disturbed or injured. Cornicles are glandular openings on the rear of the aphid's back that often look like a pair of pipes emerging from the abdomen. It is probable that the usual stimulus for release of the pheromone is when an individual is grabbed by a predator (Fig. 6.9). The pheromone stimulates other aphids to jump off the plant or to crawl around to the other side of the leaf blade or stem.

Communication by means of chemicals has been restricted so far to that between members of the same species or, in the case of aphids, closely related forms. Since the receiver of the message need not be a member of the same species, the kinds of communication are determined by the signal's impact on the communicator. In the case of "*kairomones*" the chemical (or mixture) is released by an individual of one species and induces a response by an individual of another species which favors the *recipient*. As an example, the complicated pheromone systems of *Ips* bark beetles also act in some species as a kairomone to the checkered beetle (*Enocleris lecontei*), which is their predator.

"*Allomones*" are those chemicals released by one species that induce a response in another species receiving them which favors the *emitter* of the chemical. The defensive secretions emitted by so many insects when disturbed are all examples of allomones as are the toxic chemicals in bee, wasp, and ant stings. Figure 6.10 shows two examples of defense by chemicals that disorient, repel, or injure upon contact

***Fig. 6.9** An aphid under attack by a predator emits its alarm pheromone from abdominal tubes. The pheromone causes nearby aphids to drop off the plant or move to another part of the leaf.*

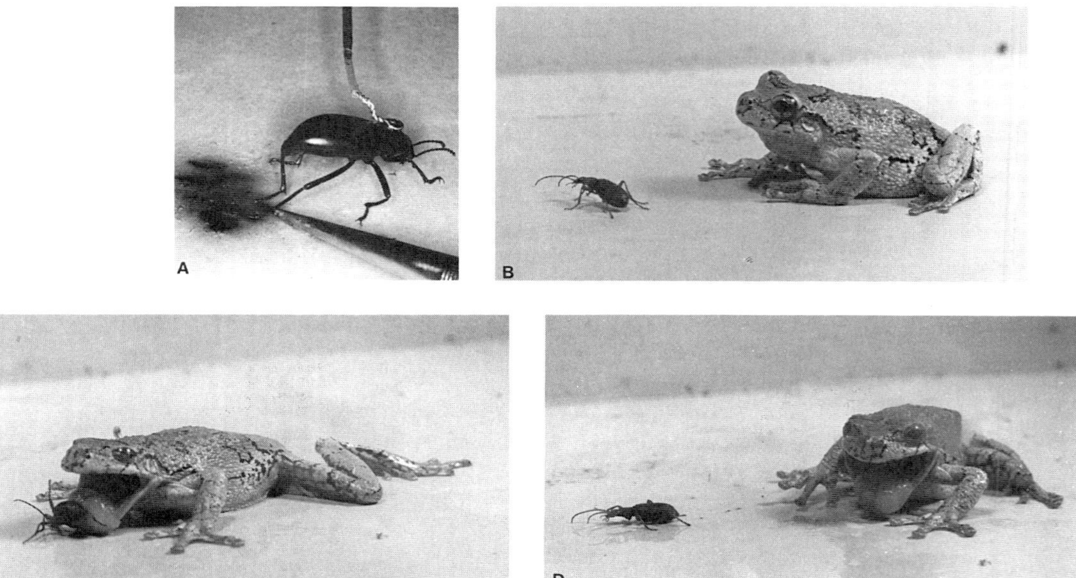

Fig. 6.10 (A) A darkling beetle sprays its defensive secretion from the tip of its abdomen when its leg is pinched. The secretion is visible in the photo only because it landed on specially treated paper. (B) A toad spots a bombadier beetle walking past. (C) The toad strikes at the beetle with its tongue, the first step in consuming the insect, but rejects the beetle (D) when the beetle discharges its defensive allomone.

with another animal. Such glandular secretions have been reported for Hemiptera, Hymenoptera, Lepidoptera, Orthoptera, Isoptera, Dictyoptera, Coleoptera, Dermaptera, and Homoptera, and include materials that could be fatal to the insect manufacturing them. One way of handling this problem is for the glands to produce the chemical components for the allomone, but to keep the components isolated, combining into the actual, often very toxic, substance only when they are released from the body.

Plants often utilize allomones to provide protection from injury caused by insects, vertebrates, and other animal groups feeding on them. These allomones are called secondary plant substances, since they are typically unimportant in the metabolism of the plant, evidently having been evolved strictly as chemical defense mechanisms. They are comparable to physical defense mechanisms such as spines, barbs, and hairy surfaces, which discourage feeding on plants with such protection.

One striking example of plant allomones that occur in the milkweed species are called cardenolides. Cardenolides are poisonous to most animals, but usually cause violent vomiting long before enough could be taken into any animal's system to kill it. Therefore, milkweeds are left alone by grazing animals and tend to grow rather abundantly in pastures. However, a number of insects attack milkweed plants and do so without suffering any ill effects from the cardenolides. It is thought that these milkweed insects (bugs, beetles, and caterpillars) had become dependent on this group of plants prior to the development of cardenolides and that the insects evolved a tolerance as the poisons became more prevalent in the insects' food supply. Thus, the insect species that can tolerate cardenolides and consume milkweed plants have a food resource that is not available to other species. Most of the milkweed insects have become specialists, that is, they only live on milkweed plants. So the milkweed allomone has not been completely effective, but it has not backfired as badly as some plant allomones have. For instance, sinigrin is an allomone of the mustard oil glycosides produced by cruciferous plants (e.g., mustard, cabbage, radish, and turnip). Although sinigrin, in addition to sinalbin, progoitrin, etc., also glycosides, do protect the plant from being eaten by general herbivores, it is a stimulant to feeding by insects such as the cabbage aphid, mustard beetle, and diamond-backed moth that specialize on crucifers. Sinigrin has recently been shown to be very toxic to black swallowtail larvae, which do not normally attack crucifers, whereas imported cabbageworms, a crucifer specialist, are not affected, even at artificially high concentrations. Cutworm larvae are bothered by sinigrin only if it is in high concentration. Thus, the utility of secondary plant chemicals as allomones has been established, but has not proved infallible.

Communication among Social Insects

The very fabric of social life depends on communication between cooperating individuals, whether the society is insect or human. Although the social insects are a major topic in the next chapter, their communication systems will be elaborated on here. In fact, some of the semiochemicals and their uses have already been presented, but their magnitude is overwhelming if we consider the scope of chemical communication for the honey bee alone.

The honey bee is the most intensively studied social insect, yet we are only at the threshold of understanding its communication. Because at least 31 different honey bee pheromones have been identified, it is safe to say that almost every facet of their complex biology is mediated by pheromones. The queen substance described earlier not only inhibits the ovary development of worker honey bees, but also inhibits queen replacement, in addition to its action as a sex attractant outside the hive—three separate functions for the same pheromone under different conditions. Other pheromones have been identified that control honey bee reactions to alarm, orientation under a variety of circumstances, and in stabilization of the swarm.

Acoustic communication is less well studied in the social insects than are visual or chemical modes. However, again using the honey bee as an example, several distinct types of sounds have been identified. Among these sounds are types known as "piping" and "quacking" produced by young queens. After several special cells have been constructed for developing queens and they are, in fact, developing there, the old queen leaves the hive with a swarm of approximately half of the workers. The first new queen that comes out of her cell produces "piping" sounds at intervals, which are answered by "quacking" sounds from the other young queens still within their queen cells. The sounds prevent premature emergence of the additional queens until after the first young queen has mated (outside the hive) and departed with an additional swarm (called the "afterswarm") of workers.

A most fascinating type of communication identified in honey bees is perhaps best categorized as tactile communication and is termed dancing. The dancing of honey bees is basically a directional message to give the location of food and water or of sites for other functions, such as establishing a new colony. The message certainly has chemical and acoustic elements, but since it is performed within the hive, there is almost no visual input.

Honey bee foragers returning to the hive from a flight in which nutrient resources have been located must communicate the good news to other foragers. The object of this communication is to recruit other field bees to the resource. However, there are several aspects to the message that must be relayed in order that those recruited are able to locate the resource. These aspects are distance, direction, quality, and character. Nobel Laureate Karl von Frisch uncovered the mysteries of bee biology, especially their "dance language." A forager returning to the hive from a nectar source close to (less than 60–100 meters) the hive relays the information by doing a "round dance." First she enters the hive and moves onto a crowded portion of the honey comb, which hangs vertically from the top of the nest. She circles as in Fig. 6.11A, while frequently offering nectar from the flower (source of the nectar) to others nearby. This round dance does not relate direction, but it does relate distance to the extent that the source is less than 100 meters away, and also character of the source by the nectar and by the flower fragrance which still clings to the recruiter's body.

At distances of 100 meters or more the round dance is replaced by the "waggle dance" (see Fig. 6.11). This dance receives its name by the way the dancer waggles her abdomen from side to side as she runs through the straight part of the dance. The straight waggling run terminates as the dancer turns around and runs back to the point of origin, usually alternating the return by the right with a return by the left. Distance of the resource from the colony is indicated by the number of turns per unit time. The excellence of the resource, its sugar content and lack of bitter substances is

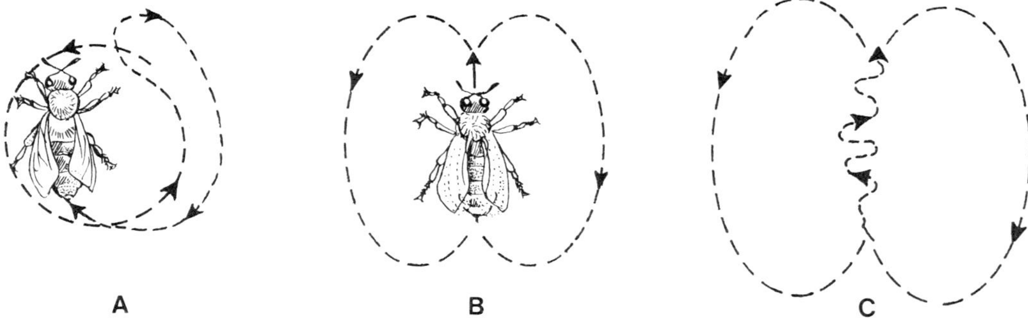

Fig. 6.11 The basic types of honey bee dance. (A) The round dance, used for resources less than 60–100 meters from the hive. (B) The waggle dance for resources at distances greater than 100 meters. (C) The waggle during the straight portion is indicated.

related by the intensity (vigor) with which the dance is executed. The character of the resource, i.e., flower type is related by fragrance as it was in the round dance. The only aspect not yet disclosed is how the direction of the resource is related.

Both honey bees and humans use a reference point with which to establish direction. We have learned that magnetized needles point toward the North pole and so we tend to orient with respect to this fact whenever we really need to know direction. Thus, even on the open ocean we can determine direction, provided we have a compass. Honey bees use the position of the sun; a ready though inconvenient reference point since it moves throughout the day with respect to the colony and the resource. In addition, the communication occurs within the hive, where it is dark. Bees overcome these orientation problems on their vertically hanging combs by translating the sun's current position into the absolute vertical. Determination of "up" is easy, even in the dark because the abdomen tends to hang down, and the body's position with respect to gravity can be detected by sensory hairs. Thus, a resource 30° to the right of the sun's current position is translated by the communicating bee as 30° to the right of "up". See Fig. 6.12 for several other examples.

A difficulty with the language is that the angle of the waggle portion of the dance changes for the same location at different parts of the day since the reference point is the sun. Another problem is created when the sun is directly overhead, as it would be at noon on the equator. This is solved by a notable foraging inactivity for a short time around noon each day. An additional problem is conversion of round dance to waggle dance at the appropriate distance. That the conversion is gradual was demonstrated

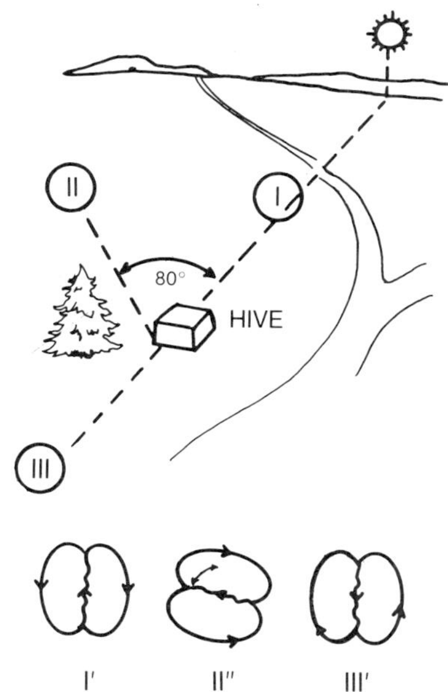

Fig. 6.12 Indication of direction from resource to hive by means of the waggle dance. Each resource I, II, and III are shown as are their dance angles (I′, II′, and III′ respectively) on a vertical comb.

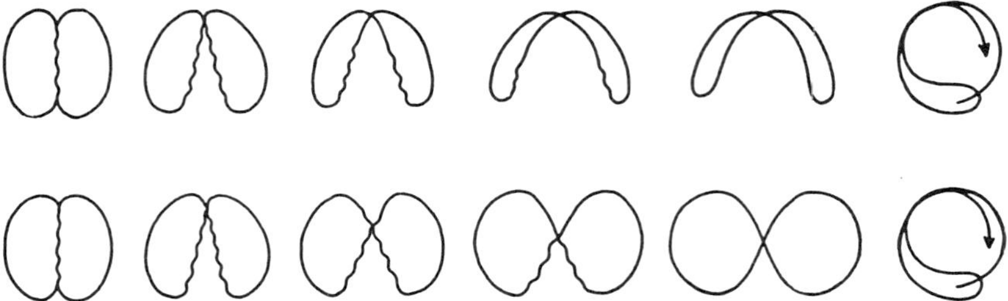

Fig. 6.13 Conversion from waggle dance to round dance as the resource is moved progressively closer to the hive. The top row shows how the Carniolan (Austrian) strain of honey bee changes its dance with distance. Other strains follow the type of changes indicated on the bottom line.

by von Frisch as he studied the dance characteristics from bees whose resource was moved further and further away from the hive (see Fig. 6.13).

Research done since that of von Frisch contradicts his work and suggests that the communication involved is primarily chemical, and that the dance may be incidental to the information exchanged.

Visual Communication

Even though the range of patterns and colors in insects is amazingly varied, their use of visual communication is generally restricted. This is due to the necessity for the sender and receiver to be exposed to view, a behavior not commonly possible in the habitats of most species. In addition, the restricted capacity of the insect eye in discerning colors and patterns limits visual communication between insects to much shorter ranges than is possible for vertebrates, or from insect to vertebrate. Therefore, insect–insect visual communication occurs over short ranges and, in many cases, following previous acoustic or chemical signals. The advantage of visual communication is that it provides accurate localization of the sender, as compared to acoustic or chemical signals.

Work on the biological significance of the details of color patterns of insects has been generally unrewarding. This is rather disturbing to the systematist who often relies on the constancy of wing or other color patterns as taxonomic characters. One area, however, in which color and pattern details have been shown to be significant is in communication between fireflies.

Fireflies (actually a specific family of beetles) predominate among the insects that can produce light. Studies over the past several years have revealed the use that the many species of fireflies make of their unique bioluminescent organs (see Fig. 6.14). Bioluminescence may be used for prey attraction, warning, intimidation of predators, illumination, camouflage, and population regulation. However, its primary use among the fireflies is one of sexual communication between partners, with almost unlimited potential for species-specific signals given the variables available. These are brightness, shape and size of the luminescent surface, timing, and move-

Fig. 6.14 Bioluminescent organs in beetles. The shape and distribution of the illustrated abdominal light organs differs among most species. This difference, in combination with differences in color, intensity, timing and movement of flashes provides for species-specific signaling.

ment, in addition to the qualities of the light produced (the wavelength frequency or combination of frequencies). For example, at a specific temperature, female *Photinus scintillans* will respond to flashes that are 0.13–0.16 seconds long, but not to flashes that last 0.20–0.34 seconds.

With all of the variables available for specific bioluminescent communication, the fireflies seem to have followed two major systems whereby the signals are used. In System I one sex, usually the female, broadcasts a species-specific signal from a stationary site, thus attracting members of the other sex to it, as with *Lampyris noctiluca.* In System II both sexes emit species-specific signals, which brings the advertiser to the correct response from its original signal, as in *Photinus scintillans.* Several species have been shown to modify their luminous behavior as the other sex approaches, apparently to eliminate attracting additional competing mates.

Insect Coloration and Visual Communication

The colors and patterns of coloration possessed by some insects are so flamboyant that they have been studied in a number of ways, one being the manner in which the colors are produced. Insect colors fall into two basic groups: colors resulting from pigmentation and those produced by structural modifications.

Pigments are chemicals that characteristically absorb a specific band of colors from a source of white light such as the sun (see Table 6.2). Thus, all of the colors not absorbed by the pigment are reflected back. We see the color that is complimentary to the color absorbed by the pigment. Pigments such as melanin may be involved in the formation of sclerotized cuticle. Other pigments are not incorporated into the cuticle itself, but lie beneath the cuticle and are seen through it.

Many metallic and iridescent colors are produced by light wave interference. Such colors are the result of light passing through a series of layers which are sepa-

Table 6.2 Pigments in Insects

Color	Pigment	Insect
Brown, black	Melanin	Walking stick
Orange, yellows, red	Carotenoids	Locusts, lady bird beetle
Greens	Combinations of yellow carotenoids and a blue pigment	Locusts, walking stick, hornworm
Whites, yellows, reds	Pterines	Butterflies and moths, *Drosophila* fruit flies
Yellows, reds, browns	Ommochromes	Dragon and damselflies, some Lepidoptera, *Drosophila* fruit flies
Reds, blues	Tetrapyroles, hemoglobin, bilins	*Chironomus* midge larvae, *Gasterophilus* fly larvae, *Chironomus* midge adults
Purple or black	Quinone-aphins	Aphids
Reds	Anthroquinones	Scale insects
Reds, yellows	Flavones	True bugs, Lepidoptera

rated from each other by specific distances, comparable to the wavelengths of light. Some of the light waves passing back are reinforced, but others are completely out of phase among themselves and cancel each other out. In addition, these colors change depending on the position of the viewer with respect to the surface. Since the distances between the lamellae (layers) vary with the viewing angle, the greens or blues may change to purples or oranges. This quality of color change is known as iridescence. The brilliance of these colors depends on the number of layers the light passes through.

Some beetles have body surfaces that possess very fine parallel grooves. These produce an iridescence by splitting white light into its spectral colors when the grooves correspond to light wavelengths. Both this diffraction iridescence and the metallic or interference iridescence can be destroyed when the grooves or lamellae are filled (as when the insect is dropped into alcohol) or the spaces between the layers change (as with drying). The colors reappear, as if by magic, when the structures are returned to their original state.

Regardless of the methods in which the multitude of insect colors and patterns are created, their presence has been shown to be of great importance to survival. Whatever value exists must lie in the particular signal being sent and who receives it. For example, predators often are camouflaged so that they can lie unnoticed by potential prey until the prey is close enough to be attacked. However, prey species are also often camouflaged so that they are overlooked by predators. Thus, color and pattern may enhance survival both as a defensive mechanism and also in offense. Another aspect of color and pattern involves our own biases about visual communication. Insects that may seem rather conspicuous to us, such as a bright green insect on a yellow flower of the same intensity, might be practically invisible to a color blind animal. Conversely, those insects that match their color background according to our own vision may be very apparent to animals with different visual ranges. Also, our own well-developed visual acuity allows us to discern insects that would go undetected by a less visually equipped organism (Fig. 6.15).

Fig. 6.15 (A) The mottled pattern of this butterfly conceals it as it rests on a tree trunk. (B) The same butterfly on a lamp post becomes conspicuous.

The Uses of Colors and Patterns

Insects have evolved colors and patterns that function in a wide variety of ways. However, the uses generally fall into those that (1) allow the insect to remain undetected by matching its surroundings, (2) distract potential enemies long enough to escape, or (3) advertise their presence and remain conspicuous. In the following subsections we will expand on these basic themes to show the variations in uses of color by insects, including the false advertising which insect mimics employ in their survival. Before discussing any of these groups, however, those insects that use no visual communication must be mentioned. Insects that live in total darkness (subterranean forms, those inside of plant structures) expend no energy in the production of pigments since there is no selection pressure from visually oriented natural enemies. Many other insects are concealed within structures for at least a part of their lives (see Fig. 6.16), although selection pressure for color and pattern may be strong due to predation when the insect is outside of its protective covering.

Concealment

Many insects blend into their surroundings, but the methods whereby they "disappear" into their background are of many types. The most common method employed probably is concealment through a matching of the insect's color and texture with its surroundings, known also as "cryptic coloration." Thus, many insects that live on plants have base colors and shapes similar to the plants on which they live (see Figs. 6.17 and 6.18). Others match their backgrounds by being transparent, or at least

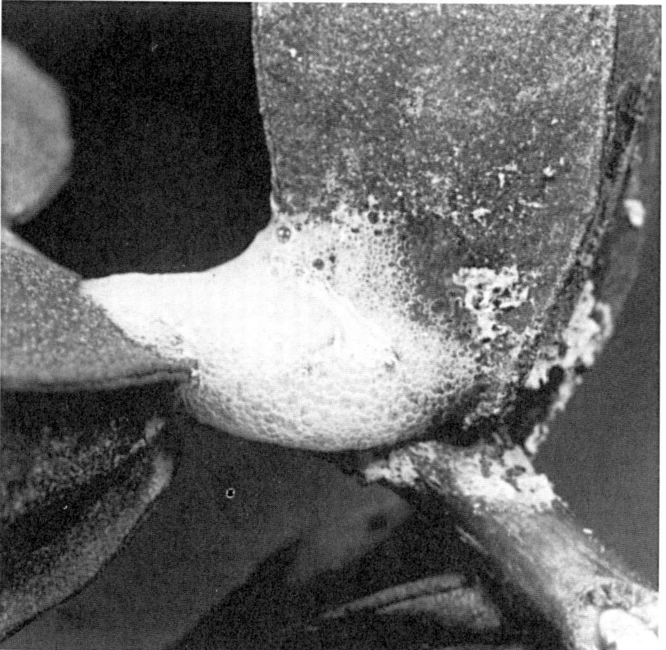

Fig. 6.16 Spittlebug nymphs conceal themselves with masses of liquid froth. Although the selection pressure for evolution of the spittle formation may not be predation, the insects are effectively hidden from sight during their stay inside the masses. Older, more mobile nymphs and the adults do not form spittle masses.

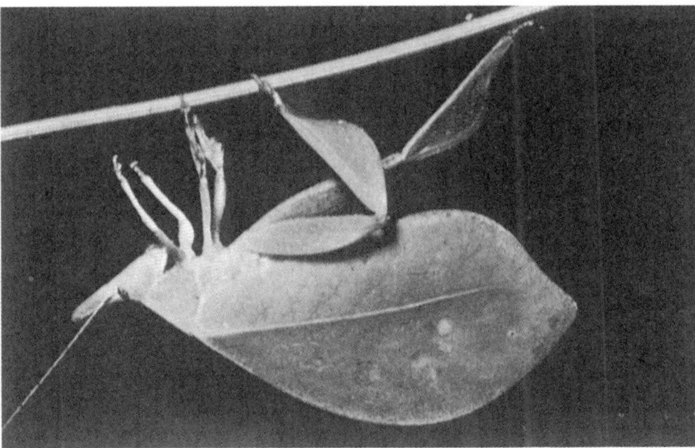

***Fig. 6.17** A katydid with leaflike forewings and expanded leaflike surfaces of the hind leg. From Tipton, Vernon J. (Editor).* **Syllabus: Introductory Entomology.** *Copyright 1973 by Brigham Young University Press. Reprinted with permission.*

having transparent wings and the rest of the body being cryptically colored (Fig. 6.19), or others assume the appearance of bird or insect droppings (see Fig. 6.20).

Predators may ignore immobile cryptically colored insects until one moves. Then the predator, such as a bird, develops a "searching image" for insects colored and shaped like the one it discovered. As long as the "searching image" is reinforced by reward, the prey continue to be taken. Thus, the pressure to remain unnoticed is great, and the insects have evolved a number of ways to increase their concealment by eliminating ways in which predatory "searching images" operate. For instance, the outline of an insect is one aspect that can be obliterated by color arrangements that

***Fig. 6.18** A walking stick insect with appendages and body resembling twigs of the plant on which it lives. From Tipton, Vernon J. (editor).* **Syllabus: Introductory Entomology.** *Copyright 1973 by Brigham Young University Press. Reprinted with permission.*

Fig. 6.19 The transparent wings of this tropical butterfly allow the background colors to show through, thus making it hard to see whether it is at rest or in its typical low flight pattern near vegetation.

break up the body form. This *disruptive coloration* is particularly effective when the base colors of the substrate are used. Many moths are disruptively colored and orient on tree trunks so that their body form is lost among the bark fissures (see Fig. 6.21).

Predators are also attracted by the shadows cast by insects, because the upper surface tends to be brighter than the lower surface. Like many other animals, insects have adopted *countershading* as a means of destroying these shadows. Countershaded insects are darker on the upper surface and lighter below, thus reflecting the same amount of light overall and losing their solid appearance. An insect such as the backswimmer has reverse countershading because it spends most of its life in water swimming upside down.

Problems in Matching a Background. Matching a background is relatively easy if the background does not vary or is widespread and the insect can remain in that particular kind of habitat. Insects that match a particular background, such as the

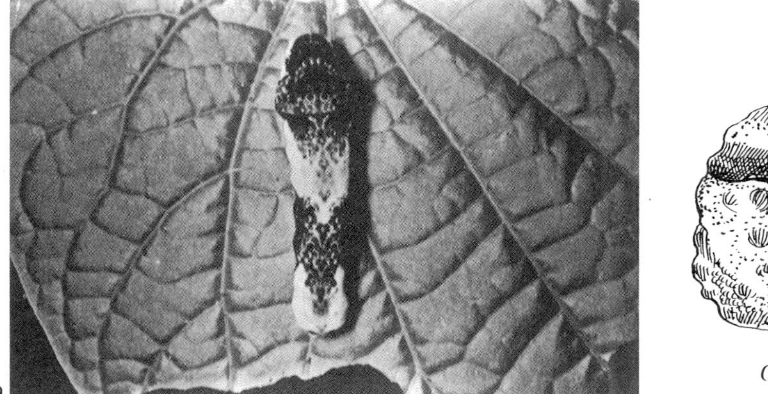

(A) (B)

Fig. 6.20 (A) Caterpillar that resembles bird droppings or (B) beetle that resembles caterpillar droppings escape the attention of predators, provided they remain motionless. (A) From Tipton, Vernon J. (editor). **Syllabus: Introductory Entomology**. *Copyright 1973 by Brigham Young University Press. Reprinted with permission.*

moth *Catocala relicta*, must rest during the day on the trunk of the paper birch tree or become extremely susceptible to predation. Since these backgrounds occur in only a small percentage of the habitat, the problem becomes one of finding the appropriate background on which to rest. In addition, many insects must also orient properly on the substrate to line up their disruptive patterns with the bark fissures or other patterns.

Some insects, such as several species of walking sticks, can undergo physiological color changes to match their surroundings. The colors change in an individual as a result of pigment movement within the epidermal cells. Other insects exhibit color differences between individuals of different generations or under different rearing conditions. For example, the phases of migratory locust species (see also, Chapter 11) not only have different body proportions and behavioral differences, but their colors differ, due to environmental conditions. Some insects change color as they age, with a young adult being one color, but fading or changing in other ways as it grows older.

A famous example of color changes in insects has been investigated by Kettlewell, among others. In England, a country with many collectors and a long-standing interest in natural history, it had been noticed that more of the darker (melanic) forms of certain moth species were represented than in past years; that is, earlier, the lighter forms of the species predominated, with a small percentage of dark forms also present. The light forms had a selective advantage because they blended better into the lichen-covered tree trunks where they rested during the day than did the dark forms. As a result, the darker forms were selectively removed by predatory birds.However, as the industrial revolution, based on soft coal, progressed, the landscape (including tree trunks) became more soot covered, and therefore darker. The previous advantage of the lighter moth forms was lost and the light forms were now selectively preyed upon because of their contrast to the dark substrate (see Figs. 6.22 and 6.23). This phenomenon, known as "industrial melanism," is reversible.

***Fig. 6.21** (A) Moth's wings are spread as they would be in an insect collection. (B) The living moth at rest folds the wings so that the light and dark patterns break up the body outline. (C) These insects are more difficult to see if they orient with the fissures on a tree trunk.*

Flicker Effect. Some insects, but especially tropical butterflies, employ a combination of dull and bright, even iridescent, wing surfaces to elude predators. Butterflies have a typical erratic type of flight. With this type of flight, a predator may only get brief, but intense, glimpses of the brilliant upper surface of the wing. The erratic flashes of brilliance tend to hold the predator's eyes at the last flash and thus to interfere with the predator's ability to intercept the moving butterfly. The iridescent

Fig. 6.22 Industrial melanism in Biston betularia cognitaria. *The darker form (swettaria) blends with soot-covered backgrounds, while the lighter form stands out. The reverse is true in nonpolluted environments.*

Fig. 6.23 Industrial melanism in an underwing moth, Catocala connubialis. *The darker or melanic form is broweri. The lighter form is cordelia.*

metallic blue morpho butterflies of the American tropics are the best examples of the flicker effect. So brilliant is the flashing from their wings that it is noticeable from a low flying airplane.

Startle Effect. Many insects that exhibit a drab or cryptic coloration at rest will react to approaching danger—such as a predator—by suddenly exposing brightly colored surfaces. Such surfaces are often hind wings, which are otherwise covered by cryptic forewings. The sudden appearance of these banded red or orange and black bright surfaces, startles the aggressor, often long enough to allow the insect to escape (see Fig. 6.24A,B). Even more effective, according to experimental results, is the sudden appearance of "eyes," in reality eyespots, which had previously been covered (Fig. 6.25A,B). The original aggressor appears to have disturbed another, possibly larger, predator, enough to cause hesitation and allow the escape of the insect. These effects are not limited to adult insects. For example, a group of moth caterpillars have the ability to take on the appearance of lizards or snake heads when they are disturbed. The posterior part of the body curves up and swells into a viperlike shape, revealing eyespots as well (see Fig. 6.26A,B).

Although the last few examples of sudden appearance of color provide an introduction to the use of active visual signaling, there are some that are even more striking. An example of the defensive repertoire employing both passive and active signaling is that of the tropical walking stick, *Orxines macklotti*. It employs concealment by selecting a daytime resting site among lichens that match its own lichenose appearance and then remaining immobile. It may also employ a slow rocking movement, common among walking sticks, and considered by entomologists to be another concealment technique because predators tend to ignore slow rhythmic movements. When disturbed, the stick insect performs a startle display by raising its cryptic forewings and exposing the expanded and brightly colored hind wings. This display is often accompanied by violent rocking. After one or more displays in response to repeated stimulation the stick insect jumps. The smaller males tend to jump more easily than the females, apparently due to the increased startle efficiency of the female's larger wings. The insect holds its wings erect as it falls after jumping from its

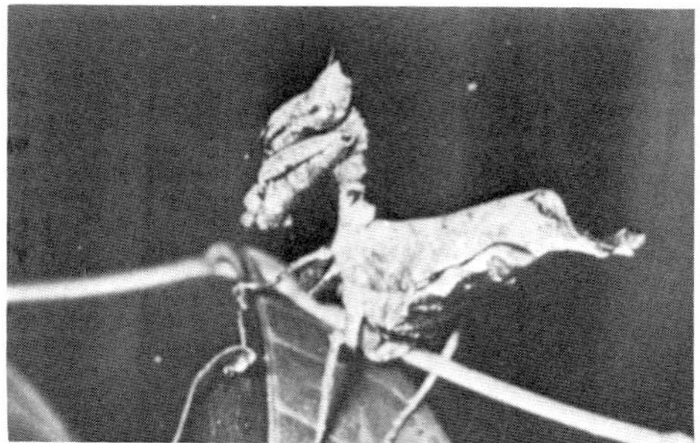

***Fig. 6.24** Flash coloration:* (left) *Mantis in repose and* (right) *revealing its brilliantly colored abdomen after being disturbed. From Tipton, Vernon J. (editor).* **Syllabus: Introductory Entomology.** *Copyright 1973 by Brigham Young University Press. Reprinted with permission.*

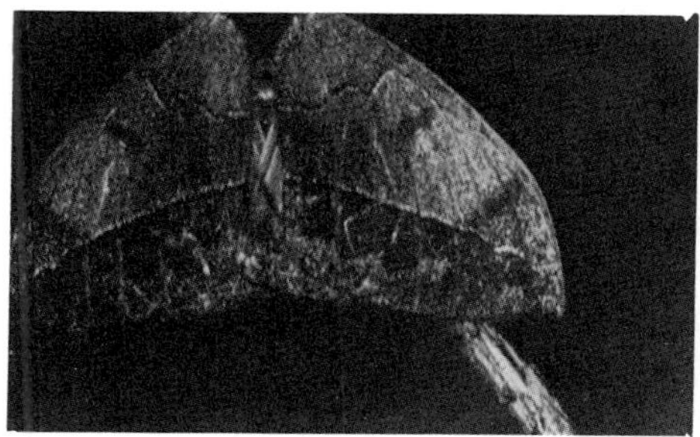

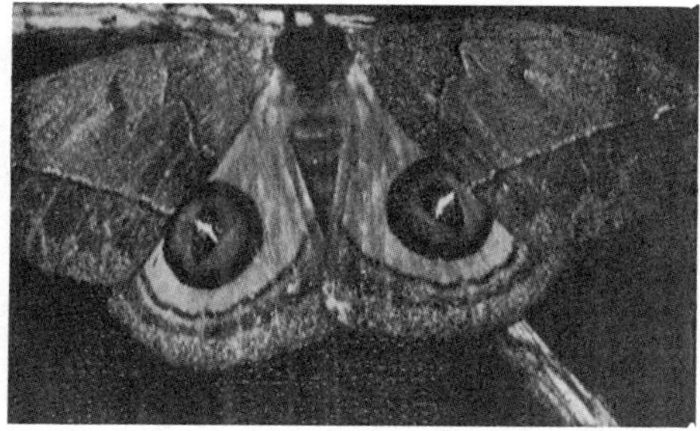

***Fig. 6.25** Eye-spots: (left) A moth in repose and (right) revealing its eye-spots after being disturbed. From Tipton, Vernon J. (editor).* **Syllabus: Introductory Entomology.** *Copyright 1973 by Brigham Young University Press. Reprinted with permission.*

perch, so that it is escaping rapidly, through gravity, from the disturbance while holding the attention of the disturber. However, the instant it hits the ground it snaps its wings shut, thereby becoming completely cryptic again and, in effect, disappearing from the view of the disturber (potential predator). In addition to this array of defensive behavior, *O. macklotti* also possesses glands that secrete a repugnant chemical when it is touched.

Advertising

While it is true that many insects are so well camouflaged that they escape detection, others apparently have evolved colors and patterns that are not in the least cryptic. Examples are lady bird beetles (often red with black spots, or the reverse), yellow jackets, and honey bees. Not only do they not blend into their backgrounds, but their behavior is one of almost flaunting their flamboyant colors. Insects with this advertising color and pattern characteristic are said to have *aposematic* coloration. These insects, in fact, are advertising their presence, because they often have something to advertise. For example, the honey bee advertises its presence because of its sting, as

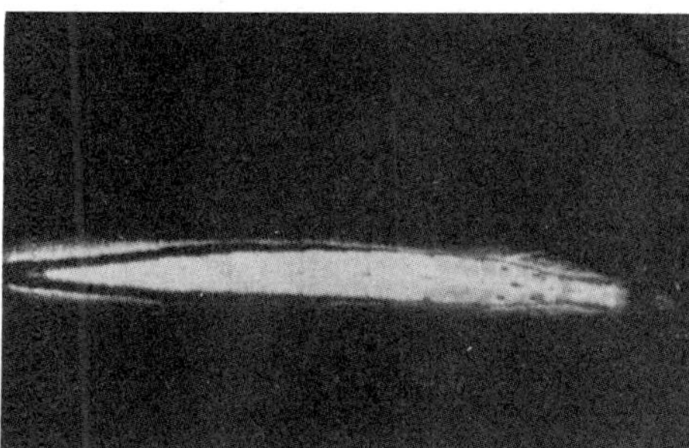

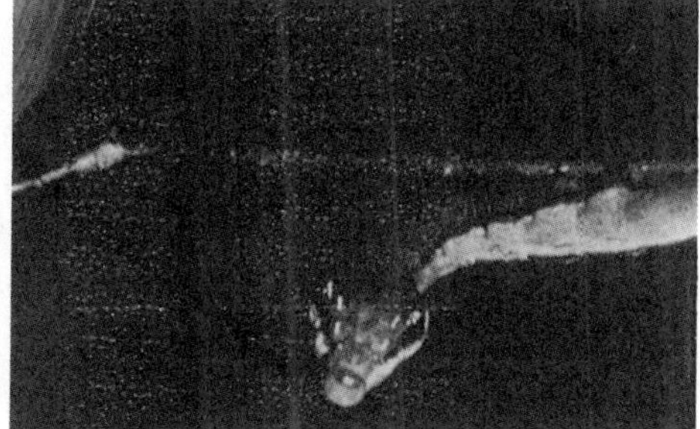

***Fig. 6.26** (left) The "harmless" sphinx moth caterpillar takes on the appearance of a snake (right) when disturbed. From Tipton, Vernon J. (Editor).* **Syllabus: Introductory Entomology.** *copyright 1973 by Brigham Young University Press. Reprinted with permission.*

does the wasp, hornet, and yellow jacket. Thus, unpleasant insects have developed aposematic colors to warn others of their bite, sting, pinch, spray or vapor (or injection) (see Fig. 6.28). An interesting characteristic of unpleasant insects is that their unpleasantness is seldom fatal to the animals that challenge them. This is an evolutionary advantage, however, because the harmed predator, at least if it is a long-lived vertebrate, learns after one or a few encounters to leave a species colored in a particular way alone. The avoidance of aposematically colored animals is a learned rather than inherited response. This has been shown by experiments with the monarch butterfly.

The caterpillar of the monarch butterfly lives on milkweed, which as mentioned earlier, has evolved secondary chemicals (cardenolides) for protection. The caterpillar is not bothered by the cardenolides in its food supply, but instead of detoxifying the cardenolides, it sequesters (assimilates into its body in the same form) them, thus becoming poisonous to some degree. Cardenolides are carried over to the adult butterfly. Avian predators, such as blue jays, have learned to avoid monarch butterflies because the cardenolides cause the jays to vomit. In the laboratory, jays that have been starved for a few hours will nonetheless try to eat monarchs and will quickly learn to accept specimens that contain no cardenolides. However, after tasting or consuming several monarchs *with* cardenolides, the jays will again avoid the showy, orange and black butterflies, provided there is an alternative food supply available (see Fig. 6.27).

This avoidance training has been employed by naturalists for other predator–prey problems, such as sheep slaughter by coyotes. Research on avoidance training with coyotes has shown that emetic-treated carcasses are initially consumed by coyotes. However, after becoming violently and repeatedly ill because of the bait, the coyotes are learning to avoid sheep. The same technique is being tried in Canada where bears traditionally cause heavy losses to bee keepers. Hives salted with emetics have made the bears sick enough that they leave the bee colonies alone.

False Advertising

Although aposematic insects such as those in Fig. 6.28A–E advertise that they are protected by a defense mechanism, not all insects have evolved warning colors along a truth-in-advertising pathway. One strategy is for the insect to conceal its most vulnerable structure, the head. A few species such as the butterfly illustrated in Fig. 6.29 draw attention away from the head by concealing it and presenting a false head to the world. Predators striking at the butterfly's "head" in reality only hit nonessential wing parts and the insect has a chance to escape without serious injury to its real head. The second type of false advertising involves various types of mimicry.

Since avoidance of aposematically colored or patterned species is a learned response, the species usually loses one or several individuals from the population each time a predator is "trained." Provided that the predator is relatively long-lived and the numbers sacrificed are few, the training is evidently an advantageous strategy, because many species have developed along these lines. The strategy depends on the predator being able to distinguish the chemically or physically unpleasant species by

Fig. 6.27 ***(A) Starved jays will eat monarch butterflies and will continue to consume them if they contain no cardenolides. (B) Shortly after eating a cardenolide-containing monarch the jay reacts to the chemical, (C) vomits repeatedly, and (D) shakes its head and drinks water, but before long regains its composure. Courtesy of Lincoln Brower.***

its coloration from other species. However, many insects have developed a different survival strategy whereby they imitate the unpleasant, aposematic insects. Such imitation is referred to as *mimicry* and is treated as a separate strategy from the imitation of a background or other cryptic color and pattern usage previously discussed in this chapter.

Monarch and viceroy butterflies provide a good example of mimicry. The monarch, a chemically "unpleasant" species is aposematically colored, thereby deriving

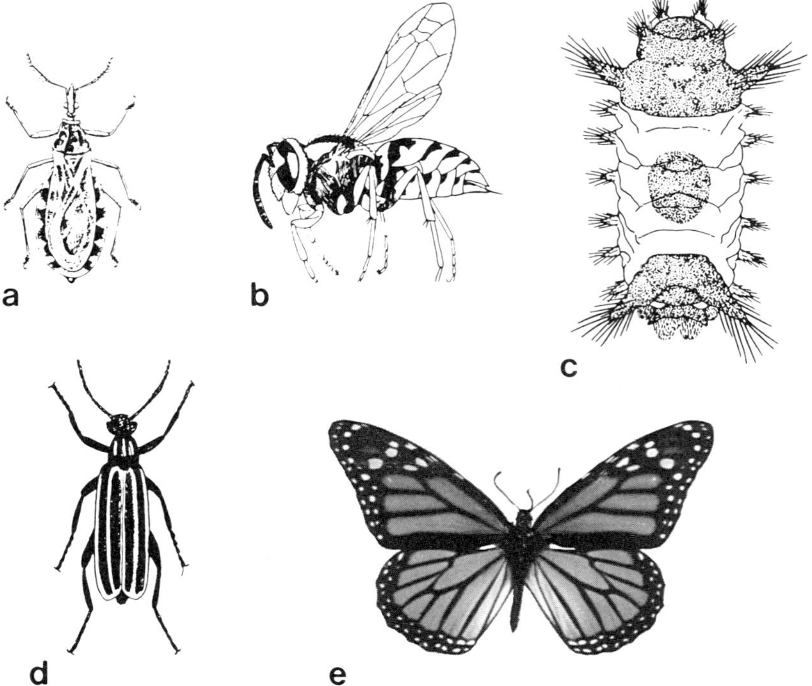

Fig. 6.28 Aposematically colored "unpleasant" insects: (a) kissing bug (bites), (b) stinging wasp, (c) saddleback caterpillar with stinging hairs, (d) blister beetle (with irritating internal chemicals), (e) monarch butterfly (cardenolides).

Fig. 6.29 False advertising. (A) The imitation head, in reality the modified structures on the trailing edges of the wings, draws a predator's attention away from the thecla butterfly's real head. (B) A related species orients head downward so that the false head is more conspicuous. The wing appendage "antennae" are moved while the insect is at rest, thereby increasing the chance that a predator's attack would minimize injury to the butterfly and allow it to escape.

protection because experienced predators recognize the monarch color, pattern, and behavior. The viceroy butterfly has evolved a very similar appearance to the monarch (see Fig. 6.30). This type of mimicry in which a palatable species derives protection by mimicking an unpalatable species is termed *Batesian mimicry*, after Henry Bates, an English naturalist who first disclosed the phenomenon from his studies of Brazilian butterflies. Figure 6.31 shows several examples of Batesian mimicry.

Density of the model species population limits the abundance of the mimetic (mimicking) species because the predator's avoidance reaction is built on several repeated adverse contacts with the model. If too many of the initial contacts are with a palatable or otherwise unprotected species, then the predator may not learn to leave the aposematically colored complex of model and mimic alone, without at least tasting the insect and thereby severely injuring or killing it. Some mimetic species have evolved a strategy that circumvents the population restriction imposed by the model population through *mimetic polymorphism*. In this phenomenon, combinations of color, pattern, and behavior are so closely linked in the *genome* (the complete set of hereditary characteristics of the species) that a group of characteristics are always expressed together. Thus, two or more sets of closely linked characteristics have evolved to the point where individuals with one set mimic one model and individuals with other sets mimic other, different, models (see Fig. 6.32).

Fritz Müller, another scientist working with Brazilian butterflies, provided the explanation for another type of mimicry, called *Müllerian mimicry*. He explained how two or more species of unpalatable butterflies living in the same area have evolved an essentially identical appearance. The unpalatable insects serve mutually as models, so that they reinforce the contact between the predator with the characters they share. In this way, each species benefits because neither has to sacrifice many individuals in the training of predators. Müllerian mimicry complexes may involve several species and are more efficient the greater the number of species that are involved.

As one might expect, a Müllerian mimicry complex provides an ample, indeed ideal, model for Batesian mimicry. In fact, several cases of mimicry complexes have been described in which both Müllerian and Batesian mimicry occur, along with

Fig. 6.30** The monarch and the viceroy. In this case of Batesian mimicry, the monarch (A), with its sequestered cardenolides from the milkweed plant, serves as the unpleasant model for the viceroy (B). The viceroy derives some protection from predators because it is easily mistaken for the monarch, due to the similarity in coloration, patterns, and behavior. From L. P. Brower,* **Ecological Chemistry**. ***Copyright 1969 by Scientific American, Inc. All rights reserved. Reprinted with permission.

***Fig. 6.31 Batesian mimicry. The wasps and bees among the stinging Hymenoptera* (left) *often serve as models for flies of the Order Diptera* (right).**

"aggressive mimicry" where a predator mimicks its prey, thereby enabling the predator to approach without alarming its prey. This is the classic "wolf-in-sheep's-clothing."

An interesting type of aggressive mimicry exists among the fireflies discussed earlier. These photocommunicators have evolved a wide variety of sizes and shapes of light organs, as well as various flashing patterns that allow each species to evolve its own specific signal to attract mates. In fact, some of the tropical species gather in trees and synchronize their flashing, thereby creating a magnificent display. Some of the predatory species have evolved behavior whereby they imitate all of the characteristics of another firefly so that they attract members of the other species. When the insect approaches the "mate" it has been communicating with, instead of the expectant copulation with the stationary signaler, it is grabbed and consumed by the predator.

In addition to the previously mentioned types of mimicry, one other has been described for insects. This is "automimicry," in which one member of a species derives protection by looking like protected members of its own species. For example, the males of stinging Hymenoptera are protected by being colored like the females.

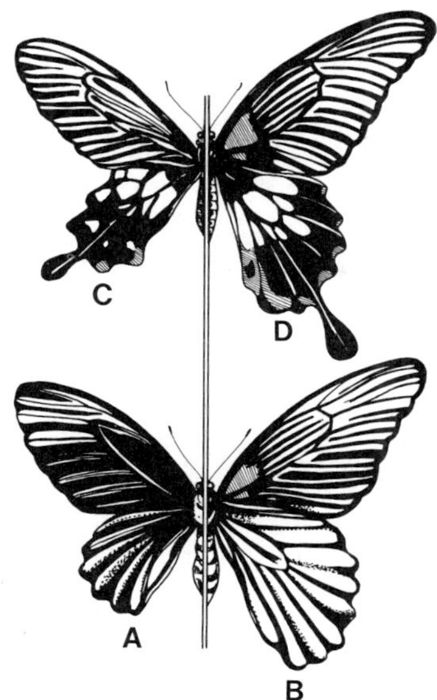

***Fig. 6.32** (A, C) The left half of two unpalatable swallowtail butterflies, (B, D) the right halfs of swallowtail butterflies that are palatable, but mimic A and C. Note that B and D are different morphs (forms) of the same species* **(Papilio memnon)**, *so that* **P. memnon** *form B mimics* **Atrophanura nox** *and* **P. memnon** *form D mimics* **Atrophanura coon.**

Since the female's ovipositor is the defensive weapon and it is not possessed by the male, his only protection lies in being so like the female that a predator cannot distinguish between them.

SUMMARY

Insects tend to show stereotyped rather than learned behavior. This is due, in part, to their reduced nervous system, and to their short lifespan.

Studies of behavior, while plagued with the scientist's tendency to describe behavior in anthropomorphic terms, and the rapidity of behavioral sequences, follows either the mechanistic description of the activities involved, or the consequential—what the result of the behavior is. The stereotyped behavior may follow one or more fixed action patterns, depending on the condition of the insect. These patterns are composed of simpler acts successively induced. When occurring between two individuals, the patterns may be a reciprocal behavior chain of actions in a particular sequence. This explains such complex behaviors as the ritualized mating typical of many insects. While most insect behavior is innate, insects have been shown to be capable of learning and to have a memory.

The primary orientation of insects is usually a response of body position to gravity (ventral earth reaction) or light (dorsal light reaction, light compass reaction). Secondary orientations involve undirected (kineses) or directed (taxes) responses to a variety of stimuli.

Insects exhibit rhythms in response to the daily and seasonal rhythms of the environment, e.g., photoperiod. Such insect rhythms have been shown to be controlled primarily by interval timers, or physiological clocks.

The sequence involved in both feeding and ovipositional behaviors usually involves locomotion to the site, recognition, feeding (or oviposition), and termination. The variety of stimuli differs among species, but may involve visual cues (wavelength of light, motion, shape), chemical cues (odors, tastes), textures, and sounds. In feeding, token stimuli may be as effective as true stimuli in eliciting the behavior.

In communication between two organisms, one must provide a stimulus for the second. The stimulus alters the second organism's ability to respond, which results in a reaction by it. Communication may be acoustic, as with the singing of crickets as they call mates, court and mate, or defend territories.

One of the most common modes of communication is through the use of chemicals. This allows for long-distance signaling without the insect being exposed, with potentially great specificity due to the number of chemicals that could be employed. The earliest known semiochemicals are the pheromones used between individuals of one species to attract mates, warn others, etc. Kairomones and allomones are chemicals released by one species that benefit the recipient or emitter, respectively. In these cases the recipient is a different species than the receiver.

Among social insects communication has reached a high level of complexity. Pheromones are widely used to control social structure and maintain security. The bee dance provides an interesting example of how complex messages are communicated.

Communication through visual signals by various fireflies involves differences in the color, pattern, and flashing rhythms employed. Color patterns and behavior also serve most insects in defense, both passive and active. Many insects conceal themselves through cryptic colors and patterns. Common patterns are disruptive, countershading, and matching the background in color and texture. Crypsis is a dynamic characteristic as shown by the phenomenon of industrial melanism. Active defenses may employ eye spots or bright colors to startle intruders.

The bright colors and patterns of aposematic insects are associated with protective characteristics (stings, bad taste, bites) that natural enemies learn to avoid. Mimetic forms (Batesian) benefit from such avoidance training. Müllerian mimics mutually reinforce the training. Automimicry, aggressive mimicry, and mimetic polymorphism are additional phenomena that increase the insect's survival potential.

SUGGESTED READINGS

Atkins, M. D. 1980. "Introduction to Insect Behavior." Macmillan, New York.

Fraenkel, G. S., and Gunn, D. L. 1961. "The Orientation of Animals, Kineses, Taxes, and Compass Reactions." Dover Publications, New York.

Matthews, R. W., and Matthews, J. R. 1978. "Insect Behavior." J. Wiley and Sons, New York.

Saunders, D. S. 1976. "Insect Clocks." Pergamon Press, Elmsford, New York.

Wickler, W. 1968. "Mimicry in Plants and Animals." McGraw-Hill, New York.

Wilson, E. O. 1971. "The Insect Societies." Harvard University Press, Cambridge, Massachusetts.

7

Interactions with Other Organisms in the Environment

HOW POPULATIONS OF TWO SPECIES MAY INTERACT

Insects live in a world of living and nonliving elements—not in isolation. The mere fact that an insect lives in physical association with another organism does not necessarily imply a relationship, however, although it is a factor in considering the kinds of interactions that insects may have with other organisms in their immediate environment.

At least hypothetically, the kinds of interactions between populations of two organisms are defined by all permutations of each consequence. That is, the interaction can have a neutral (0), positive (+), or negative (−) result for the insect. Likewise, the interaction can have a 0,+, or − result for the other organism. Thus, the combinations become 00, 0+, 0−, −0, −+, −−, +0, +−, and ++. These interactions have been given different names, as presented in Table 7.1. Examples including those for insect–plant interactions are summarized in Tables 7.2 and 7.3.

Table 7.1 Interactions between Populations of Two Species[a]

Interaction	Species 1	Species 2	Nature of the Interaction
Neutralism	0	0	Neither population affects the other.
Amensalism	0	−	One population inhibited, one not affected.
Commensalism	0	+	One population not affected; one, the commensal, benefits.
Competition	−	−	Each population is inhibited by the presence of the other.
Parasitism	+	−	One population, the parasite, is generally smaller than the other, the host
Predation	+	−	One population, the predator, is generally larger than the other, the prey
Mutualism	+	+	Interaction favors both; may or may not be obligatory

[a]0, No significant interaction in the presence of the other species
−, Population inhibited by the presence of the other species
+, Population benefited by the presence of the other species

Many of the interactions listed in Tables 7.2 and 7.3 have been introduced elsewhere in the text. For example, the mutualistic and commensalistic aspects of pollination comprise a major section of Chapter 8. The range of insect parasitism on plants is detailed in Chapter 9, but some of the interactions between the potential host and parasite remain to be discussed here. Also, insect predation on plants and some cases of mutualism between insect and plant, other than pollination, need to be presented. One case of mutualism between insects and other animals will also be explored. In addition two kinds of interaction receive special attention here because of their interest to entomologists and to other biologists. They are the roles of plants as predators of insects (a role that is the reverse of the "typical" situation) and social contact, that particular kind of interaction between insects and other organisms of the same kind.

The interactions between an insect species and the plant that it feeds on amounts to an eternal conflict between the plant's goal of securing enough energy to ensure its survival to reproduce and its minimization of energy losses suffered because of the attacks of its natural enemies. The semiochemicals discussed in Chapter 6 under allomones in the milkweed present a good example of the role that "secondary plant chemicals" play in the biochemical warfare waged between plant and parasite. In the

Table 7.2 Examples of Insect–Plant Interactions

Interaction	Insect	Plant	Habitat
Neutralism	Red oak borer	White pine	Mixed red oak–white pine stand
Commensalism	*Eucera* bee	Bee orchid	Bee pollinates orchid in attempt to copulate with the bee-mimicking flower (see Chapter 8)
Competition	Granary weevil	Grain mold	Both organisms utilize nutrients in stored grains
Parasitism	*Diplolepis* wasp	Rose	*Diplolepis* lives in galls formed on the rose
Predation	Acorn weevil	Red oak	Weevils locate and oviposit in acorns; the larvae kill and consume contents
Mutualism	Ant	Acacia plant	See discussion in text

Table 7.3 Examples of Insect Interactions with Other Animals Including Insects

Interaction	Insect	Animal	Habitat
Neutralism	Red oak borer	White pine aphid	Mixed red oak–white pine stand
Commensalism	*Musca sorbens*	Domestic dog	*Musca* breeds in dog feces
Competition	Saw-toothed grain beetle	Rat	Stored grain products
Parasitism	Cat flea	Cat	Domecile
Predation	Dragonfly	Mosquito	Waterways
Mutualism	Ant	Corn root aphid	Corn field

milkweed, a number of interactions must have occurred. First, the plant began to produce toxic metabolites (secondary plant chemicals) and to concentrate these metabolites in its tissues. However, each advance in allomone production (and concentration) must have been met with a modification in the parasite (monarch ancestor) that allowed it to tolerate the presence of allomones. An additional spinoff of the biochemical warfare between monarch and milkweed occurred as the monarch stored the plant-produced allomones in its body, thereby providing protection from its own natural enemies. Thus, the interaction between host plant and its insect parasites is a series of reciprocal changes by each species involved in the interaction. This cycle of attack and counterattack is known as *coevolution*.

Mutualism

One of the most fascinating examples of coevolution is the high level of mutualism found between certain acacia plants and ants. In this circumstance, the plant provides shelter and food to the ant while the ant protects the acacia from herbivore attack and from competition with neighboring plants for sunlight.

Acacia cornigera, the bull's thorn acacia, is a leguminous low-growing tree in the new world tropics. Its name is derived from the pair of large thorns produced at the base of the leaves.

When the thorns are developing and are still succulent, they are attacked by *Pseudomyrmex* ants. The ants bore a hole into the center of the thorn, enlarging and maintaining it while the thorn matures into its stiff woody protective final form (Fig. 7.1). The ants utilize the hollow thorns as cells to hold developing young and other members of the colony, with several to many thorns required to contain all of the colony members.

The ants leave their thorny homes to forage for food, particularly the small "Beltian bodies" (ovoid structures produced at the tips of new leaves) that are highly nutritious (containing protein) and produced in large numbers by the acacia. In addition, the ants collect nectar produced on the leaf petioles.

The ants provide protection for the bull's thorn acacia by patrolling the vegetation, removing or at least attacking caterpillars and other plant feeders. Among the plant feeders are vertebrate browsers like deer and domestic livestock, which are severely stung about the muzzle by the ants as they try to feed and soon abandon the plant.

(A)

(B)
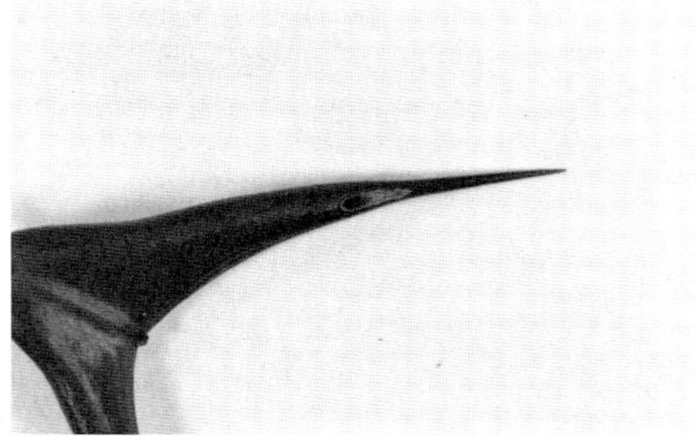

***Fig. 7.1** (A) Acacia thorns occupied by Pseudomyrmex ants. The colony may be spread among several of the thorns. (B) The entrance hole near the thorn's tip.*

Acacia cornigera is also protected from contact with other living plants, especially the vines so common in the tropics. When a vine comes in contact with acacia, the ants repeatedly attack the part in contact, chewing at it until the shoot dies so that this acacia species, which is not shade tolerant, can grow without being encumbered with competing vines.

Associations of mutual benefit whether *facultative* (optional) or *obligatory* (required) are rather unusual among animals. An exception is the tie between ants and the suborder Homoptera because both of the animals involved are insects and because the relationship is so widespread that it is frequently encountered in the field. Several insects in the suborder Homoptera (see Chapter 11) including sap-sucking aphids produce copious amounts of sugar-laden fecal liquid. Because the plant sap has only small amounts of proteinaceous material needed for growth and reproduction, the insect must process a great deal of plant sap. This results in fecal matter, known as honeydew, that is rich in carbohydrate, because the aphid takes in more sugar than it can use.

Many insects seek honeydew, and aphid-laden plants often are visited by flies, bees, and other sugar feeders, including ants. Even a fungus called sooty mold grows exclusively on honeydew.

Although many organisms seek homopterous honeydew, the relationship between the honeydew source (the aphid) and the ant is often the most sophisticated, involving more than simply the ant's feeding on the aphid's excretory products as is typical of the other honeydew feeders. *Aphis fabae*, the bean aphid, is sometimes tended by ants of the species *Lasius niger* (L). The ant strokes the aphid's abdomen with its antennae, thereby stimulating the aphid to extrude a droplet of honeydew and to hold it at the anal tip of its abdomen until it is taken by the ant. When not attended by ants *Aphis fabae* forceably ejects its honeydew as it is produced.

Ants tending honeydew-producing Homoptera (see Fig. 7.2) have evolved a considerable level of attendance, including aggressive behavior toward natural enemies of their "livestock." Ladybugs, aphid lions, and parasitic wasps are attacked by the ants when they are detected near the aphid aggregations. As another example, ant species such as *Oecophylla longinoda* build silken shelters over their attended coccids (scale insect types) during the rainy season.

The relationship reaches the most sophisticated levels between the corn-root aphid, *Aphis maidiradicis* Forbes and the ant, *Lasius niger americanus* Emery. The ant stores eggs of the aphid in its nests during the winter and installs the young aphids on appropriate host plant roots the next spring. Thus the stage is set for coevolution of these species into an obligate mutualism in which neither species can survive without the activities of the other.

Fig. 7.2 Ants on a leaf tending their aphids.

Plant Parasites and Predators

Most of the insects that attack plants fall into the category of "plant parasite." That is, they are much smaller than the host plant and live in rather continuous contact with a single host, deriving nutrients over a long period. Certainly, the definition of plant parasite fits for the many scale insects, aphids, and leaf-feeding caterpillars of perennials since each individual does little harm to the plant. In fact the impact of foliage or sap removal by a single insect is almost impossible to assess for a large perennial. Obvious exceptions to this last statement include those insects that transmit a disease, because even a single feeding by a contaminated insect may be fatal to the plant.

Several insects fit the definition of plant predator—those that kill seeds or seedlings. Cutworms are large compared to the seedlings on which they feed, and therefore destroy the plant by feeding on it, much like the ladybug does with its aphid prey so that size and brevity of contact with a single host are similar between some insect predators and their prey whether it be plant or animal.

Seed predation is rather different from leaf feeding, although they are often lumped together for convenience. The seed has much higher nutrient value, and is not directly replaced by the plant if it is lost, for seed death often occurs only after it is separated from its producer. The plant often reacts to foliage loss by producing additional leaves.

Among the 15 or so families of insects that might be considered seed predators are the harvester ants of the Genus *Pogonomyrmex*. These ants gather seeds of many kinds and store them in their nest as a food supply (see Figs. 7.3 and 7.4). In the sense that the ant consumes one seed then moves on to consume another, it is a predator of plants.

***Fig. 7.3** A dandelion seed, with its parachute still intact, being carried to the nest by a harvester ant. The ants consume the nutrients within the seed and cast the seed hull outside of the entrance mound. Reprinted with permission from R. Hutchins,* **The Ant Realm.** *Copyright 1967 by Dodd, Mead and Co.*

Fig. 7.4. The interconnected subterranean chambers of a harvester ant. In this partially excavated view, the higher chambers are "seed bins" and the lower, with the developing brood of ants, are "nurseries." Reprinted with permission from R. Hutchins, **The Ant Realm.** *Copyright 1967 by Dodd, Mead and Co.*

Dispersal of Seeds

The *dispersal* or spread of seeds is a function not usually attributable to insects. Since seeds are often rather large with respect to the size of most insects, the usual seed–insect relationship is one in which the insect develops within the seed, essentially as a parasitoid of plant seeds, or preys upon the seeds like the harvester ants described in the preceding section of this chapter. Seed dispersal is accomplished for the most part by wind or by the mechanical action of animals larger than insects. For example, seeds pass through a bird's or mammal's (e.g., cow) intestinal tract to germinate in the fecal material of the animal when the animal consumes the plant or its fruit as food.

Insects have been shown to be responsible for dispersal of the seeds of some plants, however. This relationship may be casual and infrequent or it may be the usual way that the plant's seeds get dispersed. When harvester ants remove germinating seeds from their nests, they deposit these seeds or seedlings into their nest dumps where seed hulls and other debris accumulates. Thus, the harvester ant mound tends to be devoid of vegetation on the mound proper, but the edges of the mound where the colony's trash is deposited usually has many plants as a result of the ants' deposit of germinating seeds.

A more bizarre and highly evolved example of insects as dispersal agents for plants involves other seed-eating ants. Ants visit several species of plants in the *Trillium* and *Primula* genera, gathering seeds and returning to the vicinity of the nest. The edible part, usually an oil-filled appendage (the elaisome) or the oily seed coat is stripped from the seed and taken into the ant nest for consumption; the seed is left to germinate outside the nest. The elaisome has evolved its oily nature specifically so that seed dispersal by these ants will occur. In some species of *Trillium* the stem with its developing fruit is upright until the fruit is ripe. Then the stem becomes prostrate on the ground, an adaptation that facilitates access by the ants.

ASPECTS OF SOCIALITY AMONG THE INSECTS

Degree of Sociality

The degree of "sociality," that is, tendency to associate with other individuals of the same species, in insects covers the widest possible extremes. Some species lead an existence so solitary that they may never come in contact with another individual of their own species. Such a case is the walking stick, *Diapheromera femorata* (Say). This insect is parthenogenetic, so there is no sexual behavior in its repertoire. In addition, its eggs are produced singly and allowed to drop to the forest floor. The female makes no attempt to protect the eggs or even find a suitable site for them. Note, however, that outbreaks of this insect, seemingly so suited for solitary life, often occur so that contact between individuals often occurs.

The most social of the insects are ants, some bees and wasps, and termites. In these cases the individual cannot have a normal existence except in the presence of others (usually many) of its own kind. At some time in the lives of any of these highly social insects, the individual may be out of contact with others of its kind, but these periods are usually very brief. For example, increase in the number of honey bee (*Apis mellifera* L.) colonies (a group of individuals, other than a single mated pair, which constructs nests or rears offspring in a cooperative manner) occurs when the old queen leaves the hive on a swarming flight, taking about half of the colony's number with her. Virgin honey bee queens spend a short time away from the attentions of their kind to go on mating flights. From the time she leaves the hive until she reaches an area high in the air where the drones (males) congregate she is alone. Then, after mating, the queen returns to her colony, never again to be separated from her own, except by death. Numerically, the most predominant honey bees are the worker bees that build and maintain the hive as well as gather pollen and nectar as food. The flights of these bees often separates them from others of their kind for hours each day. Yet, the question remains: Does the fact that honey bees spend most of their lives in contact with others of their own kind make them social animals or are there other components that might be considered?

In 1975 one of the greatest aggregations of insects was discovered high in the mountains of central Mexico. It was the overwintering sites of monarch butterflies from all of central and eastern North America. In areas of some tens of acres millions of monarchs cover the tall trees from top to bottom, overwintering under conditions that are ideal for them; frost-free areas that are still cold enough to allow their survival until they can begin their flight back north to repopulate much of North America. The point to be made here is not how these butterflies manage to fly to these few isolated locations, nor how they find the locations, nor even how scientists eventually tracked down the sites. The importance lies in the apparent lack of interaction between the individuals in these tremendous aggregations. Here then is a case of a species that spends part of its life in very close proximity to millions of its own kind without any significant interaction.

Types of Sociality

Scientists studying all aspects of social organizations and communications called sociobiologists recognize various degrees of sociality exhibited by insects. Those groups that are said to be eusocial (truly social) share several characteristics that have been used to define their eusociality:

1. Reproductive division of labor
2. Cooperation between individuals of the same species in caring for the young
3. An overlap of generations (at least two generations), so that offspring help their parents at least a part of their life in contributing to colony behavior

In eusocial insects some individuals have a greater capacity for reproduction than do others of their kind. The more or less sterile individuals work at tasks other than reproduction, including helping to care for the young of the species. The three traits used to define true sociality have evolved independently so that many levels of sophistication and combinations of behavior leading to eusociality have been described. A typical eusocial insect, the honey bee, will be described subsequently.

Presocial Behavior

Presocial behavior, that is, social behavior beyond sexual contact, but not expressing all three characteristics that define true sociality, can be categorized as subsocial, communal, quasisocial, and semisocial. Subsocial insects (such as some of the crickets, earwigs, sawflies, and stinkbugs) are those in which one of the parents guards the eggs and/or newly hatched young for a short period before abandoning them (see Fig. 7.5). Communal insects (like the web-spinners of the Order Embiop-

***Fig. 7.5** A sawfly guarding her newly hatched young. Her protective behavior is temporary, but typical of subsocial insects in the Orders Hemiptera, Dictyoptera, Dermaptera, and Hymenoptera.*

tera, shown in Fig. 7.6) produce offspring in the same nest but without generations overlapping or cooperation in care of the young. In quasisocial insects (some kinds of bees), a composite nest is used by various members of the same generation. They cooperate in caring for the developing young. In semisocial insects (some wasps), not only does the same generation use a composite nest and cooperate in care of the young, but some of the individuals are sterile and act as nurses for the developing offspring.

Eusocial Behavior

In the honey bee (*Apis mellifera* L.), a "typical" eusocial insect, a healthy mature colony has between 60,000 and 75,000 members during the peak of annual activity. Three distinct forms of adult bee make up the colony. The most abundant type (many thousands) is the *worker* (a sexually immature, but adult female). There is a single *queen* (a sexually mature and egg producing female). The *drones* (sexually mature males) are present during part of the year (see Fig. 7.7).

In a normal colony only the queen is a functionally reproducing female. That is, she is the only egg producer in the colony. The worker females present are inhibited from completing their ovarian development by a pheromone (the "queen substance") from the queen. If no queen is present then the workers begin to develop their ovaries. However, since the worker has not been mated, all of her eggs are unfertilized, resulting in drone bees. Thus, the unusual genetic control of sex in Hymenoptera is disclosed. Males are the result of development of an unfertilized egg; females from

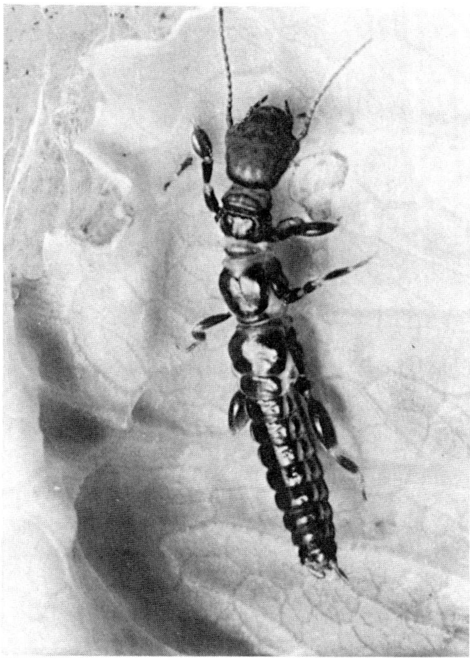

Fig. 7.6 A webspinner, typical of the communal Order Embioptera. Galleries are spun from the silk extruded by the enlarged front tarsal glands. Female webspinners guard their eggs and newly hatched young within the common silk gallery until the young have encased themselves in small silken tunnels of their own.

fertilized eggs. The fate of the fertilized egg is determined by the nutrition supplied to it by the colony. Continued feeding of royal jelly throughout larval development results in a queen (usually in conjunction with a much larger cell), whereas a switch on day 3 to honey and pollen as food will produce a worker.

A mature hive will have the maximum numbers of bees present in early summer. This is the time when the colony divides, by a process termed "swarming." The old queen and about half of the worker population leave the hive in a cloud of bees. They settle on some structure (fence, tree limb, etc.) for a few hours (see Fig. 7.8). Eventually they settle in a new site, rapidly constructing a comb made from wax produced in their dermal wax glands. The old queen starts laying eggs and the newly formed colony establishes itself by securing nectar for the winter ahead. Without a sufficient supply of nectar, the colony will not survive the winter.

In late summer the workers eject the drones from the colony and final stores from late summer flowers are gathered. The colony becomes less and less active as environmental temperatures drop. When the hive temperature falls from the rather constant 34°–35°C (93°–95°F) during the growing season and necessary for egg production by the queen, to about 14°C (57°F) or cooler, the bees in the hive form a tight mass about the queen. This cluster maintains an internal temperature of 14°C regardless of ambient temperatures because each bee adjusts its rates of metabolism so

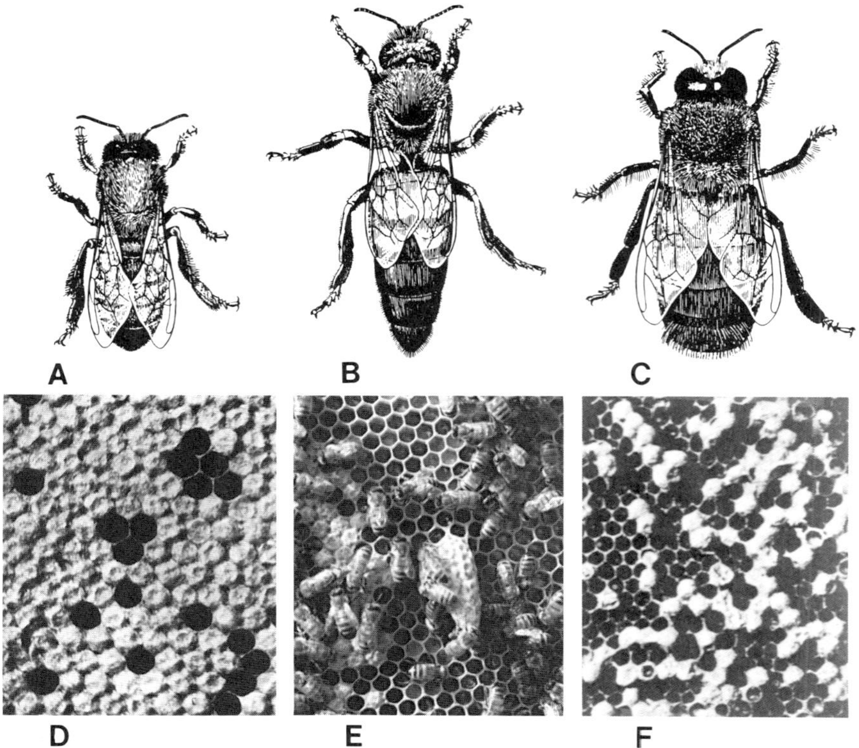

*Fig. 7.7 The three forms of adults found in the honey bee (*Apis mellifera *L.) colony: (A) worker female, (B) queen female, (C) drone (male). (D) Standard comb cells covered with flat caps within which workers are pupating. (E) The hanging, large queen cells. (F) The bullet-capped drone cells.*

Fig. 7.8 A swarm of honey bees issuing from a hive. In several hours they will move from a cluster site to establish a new colony in a place located by scout bees.

that its body produces heat, contributing to the cluster's warmth. The cluster becomes even more tightly compressed as external temperatures drop.

The bees feed on stored honey during the winter to maintain their high rate of metabolism. The bees are not hibernating and a disturbance of the hive can produce angry bees even on very cold days. The bees usually emerge from the hive on warm days of winter to go on "cleansing flights" since they will not defecate inside the hive.

The queen starts to lay eggs in the spring when the hive temperature increases and the colony becomes active. She lays one egg into the bottom of each hexagonal waxen cell. The newly hatched larvae are fed royal jelly for at least 2 days. Then they are fed a less nutritious food at regular intervals by nurse bees. On the ninth day, after a tremendous increase in size, the larva or brood is sealed into the cell by a waxen cap. It ceases feeding and molts into the pupa, spending a total of 12 days in the capped cell before chewing its way out as an adult worker bee. Drones and queens follow a similar pattern of development with differences in (a) size, position, and structure of the waxen cells, (b) nutrition, and (c) duration of each stage (see Table 7.4).

The newly emerged adult worker bee begins its life of work by polishing and cleaning cells for the queen to lay eggs in or for storage of food. After about 3 days at this activity, the worker spends about 4 days feeding the older larvae; then 7–14 days secreting royal jelly for feeding to 1- and 2-day-old worker larvae, developing queen larvae, and to the functional adult queen. This same period is when the worker's wax glands are most active, so that the wax scales produced are pulled off and used in comb construction and maintenance. At about 14 days the worker begins to forage outside the colony. Her first week as a forager may be primarily devoted to gathering pollen, but the rest of her 5 week life (as an adult worker in the summer) is spent as a field bee gathering nectar. In other seasons, bees follow a different timetable, but the sequence of duties is the same. If there is a shortage of a particular worker age class (as through accident), functions out of the normal sequence are performed by workers of various ages.

Polymorphism is the existence of two or more functionally different castes within the same sex. A caste is a set of individuals in a given colony that are both morphologically distinct (morphs) and have their own specialized behaviors. Thus, the worker caste and queen caste, both being female *Apis mellifera* L., exemplify polymorphism.

Table 7.4 Days Spent in Each Developmental Stage by the Honey Bee, Apis mellifera L.

	Queen	Worker	Drone
Egg	3	3	3
Larva	5½	6	6½
Pupa	7½	12	14½
Total	16	21	24

An example of polymorphism in termites is given in Fig. 7.9: workers (males and females), mandibulate soldiers (males and females) that protect the colony with defensive biting, nasute soldiers (males and females) whose defensive role is chemical, primary reproductives (males and females that swarm out from colonies to establish new colonies), and secondary reproductives (males and females that replace senescent or injured primary reproductives).

Production of morphs is controlled in part by the genetic makeup of the egg and in part by environmental influence, basically nutrition. Here in the insect world is a case where characteristics are "part nature, and part nurture." In ants, the primary determinant of castes, always limited to the female sex, is nutrition. In many ant species worker larvae will turn into minor, media, or major workers according to the amount of food supplied. Of course, those fed less become smaller adults than those fed abundantly. However, the difference between the smaller and larger workers is more than just in size. As discussed previously in Chapter 5 and shown in Fig. 5.5, small workers, or "minors," are clearly distinct from the larger workers. The larger ant has a head disproportionately larger than small ants, and, therefore, larger, more powerful jaws. The largest of these large ants are soldier ants. In some of the more advanced ants, hormonal control of subcaste (minor, media, or major worker) is the determining factor.

The leaf-cutting ants of the new world tropics grow a fungus for food on a

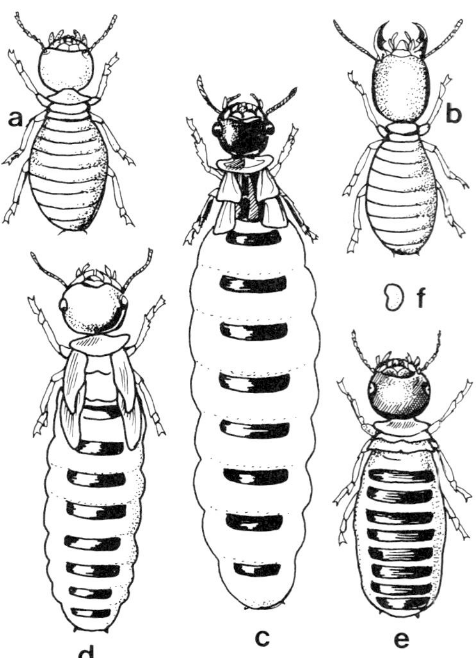

Fig. 7.9 The castes present in colonies of* Amitermes hastatus: *(a) worker (b) mandibulate soldier, (c) primary queen after 5 years of egg production, (d) secondary queen, (e) tertiary queen, (f) egg.

compost pile of plant parts, mostly leaves in an underground "garden." Media worker ants bring large pieces of leaf to the nest to be added to the pile for the fungus substrate. While on the return trip, when the media worker is occupied holding the piece of leaf, it is vulnerable to the attack of parasitic flies. Minor workers often accompany the medias on their leaf foraging and ride the leaf piece back to the nest, protecting the media worker from the parasitic flies by snapping at the flies and fencing with their hind legs (see Fig. 7.10).

The mandibulate and nasute soldiers of the termite world have their Hymenopteran equivalents among the ants. Mandibulate termite soldiers (*Zootermopsis*) have jaws designed to shear objects (usually the termite's enemy, ants) in two. Ants have shearing soldiers as well (for example, *Pheidole militicida*). Termites have stabbing soldiers with scimitar-shaped jaws (*Macrotermes*) and the ant version (*Eciton*) are saber-jawed, both forms incapable of survival without being fed by others in the colony. Blocking ant soldiers (European *Camponotus*) are somewhat different from *Cryptotermes* soldiers since the ant acts as a living door (see Fig. 7.11), whereas the termite allows itself to be cemented by workers into breaks in the colony wall. It is cemented in with its enlarged rugged head directed outward. The termite nasute soldier which has a snout used to eject a sticky or poisonous fluid at intruders, has as its analog the soldier ant of *Pheidole fallax*. The soldier ant has an enlarged poison gland in its abdomen filled with the foul-smelling chemical skatole, which is used for defense outside of the nest. Dufour's gland is the source of the trail-marking chemical. It is absent in the soldier ant but is enlarged in the worker ant, who emits a trail of chemical as it moves along the substrate. Other ants follow these chemical tracks.

Termites have exploited the abundance of cellulose in the environment as a food supply. The primitive termites, such as *Reticulotermes flavipes* (the eastern subterranean termite of North America), employ "symbiotic" protozoans to digest the cellulose taken in by the termite. The relationship between the termite and the protozoan is symbiotic because the termite could not exist without the cellulose digestion performed by the protozoan, while the protozoan requires the anaerobic (oxygen-

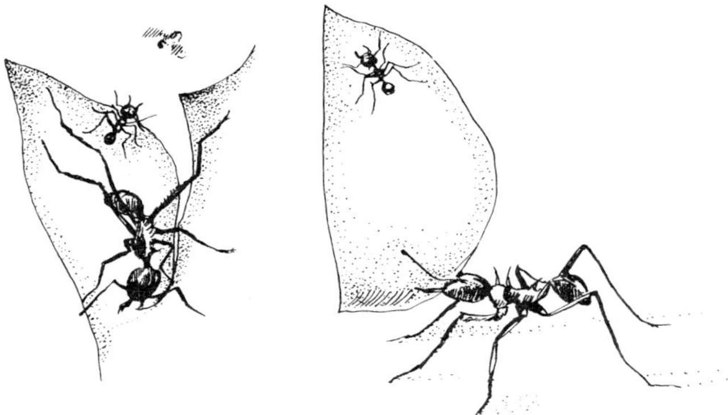

***Fig. 7.10** Minor workers of* Atta *fungus-growing ants in the process of guarding media workers as they harvest pieces of leaf. The minor is "fencing" with a parasitic phorid fly that would attack the occupied media worker, except for the efforts of the minor.*

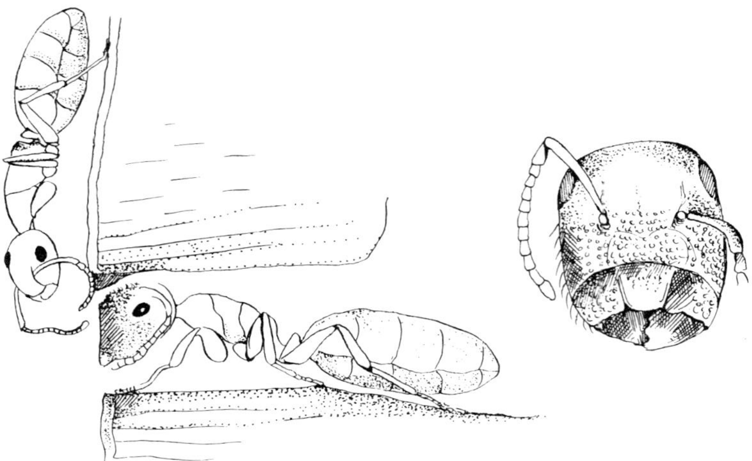

***Fig. 7.11** The "living door" soldier of* Camponotus truncatus. *The soldier plugs the nest entrance with its armored head, refusing to back away unless receiving the proper cues from an ant on the outside.*

free) environment provided by the termite's gut. Oral–anal "trophallaxis" (exchange of gut liquids between colony members) so typical of termites ensures that young pick up the gut protozoans so essential for survival.

Reticulotermes flavipes nests in the ground because it requires moist conditions to survive. Wood in contact with the soil is its normal food, but it can attack paper, wall board, and other materials containing cellulose, even if not directly in contact with the soil. In these cases, the termites construct mud tubes from the soil to the food supply so that they may reach it without exposure to predators and other dangers, including desiccation. (see Fig. 9.20). Once the colony has consumed the wooden structures in its immediate vicinity, the colony probably dies out.

Some of the very advanced termites create huge "termitaria" or termite nests that may reach over 20 feet in height (see Fig. 7.12). These structures are often very elaborate, with ventilation networks, a royal cell (for the reproductive pair), fungus gardens, nurseries, and so on. These advanced termites raise fungi for food on a substrate of cellulose from various sources. The fungus grown is unique in that it exists only in the colonies, not by itself in nature. The termites consume the substrate and termite feces along with fungus growing in the fungus gardens. In social wasps (bald-faced hornets, for example) of the temperate zone often only the mated queen overwinters. The queen selects a site in spring and constructs a nest of paper made by chewing up fragments of wood. After constructing several cells, she deposits eggs and feeds the developing larvae daily on insects she catches and chews. After these first young workers emerge as adults they take over the foraging and nest construction. The queen focuses on egg production and the nest grows rapidly. In late summer a crop of queens and males is produced. The old queen dies, and the colony begins to deteriorate as the new queens and males leave, since no more eggs are produced. The

Fig. 7.12 Some termitaria common in Australia. Such structures, constructed by termites, may be a dominant form in the landscape.

queens, once mated, seek an overwintering site and the other colony members die out.

Some ants of the more primitive sort (e.g., the carpenter ant) produce swarms of winged reproductives from mature colonies that mate and then the mated queen founds the colony on her own. However, other life cycles are also common. Several score of pharaoh's ant workers carry brood off to a new site. If they are without a queen, they rear one from the brood, and she is mated with males they also rear. Another species grows fungi for food in much the same way as do the termites. The vegetation removed by colonies of these leaf-cutting ants of the new world tropics is often so extensive that their action causes agricultural damage.

Among the fascinating ant life histories are those of the slave-making ants *Formica sanguinea*. Slavery probably evolved from the raiding of one ant nest by another species of ant, a common occurrence. Eggs, larvae, and pupae are carried off by the raiding ants and may be stored in the raiders' nest prior to consumption. If one of the "foreign" ants emerges as an adult after a few days, it may have picked up enough of the new colony's odor to be accepted by the slave-making species. Also, since the foreign ant is unaware of its capture and movement, it acts as if it were a member of the slaver colony and begins to perform work for the colony just as it would have in its own nest.

Army ants have a different life history for they have no permanent home. They alternate between a *statary phase* in which the colony stays at the same site for a 2–3 week period, and *nomadic phase*, in which the colony moves to a new site each day for 2–3 weeks. The colony forms a temporary site under protective tree roots or downed tree trunks. The ants form into a solid mass around the queen and the developing young, forming layer upon layer of interlinked bodies. This mass is termed a "bivouac." The statary phase is the time when the queen produces her eggs and the larvae begin to grow. It is also when the pupae from the previous statary phase of the cycle emerge. During the nomadic phase no eggs are laid, but the larvae already present grow rapidly. The ants make daily raids from the bivouac site and capture any and all animals they encounter that they can subdue. Most prey consists of other insects along with an occasional small or sick vertebrate (bird, lizard, etc.). Some army ant species (e.g., *Eciton hamatum*) make raids in the form of several narrow columns of raiders. Others, like *Eciton burchelli*, have broad fronts of raiders so that their foraging pattern is termed a "swarm raid." The breadth of the swarm raid in some army ants is up to 15 meters wide and effectively cleans out almost all other invertebrates within the area covered.

Insects that have evolved social behavior don't seem to regress to less sophisticated social behavior, but move on to further permutations of sociality. For example, the ant, *Teleutomyrmex schneideri*, is known only in the reproductive forms, that is, it has no worker caste. Instead, it is a social parasite. Queens invade colonies of another species of ant, *Tetramorium caespitum*, and climb onto the back of the *T. caespitum* queen. They live on her body, and in some way, suppress the host ant from producing any forms other than workers. Thus, the *Teleutomyrmex* employs the workers that it allows the *Tetramorium* colony to produce so that its own sexual offspring have care.

An overall comparison of social biology as it occurs in the termites (Order

Isoptera) and in the higher Hymenoptera (the ants and the social wasps and bees) is presented in Table 7.5.

Communication, the key to social existence, is a topic that will not be covered here. The essentials of communication insofar as they pertain to social insects have already been presented in the chapter on behavior. Included is information about chemical communication, with examples of pheromones in the honey bee, ant, and

Table 7.5 Comparison of Sociality between Termites (Order Isoptera) and Higher Hymenoptera (Ants and Social Bees and Wasps)[a,b]

	Differences	
Similarities	Termites	Eusocial Hymenoptera
1. The castes are similar in number and kind, especially between termites and ants	1. Caste determination in the lower termites is based primarily on pheromones; in some of the higher termites it involves sex, but the other factors remain unidentified	1. Caste determination is based primarily on nutrition, although pheromones play a role in some cases
2. Trophallaxis (exchange of liquid food) occurs and is an important mechanism in social regulation	2. The worker castes consist of both females and males	2. The worker castes consist of females only
3. Chemical trails are used in recruitment as in the ants, and the behavior of trail laying and following is closely similar	3. Larvae and nymphs contribute to colony labor, at least in later instars	3. The immature stages (larvae and pupae) are helpless and almost never contribute to colony labor
4. Inhibitory caste pheromones exist, similar in action to those found in honeybees and ants	4. There are no dominance hierarchies among individuals in the same colonies	4. Dominance hierarchies are commonplace, but not universal
5. Grooming between individuals occurs frequently and functions at least partially in the transmission of pheromones	5. Social parasitism between species is almost wholly absent	5. Social parasitism between species is common and widespread
6. Nest odor and territoriality generally occur	6. Exchange of liquid anal food occurs universally in the lower termites, and trophic eggs[c] are unknown	6. Anal trophallaxis is rare, but trophic eggs[c] are exchanged in many species of bees and ants
7. Nest structure is comparably complex and, in a few members of the Termitidae (*e.g., Apicotermes, Macrotermes*), considerably more complex. Regulation of temperature and humidity within the nest operates at about the same level of precision	7. The primary reproductive male (the "king") stays with the queen after the nuptial flight, helps her construct the first nest, and fertilizes her intermittently as the colony develops; fertilization does not occur during the nuptial flight	7. The male fertilizes the queen during the nuptial flight and dies soon afterward without helping the queen in nest construction
8. Cannibalism is widespread in both groups (but not universal, at least not in the Hymenoptera)		

[a]Reprinted by permission from E. O. Wilson. 1971. *The Insect Societies*. Harvard University Press, Cambridge, Massachusetts.
[b]Since the two orders are not close phylogenetically the similarities of sociality are due to convergent evolution, not common inheritance.
[c]Trophic eggs are usually degenerate and not viable, destined for colony consumption.

termite, with an additional section devoted to communication as it applies directly to the social insects.

A question still to be answered in this discussion of sociality is "What's in it for the workers?" Why should workers in eusocial insect colonies work? Are they contributing toward the welfare of their offspring? No, for they are sterile. Why does a worker honey bee use its barbed stinger to protect the colony since the stinger pulls out of the abdomen, usually proving fatal to the bee. This behavior is altruistic (self-destructive behavior performed for others) and directly contributes to colony welfare. Although the worker does not contribute to the welfare of its own offspring (it has none), she is very closely related to the other members within the colony, so that altruistic behavior toward the colony contributes to survival of the species, if not to all of the genetic components of a specific sterile female.

That sociality appears in at least 11 different lineages in the Hymenoptera, but not in other orders, except the Isoptera, may be a consequence of the haplodiploid condition found in this order. Whereas in most other groups of animals both sexes are diploid, in Hymenoptera females are diploid (that is, they have two sets of chromosomes) but males are haploid (one set of chromosomes). Thus, workers in the hymenopterous eusocial insects are much more closely related to each other than they are to their mothers. These workers (all female) have one set of chromosomes from the sperm (all of which are identical because of the haploid condition of the father). If she got gene "C" from her father, then all of her sisters have it. If it came from her mother, then there is a 50% chance that her sisters have it too. Therefore, while the degree of relatedness between full sisters in most kinds of animals is 50%, the relatedness is 75% in Hymenoptera. These animals, then, have a bigger vested interest, genetically, in rearing and protecting their sisters than would most other animals. This interest may include sterility of workers since division of reproductive labor is so much more efficient than having all individuals perform both the reproductive and nonreproductive work required.

Advantages and Disadvantages of Social Behavior

Various conditions favor social behavior and may have led to the evolution of presocial behavior in some insects. For example, parental care by a species allows it to use physically harsh rearing sites that are otherwise unusable by solitary species. Group hunting allows a species to suppress prey that is too large for individual hunters to overcome. A group can also create a larger and more protective nest. The group also benefits in cases where a sporadic surplus of food becomes available because the group can gather and store more of a given "bonanza" than could a single individual. Thus, the benefits of sociality include greater environmental control, with reserve forces for protection of the species and for exploitation of available resources.

Disadvantages of sociality lie in the concentration of populations of a species into one or several sites per unit area. Thus, an effective predator on social wasps, for example the army ants, might easily wipe out the wasps if too much of the wasps' collective biomass were concentrated in a small area. In other words, it pays (evolutionarily) for wasps to maintain a smaller colony size in the face of the highly efficient searching capacity of other invertebrate predators. On the other hand, vertebrates also prey on social wasps, and they are best defended against by large colonies of many defenders. Thus, social insects with few very large colonies are vulnerable to invertebrate attack, while being relatively secure from vertebrates.

The size of the colony seems to depend on the various environmental factors encountered by the species. In general, there are more social insects in larger colonies in the tropics than in the more temperate zones.

Degree of sociality also seems to be a balance between the advantages which social behavior provide and its disadvantages. Thus, the "presocial" states of insect biology stated earlier in this chapter are not to be viewed as instantaneous glimpses of different insects in their evolution toward eusocial life, but as the particular level of sociality that has evolved in a species with its own unique set of environmental pressures.

INSECTS AND CARNIVOROUS PLANTS

The typical relationship in nature is that the insect is a parasite on the plant. Since plants can capture solar energy through the process of photosynthesis, the producer (plant) is exploited as a food source, directly or indirectly, by the consumer (insect). However, as has been shown in other parts of this chapter and the section on pollination in Chapter 8, other relationships sometimes do exist between insect and plant, especially as "superficial" relationships are more closely examined. One of the earliest studied is the apparent role reversal that occurs when the plant eats the insect, the topic of this section.

Carnivorous plants, which are plants that utilize the proteins from the bodies of animals, have been known since the 1760s. Carl Linnaeus, the "father of taxonomy," received a Venus's flytrap from a fellow scientist who told him that the bear-trap-like structure (see Fig. 7.16) at the tip of the leaves must obviously be a "machine to catch food" and that, in fact, the two lobes did "rise up" and "lock the rows of spines together" when "the poor animal" touched the sensitive areas with its feet. Such action by plants seemed so extraordinary to Linnaeus that he presumed that the trap later opened to release the captured bug, rather than actually digesting and utilizing its body contents. Subsequent work, especially an elegant series of experiments by Charles Darwin, proved that the trapped insects were eventually digested by the plant.

Evidence for digestion, as noted by Darwin, was the dissolution of solid proteinaceous material, whether it be albumin, muscle, or gelatin. He showed by stimulating the plant with pieces of glass (so that the plant "captured" the object), that the digestive secretion normally produced by the plant could only be triggered by the presence of proteinaceous matter, not glass, water (hot or cold), various salts, heat, and other stimuli he applied. Final irrefutable proof that carnivorous plants actually assimilate materials from the bodies of their victims has been supplied through radioactive isotope studies. In these experiments, a radioactive carbon (^{14}C) isotope is fed to a fly in its food. The fly's muscle tissue eventually contains ^{14}C as an integral part of its protein structure. This radioactive muscle is dissected out and placed onto the digestive surface of the supposedly carnivorous plant. If subsequent tests of the leaf show that radioactivity is spreading and if cleansing the leaf's surface does not remove this radioactivity, then the ^{14}C-labeled fly protein must have been digested and absorbed into the plant tissue (see Fig. 7.13).

Carnivorous flowering plants can be found growing in acid bogs, heavy clays, chunks of wet drippy rock, deserts, or strictly aquatic environments. Some are even epiphytes, that is, they can grow without contact with the soil, as, for example, on other plants. Although these environments differ greatly in many aspects, they each share one characteristic—low nitrogen availability. Nitrogen is a substance required for growth. Availability of additional nitrogen would give plants in this low nitrogen environment an advantage over those that could not tap this extra source of required material. A plant could gain the additional nitrogen from the bodies of trapped insects, but this utilization would require that the insect be digested and that the products of digestion be absorbed. In the selection that led to carnivory in plants in the beginning, digestion may have originally been accomplished by bacterial activity, but most carnivorous plants that have been tested actively secrete enzymes to digest their victims.

About 400 species of angiosperms (flowering plant) are carnivorous. Since they include members of five different plant families, it is not surprising that a variety of trapping techniques have evolved. A survey of the mechanisms employed shows that they fall into two broad categories, with several variations within each major category. A classification of trapping mechanisms is presented in Table 7.6.

The Venus's flytrap has a trap at the end of each leaf, which consists of two lobes that fold together when trigger hairs on the upper surface are disturbed (see Fig. 7.14). Closure occurs as the rows of peripheral spines on each lobe intermesh. Small prey often escape "through the bars," and large strong prey can force its way out. Others not as fortunate, however, are squeezed tightly by the two lobes, which begin to pump out digestive fluid onto the victim. Digestion and absorption usually takes

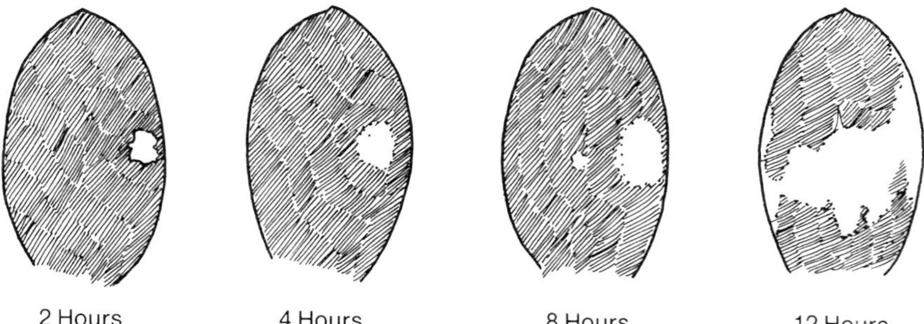

Fig. 7.13 A leaf of **Pinguicula grandiflora** *at 2, 4, 8, and 12 hours after application of ^{14}C-labeled protein to the digestive surface of the leaf. By 12 hours the plant has not only digested the radioactive protein, but absorbed the components. To examine this process, treated leaves are placed in contact with X-ray film at the specified intervals and the radiation emitted by the radioactive carbon wherever it is located in the leaf causes emulsion of the film in contact with it.*

Table 7.6 Methods of Capturing Prey by Carnivorous Flowering Plants

Trapping Mechanism	Example
Active	
Snap trap	*Dionaea* (Venus's flytrap)
Suction trap	*Utricularia* (Bladderwort)
Passive	
Pitfall trap	*Sarracenia* (Pitcher plant)
Sticky trap	*Drosera* (Sundew)
Lobster pot trap	*Genlisea* (No English common name)

one to two weeks, at which time the lobes reopen to reveal a hollow exoskeleton, since chitin cannot be digested by the plant's enzymes.

The bladderworts of North America are aquatic plants with teardrop-shaped, flattened bladders. When the sensitive hairs around the external opening of the bladder are disturbed, as by a swimming mosquito larva, the bladder walls suddenly spring out and anything near the opening is sucked into the bladder's interior (see Fig. 7.15). The action involved is much like sucking up material with an eye dropper.

(A) ***(B)***

Fig. 7.14 Venus's flytrap. (A) The plant in nature, with the bilobed leaves and fringe of spines in receptive position. (B) After the trigger hairs are touched the two lobes suddenly close, trapping the prey by means of the crossed spines.

Fig. 7.15 The bladderwort.

The trapping mechanism of the pitcher plant, as well as cobra plant, *Nepenthes*, and some others, is a body of liquid contained by the plant into which insects fall, eventually drown, and are then digested. The leaf of the pitcher plant forms the structure that holds the liquid. Insects are attracted to the entrance of the trap by the colored walls around the opening, by scent, or by nectaries that secrete sugary, fragrant liquid. One species of pitcher plant even has a row of nectaries running from the bottom of the plant up the side of the pitcher and into the opening. Some plants employ downward projecting hairs and slippery walls to discourage potential prey from crawling out of the trap (see Fig. 7.16). Others have covers over the pitfall, which act as umbrellas during rain thereby preventing overflow of the trap and dilution of enzymes and digested material (see Fig. 7.17). Cobra plants have windows of transparent tissue in the cover that admit enough light to draw the victim away from the dark trap entrance once it has climbed in and toward the light coming through the plant window.

The sticky trap of the sundew (*Drosera* spp.) consists of the leaves with their glandular hairs (see Fig. 7.18). Each hair terminates in a brightly colored bulbous tip, which is covered by a large drop of viscous fluid. In bright sunlight the droplets glisten and attract prey, which upon lighting, adhere to the droplets. The prey's struggling stimulates other hairs to bend toward the victim, thus further entangling it. The same principle of sticky droplets is employed by *Pinguicula*, but on a much smaller scale. Each of the plant's leaves is covered on the upper surface with very large numbers of minute, stalked glands that produce sticky secretions (see Fig. 7.19). Thus, the leaf becomes a kind of living fly paper to which aphids, ants, and other very

(A)

(B)

*Fig. 7.16 The pitfall trap of the northern pitcher plant (*Sarracenia purpurea *L.). (A) The whole plant and the leaf (B) with one side removed. Courtesy of Durland Fish.*

small insects stick. After an insect becomes entangled, other stalkless glands on the leaf surface secrete digestive fluid and the leaf margin rolls up, covering the victim and creating a temporary "stomach."

The lobster pot of the aquatic *Genlisea* is a modified leaf. The leaf is tubular and hollow, with hairs projecting into the lumen of the tube. The hairs occur in rings and project at an angle away from the entrance. Any animal entering the tube has a much easier time moving further into the trap than out of it (see Fig. 7.20).

Fig. 7.17 Pitcher plants with umbrellas prevent rainfall from completely filling the trap and thereby allowing prey to escape. Such protection also prevents dilution of the trap contents. One genus of mosquito lives its larval and pupal life in the liquid contents of pitcher plants. Not only does it have an abundant food supply there, but only a few other kinds of animals have adapted to this niche so that interspecific competition is minimized.

Fig. 7.18 The sundew. Insects landing on the leaves become entangled in the sticky droplets at the ends of the glandular hairs. The struggles stimulate movement of other hairs toward the victim.

Fig. 7.19 The leaf surface of **Pinguicula**, *drawn from an electron photomicrograph. The stalked mucilage glands are tipped by a droplet of their viscous secretion, awaiting a victim. This highly magnified (350×) view also shows the more numerous stalkless glands, which secrete the digestive fluid after prey has been caught.*

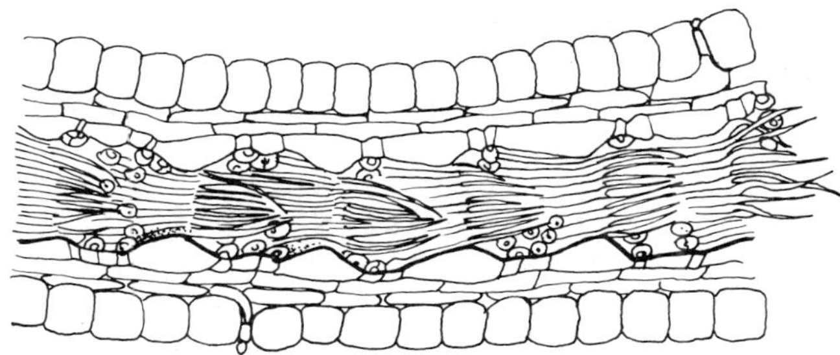

Fig. 7.20 The tubular leaf of **Genlisea**, *cut longitudinally to disclose the bands of inward projecting bristles that encourage further entry of prey into the "lobster-pot" trap.*

SUMMARY

The possible interactions between two species of organisms can be positive (+), negative (−), or neutral (0) for each interacting species. Combinations are expressed as neutralism (00), amensalism (0−), commensalism (0+), competition (− −), parasitism and predation (+ −), and mutualism (+ +). Examples of such insect–plant and insect–animal interactions are given. In the example of the bull's thorn acacia–ant mutualism, the plant provides shelter and food for the ant, while the ant protects the acacia from herbivores and competition for sunlight—a case of coevolution between two mutualistic species.

Insects that produce liquid fecal matter high in carbohydrates (honeydew) are often guarded by ants and may be protected during adverse conditions in ant nests, an example of mutualism between animals.

Most insects that live on plants are parasites, that is, the insect lives at the expense of the larger plant without destroying its host. Seed and seedling-eating insects act as predators by destroying the seeds. Ants feeding on the oil-filled appendages of *Trillium* and *Primula* seeds disperse the seeds.

Insects exhibit the entire spectrum of sociality, from absolutely solitary existence to eusociality, with reproductive division of labor, cooperation in rearing young, and overlapping generations. The honey bee is a typical polymorphic eusocial insect, with a caste system of workers, queen, and drones. As with most social insects, caste determination is controlled by genetics and nutrition. New colonies are formed by swarming.

The eastern subterranean termite is a relatively primitive form that nests in the ground, digesting cellulose by way of symbiotic gut protozoans. More advanced termites tend fungus gardens grown on cellulose substrates.

Some of the more primitive ants produce swarms of winged reproductives. The mated females establish new colonies. Other ant species produce new colonies by budding. Some ant species (e.g., leaf cutters) grow fungus gardens for food. Army ants alternate between statary and nomadic phases, depending on the age of the developing young. Slavemaking and socially parasitic ants are described.

The contribution of sterile workers to the colony is explained in terms of increased survival of the species rather than the individual. Whereas eusociality is relatively uncommon in the class Insecta, it has evolved a number of times in the Hymenoptera. This is explained in light of the haplodiploid chromosome condition found in the Hymenoptera. The social biology of termites and eusocial Hymenoptera is compared.

The advantages of sociality are use of harsh environments, group hunting, increased protection, and increased exploitation of bonanzas. Disadvantages include greater losses to efficient predators. Variations in colony size among different eusocial insects may be a result of predatory activity by vertebrates and invertebrates.

Carnivorous plants are confined to environments with low nitrogen availability, but they survive in these settings by supplementing nitrogen deficiencies with materials extracted from captured insects. Such utilization of insect material by plants has been demonstrated by radioactive tracer studies. Trapping mechanisms of plants are categorized as active (snap and suction traps) or passive (pitfall, sticky, and lobster pot traps).

SUGGESTED READINGS

Darwin, C. 1875. "Carnivorous Plants," John Murray, London.

Gilbert, L. H., and Raven, P. H. (eds.) 1975. "Coevolution of Animals and Plants." University of Texas Press, Austin, Texas.

Heslop-Harrison, Y. 1976. Carnivorous plants a century after Darwin. *Endeavour* **35,** 114–122.

Lloyd, F. E. 1942. "The Carnivorous Plants." Chronica Botanica Co., Waltham, Massachusetts.

van Emden, H. F. (ed.) 1973. "Insect/Plant Relationships." Blackwell, Oxford, England.

Wilson, E. O. 1971. "The Insect Societies." Belknap Press, Harvard University, Cambridge, Massachusetts.

8

Utility of Insects

The "usefulness" of insects is, in many ways, a value judgment concerning a particular activity performed by an insect or the result of that activity. For example, a general predator, like one of the praying mantis species, would be considered (1) beneficial if it were feeding on pestiferous grasshoppers, (2) neutral if its prey were a crane fly, but (3) a pest if it were consuming honey bees. In fact, at first, the Colorado potato beetle had neutral status before European man came to North America, then it was beneficial since it attacked a plant that was a nuisance to livestock in a horse-oriented culture. Later, when it switched hosts from native buffalo burr to potatoes, it "suddenly" became a pest.

The ways in which insects may be envisaged as being "useful" fall into four broad areas. The first involves insect action in recycling the chemical components upon which our planet's life depends. Included here is the insect's role in maintenance of the "balance of nature." The second area encompasses the specific and almost unique role of insects in the reproduction of plants by acting as pollinators.

The third is the products derived from insects. The last is the role of insects in scientific investigation, in part because of convenience, but also because of their special qualities.

Each of the subject areas mentioned above will be covered subsequently, reflecting the relative importance of each type of utility. However, the use of insects in science, while substantial enough to warrant placing it ahead of the role of insects as producers of materials, is placed last because this "use" differs somewhat from the other ways in which insects are utilized.

HOW INSECTS FACILITATE BREAKDOWN OF ORGANIC MATERIALS

Both animals and plants require chemicals in order to maintain themselves. Plants require the basic CO_2 and H_2O which they use in the photosynthetic process to convert solar energy into chemical energy. Then, both plants and animals require O_2 to release the energy bound up in the products of photosynthesis so that they can carry out the work of growth, repair, and reproduction necessary for perpetuation of the species.

Carbon, nitrogen, oxygen, hydrogen, and phosphorus are all essential to the maintenance of life. Though rather abundant, these substances often occur predominantly in forms that are not directly usable by most animals or plants. For example, the carbon bound up in plant cellulose is not directly available, since humans cannot digest cellulose (see Fig. 8.1). In fact, most animals and plants cannot use cellulose directly. Thus, the cellulose in a leaf that falls to the ground is not usable until the carbon it contains is released. This can only occur after a whole series of events take place—weathering, fragmentation, and digestion by a "saprophage" (dead organism feeder). Such common and difficult to break down chemical compounds as cellulose (plant cell walls), lignin (part of wood), keratin (protein in horn, claws, and scales), and scleratin (plasticized insect chitin) make up a significant proportion of the bodies of either plants or animals. These materials could accumulate in the environment unless a balance is developed between the decomposer organisms and other organisms. For example, in the past when decomposition was inhibited, vast beds of coal and petroleum developed, and chemical constituents were scarce. Today the opposite is true, and there is a dearth of dead organic accumulation in the tropical forests. This lack of nutrients is a limiting factor, for materials get recycled very quickly when they reach the forest floor.

While the final breakdown of cellulose into its basic components is usually accomplished by bacteria, protozoa, and fungi, insects facilitate the action. For example, a newly fallen tree will be attacked by beetles whose larvae will feed on the cambial layer, or in the wood itself. Although they only use the sugars and starches they consume and pass the cellulose through their digestive systems without acting on it, the downed tree becomes riddled with galleries or tunnels and the bark loosens. This allows the next insect to attack, like carpenter ants or termites. Carpenter ants do not feed on wood, but they make nests in wood by chewing off pieces and carrying

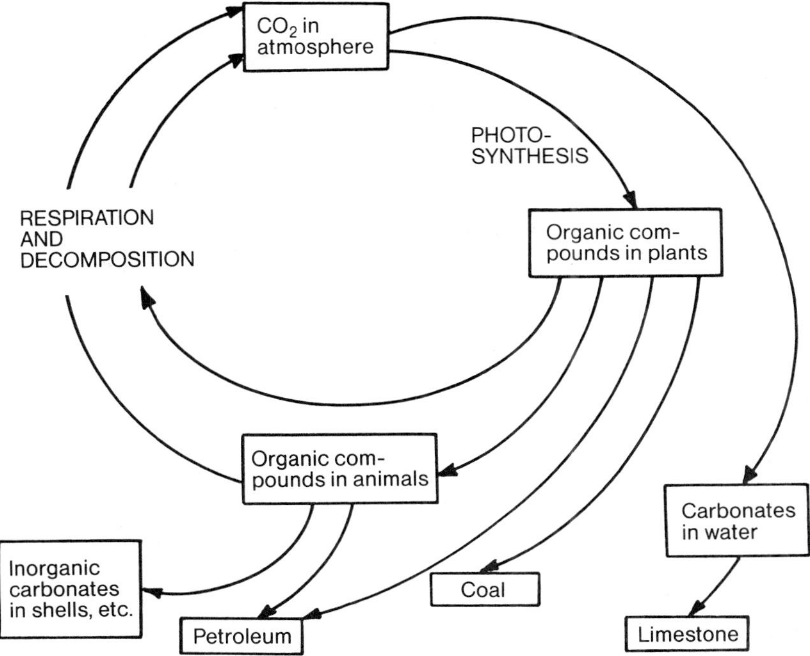

***Fig. 8.1** Essentials of the cycling of carbon in nature. Respiration of plants and animals is responsible for recycling CO_2 in the air.*

them out until a complete nest is excavated. This action again provides an avenue of entrance for bacteria and fungi and thereby contributes to the decomposition of the dead tree (see Fig. 8.2).

Some insects are capable of using the energy in cellulose. Termites, for example, employ protozoa that reside in their intestines to do the actual cellulose digestion for the insect. Thus, termites can rapidly consume even large trees. Whereas termites in temperate North America usually attack dead wood in contact with the soil (the termite nest is in the moist ground), some tropical forms attack growing trees and woody crops. Other termite forms use almost any cellulose product as a substrate to grow a peculiar kind of fungus on which the termite colony feeds. Powderpost beetles breed continuously in seasoned wood. In other words, without the activities of such groups as termites and beetles, the rate of decomposition of wood by bacteria and fungi would be insufficient to keep pace with utilization, and the world would have a large percentage of its carbon, hydrogen, nitrogen, and oxygen tied up in dead plant material.

Insect exploitation of animal remains is so efficient, it is awesome. A rabbit hit by a car and left by the side of the road will be the target of oviposition by blow flies in a few minutes. The maggots that hatch feed on the softer body parts, reducing the cadaver to skin, hair, and bones in a matter of days. Carrion beetles bury small carcasses so that their larvae may consume the food supply without disturbance.

Fig. 8.2 Galleries made in an oak log by carpenter ants. Although the wood is not eaten by the ants, it is removed as the ants initiate and expand their nests. It provides easy access for fungi and bacteria to penetrate the inner wood.

Dermestid beetles, the same types that attack wool rugs and are called carpet beetles, eat skin, hair, and fur. Thus, insects destroy animal remains even more rapidly than plant remains.

The rapidity and efficiency with which carrion insects find dead animals and lay eggs and the minute size of the eggs deposited contributed to the early theory of spontaneous generation. According to this theory, organisms arose *de nova* from mud or dead organisms. Experiments by the Italian scientist Redi in the 1700s proved to the contrary that the only time meat became wormy was when it was allowed to be contaminated by flies. That is, advanced organisms do not form spontaneously from previously nonliving material; only living forms can generate life.

INSECTS AS NATURAL ENEMIES

At the start of this chapter we discussed the change in the status of Colorado potato beetle from neutral to beneficial to pestiferous, according to the organism attacked, from the human standpoint. On this basis we will consider whether insects may be useful as natural enemies.

For almost every animal and plant that lives on land or in fresh water there is an insect (or sometimes many) that attacks it. Vertebrates are attacked by fleas, lice, and biting flies. Invertebrates have their own contingent of insect predators, parasitic wasps, and flies. Plants are attacked by borers, leaf chewers, sapsuckers, and seed feeders. We will discuss this in considerable detail in the following chapter, but the significant concept to be derived here is that if the animal or plant being attacked by insects is considered a pest by humans, then its own natural enemies can be regarded as beneficial to humans. Any suppression of the pest by its insect enemies will contribute to human welfare by suppressing the effect of the pest on humans.

This concept is the basis of applied biological control. Weeds like purple nutsedge (*Cyperus rotundus* L.), hedge bindweed (*Convolvulus sepium* L.), and *Carduus* spp. thistles are being studied to see if the insects that attack these plants in one area may be successfully introduced into other areas to provide weed control. Similar studies on pest insects have determined that their natural enemies can also be used to better advantage. Biological control specialists have successfully suppressed many pests in this manner. There is also a "natural control" (pest suppression without human intervention) that exerts an overall effect of staggering magnitude. It may be unnoticed by humans until the balance is disturbed by pesticides, pollution, or other factors and the resultant population explosion of pests ensues.

POLLINATION

Insects are often intimately related to plants, as has been discussed in Chapter 7. However, of all of these relationships, that of the insect's role in pollination seems to be most important, at least in our consideration of the ways in which insects are useful to humans.

Pollination as broadly defined, is the placing of appropriate pollen onto the receptive stigma of a flower. Since pollen grains are microgametophytes produced in the anthers, they are termed the male floral products. In order for fertilization of the macrogametophyte (ovule) within the female flower structure (pistil) to occur, pollen has to be transferred from its site of production (the anther) to the stigma (the receptive, sticky tip of the pistil). If the proper chemical and physical cues are present, the pollen grain germinates into a pollen tube, which grows down through the style and penetrates the ovary where fertilization (fusion of haploid male and female nuclei) occurs. The role played by insects is in the transfer of the pollen to the stigma. This operation can be performed in one of two ways: (1) by wind-blown transfer, and (2) by mechanical carrying of the pollen by some mobile object (see Fig. 8.3).

Wind-pollinated plants often have two kinds of flowers: Some with male (staminate) parts, others with female (pistillate) parts. Both types of flowers may occur on the same plant. The pistillate flowers have elongate sticky stigmatal surfaces exposed to the air in order to pick up one of the huge number of pollen grains typically produced by male flowers of this type. A good example of the elongate pollen traps of

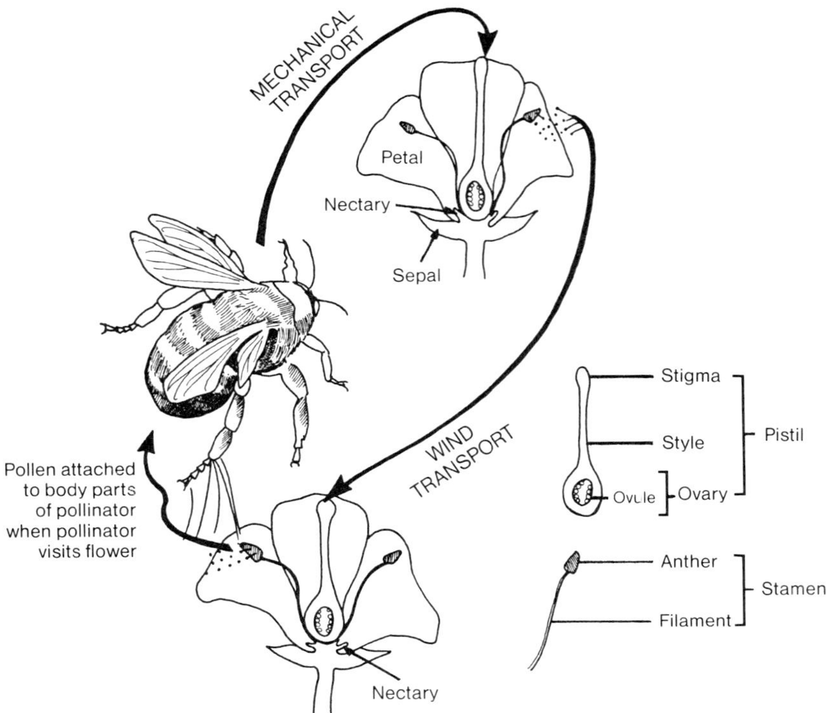

***Fig. 8.3** In order for fertilization of a plant's seed to occur, pollen must be transferred from its site of production (the anther) to the receptive stigmatal surface of the pistil. Pollen grains from wind-pollinated flowers are often small and dry, whereas insect-pollinated plants produce large, sticky pollen grains. The benefits of cross-pollination (hybrid vigor) are encouraged evolutionarily by structural and chemical incompatibility between pollen and stigma of the same flower.*

pistillate flowers is the silk on an ear of corn. The pollen-producing staminate flowers tend to occur in groups along the tips of a stem or branch so that they wave back and forth in the wind, thereby launching the many pollen grains. Common wind-pollinated temperate zone plants are many of the trees, especially the conifers, and the various grasses. Wind-pollinated types usually occur in fairly dense populations for as the distance is increased the pollen density in air becomes diluted.

Insect-pollinated plants (and plants pollinated by other organisms) have flowers that are remarkably different from those of wind-pollinated plants. They tend to be showy and smelly, with very characteristic shapes, and often possess nectaries within the flower, which produce sugary nectar. Relatively small amounts of pollen are produced and the grains are large and sticky.

The earliest insect pollinators were probably beetles that visited the plant and fed on the sticky ovules covered with pollen. Any chance recurrence of feeding by a previously contaminated beetle on another sticky ovule site of the same species would greatly increase the chance for pollination, in addition to selecting for those plant characteristics that originally attracted the beetle. Thus, pollinator attraction would be a strong selective force, with rapid evolution of dependence of the plant on a successful pollinator group.

The original pollinators no doubt fed on ovules, excess pollen, other edible flower parts (petals, sepals, etc.) and also on nectar produced by glands in the area of the flower. Since the flower is a foreshortened stem and the petals, stamen, and other flower parts are modified leaves, some of the nectar glands scattered over the plant's body would occur within the evolving flower. If the original function of the nectar gland (as suggested by some botanists) was to regulate carbohydrate content in the plant's phloem, this function soon was overshadowed by its attractiveness to insects and other potential pollinator groups.

Another evolutionary trend seen in insect-pollinated plants is toward "complete" flowers, that is, flowers with both pistil and stamen in the same flower. This trend encourages cross-pollination by both supplying and being receptive to pollen with each insect visit.

To avoid the problem of self-pollination, which results in a lack of hybrid vigor, various physical and chemical means have evolved. For example, in the flower of the avocado (Fig. 8.4) the female organs of a given flower are receptive at a time when that same flower is incapable of producing pollen. Then, after the stigma has senesced, the anthers open and release their load of pollen; a temporal separation of functions. Structural modifications such as that found in the flower of the French prune (see Fig. 8.5) minimize self-pollination because the style is longer than the stamen's filament. Thus, pollen released from the anthers tends not to reach the stigma. A mechanical means (e.g., insect) must be employed. In Mariposa plum flowers (Fig. 8.6), the lack of structural separation between anthers and stigma is offset by the chemical incompatability between pollen and stigma of the same plant, which prevents pollen tube formation.

Insects, like the other animals involved (Table 8.1), do not pollinate plants because of any conscious good will they bear for the plant, but rather because the cues emanating from the flower attract the insect and elicit some type of behavior once the insect is in close proximity to it. The two basic behaviors elicited in the insect are those in response to feeding or reproductive requirements of the insect. The last incorporates either false signals that a mate is present or that the site meets the oviposition requirements of the insect.

Food is the most common inducement to the pollinator; it may be as nectar, pollen, or flower parts. Pollen contains from 7 to 35% protein and is also rich in lipids. Therefore, it is a suitable protein source for small organisms, provided they are either small enough to feed on individual grains (like the Thysanoptera) or have an efficient way of collecting and concentrating this rich, though scattered, nutrient source. Figure 8.7 shows the structural modifications that have evolved in the honey bee, which aid in the gathering of pollen, the primary source of protein used in the rearing of young honey bees.

The nectar supplied by plants to pollinators is essentially a solution of sugars (see section on Honey). There is some variation in sugar concentration and other components. The plants exhibit differences in quantity and quality of nectar, as well as in timing of floral nectary production, which aids in restricting the type of potential

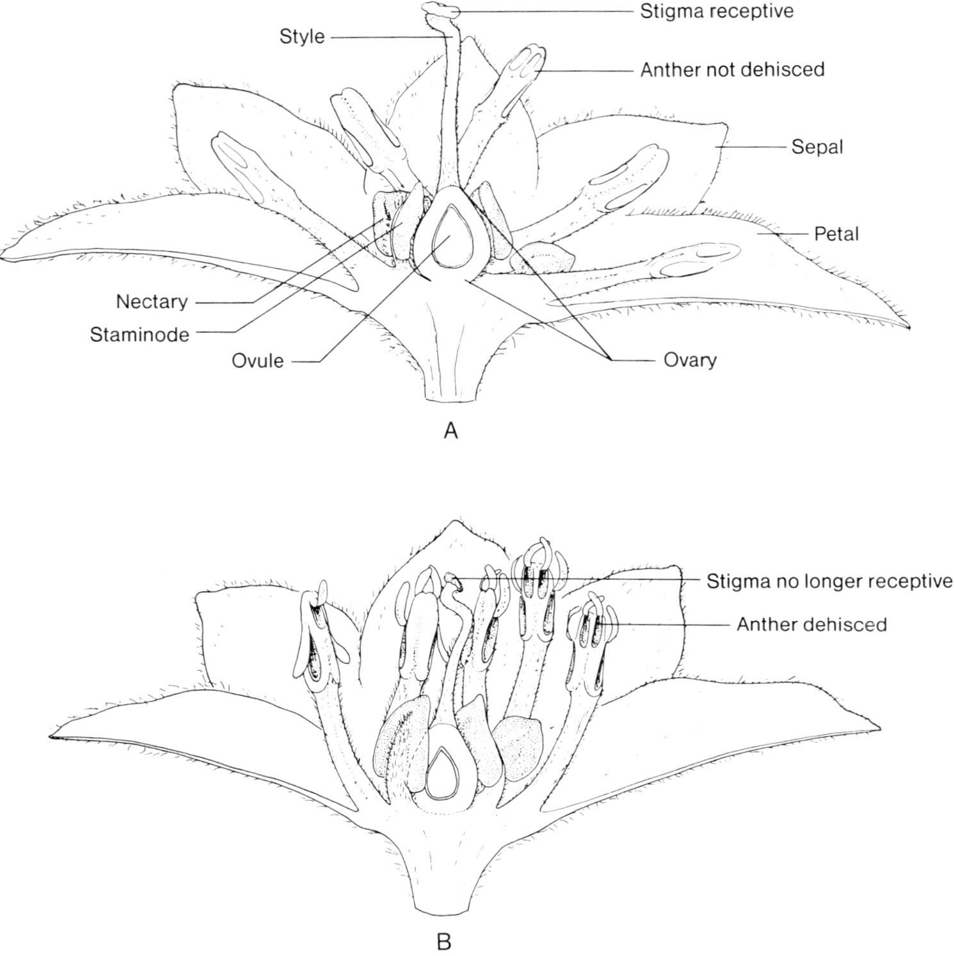

Fig. 8.4 The flower of the Fuerte avocado. (A) On the first day the stigma of the open flower is receptive to pollen, but the stamens are bent away from the pistil and the anthers are closed. After a few hours the flower closes. (B) When it opens the next day, the pistil is shriveled and is no longer receptive to pollen. The stamens are erect, the four pollen sacs of each anther are open, and the valves are hinged back to release the sticky clumps of pollen. The second day the nectar production is even greater than on day one. Thus, the avocado is structurally a bisexual flower, but temporarily it functions as a unisexual flower.

pollinators that visit the flower. For example, golden saxifrage flowers are so tiny and offer such minute amounts of nectar that only a small fly could benefit by visiting them. At the other end of the spectrum, the bird-pollinated flowers such as red columbine, fuschia, hibiscus, and passionflower offer copious, though dilute nectar. But the flower is in a very long narrow tubular shape that restricts access to this nectar; an insect could not easily tap the supply. The advantage for such restriction is that the amount, quality, and availability of nectar have all evolved to induce pri-

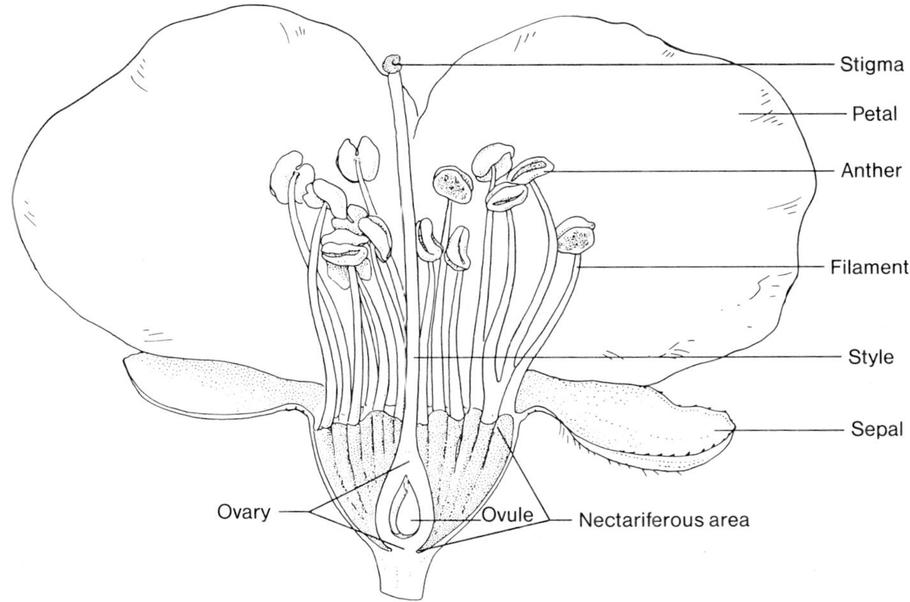

Fig. 8.5 The longitudinal section of the French prune tree flower shows that the height of the stigma is sufficient to minimize the chance of pollen from the shorter anthers contacting the stigma. The honey bee is the prime transporter of pollen in this crop.

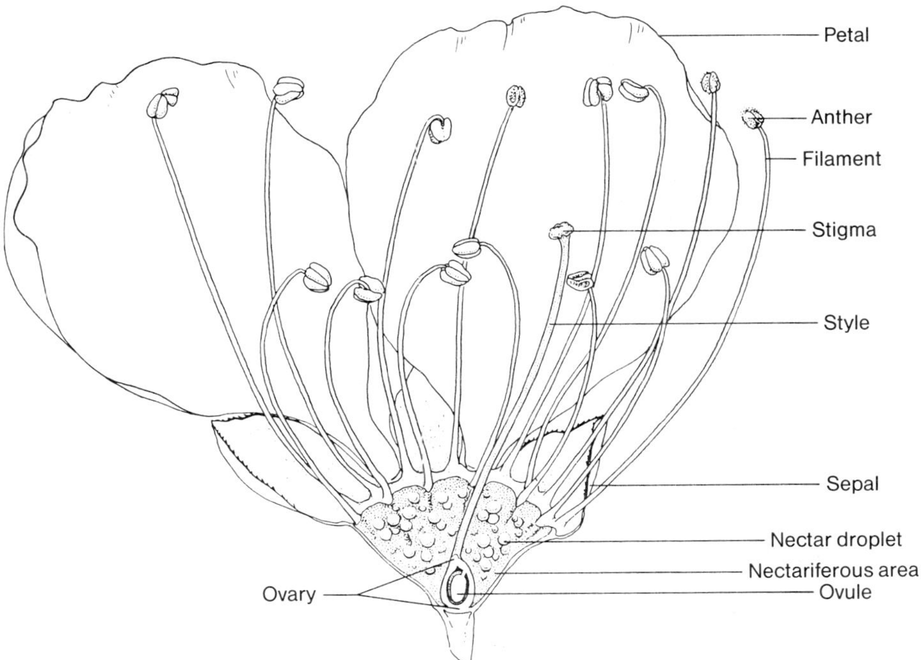

Fig. 8.6 In the flower of the "Mariposa" plum, most varieties of plum are self-sterile so that the lack of structural separation between anther and stigma is not crucial in maintaining out-crossing (hybridization).

Table 8.1 Animal Groups That Include Pollinators

	Common names
Major groups	
Coleoptera	Beetles
Diptera	Flies, mosquitoes, midges
Hymenoptera	Ants, bees, wasps
Lepidoptera	Moths and butterflies
	Birds
	Bats
Minor groups	
Thysanoptera	Thrips
Orthoptera	Crickets and grasshoppers
	Snails and slugs
	Squirrels
	Rats
	Many other arboreal species

marily those organisms that can aid in cross-pollination. If insects had access to large amounts of nectar in one plant, the insect would not have to visit additional flowers to satisfy its energy needs, with a resulting failure in cross-pollination for the plant. The bird (e.g., hummingbird) requires large amounts of energy because of its high metabolic rate and is thus required to visit more than one blossom (or plant) to feed, thereby accomplishing cross-pollination.

Some species of orchids, known as "fetish" flowers, practice a type of sexual deceit to attract pollinators. The form and coloring of the female of the pollinating wasp is imitated by these orchids (Fig. 8.8). The male wasps often emerge as adults somewhat earlier in the season than do their female counterparts. The flowers also mature before the female wasps appear. The males attempt to mate with the floral imitations and during repeated pseudocopulations pollen is transferred. When the

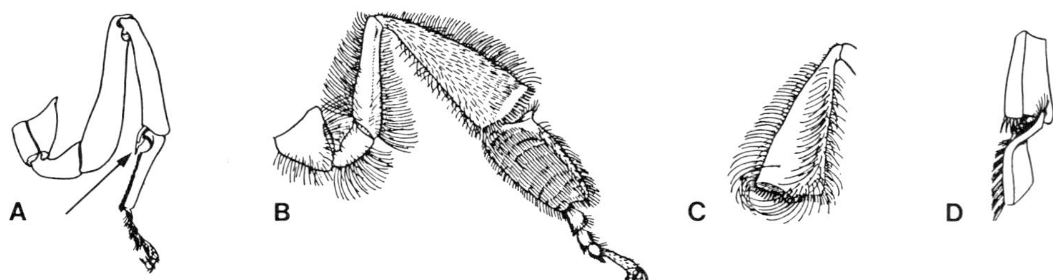

Fig. 8.7 The legs of the honey bee are adapted to the job of pollen concentration. (A) The antenna cleaner of the foreleg (drawn without hairs), a structure possessed by most insects to clean the sensory surfaces of the antenna. (B) The inner surface of the hindleg illustrates the pollen collecting brush and pollen press. (C) The outer surface of the hind tibia possesses the pollen basket, pollen rake, and auricle, two structures. (D) Rake and auricle magnified even more.

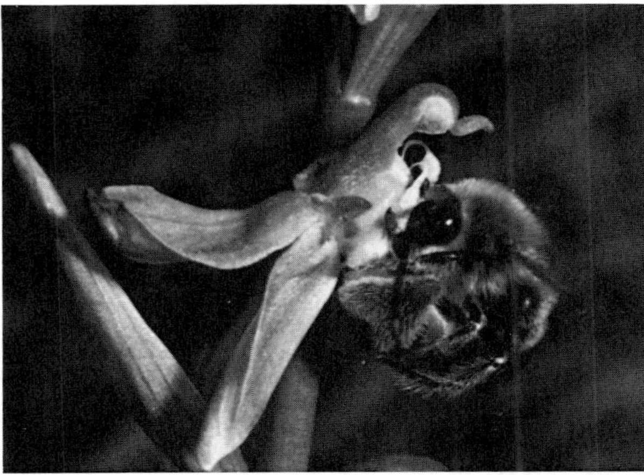

Fig. 8.8 ***Pseudocopulation between the male bee,* Eucera longicornis, *and the bee orchid,* Ophrys apifera. *Males are attracted by species specific odors, for the plant produces no nectar.* Pollinia *(masses of sticky pollen) become attached to the bee and are transferred when copulation with another blossom is attempted.***

real females finally emerge, they are able to successfully contend with the floral imitators for the attentions of their potential mates.

Characteristics of Insect-Pollinated Flowers

The combination of shape, size, color, and smell serve to advertise the presence of a particular plant's flowers. These characteristics, in combination with changes in their color patterns and synchronization of blossoming, a phenological event, increase the possibility of species-specific cross-pollination.

Flowers have evolved a variety of forms that increase the probability that pollination will occur. Many bee-pollinated flowers have evolved a landing platform for the pollinator (Fig. 8.9). The colors of the flowers are predominantly yellow or blue, and never pure red, for bees do not see red. Examination of such flowers with ultraviolet-sensitive equipment or film shows patterns are not visible to humans (see Fig. 8.10). These color patterns enable an insect to distinguish one part of the flower from another and serve as nectar guides. Nectar in bee-pollinated flowers often is located at the base of an elongate tube so that only those insects with long sucking mouthparts like the bee may utilize this nectar. Bees are the largest group of pollinators.

Moth-pollinated flowers are usually white rather than other colors, for colors fade during the diminished light at twilight. Many of these flowers only open at dusk. They have elongate tubular corollas (fusion of petals), but lack a landing platform since the moths hover like a hummingbird rather than landing to feed as a fly or bee does. The moths of the family Sphingidae with their narrow wings and large body are often mistaken for hummingbirds.

Beetle-pollinated flowers are either very large and cup-shaped like the magnolia,

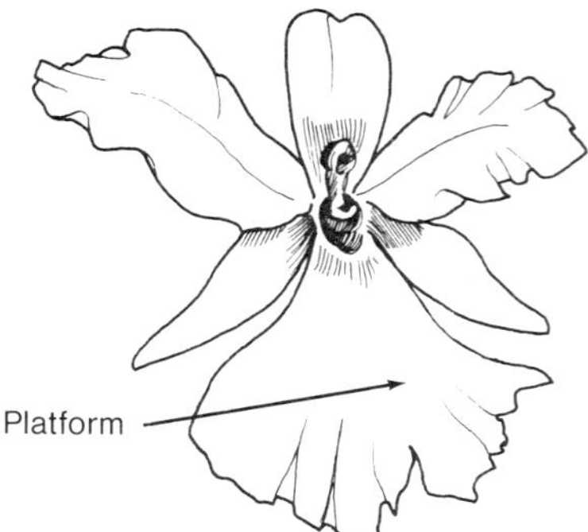

Fig. 8.9 The landing platform commonly evolved by flowers that employ bees in their pollination. The platform, or lip, is a modified petal.

or have tight aggregations, like spirea. Beetles are less discriminating visually than chemically, so that the flowers they visit tend to be white, but with strong odors. The ovules are often buried in tough tissue to protect them from chewing by pollinating beetles.

Some flowers trap pollinators and eventually release the insects only after they have been covered with pollen. For example, the *Arum* lily (Fig. 8.11) traps flies that are attracted by the fecal odor emitted from the spadix. The stigmas pick up any *Arum* pollen already on the visitors. The guard hairs keep the insect in the trap because they bend down easily, but not up. The pollinators are able to leave only after the anthers open to cover them with pollen and the guard hairs have withered,

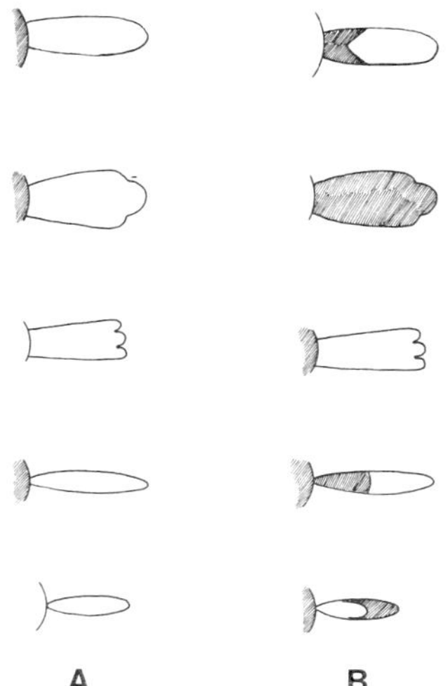

Fig. 8.10 Petals of five yellow flowers as viewed in visible light (A) and ultraviolet (B). Note that each has a distinct pattern making it easily discernible. In some species the UV reflectance does not appear until the flower is receptive to pollination.

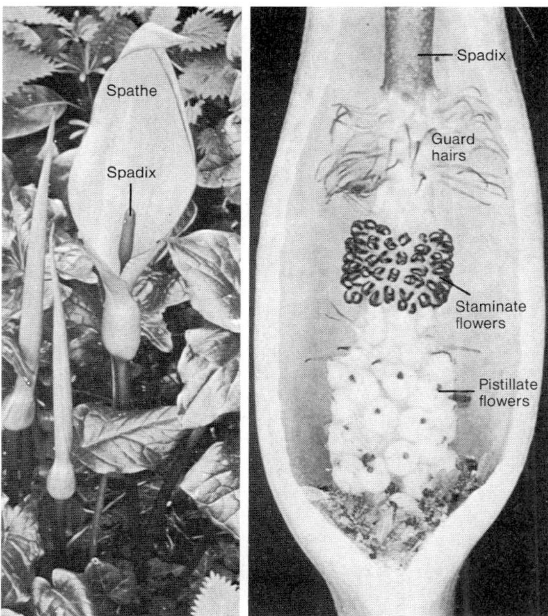

Fig. 8.11 ***Flies attracted to the fecal odor of the* Arum *lily emitted from the spadix crawl down through the guard hairs to the lower chamber, where receptive stigmas and pollen-producing anthers are present. After some time (several hours to days) the guard hairs wither and the pollen-coated flies are released.***

providing an avenue for escape. Other trap flowers utilize carrionlike odors to attract flies and beetles.

Bees of the genus *Euplusia* are attracted to the bucket orchid, *Coryanthes*, by a strong scent. They approach the base of the lip and scratch the area with their forelegs, apparently attempting to gather the liquid that oozes from the tissue (Fig. 8.12). This fluid has an intoxicating effect and the exposed bee gradually loses its grip and falls into the bucket. Since the bucket is filled with water secreted by a pair of glands, the bee becomes thoroughly saturated. Most of the sides of the bucket are too steep and slick to climb so that after the bee starts to orient properly again it eventually finds the one tightly fitting tunnel through which it can crawl, picking up pollinia on the way out.

Several flowers have evolved which take a very active part in applying pollen to the visiting pollinator. *Salvia* blossoms have hinged stamens that swing down to deposit pollen onto the pollinator's back (see Fig. 8.13). Others, like broom (*Sarothamnus* sp.) have stamens and pistil within a keel (modified landing platform). When the bee alights on the keel to probe the base for nectar, the keel abruptly releases the springlike stamens and pistil. These then coil up violently, striking the bee and depositing pollen onto its back and simultaneously picking up pollen on the stigma.

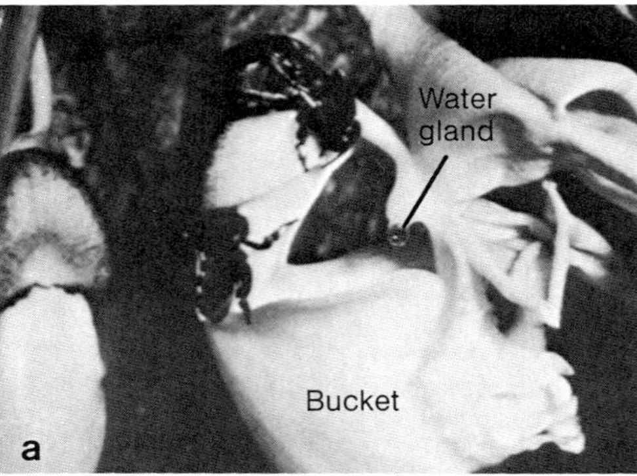

Fig. 8.12 (a) **Euplusia** ***bees land on a bucket orchid blossom. Scratching at the area that secretes the intoxicating material causes the bees to fall into the water-filled bucket. (b) A bee crawls out of the bucket through the narrow opening, thereby picking up pollinia, which it may later transfer to another blossom.***

Current Trends in Pollination Research

The descriptive phase (the mechanics of pollen transfer) in pollination research may be said to have run from the 1750s, when the role of insects in cross-pollination was first described, to the mid-1960s. Since that time the scientific aspects of pollination research have been pursued using several rather new techniques. Visual studies using color television analyze colors and ultraviolet components faster than the complicated photographic efforts previously used. Sophisticated sound analyzing equipment has identifed characteristic flight changes in some pollinators. Electrophysiological studies of odor perception by pollinators now employs such techniques as the electroantennagram, in which the insect sensor (usually the antenna) is tapped with microelectrodes to determine when the insect's sensory apparatus reacts to various chemicals blown over the isolated antenna. Also, highly discriminatory microchemical analyses are helping to explain pollen–stigma interactions and other chemical puzzles in the pollination process.

The Importance of Insect Pollinators in Agricultural Production

Approximately 80% of the flowering plants in temperate zones are pollinated by insects. Examination of Table 8.2 reveals that insect pollination is crucial to the variety of our food resources for if insects were not around to pollinate those fruits and vegetables that make up such a large percentage of our diet, we would have to depend more on such wind-pollinated plants as the grains. Humans might be able to survive without insect-pollinated plants, but our diets would be far less interesting and the world would not be as aesthetically pleasing a place.

Greater crop yields result when pollinators that naturally occur in an area are

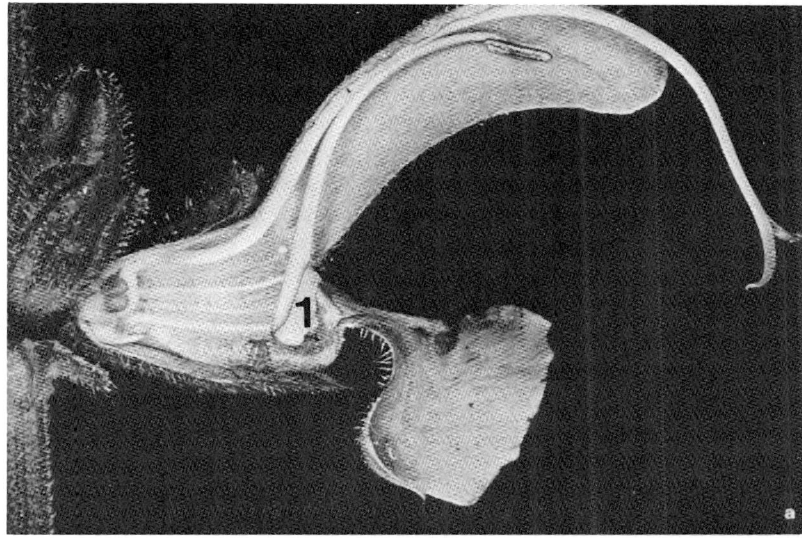

***Fig. 8.13** (a) Part of the* **Salvia** ***flower** has been removed for examination. The moveable stamen, when pushed at point 1 by the bee seeking nectar at the base of the flower, swings down onto the back of the bee. Note the open stigma directly in the path of an entering pollinator. (b) The pollinator is contacted by the anther as the base is pushed.*

augmented with colonies of honey bees. Honey bees seem capable of pollinating most of the crops grown by farmers in temperate areas. In one experiment, yields of blueberries increased from 0.5 tons per acre to 2.75 tons per acre by providing two strong honey bee colonies for each acre grown. Thus, a tremendous increase in yield is sometimes possible if pollinators can be supplied. One difficulty is that the pollinators are only needed for the short blooming period, and then not again until the next year.

Table 8.2 Representative Commercial Crops That Require or Benefit from Bee Pollination[a]

Crop
Alfalfa (seed)
Almond
Artichoke
Asparagus
Avocado
Berries (most types)
Carrot
Celery
Citrus (most types)
Cole crops (cabbage, broccoli, related types)
Cotton
Cucumbers
Eggplant
Flax
Lima beans
Melons (most types)
Onions
Peppers
Pome fruits (apple, quince)
Squashes (most types)
Stone fruits (plum, peach, apricot, related types)

[a]For a complete listing consult Insect Pollination of Cultivated Crop Plants. USDA Agric. Res. Serv. Agric. Handbook No. 496.

So the blueberry grower would also have to become a beekeeper in order to maximize production unless some alternative were available. One alternative common in agriculture is the rental of bee colonies.

Colonies of bees are rented at the rate of one colony or more per acre by producers of many different crops. Honey bee colonies have been provided for apples, almonds, avocados, cranberries, melons, and many crops grown for seed. For example, California, probably the greatest user of rented pollinators, used over 400,000 colonies in 1974, according to a survey of pollination practices. Costs of rental vary with location, length of time needed, and hazard (i.e., pesticidal exposure of the colony), but range between $5 and $22 per colony for just a few weeks. Many pollinator services truck their bees from one area of the country to another, renting them several times each season.

PRODUCTS DERIVED FROM INSECTS OR THEIR LABORS

Insects are the source of several types of products that are useful to humans. Among these are silk, waxes, dye, and honey. Since the most significant products, honey and beeswax, are derived from the honey bee, we will discuss the products from this insect first. Most other insect-derived products have been largely supplanted by synthetic materials, for example, nylon, polyester, rayon, and other synthetic fibers have largely replaced silk and polyurethane has replaced shellac. However, the recent increases in cost of petroleum-based raw materials may foster a return to some natural products.

Honey

Honey is the sweet viscous fluid elaborated by bees from plant-derived nectar, which is modified and stored by honey bees for use as food. The "raw materials" of honey are of two types, both derived from the phloem tissues of plants. The first is nectar, the aqueous, sugar-containing secretion of plant glandular structures called nectaries. Nectaries are often, though not always, associated with flowers and may have originated as sugar valves to regulate the amount of sugars in the phloem sap of plants. Nectar is actively sought by many insects as an energy source of readily utilizable carbohydrates.

The second major raw material in the production of honey is honeydew. Honeydew has been called the "sweat of heaven," or "plant perspiration." It is the term applied to "those clammy drops that glitter on the foliage of many trees in hot weather." For many years the source of this material was unknown. It was finally determined to be the carbohydrate-rich excretory fluid of plant-sap feeders, especially aphids, but always limited to plant feeders of the Suborder Homoptera of the Order Hemiptera.

Nectar and honeydew are collected by the honey bee and deposited into its honey sac, where solids are actively filtered off by the gut. When the collecting bee returns to the hive she regurgitates the saliva-diluted nectar, which is drunk in by another bee and eventually, after one or several more bee to bee transfers, is deposited into one of the waxen cells in the colony. This material would quickly ferment because it is dilute. Therefore, the bees actively spread the nectar, thereby exposing it to the dry, warm air of the hive. Eventually the water content drops to 20% or less, the condition of ripe honey. Honey-filled cells are capped (sealed) by the bees with wax for preservation until it is needed by the colony.

Honey bees can collect and store up to 400 or more pounds of ripened honey in a single season, depending on environmental conditions and the kind of colony managment employed by a beekeeper. Since the colony requires only 50 lb of honey to survive the winter, as much as 350 lb of honey may be available for removal by the beekeeper.

The composition of honey varies for several reasons, the primary one being the source of nectar (or honeydew). Other factors are the location, external conditions, and method of extraction. The average composition of honey is presented in Table 8.3. Note that there is an extreme range of sugars due to differences in nectar components of different plant species. The three types of nectar include (1) those in which sucrose dominates or is the exclusive sugar, (2) those with about equal parts of sucrose, fructose, and glucose, and (3) those with high glucose and fructose, but very little sucrose. Total sugar content of nectar varies from 5 to 80%.

The acidity of honey occurs toward the end of the ripening process and contributes to the honey's flavor as well as to its stability in the presence of microorganisms.

Table 8.3 The Average Composition of Honey

Component	Average %
Moisture	17.2
Levulose[a]	38.2
Dextrose[a]	31.3
Sucrose	1.3
'Maltose'	7.3
Higher sugars	1.5
Total acid (as gluconic)	0.57
Ash	0.169
Nitrogen	0.041
pH	3.91
Diastase (amylase) value	20.8

[a]The result of the action of invertase on sucrose

Enzymes present in honey include diastase (amylase), a contaminant added by honey bees, since it is an enzyme for starch (not honey), and invertase, the enzyme bees use to split sucrose into levulose (fructose) and dextrose (glucose).

Honey, essentially a saturated water solution of dextrose and levulose, typically crystallizes some time after it is formed. The conditions involved in crystallization include water loss (which raises the concentration of dissolved sugars), temperature (more sugar can remain in solution at higher temperatures than at low temperatures), and the presence of colloidal materials or other objects in the honey used as starting points for crystal growth.

Various vitamins present in honey are attributed to the pollen contaminants almost universally present in honey. Pure honey contains several vitamins, but in concentrations too low to be a good source.

In the preface to his "Treatise on the Management of Bees" (third edition) published in 1778, Thomas Wildman very adequately describes the change in the status of honey as a sweetener in the diet of Europeans over the past 400 years:

> BEFORE ſugar became ſo plentiful as it has been ſince the Europeans have got poſſeſſion of the Weſt India Iſlands, honey was much more valuable than it is at preſent, being then the chief ingredient in general uſe for ſweetening every article of food.

As Wildman points out, the insect-derived sweetener has been supplanted by another cheaper substitute. However, honey is still a valuable commodity on the world market and brings a rather high price.

Much of the honey humans extract from bee hives is consumed directly as a sweetener for breads, beverages, or other foods. A large percentage is also used as an ingredient in prepared foods. This latter use is the primary method for utilization of the heavier, darker, and stronger tasting honey types not suitable for table use. More of the lower grades is consumed in bakery products than in other processed foods. Bakers find that their products have a better and more moist quality than when other sweeteners are used. Not only is the product more moist initially, but it retains its moisture for a longer time, too. However, honey is not often used in candy making because the high temperatures involved destroy the flavor.

Another use of honey is in the production of mead (honey wine). In *The Honey-bee* Edward Bevan (1827) states:

> Mead was the ideal nectar of the Scandinavian nations, which they expected to quaff in heaven out of the skulls of their enemies; and, as may reasonably be supposed, the liquor which they exalted thus highly in their *imaginary celestial banquets*, was not forgotten at those which they *really* indulged in *upon earth*. Hence may be inferred the great attention which must have been paid to the culture of the bee in those days, or there could not have been an adequate supply of honey for the production of mead, to satisfy the demand of such thirsty tribes.

In the singular paradise which the Scandinavian religious figure, Odin, sketched for his followers, the principal pleasure was to be derived from war and carnage; after the daily enjoyment of which, they were to sit down to a feast of boar's flesh and mead. The mead was to be handed to them in the skulls of their enemies, by virgins somewhat resembling the houri of the Mahometan paradise, and plentiful draughts were to be taken, until intoxication should crown their felicity. Hence the poet Penrose thus commences his "Carousal of Odin."

> "Fill the honey'd bev'rage high,
> Fill the skulls, 'tis Odin's cry!
> Heard ye not the powerful call,
> Thundering through the vaulted hall?
>   Fill the meath and spread the board,
>   Vassals of the grisly lord!—
> The feast begins, the skull goes round
> Laughter shouts—the shouts resound!"

A recipe for this most interesting of beverages is presented in Thorley's *The Female Monarchy*, published in 1744:

How to make Mead, not inferior to the beſt of foreign Wines.

Put three Pounds of the fineſt Honey to one Gallon of Water, two Lemon Peels to each Gallon; boil it half an Hour (well ſcummed) then put in while boiling Lemon Peel. Work it with Yeaſt, then put it in your Veſſel with the Peel, to ſtand five or ſix Months, and bottle it off for your Uſe.

N. B. If you chuſe to keep it ſeveral Years, put four Pounds to a Gallon.

* Pſal. cxviii. 12. † *Fas eſt et ab Hoſte doceri.*
‡ Prov. vi. 6, 7, 8.

F I N I S.

Beeswax

Beeswax is a true wax secreted by epidermal wax glands on the abdomens of worker bees. Since wax is a normal constituent of the insect integument, the specialization of wax-producing glands in the cuticle of the honey bee is not surprising. Bees use wax to build the comb of individual cells for storage of honey and pollen and for the rearing of individual larvae. Waxen caps are formed over ripe honey, and because the caps must be cut off when the cells are opened for honey removal by the beekeeper, some beeswax is harvested along with the honey.

Beeswax has an agreeable odor and a very slight, but not unpleasant taste. It is completely insoluble in water and melts at about 147°F. Because of these characteristics, beeswax is used predominantly as an ingredient in cosmetics. It can be found in lipsticks, lotions, pomades, ointments, cold creams, and rouges. A secondary use for beeswax is for candles. This use is of long standing, as can be seen in the following quote from Wildman (1827), in which he elaborates on the importance of bees in general, and in their uses in commerce.

> As the value of honey has, however, leſſened, luxury has increaſed the price of wax, which is now become the greateſt ſupply of light in all polite aſſemblies, as well as in the Romiſh churches, in which wax-candles are kept conſtantly burning. By this means, wax is become a conſiderable article of commerce, and as ſuch, is now become the chief inducement for the care beſtowed on bees, eſpecially in the warmer climates;

The third major use of beeswax is by the beekeeping industry itself. Beeswax is used to produce foundation, the base sheet of beeswax placed in wooden frames (10 frames to the hive box) to be drawn out by the bees into a double-sided "honeycomb" (Fig. 8.14). By reusing beeswax to encourage comb construction, the beekeeper allows the colony to devote more energy into storage of honey than into wax production, for bees must consume 6–7 pounds of honey for every pound of wax secreted.

Beeswax is used to some extent in dentistry (impression wax), pharmaceuticals (salves, ointments, pill coatings), and in polishes for wood and leather.

Royal Jelly

Royal jelly is a substance fed to all honey bees after hatching from the egg. It is discontinued as a food for worker and drone bees after the first two days, but is fed to developing queens throughout their larval development. Royal jelly is secreted by the

Fig. 8.14 A wooden frame with the central layer of wax foundation drawn out by the bees into the comb. The frame hangs with nine others inside the hive body (box) resting on an internal ridge by the extensions of the upper frame member. Such frames can be easily removed by the beekeeper without damaging the comb or its contents.

hypopharyngeal glands of adult worker bees when they are between 5 and 15 days of age. A complex of nutrients, royal jelly has been exploited as an ingredient with miraculous powers in face creams and health foods, despite lack of scientific proof of efficacy.

Other Materials Produced or Gathered by Bees

Studies on bee venom indicate a possible therapeutic value as an antirheumatoid-arthritic treatment. The low incidence of rheumatoid arthritis in beekeepers has lent credence to these ideas, but definitive, unbiased studies still need to be conducted. More importantly, venom may prove to be valuable in the treatment of humans who have become sensitized to the proteins in bee venom. Persons known to be allergic to bee stings are often urged by their doctors to undergo desensitization. This involves injection of minute quantities of the allergin (venom) into the patient's bloodstream. Usually minute quantities of whole-body homogenates of the stinging insect are administered over a number of months with increasing quantities. A newer approach overwhelms the antibody system through a short, intensive injection series. One difficulty with this treatment is that the patient becomes desensitized to a single species of stinging Hymenoptera, to the exclusion of other species.

Bees gather pollen from flowers for use in rearing their brood (young). Several morphological and behavioral features have evolved in honey bees that make them highly efficient pollen collectors. Among these are the branched body hairs, called "bee feathers," that hold pollen, and the series of combs, spines, and storage structures that allow the bee to comb pollen from its body surface while in flight, compact the sticky pollen into a solid pellet, and carry the pellet back to the hive. Pollen can be collected at the hive's entrance by the beekeeper. Pollen has been considered a source of energy by some, but the physiological effect of pollen as a nutrient for humans has not been thoroughly studied, although the protein content of pollen compares favorably with that of beans or peas.

As a part of a "back to nature" move, many people are investigating the possibility of keeping bees for the products they may derive, essentially the honey and wax. Each hive can produce 40 to 100 pounds of honey per year (quantity and quality depending on location, season, and management). In addition, the presence of bee colonies promotes fruit and vegetable production so that gardens benefit, too. An additional aspect of beekeeping is an appreciation of the order that exists in nature. Beekeeping also fosters the qualities of patience and precise handling, for movements that are too rapid or general clumsiness are soon rewarded by stings from irate bees. These, however, can generally be avoided and it is possible to work four or five colonies without being stung more than once or twice a year.

Silk

Silk is made from the fine, glossy strands extruded by moth larvae as they construct the cocoon in which they pupate. Although silk is produced by many kinds of insects (caddisworms, black flies, katydids, lacewings, sawflies, and fungus gnats) and other arthropods (mites, spiders) only the silk of a few Lepidopterous caterpillars is used in commerce. The moth species predominantly used for silk production is *Bombyx mori* L. Others are *Antheraea pernyi* (Guerin-Meneville), *A. yamamai* (Guerin-Meneville), *A. mylitta* (Drury), and *Philosamia cynthia ricini* (Boisduval).

Silk, as fibrous protein, is not foreign to animal cells, since fibrous proteins also comprise muscle fiber, connective tissue, and keratin. Silk proteins are like connective tissue in that the cell produces it exteriorly, that is, it is used by those that produce it for external functions, as in attachments, webs, and cocoons.

Silk is produced by different organs in various insects. Those few adult insects that produce silk (e.g. *Chrysopa* lacewings) are always females and the silk originates from colleterial glands that are part of the reproductive system. Other insects use Malpighian tubules for silk production, and some extrude the peritrophic membrane as silk. However, most insect larvae produce silk from the salivary glands, through a "spinneret" (hollow spine with the opening of the duct from the silk glands).

Silk from *Bombyx mori* L. is extruded by the larva from a pair of silk or salivary glands. One gland produces fibroin, the main component of the silk, while the other gland secretes sericine. These two proteins consist of six predominant amino acids. X-ray studies have revealed that salivary gland silk, secreted as small globules which coalesce as ill-defined masses in the gland lumen, have a pleated-sheet structure of protein chains.

B. mori silkworms eat mulberry leaves, especially the white mulberry, *Morus alba* L. During the 4500 years or more that humans have cultured silkworms the insect has become so domesticated that it could not survive without human assistance. Its thoracic legs are too weak to climb to where leaves would be found in nature and its jaws are so weak in the first instar that tender mulberry leaves must be finely shredded before they can be eaten. Fresh leaves every two hours for the first few days and then fresh leaves every 4 hours for the rest of its 30–40 day larval life are required. Provided its surroundings have been airy, bright, and cleaned frequently, it will reach larval maturity, become restless and spin a cocoon over a 3 day period. In spinning the cocoon the larva swings its head from side to side (at about 65 oscillations per minute) extruding a single continuous filament of from 800 to 1200 yards in length.

The first silk secreted is likely to be coarse, while the last part is broad and flat, but the middle 400 to 800 yards are suitable for use by humans. Although the larva is said to "spin" silk, the silk is not twisted as it is extruded.

Over 1000 strains of *B. mori* are known. The caterpillars used for silk production are first generation hybrids between strains, the hybridization yielding heterosis (hybrid vigor) and providing greater yields of product much as does hybrid corn provide greater yields than either parent strain.

Cocoons produced for silk are treated with heat to kill the pupae about 7–8 days after spinning. If the pupa is not killed, the adult dissolves part of the cocoon when it emerges and destroys the prime value of the silk, since it no longer is one continuous strand.

The 8-day-old cocoons are boiled to loosen the cement impregnating the fiber, then one or more cocoons are unwound together in a process unique to the silk industry called reeling. The ends of the filaments are caught, unwound, and kept together by coagulation of the natural gums to form a single uniform filament.

Depending on the treatment of the silken filament (number of filaments combined, sequence of weaving and dyeing, and degree of twisting) it is produced into one of several types of fabric. Those commonly associated with silk (at least in the past) are taffeta, satin, chiffon, crepe, shantung, organza, and faille. All of these are very high quality silk fabrics that are relatively expensive because of the labor involved in the creation of magnificent materials of unique quality.

Synthetic fibers have replaced silk for some uses. Ladies stockings were made of silk before World War II, but are now made of nylon. However, silk still has enough unusual quantities to allow it to compete for other uses. Silk has almost the tensile strength of iron wire (iron being 90,000 lb/in^2, silk 64,000 lb/in^2) and has an elasticity of about 20%. Thus, it can be stretched about one-fifth of its length and return to the original dimension, making it a very wearable material. It also has very low thermal conductivity.

Silk is highly absorptive, thus enabling it to be dyed at much lower temperatures than fibers like wool. In addition, this greater capacity for absorption permits silk to take on colors more brilliantly than other fibers.

The true origin of "sericulture" (the production of silk) and the resulting domestication of the silkworm is lost in antiquity. The Chinese dynasty of Hyang Ti (2640 BC) and more particularly the empress, Ling Shi, is attributed with the development of sericulture. Silk production (for the use only of the nobility) was a tightly guarded secret of the Chinese for centuries until the secret was stolen (according to tradition by Koreans) for the Japanese. The secret was spread to India by a Chinese princess, who hid silkworm eggs and mulberry seeds in her headdress as she left for her marriage in India. Centuries later, the emperor Justinian acquired silkworm eggs and mulberry seeds smuggled out of China in the hollow bamboo staffs of two Persian monks.

Sericulture was first introduced to the New World in 1522 by Hernando Cortes, but has never been successful here because it requires cheap, abundant labor. California has many areas where mulberry was planted along roads in earlier efforts to promote a California silk industry. Indeed, the introduction of the notorious gypsy moth into North America is the result of an abortive hybridization scheme in which a Frenchman sought to bring together the vigor of the gypsy moth with the silk-producing qualities of *B. mori*. Unfortunately for much of North America, the gypsy moth escaped. If a hybrid had been produced, the northeastern United States would have taken over the world leadership in silk production while the northeastern forests were being stripped by gypsy moths.

The leading producers of silk remain Japan, China, and Russia with India and South Korea contributing lesser quantities to the annual world production of nearly 33,000 metric tons.

Other Insect Products

Other products from insects include a dye and shellac. Neither of these products is able to compete successfully with synthetic products, either because of greater costs of the natural product, or because the synthetic compounds have fewer undesirable characteristics.

Shellac is made from the waxy body coverings of a scale insect, *Laccifer lacca* (Kern), that lives in Southern Asia. The resin is purified, then pulverized and dissolved in a solvent. Shellac is used as a finish for wood, often imparting an attractive orange tinge with a smooth, glossy finish. One reason shellac is being supplanted by synthetics like polyurethanes is that liquids spilled on shellac will cause the shellac finish to become white and opaque, whereas it will not affect polyurethane. Thus, a cocktail glass will permanently mark a shellacked coffee table with a white ring if any of the beverage is spilled.

Scale insects of another species, *Dactylopius coccus* Costa, were grown on cactus in Central America and the Canary Islands for the purpose of extracting a brilliant red pigment known as cochineal. The culture of this material is one of the prime reasons why the Spanish conquistadors enslaved the Indians of Mexico. Cochineal dye has long been used in cosmetics and as a food coloring. However, two factors have led to the near demise of the cochineal industry. One is that cochineal has been shown to be carcinogenic. The second is that aniline dyes can be produced less expensively and give satisfactory results.

INSECTS AS A SOURCE OF NUTRIENTS

All major vertebrate groups have representatives that consume insects on a regular basis. In fact, at least one representative in each group may be found that lives exclusively on insects and other arthropods (see Table 8.4).

Table 8.4 Insectivory among Vertebrates

Groups	Completely insectivorous	Partially insectivorous
Pisces	Mosquito fish	Trout, sunfish
Amphibia	Toad	Salamanders, frogs
Reptilia	Anole, many lizards	Gecko, garter snake
Aves	Martins, swifts, swallows, woodpeckers	Robins, thrushes, shrikes, creepers, sparrows, starlings
Mammalia	Bats, ant eaters, aardvarks	Moles, skunks, shrews, bears, foxes, rodents, primates

Note that primates, including humans, are listed as being partly insectivorous. In many cultures insects are a normal part of the human diet. For example, the nkhungu fly is consumed with other ingredients as a snack in Malawi, Africa.

The nkhungu fly is a gnat that lives as a larva in Lake Malawi (Lake Nyassa). Like its relative in Clear Lake, California, it emerges as a nonbiting adult to mate and lay eggs for the next generation. Those in Africa are called *nkhungu* (fog, cloud) because the vast numbers look like a cloud. Nkhungu are gathered in baskets, pressed into cakes and dried for later consumption. Similar insectivory may be found in most other cultures.

The nutritional value of insects is rather high compared to other foods consumed by humans. In Table 8.5, several types of insects are compared with beef and fried fish, showing that the insects are the richer source of nutrients, overall.

The point to be made is not that we should all start eating insects, just because other peoples consume them, but that they do constitute an acceptably nourishing material. The food industry has made great prógress in increasing the palatability of many nutrients. A food technology sophisticated enough to make spiney dogfish flesh into such appetizing and acceptable foods as "hot dogs," "corn curls," "shrimp roll," and "shrimp" should be able to meet the challenge offered by insect proteins without difficulty.

Table 8.5 How Insects Compare to Other Nutrient Sources

Food source	Protein (%)	Fat (%)	Ash (%)	cal/100 g
House fly pupae	63	16	5	—
Termites (live)	23	28	—	347
Termites (fried)	46	36	5	508
Locust	61	10	5	—
Beef	17	7	1	127
Dried salt-fish	44	3	21	203
Peanut oil	28	50	3	598

INSECTS AS MODEL SYSTEMS FOR BIOLOGICAL RESEARCH

Insects possess a number of characteristics that distinguish them from other types of organisms. Such aspects as body covering, number and grouping of appendages, and organizational plan of the organ systems are among those that make insects different in some degree from plants and the other animals. Despite these many differences, however, there are enough fundamental similarities so that insects can be used as model systems. Model systems are living systems used to ask particular questions in experimentally simple ways. The trick is for the investigator to select the particular model system in which the organism's mode of living makes it particularly amenable for experimental use.

That insects are ideal model systems for investigation in many areas of biology is shown by the breadth of scientific disciplines employing insects as research organisms. A search through college biology (or zoology) departments reveals that insects are being used in genetics, sociobiology, behavior, ecology, systematics, evolution, physiology, toxicology, and endocrinology. They are used in these disciplines for two basic reasons. First, information gained from studies of insect model systems can usually be generalized to other organisms because of the fundamental life process similarities among organisms. Second, insects are convenient and inexpensive models. Before presenting examples of areas in which insect model systems are used, a look at the second reason, convenience, is appropriate.

Convenience is a matter involving emotional as well as technical aspects, that is, the investigator's relative empathy with the test organism. We have little empathy with plants because we are confident that they "have no feelings." Thus, experimentation with a cabbage plant that requires dissection bothers us about as much as making cole slaw. At the other extreme, legal actions have been brought against scientists using dogs, rodents, or primates for research. The use of insects in research, on the other hand, is ignored by the non-scientist, and the scientist experiences few qualms when insects are the model systems employed.

The technical aspects that make insects desirable for use are many. Generally, little room is required for maintenance of laboratory colonies of experimental animals. Whereas a colony of beagles would take a whole building to produce 100 experimental animals, the space of an equally productive *Tribolium* flour beetle or *Drosophila* fruit fly colony might be a single one-gallon jar. Concurrent with the differences in space required are those costs involved in maintenance. Food and labor costs for the beagle colony would be many thousands of dollars per year, for each cage requires a watering system, one for waste disposal, and labor for feeding. The *Drosophila* require only a single banana (or some custard-like growth medium) every week or two. The *Tribolium* is even easier for it lives in wheat flour, without the requirement of a free water supply or of a liquefied medium with concurrent mold problems. Thus, a lab colony of four insect species can be maintained at an annual expense of under $5.

Another factor is the speed of development and short generation time of insects. A study over 10 generations would take 20 years with dairy cattle and several years with beagles, but only 6–8 months with the fruitfly, house fly, or mosquito. Although bacteria have shorter lifespans, they require compound microscopic techniques that are not necessary for many insect model systems.

Special characteristics of certain insects make them organisms of choice for scientists. T. H. Morgan chose *Drosophila melanogaster* as an organism to study because it could be reared in ½ pint milk bottles on mashed fruit and yeast. However, subsequent events proved how fortuitous this selection was, for *Drosophila* has other characteristics that makes it a special model in genetics. *Drosophila* has four pairs of chromosomes, as well as polytene chromosomes in some tissues. These "giant" chromosomes exist in the fruit fly's larval salivary gland and consist of chromosomes that replicate, but do not separate, so that the polytene chromosome may be 1000 or 2000 chromosomes packed side by side. The regularities in their structures are emphasized and the banded chromosome is visible under the microscope. The linear arrangement of genes on the chromosome has linked directly the phenotypic characteristic (that is, the external expression of genetic traits) with the position of the gene on a particular chromosome. Further research in this area has led to "chromosome mapping," the determination of the sequence of genes on a chromosome, which, in turn, has facilitated research on the evolution of drosophilic flies. The genetic structure of these flies is shown in Fig. 8.15.

Drosophila melanogaster is still the most common model system used to introduce science students to classical genetics (see Table 8.6). Using *Drosophila*, students can do experiments in genetics within a period of a few weeks because of the ease with which *Drosophila* flies can be manipulated and their relatively distinct phenotypes (see Fig. 8.16).

A common ecological model system is *Tribolium confusum*, the confused flour

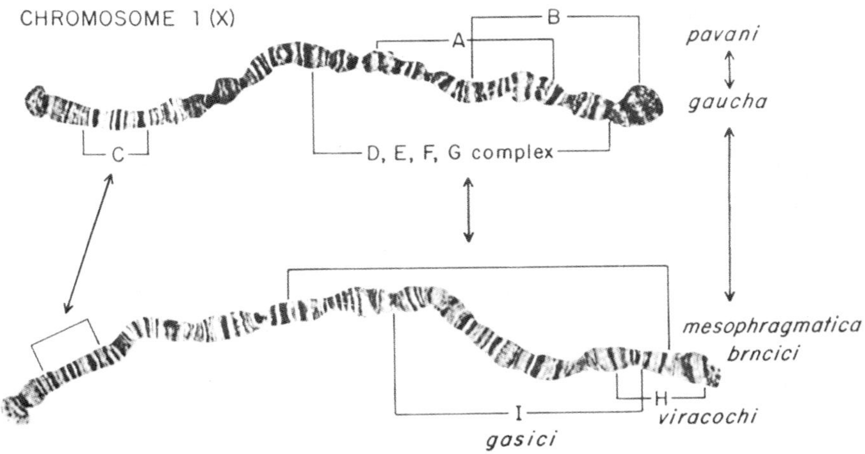

***Fig. 8.15** Polytene (giant) chromosomes of drosophilid flies. Shown is a composite photographic map of the X chromosomes of* **Drosophila gaucha** *and* **D. mesophragmatica**, *two drosophilids from South America. The brackets indicate chromosomal inversions (aberrant events in which pieces of chromosome have broken out, turned around, and become fused again). Particular inversions, characteristic of the chromosomal arrangement of each species in the species group, are identified by brackets.*

Table 8.6 Experimental Organisms Used in Lab Exercises of College-Level Introductory Genetics Courses[a] in the United States[b]

Organism	Scientific name	%
Fruit fly	*Drosophila melanogaster*	84
Corn	*Zea mays*	48
Bacterium	*Escherichia coli*	41
Bread mold	*Neurospora crassa*	34
Bacterial virus	Phage T_4 and/or T_2	24
Fungus	*Sordaria fimicola*	18
Tobacco	*Nicotiana tabacum*	10
Bacterium	*Bacillus subtilis*	10
Bacterium	*Diplococcus pneumoniae*	3
Green alga	*Chlamydomonas reinhardi*	3

[a]Many employ more than one organism.
[b]Committee of Undergraduate Education in the Biological Sciences.

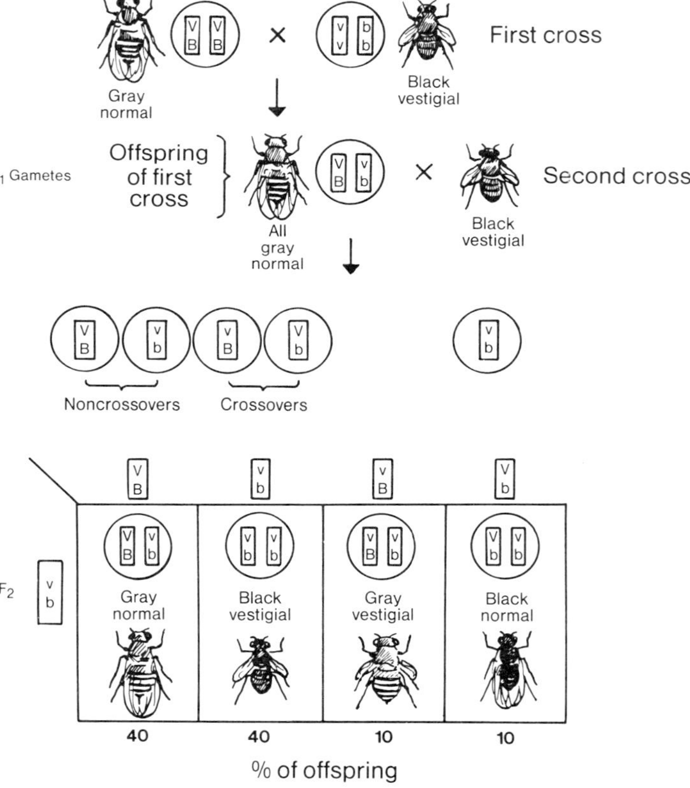

Fig. 8.16 Diagram of two sequential crosses (matings) between* Drosophila melanogaster *strains. The crosses show linkage and crossing-over. Genes for vestigial versus normal wings and black versus gray body in fruit flies are linked, that is, they are located on the same chromosome. The percentage of offspring in the second cross with each combination of characters provides the evidence that the wing and body characteristics are not always inherited together, but that some crossing-over has occurred.

beetle. The name refers not to the mental state of the beetle, but to the entomologists who try to separate this species from its near relative, *Tribolium castaneum*, the red flour beetle. *Tribolium* was chosen by R. N. Chapman to quantify ecology, formerly known as natural history. He and subsequent workers found *Tribolium* to be ideal because an entire population could be maintained in a small jar with just a few grams of flour, since the medium is both food and habitat for all life stages. With a generation time of about one month, each little experimental universe (the jar–flour–insect unit) could be sieved and the numbers of individuals in each stage recorded. Thus, evaluations of group responses could be made to such isolated factors as temperature, humidity, pollution, and competition with other similar forms (see Table 8.7).

Insects are used in sensory physiology because the small numbers of neurons innervating the sensory receptors allow the researcher to monitor the pattern of impulses passing down the nerve as the sensory receptor is stimulated. The sensory hairs on fly mouthparts are both chemical and mechanoreceptors, with five primary neurons. It is far easier to tap the same neuron with a microelectrode in several flies to get an electronic record of the response to a particular condition than to do the same kind of work in a rat or rabbit where thousands of neurons follow the same pathway and the chances of hitting the same neuron in several preparations are very slim.

Insects have been the source of curative chemicals and have also been used alive to treat specific problems. Ground up roaches were thought to have diuretic properties. The Spanish fly, a blister beetle, is the source of cantharidin, a vesicating irritant, especially on urogenital tissues. This material previously was used in human medicine, but its use is now restricted to veterinary medicine.

Intentional infestation of human wounds with fly maggots was popular in the 1930s and is now being practiced again in several of North America's leading medical centers. The early work established that the maggots of certain species of flies were so selective that they would feed only on dead tissue and leave the healthy tissue alone.

***Table 8.7 Competitive Exclusion as Demonstrated by Interspecific Competition between* Tribolium confusum *and* T. castaneum**[a]

			Results of Interspecific Competition[b]	
Climate	Temperature (°C)	Relative Humidity (%)	*Tribolium castaneum* wins	*Tribolium confusum* wins
Hot-wet	34	70	100	0
Hot-dry	34	30	10	90
Warm-wet	29	70	86	14
Warm-dry	29	30	13	87
Cool-wet	24	70	31	69
Cool-dry	24	30	0	100

[a]Although each species can live alone under each of the experimental conditions, when the two species are mixed only one survives over the long term.

[b]The % of time (replicates) each wins.

In addition, the maggots secreted substances (allantoin, urease) into the wound, which aided healing. Although the treatment of intractible osteomyelitis with maggots proved remarkably successful under the supervision of the treatment's original user (W. S. Baer of Johns Hopkins), it so traumatized patients that it was discontinued. The treatment has recently been used again to treat intractible cases of abscesses, burns, gangrene, and ulcers.

Insects are used in many other biological disciplines because of some or all of the characteristics mentioned above. Of particular interest is the relatively new area of sociobiology, in which much of the foundation has been established with work on ants, bees, termites, and other social insects.

SUMMARY

The action of insects that breaks down dead organic matter greatly accelerates the rate at which these materials can be recycled in the environment. The insects that attack cellulose are particularly important since cellulose and some other plant materials are slowly decomposed by microorganisms.

Insects are categorized as beneficial if they attack weeds or animals that we consider to be pests.

Flowering plants are pollinated either by wind or mechanical means. While other animals may be involved in mechanical cross pollination, this process of genetic mixing is performed predominantly by insects. Most insects visit flowers seeking nectar, pollen, or edible flower parts. The sticky pollen, characteristic of insect-pollinated plants, adheres to the insect's body and is transfered when the insect brushes the sticky stigma of the same flower or another flower of the same kind. In some cases, pollinators are intoxicated by floral chemicals or attempt to mate with insect-mimicking flowers.

Cross-fertilization is facilitated by the unique combinations of form, color, shape, and smell of each type of flower, as well as synchrony of bloom. In "passive" flowers, the insect brushes against the pollen source (the anther), but in "active" flowers, both anther and pistil may move, sometimes violently, to contact the insect.

Since insects pollinate most of the fruits and vegetables we eat, they play an important role in agriculture. Growers rent beehives or supply other pollinators to increase crop yields.

Honey and silk are the two most valuable products we derive from insects. Honey is modified condensed nectar or honeydew. It is consumed directly by humans to some extent, but much of it is used in bakery goods and other processed foods. In earlier times, it was often the sole sweetener in the human diet and was the basis for mead, honey wine. Bees also produce beeswax, used in cosmetics, candles, and in the beekeeping industry. Royal jelly, pollen, and bee venom have some limited uses.

Silk is spun from salivary glands by the mature silkworm as it forms its cocoon. Of the five moth species used commercially, *Bombyx mori* is the most common. Silk is processed and treated in various ways to produce several types of material. Its qualities of strength, elasticity, low thermal conductivity, and dye absorption make it competitive with synthetic fibers.

Insects are eaten by many kinds of vertebrates because they are plentiful and nutritious. Analysis reveals that they compare favorably with beef and fish in protein content.

Insects are used as model systems for research in many areas of biology. This is possible because so many functions in insects are the same as in other kinds of animals. Also, insects reproduce rapidly, are small, and relatively easy to rear in large numbers. *Drosophila* fruit flies are used widely in genetics, in part because of their giant chromosomes. *Tribolium* flour beetles are used in ecology. Medicinal maggots are used in the treatment of some types of human disease.

SUGGESTED READINGS

Crane, E. (ed.) 1975. "Honey, a Comprehensive Survey." Heinemann, London.

Dadant and Sons (eds.) 1975. "The Hive and the Honey Bee." Dadant and Sons, Hamilton, Illinois.

Essig, E. O. 1945. Silk culture in California. Circ. 363. University of California, Berkeley, California.

Faegri, K., and van der Pijl, L. 1971. "The Principles of Pollination Ecology." Pergamon Press, New York.

Gojmerac, W. L. 1980. "Bees, Beekeeping, Honey and Pollination." AVI Publishing Co., Westport, Connecticut.

Manchester, H. H. 1916. "The Story of Silk and Cheney Silks." Cheney Bros., South Manchester, Connecticut.

Morse, R. A. 1980. "The Complete Guide to Beekeeping." Dutton, New York.

Proctor, M., and Yeo, P. 1972. "The Pollination of Flowers." Taplinger Publishing Co., New York.

Richards, A. J. 1978. "The Pollination of Flowers by Insects." Academic Press, New York.

Root, A. I. *et al.* 1975. "The ABC and XYZ of Bee Culture." Root Co., Medina, Ohio.

Taylor, R. L. 1975. "Butterflies in My Stomach: Insects in Human Nutrition." Woodbridge Press, Santa Barbara, California.

9

Negative Insect–Human Interactions

The title of this chapter might be interpreted in two ways. One is the negative effect that humans exert on insects, and of course this is extensive as humans modify the physical and biotic aspects of the world. For example, irrigation of drylands may aid humans (and the insect pests of irrigated crops), but at the same time it is often detrimental to the dryland insect fauna. The second interpretation of the title is the topic of concern in this chapter. It is the negative effect that insects have on human survival.

The human population needs certain requisites from the earth's resources in order to survive. These are a place to live, food of adequate quality and in sufficient quantity, a minimized effect from other organisms (predation, parasitism, disease, competition), and physical and chemical conditions within the range of tolerance for humans (see also Chapter 1). Of these requisites, the most important interactions between insect and human are (1) the direct competition with insects for resources, and (2) the attack of insects on humans, that is, the medical problems caused by insects. In the temperate climates of the world, especially Western European and North American areas, the current primary insect problems are competition for foods and other natural products. The more tropical climates have the increased threat of insect-borne diseases, a topic covered later in the chapter. As a consequence of the severity with which humans and insects interact we continue to try to minimize losses caused by insects. This effort usually falls within the scope of insect control, which will be covered in the next chapter.

COMPETITION FOR FOOD, FIBER, AND STRUCTURAL MATERIALS

As discussed in Chapter 3, insects are the super specialists of the animal world. Therefore, it should not be surprising to find that almost every plant and animal used by humans has insects that attack it. The few exceptions are marine animals and plants used by humans, but in these cases the place of insects is taken by their marine equivalents, the Crustacea. An example of the range of insect damage to a single useful natural product is the oak tree (Genus *Quercus*), which is useful for its high quality wood and as an ornamental plant. In addition, the acorns produced by oaks in the western United States were such an important food resource for the Indian tribes in that area that battles used to occur over prime oak stands. Table 9.1 lists the various parts of an oak or its stages of development, along with the types of injury, insect group involved, and the number of species of insects involved. The total, 208 species, is probably not unusual for an ecologically significant genus of plants such as *Quercus*. Other important plant species for which the insect fauna are well-known tend to be agriculturally important crop plants such as cotton, with a worldwide fauna of 1326 insect species that live on this plant. In the United States alone, about 100 species of insects and mites live on cotton. Worldwide, apples support about 500 phytophagous (plant feeding) insects, with over 100 species found in one single area studied (Wisconsin). A study of cabbage in Minnesota yielded 125 species living directly on cabbage. Alfalfa, a perennial crop, supports 312 herbivorous insect species in New York state, although on a worldwide basis only 43 herbivorous species are important enough to be considered alfalfa pests. In Chapter 10 we shall see that even those crops, like alfalfa, with many important pest species usually have very few in any single geographic region.

Damage to Plants and Structures

Insects reduce yields of crop plants or lower the value of the crops through contamination and reduced marketability. The insect activity most commonly responsible for this crop damage is the result of feeding by the insect.

Feeding Injury

Such insects as grasshoppers, caterpillars, beetles, and sawflies injure the plant when they remove pieces of plant material with their chewing mouthparts. Leaves of plants

Table 9.1 Insect Attack on Oaks in the Eastern United States

Plant part attacked	Type of injury	Insect group involved	Number of species on oak in U.S.
Acorn	Borers	Beetles and weevils	9
		Moths	2
Bark and phloem[a]	Borers	Beetles	9
		Moths	1
Bud and blossom	Chewers	Beetles	1
		Moths	2
Twig and shoot	Borers, sapsuckers	Beetles	3
		Aphids	1
		Scales	12
Roots	Borers	Beetles	3
Wood	Borers	Beetles	51
		Carpenter bee	1
		Horntail	1
		Moths	4
Leaf	Chewers	Beetles	20
		Butterflies and moths	57
		Grasshopper	1
		Sawflies	3
	Sapsuckers	Aphids	3
		Lacebugs	2
		Leafhoppers	4
Galls[b]	Various	Beetles	1
		Midges	3
		Wasps	14

[a]Vascular tissue of plants that transports organic nutrients
[b]Galls are abnormal plant growths.

so affected either show a ragged appearance such as that in Fig. 9.1, or a lacy appearance, known as skeletonizing, when the harder vascular tissue is left behind (Fig. 9.2).

The leaves of most green plants are the primary or sole site of photosynthesis. It is the leaf tissue that captures solar energy in the form of high-energy chemicals for use in the plant's maintenance, growth, and reproduction. Thus, any reduction in leaf surface area, such as occurs when pieces are cut off by an insect, reduces the total photosynthetic potential of the plant. A plant with reduced photosynthesis has less energy available for growth over and above its maintenance requirements. For plants in which the leaf is the commodity to be sold, as with lettuce, spinach, tobacco, or tea, leaf consumption directly reduces the amount and/or marketability of the crop.

Leaf chewing may also affect the nutrient quality of the remaining plant material, even if it is still marketable. For example, alfalfa is almost exclusively grown as forage—food for livestock in general—but more commonly only for dairy animals. The alfalfa is harvested when it begins to bloom (1/10th bloom is the usual level), thus assuring an optimal balance between percentage of protein content and fiber. If the plants become too mature, the digestability of the resulting hay drops, because of the increased fiber content, while protein content declines. If the plants are too early in the pre-bloom stage, not enough nutrients are present in the roots to ensure

Fig. 9.1 Chewing injury to potato leaves by the Colorado potato beetle.

regrowth after the tops have been harvested. Since fields in the Northern half of the United States are cut from three to four times each year depending upon location, maximum productivity at each cutting increases production efficiency while ensuring that the plants can overwinter successfully. Many alfalfa fields are kept in continuous production for 5 to 7 years before they are rotated to another crop. In warmer areas of the Southwest fields never go dormant as they do in the North and may be harvested

Fig. 9.2 A skeletonized leaf after tissues are removed by the chewing insect, leaving the more rigid vascular tissue behind.

eight times each year. When insects defoliate alfalfa by chewing, they not only remove photosynthetic surface, but reduce the protein content of the hay. Since alfalfa is grown precisely because of its high protein content, defoliation greatly affects hay quality.

Insects with piercing-sucking mouthparts can also feed on plant foliage, but because they withdraw the plant sap rather than pieces of tissue, the damage may be less obvious. Severe injury to leaf tissue results in a necrotic (dead) area that changes color. If the tissue is killed while the leaf is developing, the leaf form becomes distorted because some parts continue to grow around the affected area (Fig. 9.3).

Both chewing and piercing-sucking insects also attack buds and the tender young growth of plants. In fact, many insect species restrict themselves to new growth since it has more available nutrients than mature plant parts. Removal of even small portions of the buds and developing plant parts by chewing insects will result in severe damage to the remaining portion of the vegetation as it grows out. For example, the billbug is a weevil (snout beetle) that attacks corn both in the adult stage and as a larva. The adults (Fig. 9.4) chew on the tender center leaves of the corn plant while it is a seedling and, if the plant survives, the fully developed leaves show a characteristic transverse row of identically shaped holes in one or more leaves.

Insects with piercing-sucking mouthparts inject saliva into the tissue on which they are feeding. In many species, the saliva causes a toxic reaction by the plant. This reaction, in conjunction with the physical injury caused by the lancing of the tissue, may result in a killed or greatly distorted terminal (Fig. 9.5). The same kind of feeding on fruit may result in "catfacing" (Fig. 9.6). Those insects without toxic saliva may stunt the plant by massive feeding. The result is still an economic loss to the grower.

Insects that tunnel within living plants cause damage in several ways. Borers that feed just beneath the bark of living trees are often responsible for death of the tree or

Fig. 9.3 Curling and distortion of leaves as the result of feeding by aphids (left and center) compared with normal foliage at right.

Fig. 9.4 Foliar damage results from billbug feeding while the leaf is forming in the whorl. The larval billbug, a legless grub, bores in the corn stalk itself.

Fig. 9.5 Distortion of terminal growth by the tarnished plant bug, an insect which produces a toxic saliva. In this photo, all of the flower buds but one have been destroyed.

Fig. 9.6 ***"Catfacing" of fruit is a result of feeding injury by such insects as the tarnished plant bug,* Lygus lineolaris *(P. de B.). The affected area dies and corks over while the surrounding tissue continues to grow, causing a dimple to form.***

branch under attack. The area just beneath the bark is the cambium, a thin layer of cells that have the capacity to grow and multiply. These are the cells responsible for annual growth ring formation in trees, and because it is highly nutritious the layer is sought out as food by many borers. If the activity of the borers results in a complete encirclement of the cambial layer, the tree is said to be "girdled" and will die because nutrient transport has been blocked (see Fig. 9.7).

The tunnels of caterpillars in the stalks of corn seldom kill the plant outright because the transport tissues of monocotyledonous (grasses) plants are scattered in discrete bundles rather than in a cylindrical tissue as in conifers (needle-bearing plants) and in dicotyledonous plants (broad-leaved plants). The boring does damage the stalk structurally though and may weaken it so much that it falls over (lodges) in the wind or even when heavily weighted down by water. These "lodged" plants are impossible to harvest by mechanical means (see Fig. 9.8). After grubs of any one of the corn rootworm beetle species attacks the roots the corn plant is easily blown over, but may subsequently start to grow vertically again.

Wood boring insects cause structural damage in the trunks of trees that have

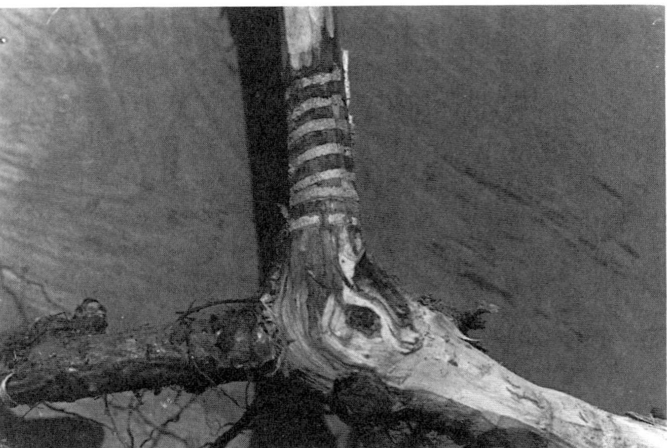

Fig. 9.7 Feeding by a beetle grub under the bark killed this young aspen. Nutrients cannot pass through the girdled trunk.

Fig. 9.8 Stalk boring, as by the European corn borer, often results in (A) breakage of the stalk or loss of the ear. (B) Lodged corn plants are lost when fields are mechanically harvested.

Fig. 9.9 Structural damage by a wood-boring beetle grub. In trees grown for timber, the resulting wood is of very low quality and strength and is not marketable as timber.

been grown for timber by reducing the strength of the resulting timbers, in most cases also thereby lowering the quality (Fig. 9.9). Anomalies exist, however, because although most people would not accept hardwood flooring that had beetle holes in it even though it had been kiln dried, they might pay a premium price for a "wormy chestnut" bureau. In fact, some wood workers "distress" furniture with fine shot from a shotgun to produce the desired effect.

Insects also provide an avenue of entrance for organisms that could not otherwise penetrate the plant under attack. Feeding injury or any other insect activity such as oviposition (egg laying) that disrupts the plant's exterior surface allows rot organisms such as fungi or bacteria in. Later in this chapter we shall cover some of the many associations between insects and the disease-producing organisms they often introduce with their feeding.

Egg-Laying Injuries

The kinds of plant injury discussed thus far show the range of ways in which insects attack the plant by feeding. Whether the result is catfacing, lodging, girdling, or skeletonizing simply depends on when, where, and how the feeding occurs. Another type of injury, but one that is not associated with the feeding process, is caused by the female when she oviposits. In some species, the female has a spear-, saw-, or sword-like ovipositor, which she uses to pierce the plant tissue, thereby making a slit or hole

Fig. 9.10 Ovipositor damage to a twig in the process of egg laying by a cicada.

into which the egg is then thrust. Figure 9.10 shows an example of such ovipositor injury. Figure 9.11 shows the characteristic crescent-shaped cut chewed into the apple by the plum curculio (a weevil). The crescent partially encircles the egg she has deposited through another cut chewed with her jaws. The crescent slows the growth of the tissue in the vicinity of the egg.

Eleven Major Agricultural Pests in North America

It is logical at this point to survey the serious pests of our major agricultural crops and to discuss their ways of living. However, restrictions on space force us to condense a potentially lengthy discussion of major agricultural crop pests into two tables. The species that are major agricultural threats are presented in Table 9.2. Table 9.3 lists the groups of general feeders among crop pests. It also includes aphids and scales, two groups that contain species that are often very host specific. They have been included because each group contains large numbers of pest species that, collectively, cover a wide variety of agricultural crop plants.

Fig. 9.11 (A) Crescent-shaped wounds to a developing apple inflicted by a plum curculio. The egg is laid into the fruit tissue through a slit chewed by the weevil. Then she chews the crescent-shaped wound which retards growth in the area of the developing young. (B) Damage later in the season, when the fruit is nearly mature.

Table 9.2 The Insect Pest Species That Constitute a Major Threat to North American Agricultural Crops

Common name	Scientific name	Crop	Damage
European corn borer	*Ostrinia nubilalis* (Hüb.)	Corn, vegetables	Tunneling larva weakens stalks; ears drop off; stalks break off
Corn earworm	*Heliothis zea* (Boddie)	Corn	Chewed kernels at end of ear; ragged leaf whorls
Tomato fruitworm	*Heliothis zea* (Boddie)	Tomato	Tunnel in developing fruit
Cotton bollworm	*Heliothis zea* (Boddie)	Cotton	Each larva tunnels in 1–12 bolls
Pink bollworm	*Pectinophora gossypiella* (Saunders)	Cotton	Buds are bored into; tunneled bolls fail to yield lint; seed is damaged, reducing oil content
Cotton boll weevil	*Anthonomis grandis* Boheman	Cotton	Grub lives within the boll, destroying the lint
Tobacco hornworm	*Manduca sexta* (Linnaeus)	Tobacco	Larvae defoliate the plant by chewing
Tobacco budworm	*Heliothis virescens* (Fabr.)	Tobacco	Larvae attack buds, thereby causing the resulting leaves to be ragged and of less value
Cabbage looper	*Trichoplusia ni* (Hüb.)	Cabbage family, many other vegetables	Defoliates and contaminates vegetables
Codling moth	*Laspeyresia pomonella* (Linnaeus)	Apple, pear, walnut	Larva destroys fruit by tunneling in it
Sugarcane borer	*Diatraea saccharalis* (Fabr.)	Cane, corn, rice, sorghum	Tunneling larva weakens stalk; reduces yield; in cane, purity of juice is also impaired

Table 9.3 Important Insect Pest Groups That Are General Feeders

Insect group	Crop	Damage
Grasshoppers	Corn, small grains, alfalfa, many others	Defoliate, devour various parts of plants
Cutworms, armyworms	Most crops	Defoliate, cut off at soil surface
Wireworms	Corn, small grains, potatoes, vegetables, root crops	Devour germinating seed and subterranean parts
White grubs	Grasses, grains, most cultivated crops	Devour germinating seeds and subterranean parts
Fruit flies	Most fruits	Larvae feed in fruit, destroying it and fouling the remainder
Plant bugs	Cotton, seed crops, fruits	Toxic symptoms at feeding site; loss of bud or seed at site of feeding
Aphids[a]	Almost all crops	Suck sap, foul with honeydew, transmit diseases
Scale insects[a]	Perennial crops	Suck sap, foul plant with honeydew

[a]Groups with many different species that collectively attack a wide variety of agricultural crops. Most species in these groups are quite host specific.

Garden Pests

The problems created by insect pests of gardens, both vegetable and fruit, and of ornamentals are similar in several ways. First the plantings of any single plant species are usually few. Second, the individual plant has high value to the grower.

Gardens are often characteristically limited to a single row of plants (e.g. tomatoes) or even one plant (e.g. parsley) of a given species. Ornamentals, especially woody perennials, are even more commonly planted singly. This means that garden areas have a high diversity of plant species and that host-specific insects may have some difficulty in locating their food plants. Unfortunately for the plant (and for the gardener) most of the pests of annual plants are insects that have evolved highly effective colonization characteristics. A part of this colonizing ability is the insect's small size, which contributes to its ability to become widely dispersed by winds. Coupled with this ease of dispersal is the very high reproductive capacity typical of these colonizers. The type of insect just described, termed an r-strategist by evolutionary ecologists, has an advantage when it comes to locating and exploiting new resources, such as an isolated, but acceptable food plant. These insects do relatively well on annual crops, but not so well on perennials. The r-strategist is not very effective when it has to compete with other insects for limited resources. On the other hand, k-strategists, the insects that usually live on perennials, tend to be poor colonizers, but are more likely to be highly efficient competitors. Thus, annual plants may not escape attack even the first year they are planted. Perennials may not have many pests the first few years, but eventually they are discovered by pests and may end up supporting several species of pests.

Mostly, however, each plant species typically has its own array of pest insects, many of which are fairly host specific. In addition, several groups of general feeders cause serious problems in gardens and ornamental plantings.

Although in some cases home vegetable gardens are used to produce foods inexpensively, most gardening efforts fall into the hobby category. Indeed, if a gardener were to calculate the costs of seed, fertilizers, pesticides, and tools as compared to the yields achieved, the cost of home-grown vegetables and fruits would be high. This would be even greater if real costs, such as labor, land taxes, and crop quality were considered. However, hobby gardeners are usually involved in the endeavor basically for the enjoyment and thus, will go to considerably more effort and expense to protect their plants than would be feasible on a strictly economic basis. Among the most common and widespread pest groups are those presented below.

Subterranean Beetle Larvae. Seeds, seedlings, and transplants of many crops may be so severely damaged by wireworms (Fig. 9.12) and white grubs (Fig. 9.13) that few plants become established. These insects are especially numerous in gardens that have been newly created from land previously in sod.

Foliage-Feeding Caterpillars. The larvae of various moth and butterfly species are most often restricted to one plant host or to a group of closely related plants. Some species, like the cabbage looper previously mentioned, can attack many vegetable

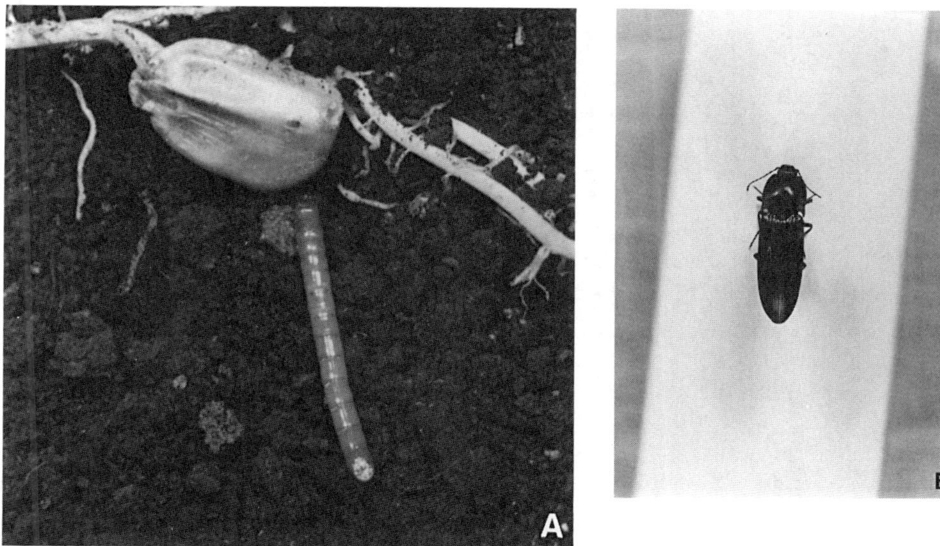

Fig. 9.12 (A) A wireworm larva and (B) the adult beetle (Family Elateridae) known as the click beetle because of its ability to fling itself into the air by flexing its body.

crops. Another group, the cutworms (Fig. 9.14), contains many species among which are some that attack crops, often cutting off the young plants at ground level.

Root Maggots. Plants attacked by root maggots first appear to be stunted, then wilt and die. The legless maggots attack the root system by tunneling in it and

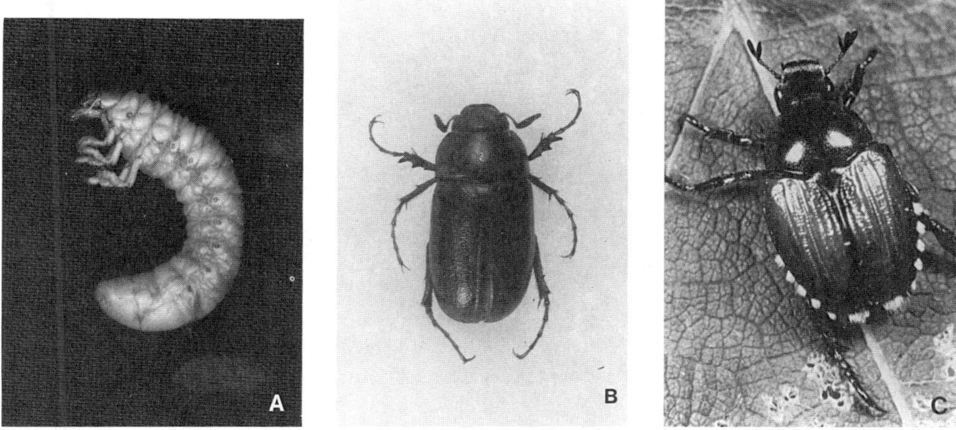

Fig. 9.13 (A) The larvae of the beetle family Scarabaeidae are known as white grubs. (B) The adults are scarabs, but are also known as June beetles. One common example is the Japanese beetle (C).

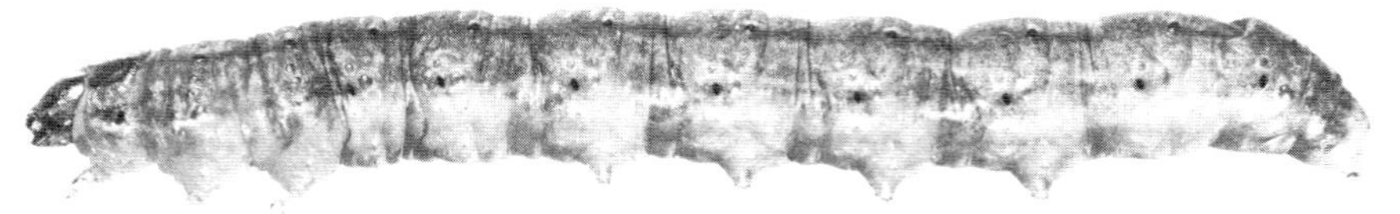

Fig. 9.14 Cutworms are nocturnal, so they can only be found during the day curled in a tight coil in the soil or litter around the base of the plant. Photo courtesy of R. W. Rings, Ohio Agricultural Research and Development Center.

introducing bacteria that cause the plant tissue to rot. Adult flies favor plants growing in soils with high organic content. They lay their eggs on the plant itself or on the soil near the roots. Cole crops (cabbage, broccoli, radishes, and cauliflower, etc.) beets, celery, onions, spinach, corn, beans, peas, and seed potatoes may be attacked by one or more of the various root maggot species.

Foliage-Feeding Beetles. Although the Japanese beetle (Fig. 9.13c), rose chafer, and several other scarabs (June beetle types, Fig. 9.13b) feed on leaves, the most destructive beetles injuring foliage are the leaf beetles (Family Chrysomelidae, Fig. 9.15). This group contains the small jumping beetles (flea beetles) that chew small round "shot holes" in leaves as well as cucumber, asparagus, grape, and potato beetles that feed as adults on the foliage of these crops. In addition, a number of species feed on foliage both in the larval and adult stages, the best example, perhaps, being that of the Colorado potato beetle, *Leptinotarsa decimlineata* (Say). It feeds on many species of Solanaceae (potatoes and tomatoes) but seems to prefer eggplant over the other members of this plant family.

Another common foliage-feeding beetle of interest is the Mexican bean beetle (Fig. 9.16). This is the only phytophagous lady bird beetle of any economic significance in North America. The other economically important "ladybugs" all are predators of small soft-bodied plant pests such as aphids, scales, and mites.

Pests of Trees and Shrubs

Many of the pest groups so common on garden crops are to be found on herbaceous annual ornamentals. However, the woody perennial ornamentals develop severe pest problems that differ from those previously mentioned and are more typical of the kind of insect problems found in forest trees. On the other hand, specimen ornamental perennials, such as trees, have certain other characteristics that make them rather unique with respect to insect damage.

Ornamental perennials, especially trees, have a high dollar value, in part due to the aesthetic value they add to the property on which they are located. They also increase privacy by creating barriers between neighboring building sites as well as

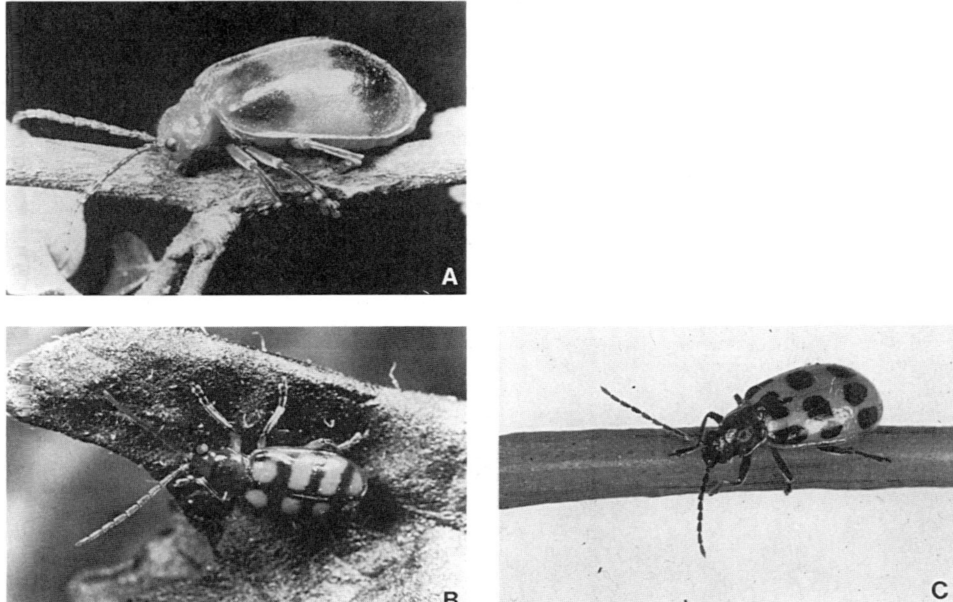

Fig. 9.15 Leaf beetles of the Family Chrysomelidae. (A) Larger elm leaf beetle, (B) banded cucumber beetle, and (C) 12-spotted cucumber beetle.

Fig. 9.16 The Mexican bean beetle is the only plant-feeding "ladybug" of any economic significance in North America. Ladybugs are usually considered carnivorous types that feed on aphids and other soft-bodied insects.

providing shade. In addition, trees, especially if they are relatively large, add to the feeling of permanence in one's home.

Other aspects of ornamental tree value are more pragmatic than those mentioned above. Once a tree is dead, whether killed by insects or some other cause, it has to be removed because of the potential for damage its uncontrolled fall could cause. Professional felling and removal of large trees may cost from $300–1200, depending on the size of the tree, its location with respect to hazards (such as wires, structures), and the region of the country. Therefore, property owners would be wise to protect their trees from severe injury if only to keep from having to pay for their removal.

Not only is the cost of tree removal high, but most property owners opt to replace a lost tree with another perennial, often another tree, adding to the cost. Therefore it is prudent for property owners to be aware of the possible problems that may occur. However, the very size of large trees often complicates insect pest problems. First of all, it is difficult to see problems and to assess their degree of severity when the problem is occurring well above a person's head. Because the canopy of large trees cannot be inspected closely without special equipment or effort, the problems often become very severe before they are noticed. In addition, even after the problem has been identified, specialized equipment is often required for such control procedures as spraying with pesticides.

The insect pests that are most common on woody perennial ornamentals, including trees, are borers, bark beetles, defoliators, and scale insects.

Scales. Because they are small, immobile, and not "insect-like" in appearance (Fig. 9.17) these sap-sucking pests can become so numerous that they kill the plant. Homeowners often bring scale-laden branches to county and university extension offices for analysis. The scales have not even been noticed by the owner, even while

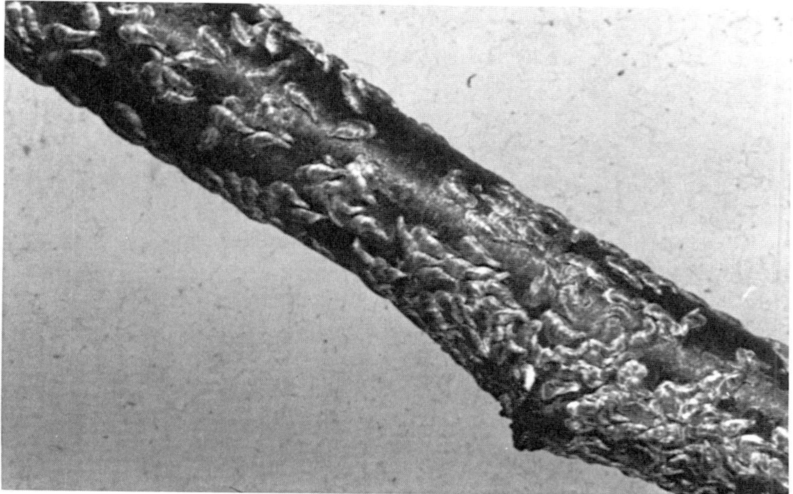

Fig. 9.17 The lack of typical insect characteristics—legs, wings, antennae, or body regions—along with their small size and obscure coloration, often accounts for oystershell scale being overlooked as a cause of plant damage.

handling the specimen. Some scales, like the San Jose scale on lilac, are so similar in appearance to the bark itself that specimens are often submitted so badly infested that only scales are visible externally, having covered every bit of the bark. Individual scale insects do little harm, but in large numbers they remove so many plant nutrients as to stunt or kill branches or entire plants.

The first instar scale nymph hatches from the egg, or is born alive. This is the dispersal stage for this type of insect—only after considerable traveling does it settle down on an appropriate host, insert its mouth parts, and molt. The "hard" scales lose their legs at the molt and stay put, developing a thick waxy covering over themselves. The "soft" scales have a rigid exoskeleton rather than the waxy covering. They retain greatly reduced legs and thus are capable of movement throughout life. Females remain in their scale form even at maturity, but males take on a winged, gnatlike appearance as adults and fly about actively seeking females for mating.

Defoliators. Three groups of insects are primarily responsible for the defoliation (foliage removal) of ornamental perennials. They are the caterpillars of moths and butterflies (Order Lepidoptera) and sawflies (primitive Hymenoptera), and the Coleoptera, both as larvae and adults. The apparent effect of defoliation is the loss of leaves or portions of them, with a ragged or tattered appearance and a scraggly canopy. Because of this loss of aesthetic value, even if it is only temporary, many home owners seek a way of solving the problem as rapidly as possible. In some cases action is warranted, but it depends upon the amount of defoliation and the plant being attacked.

Most broad-leafed perennial ornamentals are deciduous. That is, they lose all of their leaves at the end of the growing season and develop entirely new foliage the next spring. There are many exceptions to this rule, such as rhododendron, laurel, some azaleas, camillias, and citrus, but the pattern is widespread in the temperate zones of the world. A single complete defoliation of most broad-leafed ornamental perennials does them little apparent harm. If it occurs early in the growing season, the plant usually will produce another set of leaves. If the defoliation occurs later in the season, the leaves may not be replaced, but they will have served much of their usefulness by then anyway. The results of defoliation show up in the growth rings of the tree too, with an abnormally small growth ring during years of defoliation, or in immediately subsequent years, with gradual recovery stretching over the next several years.

Conifers, the needle-bearing trees, and their related ornamental forms react to defoliation in a more spectacular way than do many broad-leafed deciduous trees. Whereas complete defoliation may only stunt a broad-leafed form, complete defoliation of most conifers, unless quite late in the growing season, usually results in the death of the tree. Likewise, a completely defoliated branch of a conifer does not produce needles again unless the current year's growth is far enough along to have bud formation. Thus, the results of severe defoliation in most conifers are much more serious than in broad-leafed deciduous ornamental perennials.

Bark Beetles. The dry surface bark of most trees is a relatively inert material of poor nutritional quality. The so-called bark beetles of perennials usually feed on the cambium and vascular tissues just beneath the bark, often killing the tree by girdling it.

Initial attack on a tree is by the adult bark beetle. Healthy trees, such as the conifers used for most lumber production, resist the attack of bark beetles by means of the oleoresins (pitch) they release upon being injured. The pitch tends to flood out, drown, or entangle the beetles as they attempt to bore galleries into the tree. The same species of tree under suboptimal conditions is much more susceptible to attack. Thus, a tree weakened by drought, age, defoliation, mechanical injury, or fungus loses its ability to ward off bark beetle attack.

Many bark beetle species are called secondary invaders because they normally are capable of attacking only weakened trees. Those capable of overwhelming healthy vigorous trees are called primary invaders. Bark beetles bore a nuptial cell in the host tree and the female (of monogamous species) or females (of polygamous species) bore elongate egg galleries, placing eggs along the sides. The larvae hatch out and begin to feed. As specific as bark beetles are in the type of host attacked and in the basic form of egg galleries within a species, the pattern of larval feeding is as consistent, but varies between different species. Such differences are sufficiently consistent to use in identifying pest species by host and gallery, without actual examination of the insect involved.

The larvae of several species of bark beetles feed in a group, as in *Dendroctonus valens* Lec., the red turpentine beetle, but most larvae make individual larval galleries at right angles to the egg gallery. The larval galleries widen progressively as the larvae increase in size (Fig. 9.18). Pupae are often found in individual cells close to the tip of each gallery and the callow (new, soft, unsclerotized) beetles remain in the gallery system until they mature. In fact, they often pass the winter in their larval gallery system, emerging the next spring to initiate the next generation.

The most important types of borers are beetles (Order Coleoptera), moths (Order Lepidoptera), and horntails (Order Hymenoptera). Beetles and horntails can bore in wood in both the larval and adult stages because both have chewing mouthparts. Only the moth larva chews wood; the adult has siphoning mouthparts suited only for consumption of liquids. Although many horntail species are general feeders, that is, each attacks a wide variety of host plants, moths and beetles are more variable in their host range.

The leopard moth, *Zeuzera pyrina* (L.), attacks a wide variety of hardwoods, preferring elm, maple, ash, oak, apple, pear, and plum, but also attacks at least 25 other species of trees. Other moths are restricted to one or a few hosts; for example, *Thamnosphecia americana* (Beut.) attacks alder, *Podosesia fraxini* (Lugger) attacks ash, and the lilac borer, *Podosesia syringae* (Harr.) attacks mountain ash and ash in addition to lilac. The beetles show the same range of specialization. Some, such as the genus *Goes*, attacks many species, others are so restricted that they attack only a single host species (*Megacyllene robiniae* (Forst.) on black locust).

The attack of insects, particularly the chewing insects that bore in the growing tips and buds of plants, often modify the shape of the plant. The terminal (or uppermost) bud of a plant tends to dominate lateral bud growth so that the most vigorous development is at the tip. Thus, gardeners who wish to produce short, dense plants do so by pruning off the terminal bud or buds. This allows the lateral buds to develop. Insects can cause the same alteration in plant growth form by injuring the

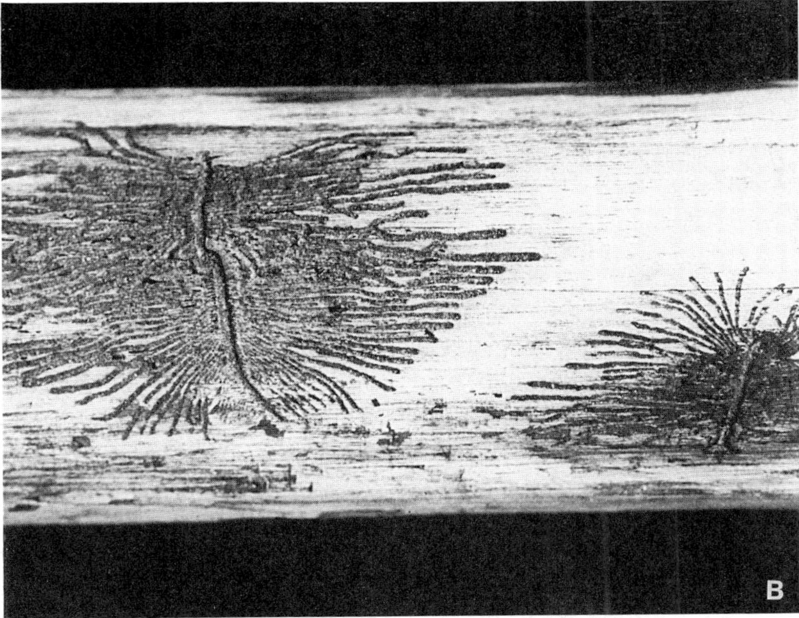

Fig. 8.18 (A) The adult beetle (only 2–3.5 mm long) and (B) a single gallery system of the native elm bark beetle, **Hylurgopinus rufipes** *(Eich.). Note that in this species the egg gallery tends to run across the grain or at a slight angle and larval galleries run with the grain.*

growing tip. A common example is the terminal destruction by the white pine weevil, *Pissodes strobi* (Peck) (Fig. 9.19).

Borers can attack ornamental plantings girdling the plant or weakening the plant structurally, thereby enhancing breakage under heavy winds. As we shall see in a later section of this chapter, they also provide an avenue of entry for disease and rot organisms. Another problem with borers is their affinity for newly killed timber being cut for firewood. They often emerge from the firewood in the house or garage where the wood is being stored causing considerable concern. Some of the species are relatively large, and their chewing can be heard from several feet away. When they emerge indoors the concern is whether or not the wood of the home and the furnishings is under attack. For the majority of these insects none of the dried structural wood in a home is suitable as a breeding site. Most of these insects require freshly killed and relatively moist wood, and are, thus, unable to survive in seasoned wood.

Pests of Structural Wood

It is easy to understand why a person might be concerned about introducing a wood boring insect into the home. After all, most people know that some insects attack wood, even if they have only heard of termites. Such concern is well-justified, considering the range of insects that do attack dry, seasoned wood—wood used in furniture and structural wood of the house. Omitting the minor groups (old house borer,

Fig. 9.19 Plant deformation by insects. (A) The terminal of a white pine killed by larval feeding of the white pine weevil, Pissodes strobi *(Peck.). (B) Growth of the lateral buds often results in a pine with more than one bole, and greatly reduced timber value.*

carpenter bees, and timberworms, plus a few others) the major pests are termites, carpenter ants, and powderpost beetles (see Fig. 9.24).

Termites. These colonial insects are the most important pests of seasoned wood in the temperate zones of the world and are even more prevalent in the tropics. North America has two basic types of these colonial insects: subterranean and damp (or dry) wood termites. Subterranean termites are so called because their colony is in the soil; only a portion of the colony feeds on the wood that is their basic food. Damp or dry wood termites are those that live in the wood and need not maintain contact with the soil. The subterranean termites are more widely distributed, extending over almost all of the United States and into some parts of Canada (see Fig. 9.22). The damp and dry wood termites are restricted to the southern states, extending from about North Carolina across the Gulf states into California.

All species of termites are colonial and have a complex social order, a subject covered elsewhere in this text. The significance of termite social behavior, in economic terms, is that they seldom occur in small numbers. Thus, if a few termites and some damage are found in a building, there is a good chance that many more are present and that there may be severe damage in other parts of the structure.

Termites often make their presence known in a building when the colony produces a swarm. The termites issuing from the colony are winged reproductive forms whose role is to establish new colonies. The wood damage is done by workers that are seldom visible.

Termites consume cellulose in a variety of forms, such as paper, plants, books, and cardboard. However, the most common cellulose material is the wood used for structures (Fig. 9.20). Attacks often result in wood with a hollow core, that causes collapse of stairs, walls, or porches.

Termite damage to wood is so common that most property transfers require termite inspection (Fig. 9.21). These inspections usually involve testing of the soundness of wood in suspect areas and searches for other signs of infestation. For example, when wood is not in direct contact with soil, subterranean termites build earthen tubes across the intervening material. The earthen tubes provide access from the colony to wood without exposure to the external environment, thus maintaining the high humidity necessary to the survival of these insects. Most modern buildings are so designed that wood does not come into contact with the soil for the very purpose of thwarting termite attack. However, termites can build tubes across or can travel up through cracks in the foundation or when a piece of lumber is inadvertently left in a cement foundation form.

Carpenter Ants. Colonies of carpenter ants are commonly encountered living in galleries excavated out of wood. This insect is not as destructive as is the termite. Whereas termites consume wood and derive their energy from the breakdown of cellulose, the carpenter ant consumes other insects, the succulent parts of growing plants, and the sugary fecal matter from aphids known as honeydew. Carpenter ants chew out the network of galleries in wood simply to create a colony site. The wood that is removed is deposited outside of the nest. Thus "sawdust" is coarse and granular, and also contains fragments of insect skeletal remains.

***Fig. 9.20** (A) The earthen tubes constructed by subterranean termites across a foundation wall to connect the colony with wood not directly contacting the soil. (B) Damage to a pine 2 × 4 by termites. (C) Extensive termite damage to a home.*

Even though carpenter ant attacks are less serious than those of termites, it is usual to try to eliminate the colonies. In fact, few people would tolerate these very large ants in the house. However, the problem is locating the colony. The ants are more active at night than during the day, and they tend not to use the same trails, as is the habit of many other ants. This, apparently sporadic appearance and "wandering" make it difficult to locate the colony, especially if the sawdust is being deposited between wall studs or in some other undetected spot.

MINUTEMAN PEST CONTROL Co. Inc.

196 N. Pleasant St.
Amherst, MA 01002
Amherst 413-549-1800
Northampton 413-586-1009

Termite and Woodboring Insect Report

Date of inspection: July 30, 1983

Inspector: Alfred Jones   Lic. No. 02506

Inspection requested by: John Smith

Location of property inspected: Owner: John Smith
Address: 3 Maple Street, Amherst

What types of buildings are included: House only

Type of inspection:
(x) Real Estate ( ) Conventional ( ) Transfer ( ) Other ______

Were there any signs of live termites or woodboring insects? If yes, what and where? Yes ( ) No (x)

Were there any signs of damage by termites or woodboring insects Yes (x) No ( )

Signs of damage from termites to basement sills and joists at several locations.

Were there any obstructions to a visual inspection, either inside or outside of buildings being inspected? If yes, what type and where Yes ( ) No (x)

Recommended Treatment or Remarks: It is our recommendation that house be treated for termite infestation.

Please Read Carefully:

This inspection was made only to determine VISIBLE evidence of the presence or absence of noted organisms. It is made only in those areas of noted structures which were readily accessible and visible. Inspection has been made in the areas in which infestations are most likely to occur.

No inspection was made in inaccessable areas which might require breaking into, breaking apart, dismantling, removal of an object, including but not limited to moldings, floor coverings, insulation in basement & attics, wall coverings, siding, ceilings, floors, furniture, appliances and/or personal possessions.

THIS IS NOT A STRUCTURAL DAMAGE REPORT, neither is it a **warranty** as to the absence of wood destroying organisms.

Neither Minuteman Pest Control Co., Inc. nor its employees claim expertise in either building construction or structural engineering. If a qualified opinion is desired as a determination of the extent of damage or structional integrity of the structure we recommend you contact a qualified representative of the appropriate building trade.

No liability will be assumed and by receipt of this inspection report, holder acknowledges same.

MEMBER NATIONAL PEST CONTROL ASSOCIATION   Protectors of Health and Property

Date: ______________

Authorized Signature: ______________

Fig. 9.21 A typical termite inspection form, often required in transfer of real estate.

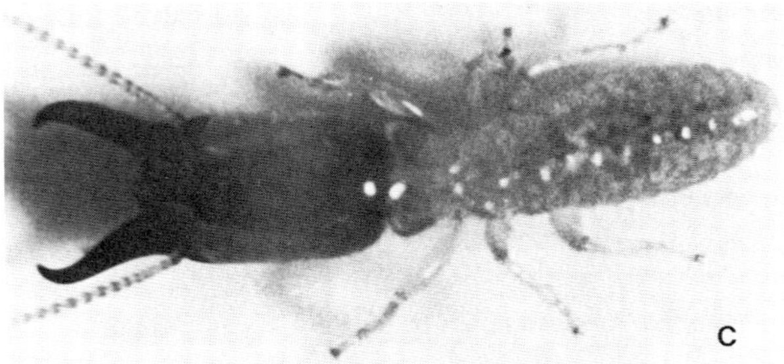

Fig. 9.22 (A) A winged reproductive termite, typical of the forms that periodically swarm out of the colony. Note the length and similar venation of the four wings. (B) The worker and (C) soldier are seldom seen unless the colony or the wood on which it is working is opened.

Carpenter ant colonies, like those of termites, periodically produce swarms of winged reproductives. The mated females locate moist rotting wood in which to start their new colony. They frequently attack wooden window sills, stairs, and other structural wood, especially in areas where the wood gets soaked with rain.

Other Pests in Homes and Commercial Establishments

The insects included in the previous section are some of the most devastating and costly insect pests that can occur in a building. However, that group is restricted to those that attack cellulose, primarily in its structural uses. If all of the other species that cause problems in houses, apartments, or commercial buildings were enumerated, the list would number well over 100. We will only discuss the three groups most representative of the common household pests.

Roaches

Although there are at least 2000 species of roaches worldwide, only a half dozen species are important in temperate zones of North America as domestic pests. Even

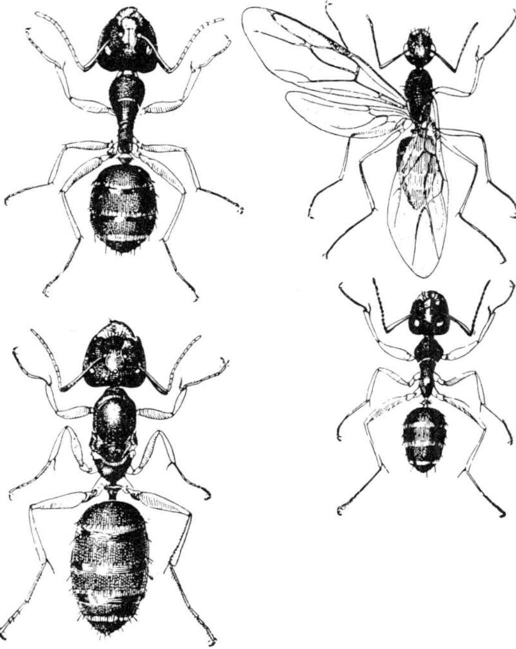

Fig. 9.23 The winged male carpenter ant (upper right) and larger de-winged female (lower left). The two other ants are workers (always wingless). Note both the relatively short length of the wings, as compared to the termite in Fig. 9.22A, and the prominent "wasp waist" or abdominal construction that distinguishes these insects from termites.

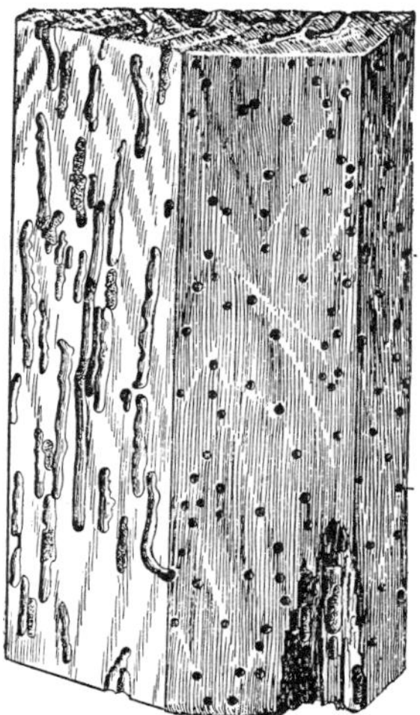

Fig. 9.24 Adult powderpost beetles emerge from holes in the wood after having completed larval development and pupating. The powdery frass (solid larval excrement) may fall or be dislodged from the emergence holes if the wood is jarred.

so, these few species are unquestionably the most significant urban insect pests. They rank along with the rat as the prime targets for the majority of urban pest control firms.

Roaches are general feeders, that is, they can consume a wide variety of materials. Open packages of foods in a kitchen, scraps in a trash can, exposed dishes of pet food, as well as crumbs and other food debris, even in small quantities, sustain roach infestations once they have gained entrance into a dwelling. In the more southerly regions, roaches may move from building to building by themselves, often living outside of buildings for most of the year and moving indoors with the onset of cold weather. In the more temperate regions these pests are essentially restricted to structures. Central heating of structures provides pathways for expansion of range by these species, and sewers, heating tunnels, or other connections between buildings are used by roaches to move into uninfested areas.

Roaches often gain entry into structures within packages or other objects being transported. The female glues the beanlike oothecae (egg cases) she produces into cracks and crevices. Thus, the seams of corrugated cardboard boxes, undersides of chairs and drawers, or other objects may contain one or many egg cases. Since a population of five pairs of roaches can give rise to 50,000 within a 7 month period, one egg case may be enough to establish a severe infestation.

Cockroaches are pests, but not especially harmful in themselves. The greatest problem is one of food contamination. This may occur in two ways. (1) The roaches may feed on human fecal material at one time and then move to human food supplies, thereby transmitting filth and with it the possibility of disease organisms. (2) The roach disgorges partly digested foods at intervals and defecates wherever it is. These body excretions have a strong and disgusting odor that is characteristic of cockroaches. Thus, contaminated foods or other products acquire a disagreeable odor and become soiled.

Populations of cockroaches often become unbelievably large in apartments or restaurants where it is difficult to detect and control them. This may be partially attributed to the secretive nature of cockroaches. They are nocturnal and spend the daylight hours in hiding. Since the early instars are small, as many as 50 may hide in a straight-backed chair. Then, once the lights go out, the cockroaches emerge.

Although the adults of most cockroaches have fully developed wings, they seldom fly. Both nymph and adult stages run very rapidly and possess very long, constantly moving antennae. The body is rather flattened, with the coloring of the pest species ranging from light brown through chestnut to black (Figs. 9.25, 9.26, and 9.27).

Pantry Pests

In the prior discussion of roaches, their ability to infest stored products was mentioned. They are not, however, the usual, or most serious type of insect that infests stored foods. Not only does their size prevent access to relatively well-packaged foods, but they require a constant and abundant supply of water, a factor often missing from areas where foods are stored. Thus, the insects that can exist in stored foods

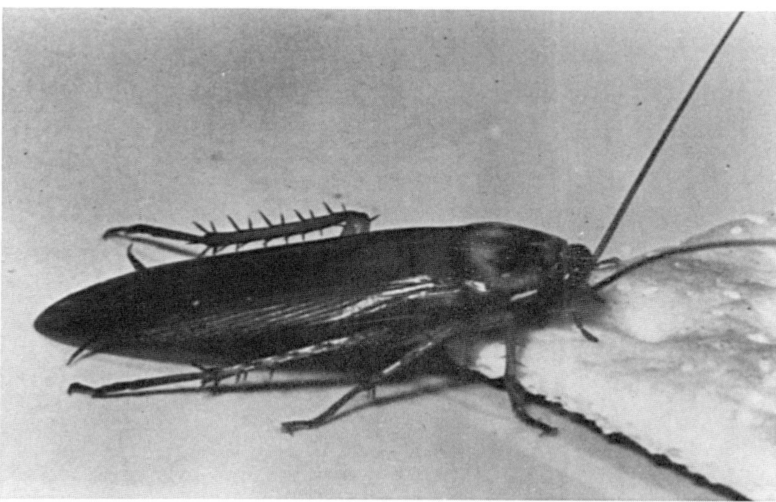

Fig. 9.25 The American cockroach, **Periplaneta americana** *(L.). As well as being a domestic pest, this species is important as a laboratory animal used for a wide variety of experiments in physiology and toxicology.*

without the need for free water are the more usual pantry pests. They enter packages prior to their sealing or burrow in after the package is in storage. Of the many species that act as pantry pests, two basic groups stand out. They are the flour-grain beetles and meal moths.

Flour-grain beetles, like the bean, rice, and granary weevils, attack whole, un-

Fig. 9.26 Only the male of the oriental cockroach, **Blatta orientalis** *(L.), has fully developed wings. In the female, wings are similar in appearance to the wing pads of a mature nymph.*

***Fig. 9.27** The adult German cockroach,* **Blattella germanica** *(L.), is much smaller than either the oriental or American species. It has two longitudinal stripes behind the head to distinguish it from the similarly sized and shaped brown-banded cockroach,* **Supella longipalpa** *(Fabr.), which has a single irregular blob.*

broken grains. Others, like the very minute saw-toothed grain beetle, *Oryzaephilus surinamensis* (L.) (Fig. 9.28), and flour beetles, *Tribolium confusum* Duval and *T. castaneum* (Herbst.) require broken or milled grain products for their maintenance. Still others like the drugstore beetle, *Stegobium paniceum* (L.), attack dried spices and the cigarette beetle, *Lasioderma serricorne* (Fabr.), attack cigars and tobacco in other forms.

***Fig. 9.28** The saw-toothed grain beetle and its larvae may infest almost any dried stored foods. Its very small size and flattened body allow it to work its way into packages that appear to be tightly sealed.*

Meal moths attack grains, usually after they have been milled. Of the several species that may be involved, the Indian meal moth, *Plodia interpunctella* (Hbn.), is probably the most common. This bronze moth with coppery wing tips (Fig. 9.29) often enters houses as larvae within stone-ground flour or in bags of pet food. Since these products are usually purchased in large quantities and opened bags are left in out-of-the-way places, infestations are allowed to grow and spread to cereals, crackers, and similar materials.

Dermestid Beetles

One of the most common insect pest groups encountered in dwellings is the beetle family Dermestidae. This large group of insects has species that attack various materials, primarily of animal origin. Thus, the common names of many species include the terms skin, hide, carpet, or buffalo, along with the term beetle.

The dried animal materials favored as food by dermestids include ham, bacon, hides or stuffed animals, felt, wool carpeting, and clothing, such as wools, or fleece-lined gloves and boots. Several species are the major pests of materials in museums, since they attack insect specimens as well as other preserved animal materials. Their efficient utilization of dried animal remains is employed by many major zoological units where colonies of dermestids are maintained to aid in cleaning up animal skeletons prior to study. A large colony can strip a carcass of all soft body parts in less than a month, leaving only the clean, odor-free skeleton behind.

Fig. 9.29 The Indian meal moth is a common pest of stored grain products. The larva is the injurious stage.

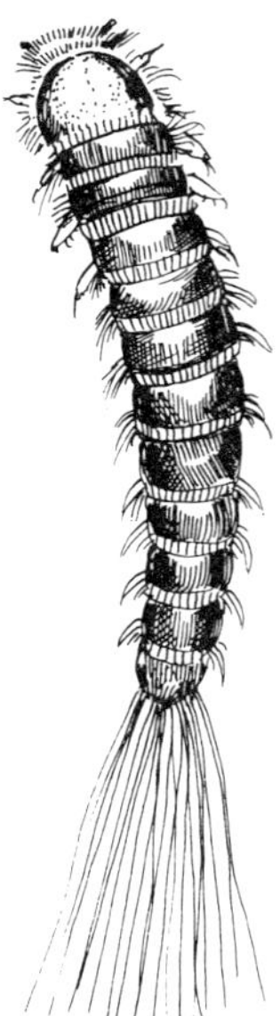

***Fig. 9.30** The small, slow-moving black carpet beetle larva is distinctive because of both its tuft of hairs at the posterior end of the body and its cinnamon-brown color. The adult beetle is black and a little smaller and narrower than a lady bird beetle.*

Dermestids can be found in almost every dwelling. Not only do they gain entry when infested materials are brought in, but many of the species occur outdoors and the adults can be found on flowers. The easiest place to locate dermestids indoors is under refrigerators and stoves. Since these appliances are heavy and awkward they are seldom moved. This, in combination with the dropped foods that inevitably accumulate there, provides an ideal environment for a culture of dermestids to develop. The most common evidence of an infestation is the slow moving larval stage (Fig. 9.30) or the shed larval "skins."

PESTS OF ANIMALS AND HUMANS

In this section we cover those insects that attack, annoy, or live on various vertebrate animals, including humans. While the numbers of species involved may be smaller than the group that feed on plants, the numbers are still substantial, and in terms of economics, at least as significant. A great part of the significance results not from the direct attack and subsequent weakening of the animal or human, but because many of these types of insects transmit disease organisms from one animal to another. This last aspect will be discussed in the final section of this chapter.

Insects That Attack Humans

The first group to be covered are those that attack animals and humans not because of a need to derive food or a place to live, but to defend themselves, or their home range (nest, colony) from intrusion and possible destruction. Here we refer to such insects as ants, honey bees, wasps and other stinging Hymenoptera, as well as some tropical termites. Also included are insects that defend themselves when approached or handled by biting, pinching, stabbing, or scratching the offender.

Stinging insects are distinguished from biting insects by how they inflict injury. Biting insects use their mouthparts to pinch or stab; a cricket could inflict a painful nip with its large chewing jaws. The mosquito pushes its long sharp stilettolike mouthparts into its victim, actually poking a hole through the skin. Most stinging insects use their egg laying appendages as defensive weapons. There are a few insects that possess stinging hairs, which break off at contact and inject irritating chemicals or mechanically irritate like fine glass fibers do. The stinging hair defense is most common among caterpillars (Fig. 9.31). With these few exceptions then, most stinging insects are found in the Order Hymenoptera, a group that has evolved into a number of life styles associated with various uses of the female's ovipositor (see Chapter 11).

The sting apparatus of Hymenoptera is possessed only by the females since it is a modified ovipositor. Associated with the physical stabbing structure is a venom-producing structure that may consist of one or more glands (see Fig. 9.32). Thus, the sting of a wasp is a dual operation in which the victim is first punctured by the sharp, hollow stinger, and then venom is injected into the interior of the wound by muscular contraction, similar to the action of a hypodermic needle.

A sting almost always causes considerable pain, due in part to the physical puncture, but mostly because of the venom. The chemical composition of venom

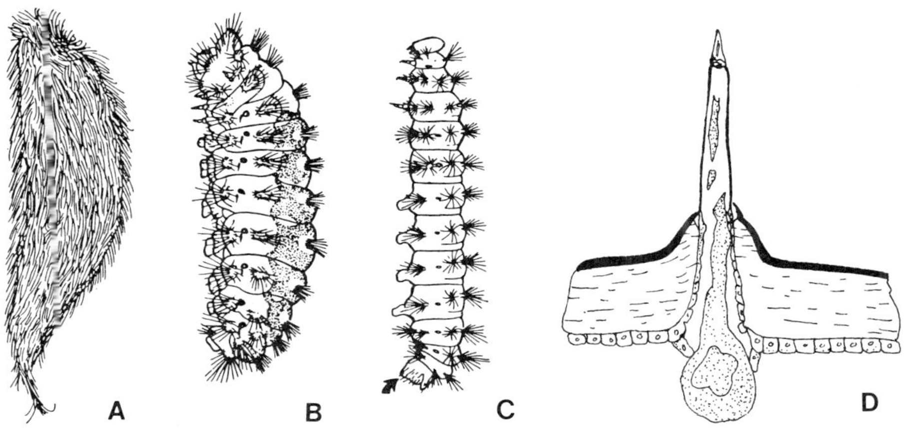

Fig. 9.31 Stinging caterpillars are usually brightly colored (aposematic): (A) a puss caterpillar, (B) the flannel moth, and (C) io moth larvae. (D) A diagram of a stinging hair.

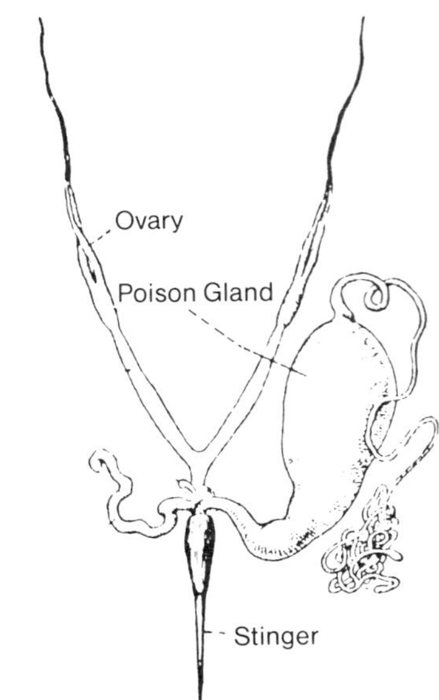

Fig. 9.32 ***The stinger and associated structures of the worker honey bee. The hollow stinger is retracted within the abdomen of the bee except when it is thrust into the victim by muscular activity. Because the tip of the stinger is barbed, the entire structure may be ripped out of the bee's body.***

varies with the species involved, but they usually have several components in common, among them several that are proteinaceous in nature. One present in honey bee venom is lecithinase, which destroys red blood cells and interferes with blood clotting. Another, hyaluronidase, destroys the cushioning fluid in the joints. Histamine is the component that is responsible for the local pain and swelling. It also causes the release of adrenalin, which accelerates the heart rate. The initial sting received by a person, while a painful experience, is rarely a severe medical problem, unless the attack is by a mass of 500 honey bees. The danger in stings occurs after a person has become sensitized by previous stings.

Of 460 deaths in the United States that could be definitely attributed to venomous animals over a 10 year period, 229 were caused by stinging Hymenoptera. These deaths were due to an allergic response. The allergic response is a manifestation of our body's immunologic defense system, a third line of defense against foreign substances. Most foreign materials are kept out of the body by its skin, or by the antimicrobial characteristics of the moist mucus membranes. The blood also contains cells whose function is to destroy microbes that may have been introduced.

The immunologic defense system protects the body from a foreign object by first recognizing that the object is foreign and then neutralizing it. The foreign objects that cause the most trouble are pathogens such as viruses, bacteria, and other disease-causing microorganisms, all of which share some common structural materials such as proteins or other organic macromolecules. The presence of these foreign substances, called "antigens," in the body stimulates the production of immunoglobulins (also called antibodies). The classic type of antibody, the B-type, circulates in the blood. This immunologic reaction is used to protect us when we are vaccinated with a weak strain of the pathogen and our bodies are allowed to build antibodies against the pathogen. However, this type of protection is only feasible prior to exposure to the pathogen. If the person already has the disease, it is not advisable to wait for antibody production. In these cases, blood serum is used from another animal which previously has been caused to manufacture lots of antibodies. The blood is processed to remove materials other than the antibodies and the antibody-rich serum is injected into the patient.

Five types of immunoglobulins have been identified in humans thus far. In addition to the B-type mentioned above, the E-type immunoglobulin is the type that causes problems in persons sensitized to the proteins present in the venom of stinging Hymenoptera. As far as is known, a person is not sensitized (that is, with E-type immunoglobulins specific for hymenopterous venom) to a particular venom unless that particular venom has been present in the blood at some earlier date, that is, unless the person has been stung by that species before. The prior sting initiates the conditions whereby the victim's body starts to manufacture E-type immunoglobulins. Then, at a subsequent exposure, the body undergoes an allergic reaction. The amount of venom antigen introduced and time between stings required to produce a violent allergic reaction is so variable that it is difficult to predict how many stings are necessary before a person becomes sensitized. Note that a high degree of sensitization to Hymenopterous venom may be produced by a single sting and that even one subsequent sting can produce the most severe type of allergic reaction, called "anaphylaxis."

Symptoms of anaphylaxis are not localized around the site of the sting. The victim develops difficulty in breathing, with obstruction of respiratory passages. Urticaria (blisters) form. The skin is itchy. Gastrointestinal disturbances such as diarrhea, cramps, and vomiting are common. The vascular system collapses. These changes may begin to appear 30 seconds after the sting occurs and in extreme cases death may ensue in as little as 16–120 minutes.

People who are sensitized to stings of Hymenoptera should seek medical advice. Short-term symptomatic treatment usually involves administration of antihistamines to counteract the swelling. However, the damage to blood vessels and other effects may continue. An alternative to symptomatic treatment is desensitization in which the patient is injected with small amounts of the allergin over a long period of time. Results of desensitization are variable so that many sensitized persons carry "sting kits" containing a tourniquet, epinephrin, and antihistamines.

Insects That Annoy

Practically any insect flying around us or crawling on us is annoying to some extent, but the insects we will discuss here are more intimately associated with humans and animals than through chance encounter. Insects that fall into this group are primarily nonbiting flies; those of the Genus *Musca* are perhaps the best examples.

The Genus *Musca* contains three serious pests of vertebrates. The best known being *M. domestica* L., the house fly. The other two are *M. autumnalis* DeGeer (the face fly) and *M. sorbens* Wiedemann (the bazaar fly). All three of these flies spend their larval lives in organic material, preferentially animal dung, but of differing types. The face fly prefers to oviposit into warm, moist (fresh) cow manure. The house fly can get along in manure of various types, but prefers horse, pig, or human feces, if available. The bazaar fly prefers dog feces as a larval habitat.

Adults of all three *Musca* species tend to rest on or near animals; the house fly is equally at home on an animal or some structure near an animal. It is easily disturbed and may fly some distance before landing after being startled. The face fly spends much of its time resting on the face, as the name implies. Fortunately, the animals involved are horses or cattle rather than humans (see Fig. 9.33). Face flies feed on the moist secretions of the nasal passages and around the eye. While the flies are nonbiting, they still cause considerable annoyance to the animal and may cause them to stop grazing and cluster together, thus reducing milk or meat production.

One might think that humans could not or would not tolerate the presence of face flies. However, humans have adapted to the presence of large numbers of flies, as shown in Fig. 9.34. The bazaar fly adult not only prefers to rest on the heads of humans, but it is considerably more persistent in remaining than its house fly or face fly relatives. If it does get brushed away, it immediately returns. This aspect, coupled with its larval habitat (dog feces) makes it an extremely undesirable insect. Another fact worthy of note about the bazaar fly is its rapidly expanding geographic distribution. At this writing it has not been detected in the Western hemisphere, but it has

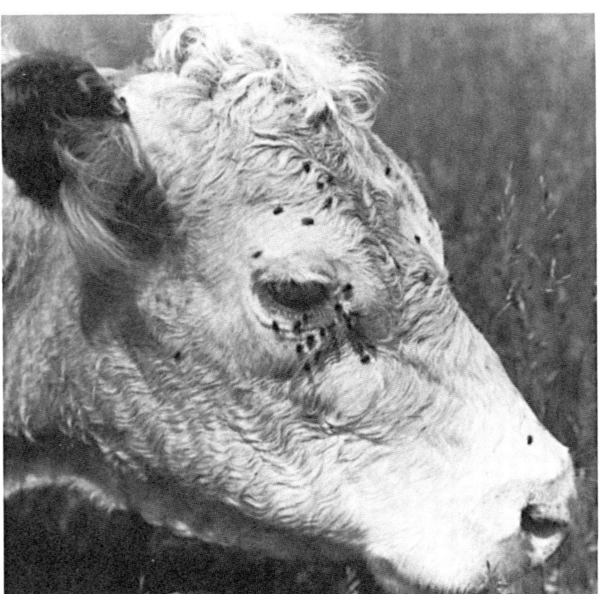

***Fig. 9.33** The face fly,* **Musca autumnalis** *DeGeer, in one of its most common locations. Adults feed on nasal and eye discharges. Larvae live in cow manure. This insect has been shown to transmit eyeworms and other eye-disease-causing organisms.*

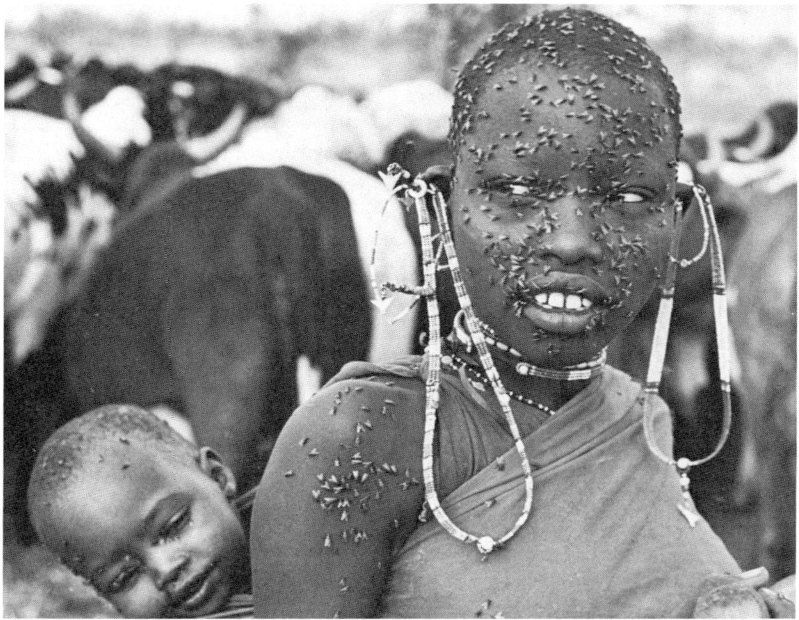

Fig. 9.34 The bazaar fly, **Musca sorbens** *Wiedemann, breeds in dog feces and prefers resting on the heads of humans, is not easily dislodged, and it returns directly after being disturbed. This fly is not currently distributed in the New World, but now exists in Hawaii and may soon make the jump to the United States mainland. Reproduced with permission from Camera Press, London. Photo by Raymond Lewis and Denys Dawnway.*

invaded the Hawaiian islands. Coastal states in the western United States run surveys to determine when and where it gains entry, if it ever does.

As covered in greater detail later in this chapter, the nonbiting *Musca* flies described above also cause problems by transmitting disease organisms, as do several other types of nonbiting flies.

Parasitic Insects

Some animal parasites spend their entire lives on the body of the host animal; others have one or more stages that are nonparasitic and sometimes quite removed from the host animal. The host is an animal at whose expense the insect lives. In other words, the insect parasite benefits from the association, the host does not. The parasite is usually rather small in size compared to the host. Thus, when the parasite is an insect and the host is a vertebrate animal, the relationship is seldom one that kills the host. We shall see later how different the results of parasitism are when both parasite and host are insects.

Ectoparasites

The two ecologically dominant vertebrate groups that are heavily parasitized by insects are the birds and the mammals. Both of these groups are warm-blooded, and therefore, they have insulating body coverings that facilitate temperature regulation. It is in these coverings—feathers for birds and hair or fur for mammals—where the external parasites or ectoparasites live.

Ectoparasitic insects that are permanently associated with a host have several problems, the most serious being a method of retaining contact with the host. In the active stages they simply hang on with their legs. The egg stage, however, is easily dislodged, so that most permanent ectoparasitic insects glue their eggs to the host's body covering. Such glued-on eggs are called "nits" (Fig. 9.35).

The feet of most insects end in a pair of claws, so that insects are fairly well equipped for grasping either feathers or hairs. Some insects have evolved even better feet or tarsi and claws. As shown in Fig. 9.36, sucking lice have only one claw. It is moveable and works almost like vice-grip pliers because of the opposed, but nonmovable spine.

Another need of permanent ectoparasites is to infest additional host animals. Parasites must move to new hosts or the parasite would soon disappear. Transfer can occur whenever two or more hosts are in close contact with each other, the most common events being mating and when the young are being cared for by the parents. Undoubtedly transfer of parasites also takes place during other social acts, as when animals herd up or nest in aggregations.

As mentioned earlier, lice are the most typical of the insect ectoparasites that remain in permanent contact with their hosts. Two rather distinct groups of lice exist (see Chapter 11), each having evolved into similar niches, but one is primarily limited to mammalian hosts, the other is more common on birds. The mammal lice (Order Anoplura) have piercing-sucking mouthparts, which permit them to feed on blood. Bird lice, also known as chewing lice, have chewing mouthparts. Actually the term bird lice is a misnomer because they sometimes occur on mammals. Chewing lice

***Fig. 9.35** Nits, or eggs, of a louse. The nits are glued to hairs of the host by the female louse as she produces them. Attachment of the eggs assures the continued contact with the host necessary for survival of the louse upon hatching.*

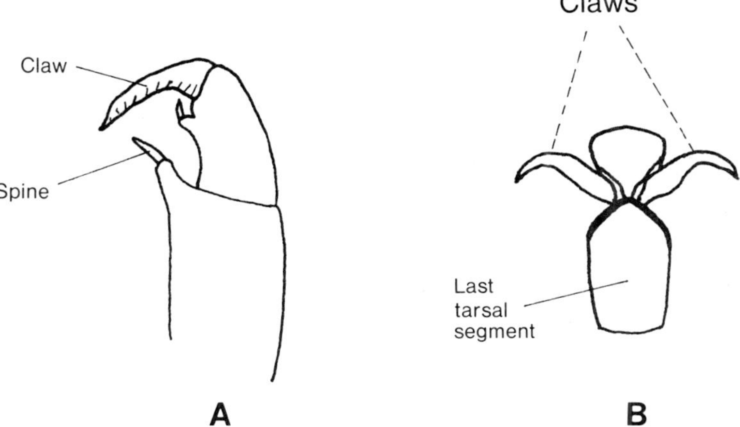

***Fig. 9.36** The arrangement of claws on the tip of a louse's leg (A) as compared to the tip on most insects (B). Note that the second claw has disappeared, but an immovable, spine-bearing peg opposes the single claw for ease in grasping hairs on the mammalian host. The fleshy pad between the two claws on a typical insect's leg is also absent in sucking lice.*

feed on skin scales, developing feathers, scabs and so on. Some species do bite the skin and then feed on the blood oozing from the wound.

Lice occur on most species of birds and mammals, but the relationship between the host and its louse parasite is highly specific. Somewhat less than 500 species of Anoplura have been described, and these occur on 840 different species of mammal, reflecting the specificity of the host–louse relationship in which a particular species of louse is often restricted to a single species of mammal host. This relationship is even tighter than the figures above indicate if we consider that some one-way transfers of lice occur from prey mammal to predator mammal. Thus, each mammal or bird species typically has its own particular complex of lice, except for some louse-free mammalian groups like bats, primitive mammals such as platypus, opossum, and kangaroo types, sloths, anteaters, the marine whale–porpoise–sea cow groups, and most terrestrial carnivores. However, more than one louse species can live on a species of mammal or bird. For example, two distinct louse species live on human bodies: the head and body louse and the pubic louse. Interestingly, even these two species show specialization.

Pediculus humanus L. has two distinct strains: the head louse is found among the head hairs and the body louse spends most of its time in the seams of its human host's clothing. If we consider that the clothing is the body covering of the essentially barren human body, then the fact that the body louse spends its time in clothing is not unique. The pubic louse, *Phthirus pubis* (L.) lives predominantly in the pubic region. A closely related species of *Phthirus* is found only on the gorilla, a bit of evidence of the evolutionary kinship between humans and closely related primates. Figure 9.37 illustrates the two species of louse that live on humans.

The term for a louse infestation is pediculosis. Considering the small size of a louse it would seem that a sparse population of lice easily could be tolerated by an

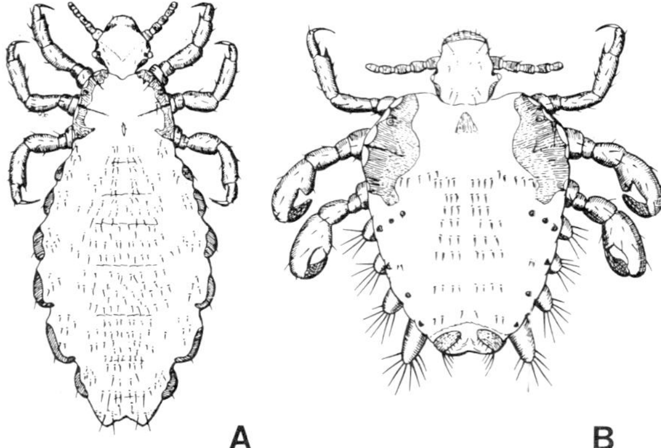

Fig. 9.37 The two types of human lice: (A) the head and body louse and (B) the pubic louse.

adult human without any ill effects. This is not true because a chemical in louse saliva produces lethargy. Thus, even the presence of a few lice makes the host human feel tired and achy all over. This is probably the origin of the term "feeling lousy."

Human lice are not as much of a problem in these times of powerful synthetic insecticides as they were before World War II. Lice also are kept at a minimum by sanitary conditions, including the ability of the human population to eat well, keep warm, have clean clothes, and get proper rest in uncrowded conditions. When these conditions are not met, for example in war-torn areas, body lice, or "cooties," become so abundant that 400–500 may easily be removed from an infested person, and some individuals may have as many as 10,000 active lice in their clothing. There does appear to be an upper limit to numbers of lice a single person can harbor, although it is high.

Under normal urban conditions lice become only sporadic problems, usually occurring in children at the elementary school level. Once lice show up in a school there usually is a general eruption because lice are so easily transferred through contact and they have a generation time of only three weeks.

Intermittent Parasites

Unlike the lice just discussed, most insect parasites of vertebrate animals spend some or most of their lives away from the host. These insects are called intermittent parasites because they approach the host, essentially for a meal, and then leave.

The two most significant insect groups are vertebrate parasites as adults, but have free-living immature stages. They are the common fleas of temperate regions, the mosquitoes, and other blood-feeding flies.

Fleas. Fleas have complete metamorphosis, so that the free-living egg, larval, and pupal stages occur out of direct contact with a host. However, since these stages are still somewhat associated with the vertebrate, and the vertebrate typically is one that has a home shelter (nest, den, burrow), the free-living stages can usually be found among the dust, debris, or litter of the home shelter. Larvae of fleas can be found in rugs, under furniture or baseboards, and in cracks between the floor boards in the homes of humans. In Roman times, rich homeowners would seal up cracks in the floor with beeswax in an attempt to reduce the available flea habitat. Larvae eat organic debris, including skin scales as well as feces produced by the adult fleas. Such flea droppings are especially high in protein content and thus are a good nutrient source for the larvae. One problem for fleas associated with the nest habitat is that, if the host is at all resourceful, the nest is rather dry. This means that the availability of water is a real problem for the larval flea. The flea has evolved the ability to absorb moisture from the air, even in the vapor state, to alleviate the need for free water.

Although some of the more advanced types of fleas, like the sticktight flea of poultry and the chigoe flea of humans, spend all of their adult life on their hosts, the adult of most familiar species spends the majority of its life away from its host, seeking the host only for a meal or to locate a mate. Fleas locate the host by its smell or the warmth from its body, although evidently not from a very great distance. Some fleas are also attracted by movement. For example, those that infest ground dwelling birds will jump when an object is inserted into the nest. They may be collected by inserting a brush into the nest, provided the brush has previously been dipped in alcohol.

Many flea species can survive through extensive periods without a blood meal. This can cause a problem when returning to a house after an extended absence. The immature fleas (both larval and pupal forms) complete their development and await the arrival of a suitable host, preferably a dog or cat. If the pet is not present when the house is first re-entered, members of the family are vulnerable to flea attack. Fleas can also become a problem in homes after a pet has died. With the loss of the favored host, the fleas only have the humans left to use as a food supply. In cases like this, where the dwelling is badly infested, insecticide is used to reduce the numbers of fleas, or another dog or cat can be introduced into the home to attract all of the hungry fleas. It can then be taken out of doors and dusted with the appropriate flea powder.

Other ways of getting rid of fleas have been devised. The collars now sold to control fleas contain resins impregnated with an insecticide that volatilizes slowly, killing fleas that are nearby. Most infested animals scratch themselves and often do dislodge the flea. The story of how foxes eliminate their fleas is one of the most interesting, even if it is not well documented. According to this story, foxes back slowly into a cold stream while holding a large piece of moss in their mouths. Supposedly the fleas keep moving up the body to avoid the cold water and eventually end up on the moss as the fox becomes increasingly immersed. Since the last dry portion is the moss, the fox releases its hold on the moss just as it puts its head under the water and the fleas drift away on the moss, leaving behind a nearly flealess sly fox.

Blood-Feeding Diptera. The Order Diptera (see Chapter 11) contains a number of groups whose members may be intermittent vertebrate blood-feeders. They fall into two principal categories: the primitive groups in which only the female takes blood (e.g. mosquito, black fly, punky, horse flies and deer flies, and their relatives) and the advanced group (stable flies, horn flies and tsetse flies) in which both sexes feed on blood.

The most ubiquitous group in the primitive dipteran category are the mosquitoes. Male mosquitoes are not blood-feeders, but feed on nectar for energy, as do females. The female mosquito uses the protein taken in at the blood meal to build eggs. This is comparable to the honey bee where the pollen is the protein source. Without a protein source neither insect could produce eggs.

Mosquitoes generally require a blood meal before producing eggs, though not always. Whereas the bird feeder *Culex pipiens pipiens* is incapable of producing a single egg unless a blood meal has been obtained, its close relative, the human feeder *Culex pipiens molestus*, can produce the first batch of eggs without requiring a blood meal. The proteins for egg development are supplied from its body reserves. It is interesting that if a blood meal is supplied, more eggs are produced. After the first batch of eggs is produced, it requires blood meals in order to produce subsequent batches.

Mating for most species is facilitated by the swarming behavior of the males. Groups of dozens to hundreds of males of a species fly in a rather tight pattern. Then, when a female approaches, the swarming males react. Their reaction is stimulated by the whine of the female's wing beat, which causes a resonance detectable by the bushy antennae of each male. Each species has wing beat frequencies that differ enough so that the antennae of the correct males resonate, rather than all of the male mosquitoes within range. The first male to reach the female grasps her, and they tumble out of the swarm with her wings stilled so that the antennae of the other males stop reacting; the pair mates without further interference.

The female mosquito lays her eggs on or near water, but, using physical and chemical cues, each species ends up ovipositing in fairly specific locations. *Culex pipiens pipiens* lays rafts of 12–250 eggs that stay together floating on the surface of polluted ground water (sewage effluents, drains, and gutters) for several days until embryonic development is complete at which point they hatch into the water. *Aedes aegypti* (the yellow fever mosquito) lays her eggs singly just above the water line in small containers such as rock pools or discarded tires and tin cans. The eggs complete their embryonic development and then the completed embryo remains inactive until particular conditions prevail. The mature *A. aegypti* egg will not hatch when it becomes wetted, as with dew, or a light shower. It is only after the combination of water and low oxygen levels occurs that the larva will hatch out. These conditions normally only occur when the container becomes filled with water to a level above the eggs and microbial action in the debris within the water has lowered the amount of oxygen. Then the larva emerges and begins to feed on minute particulate foods such as algae and detritus. After four instars, the larva (Fig. 9.38) pupates into a free-swimming comma-shaped form that takes in air by means of respiratory horns on the thorax when it is resting at the water's surface. The adult emerges from the pupal case and flies off to complete the generation. Under ideal conditions mosquitoes may go from egg to adult in 5 days or less.

The adult female finds her blood source by a combination of stimuli. In fact, most emanations that could be listed for an animal are used as cues by some species of mosquito. Among those known for the yellow fever mosquito are moisture (water

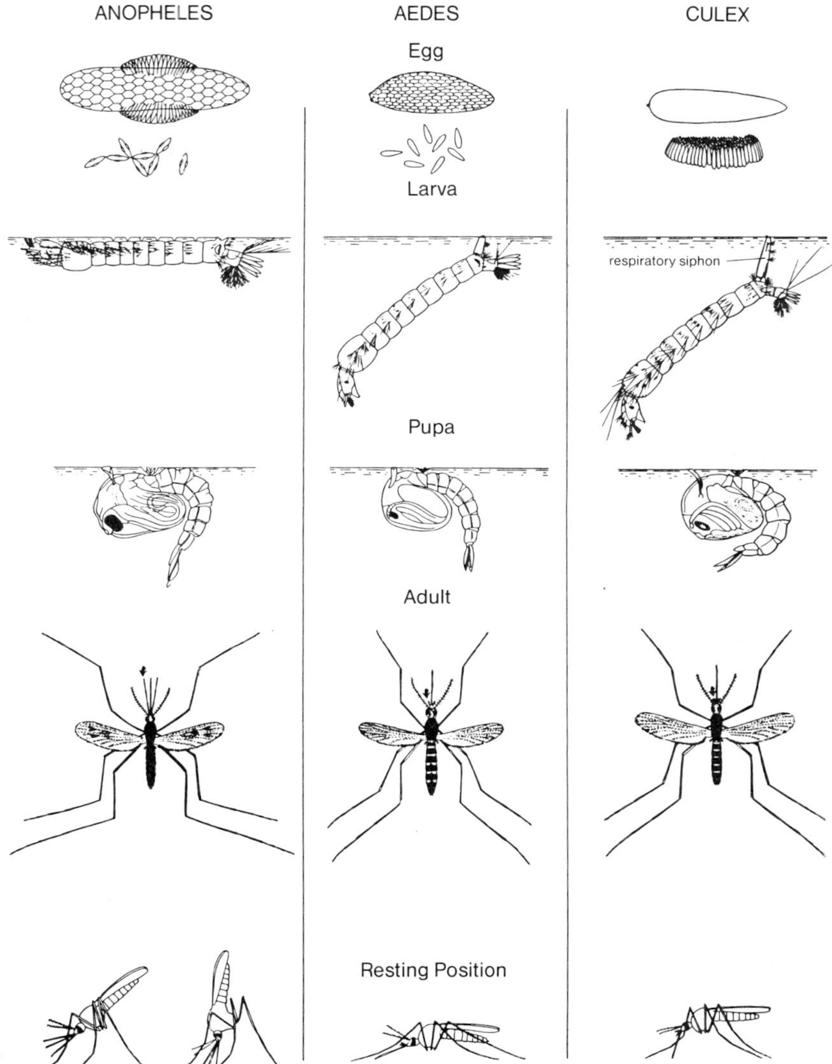

Fig. 9.38 The life stages of mosquitoes. The eggs of **Anopheles** *are produced singly, like* **Aedes**, *but they have floats on either side.* **Culex** *produces eggs in floating groups called rafts. Larval* **Anopheles** *lie at the water's surface, filtering out surface debris with their mouth brushes, whereas both* **Aedes** *and* **Culex** *hang from the surface by their respiratory siphons. All three take in air through spiracular openings as do the pupae, except that the pupal openings are on the anterior part of the body. Only the female of the flying adult is illustrated. Males have much bushier antennae.*

vapor), carbon dioxide, heat, color, motion, and lactic acid. Water vapor, carbon dioxide, and heat are expelled from the human lungs. Moisture and lactic acid are ingredients of perspiration. Thus a person undergoing intense exercise is particularly attractive to mosquitoes. Also, since both male and female mosquitoes visit flowers to feed on nectar, sweet fragrances such as those in deodorants, after shave lotions, and perfumes attract mosquitoes.

Mosquitoes are seldom so numerous as to cause great injury to their hosts. The annoyance may be severe and reactions to bites can cause considerable discomfort. It is of interest to note that long-term repeated biting exposure often results in a marked decrease in sensitivity to the bites. Sensitivity to bites varies with the type of species so that a person may become almost insensitive to one type, but react violently to another species. See the last section of this chapter for a discussion of mosquitoes as disease transmitters.

The North American species of black flies (Fig. 9.39) are restricted to well-oxygenated, rapidly moving fresh water. The immature forms are aquatic; as adults they are primarily daytime feeders. Their bite is especially painful and may leave the victim with a bleeding wound at the site of the attack. They are persistent feeders and sometimes occur in such huge numbers that livestock have died as a result. Conflicting reports attribute the livestock deaths to massive shock reactions or to clogged respiratory passages due to the clouds of insects, but not to loss of blood.

Punkies or no-see-ums are biting gnats of such small size that they are not easily detected even when they are biting. The bites are painful and appear to be without a cause until very close examination reveals a minute fly with its mouthparts inserted. The immature forms are aquatic. Their small size often permits punkies to crawl through the mesh of window screens and to become nuisances in houses.

Female horse flies and deer flies (Fig. 9.40) attack a variety of large mammals and in taking their blood meals, inflict a painful bite. Both males and females feed on

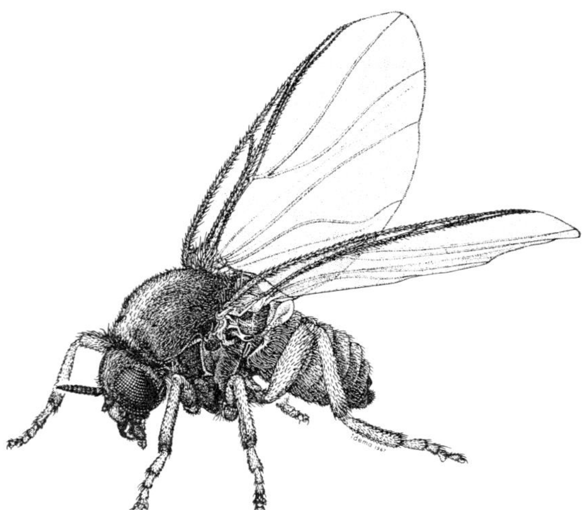

***Fig. 9.39** An adult black fly. The aquatic larvae cling to rocks in fast-moving water by means of silk they produce and attach to the substrate. Black fly larvae are filter feeders.*

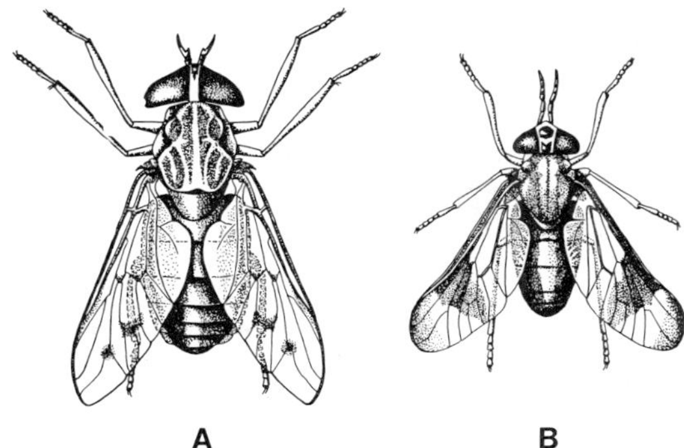

Fig. 9.40 The horse fly (A) and deer fly (B) are often a combination of dark and yellow colors. However, the compound eyes of many species offer a kaleidoscope of metallic blues, greens, and reds.

nectar. The female lays her eggs in shingled masses on vegetation overhanging water or damp land and the larvae fall down after hatching to exist as carnivores or scavengers beneath the surface. They pupate in drier soils and, in temperate zones, the pupa is the overwintering stage, with most species taking at least a year to complete a generation.

Deer flies are more commonly pests of humans than are horse flies, probably because the attack of deer flies is more subtle. For most species the severity of attack is greatest on hot, sunny, humid days, with reductions in attack caused by lower humidity or drops in temperature and sunlight. These pests certainly interfere with the tourist trade in many ocean beach communities near salt marshes. In addition, these insects of the family Tabanidae are large enough in size and often so numerous that irritation and blood loss to livestock reaches significant economic levels.

Tsetse flies are rather similar to house flies in size and shape, except that their wings overlap each other on the back. However, unlike the cosmopolitan house flies, tsetse are restricted to Africa. There are a number of species of tsetse, each with somewhat different characteristics, but they all share the basic biological aspects of bloodfeeding and development. Both sexes feed on blood, a characteristic that distinguishes these flies, as well as stable flies and horn flies (Fig. 9.41), from all of the more primitive groups previously mentioned. Their larval development is also rather unique in that it takes place within the female's body. Only one larva develops at a time, hatching within the female's body and feeding on fluids from special "milk glands." Upon completion of the larva's development its parent larviposits (gives birth to the fully developed larva) in shady areas, and the larva immediately burrows into the soil and pupates. The significance of tsetse lies not in their painful bite, but in the ability to transmit disease (covered later in this chapter).

Another group of parasitic flies resemble tsetse to the extent that they too have internal development of larvae which pupate upon emergence from the female. All of this group are external parasites of mammals and birds. Many are entirely wingless, like the sheep ked, *Melophagus ovinus* (L.), but some are winged throughout their adult life or only lose their wings later in adult existence.

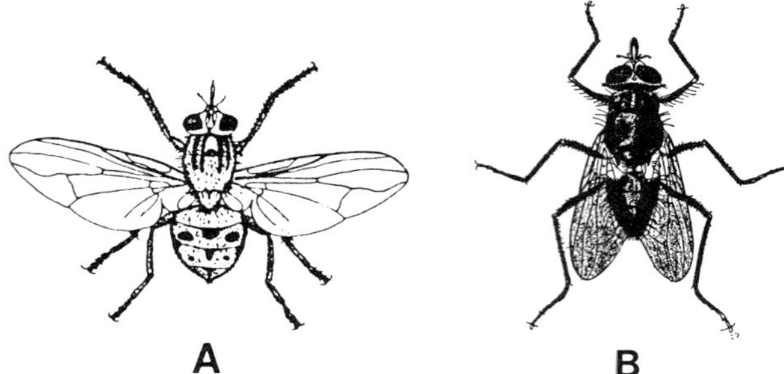

Fig. 9.41 ***(A) The stable fly breeds in plant debris like wet hay or straw to emerge and attack horses or cows around the legs, causing the animals to stamp their feet. (B) The much smaller horn fly will only oviposit in very fresh cow manure. They are found on the body or about the top of the head.***

The advantage of internal larval development and pupation upon emergence is the increased survival of offspring, as well as being able to retain constant contact with the host. The sheep ked larva secretes a sticky material that glues it to the sheep's wool throughout the pupal stage. Some louse fly parasites of birds produce larvae that drop off the host into the bird's nest to pass the pupal period. The most bizarre looking louse flies are those associated with various bats. One group strongly resembles a spider in body form. The more common louse flies may be distinguished from other flies by the great distance between the bases of each pair of legs—the legs are at the sides of the thorax rather than at the bottom.

Bed bugs. The human bed bug, *Cimex lectularis* L. has piercing-sucking mouthparts in both the nymphal and adult stages making it capable of feeding on human blood after it hatches from the egg. These bugs hide in human habitations during the day, most often in the sleeping quarters, where they may be found in mattress ticking, bed frames, under wallpaper, behind baseboards, and so on. They come out at night, find a host, and insert their mouthparts, withdrawing a small amount of blood. After becoming engorged, they return to their hiding places. They usually feed once between each molt and again prior to reproducing.

Bed bugs are small, flat, with rudimentary wings, and reddish-brown, with some black, if they have recently fed (see Fig. 9.42). Like many other members of the Order Hemiptera, they possess scent glands and give off a peculiar, pungent, and distinctive odor that is noticeable if they occur in fairly large numbers.

Ticks. Another common animal parasite similar in parasite–host relationships to the bed bug is the tick group. Ticks are not insects at all, but rather in the Class Arachnida along with the mites, spiders, and scorpions (see Fig. 9.43). All ticks feed on vertebrate blood, but use hosts in different ways. Some spend their entire lives on a single host once it has been located, others, like the bed bug, must approach a vertebrate for a blood meal each time it molts. The first two active stages of the tick often derive blood from small mammals, like rodents or rabbits, whereas the adult stage attacks larger mammals like deer, livestock, and humans.

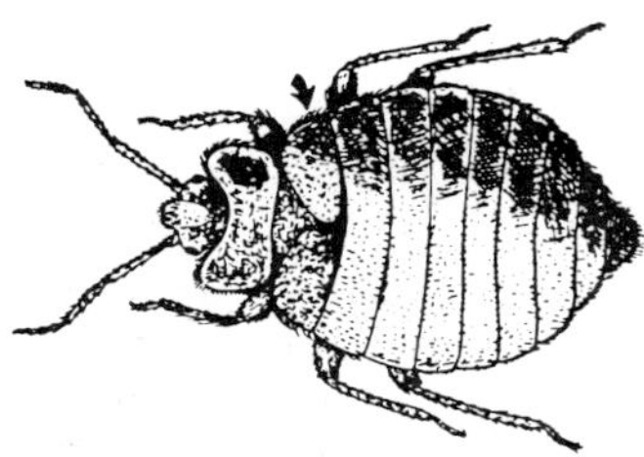

***Fig. 9.42** The bed bug is a wingless, flat blood feeder that approaches its host while the host is resting. Forms with very similar appearance and biology are intermittent parasites of bats, barn swallows, and poultry.*

Protelean Parasites. The groups of parasitic insects discussed thus far have been either those that spend their entire lives on the host (e.g., lice) or those that only visit the host to acquire a meal (e.g., fleas, mosquitoes, and tsetse). The remaining group is made up primarily of Diptera species whose larvae are parasitic, but whose adult stage is free-living. Insects with this type of life history are termed "protelean" (pro = before, teleo = the end) parasites, and collectively present some of the most bizarre examples of insect adaptation. With a few exotic exceptions to be discussed later (Tumbu fly, Congo floor maggot) the species may be grouped into three broad categories. The first group is associated with the vertebrate's digestive and/or respiratory system, the second group causes skin cysts, while the last group is primarily associated with wounds.

Flies in the genus *Gasterophilus* attack livestock and are known as bot flies. The eggs of the bot fly, *G. intestinalis* (De Geer), are attached to the hairs on a horse's legs and are much like the nits of lice. After developing, the embryonic larva hatch out given the proper stimuli (a combination of warmth and moisture), normally provided when the horse licks its legs. The larva (Fig. 9.44) hatches instantly and attaches to the host's tongue, burrowing into the flesh. It migrates to the stomach and attaches to the stomach lining by means of its mouth hooks, where it remains until ready to pupate. Intestinal bots cause damage by injuring the intestinal lining and causing blood loss and irritation of the bowel, and are a possible source of infection. They

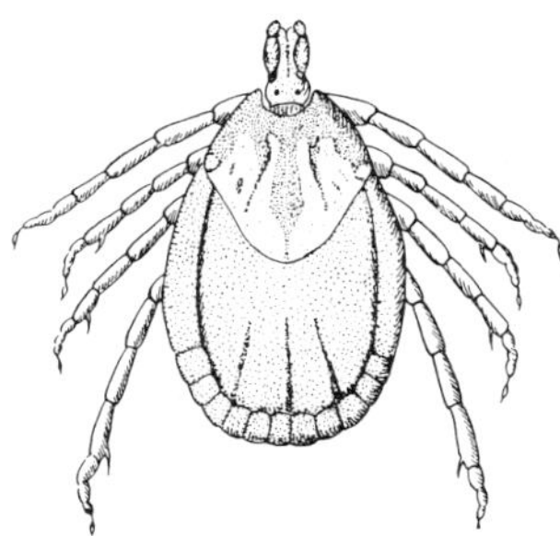

***Fig. 9.43** A typical representative of the "hard" tick group. The "soft" ticks have a more wrinkled appearance. After the first molt, ticks have four pairs of legs rather than the three common in insects. Note also the absence of antennae.*

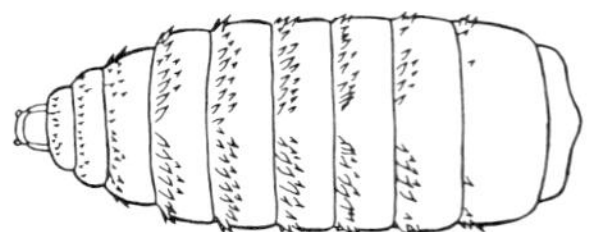

***Fig. 9.44** The horse bot,* **Gasterophilus intestinalis** *(De Geer). The first instar hatches from the egg immediately upon contact with saliva. With properly conditioned eggs the reaction can be seen under the stero microscope by moistening the egg with saliva. The rows of elongate recurved spines aid the larva in clinging to the host's tongue and thus to begin the migration to the stomach.*

sometimes become numerous enough to block the digestive tract. Mature larvae pass out in the horse droppings and pupate where they fall, either in the droppings or soil. Each of the six species of *Gasterophilus* that attack horses or donkeys differ slightly as to where the eggs are laid and where the larval stage occurs within the digestive system. Other flies in this family attack a variety of large mammals.

Skin cysts are caused in rodents, rabbits, and humans by bots, and in cattle by warbles. The adult flies are very large and often resemble bumble bees in coloration and hairiness, especially those attacking cattle. The rodent and rabbit bot larvae attach themselves to a passing host after having hatched from eggs that were deposited in areas frequented by hosts. The larvae crawl into the host through one of its orifaces and eventually locate beneath the skin, punching a hole through to the exterior for air and dropping out to pupate.

Warble flies on cattle oviposit directly onto the host's hairs and the emerging larvae penetrate the skin and migrate through the body for 4 months, eventually settling alongside the backbone and puncturing the hide making an air hole to the exterior (see Fig. 9.45). When mature the larva squeezes out through the hole and drops to the ground to pupate. Warbles on cattle reduce their market value in two ways. First, the hide is much reduced in value because of the holes made by the warbles. Second, the area usually infested yields some of the more expensive cuts of beef, which must be discarded.

The human bot, *Dermatobia hominis* (Linn., Jr.), also known as the torsalo fly, occurs in the tropical Americas. It is quite unique in its method of contacting the host. Rather than laying eggs directly on the human, it captures mosquitoes and other blood sucking insects that seek humans or cows, its other primary host (see Fig. 9.46). It then glues its eggs onto the other blood sucking insect. The larvae hatch upon contact with high temperature of the human body as the mosquito feeds and enter through a hair follicle to produce a boil-like cyst at the site. Eggs may also be laid in an area where the larva might be picked up by a host. After the bot emerges, the cyst heals rapidly and without infection, due to chemicals secreted into the cyst by the bot. This is true for most of the cyst-forming Diptera.

The screwworm fly, *Cochliomyia hominivorax* (Coquerel), is probably the most severe of the wound-infesting Diptera. Screwworms are rather advanced types of blow flies, that is, the flies that attack dead animals. However, the competition for newly dead vertebrate carcasses is so severe that some species of blow fly have developed the ability to infest wounds of live, though presumably about-to-die, animals where they consume the dead or dying tissue. The screwworm fly maggots can

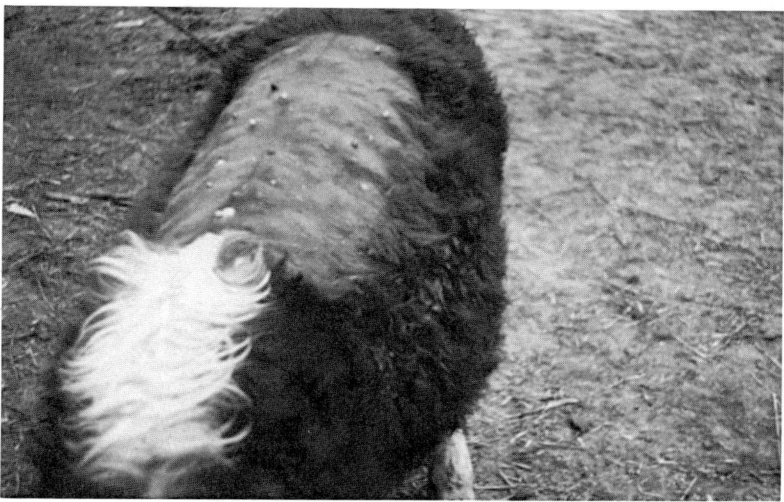

***Fig. 9.45** The damage caused by warbles, or cattle grubs, to hides and carcasses is extensive. The oviposition flights of the nonbiting adult flies can cause cattle to herd up or panic and injure themselves by jumping fences attempting to escape. The adult fly strongly resembles a bumble bee.*

attack healthy flesh, provided only that there is some break in the animal's skin through which they can gain entry.

Adult screwworm flies look somewhat like green bottle flies. They are only capable of ingesting liquids. The gravid female is attracted to wounds and oviposits near breaks in an animal's skin, even those as small as one made by a biting tick or horse fly. The larvae enter the wound after hatching and feed on the flesh, extending the injured area and causing it to supperate, becoming even more attractive to females, thereby enhancing additional oviposition and larval attacks (see Fig. 9.47).

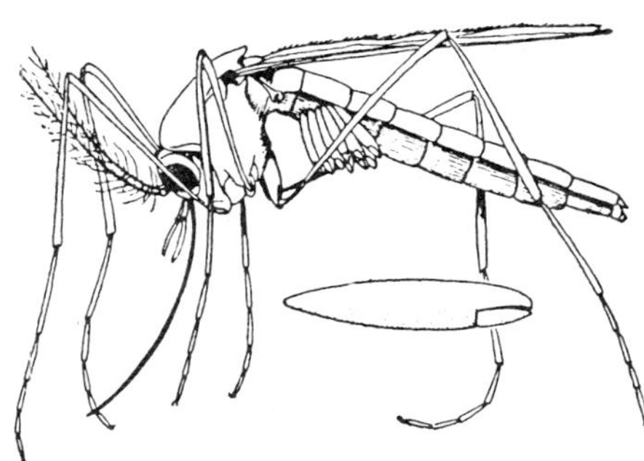

***Fig. 9.46** The eggs of the human bot fly,* **Dermatobia hominis,** *attached to the abdomen of a mosquito. When the mosquito approaches its human or bovine host for a blood meal, the human bot fly egg receives the hatching stimulus—a sharply increased temperature—and the larva immediately emerges and burrows into the host.*

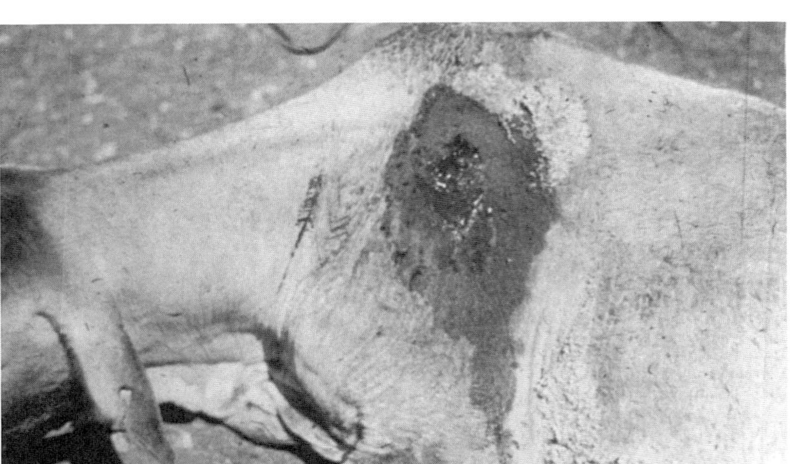

Fig. 9.47 Damage to livestock by the screwworm fly.

The animal with such a wound tends to find a thicket, enter it, and lie down. Such attacks very often result in the animal's death unless it is located and treated.

Lucilia sericata (Meigen) and other blow fly species cause "sheep strike." That is, they oviposit into wounds or soiled wool, especially around the sheep's vent. The fecal- and urine-soaked wool causes dermal inflammation, which is enhanced by the larval scratches on the skin as they attempt to feed. The animal begins to rub and soon causes larger lesions, which come under greater attack. The whole sequence often ends in the sheep's death.

The tumbu fly, *Cordylobia anthropophaga* and Congo floor maggot, *Auchmeromyia luteola* are found in tropical Africa. The tumbu fly maggot imbeds itself in a human or monkey and forms a cyst somewhat like the human bot. The Congo floor maggot attacks humans that sleep on dirt floors or mats. The larvae scrape at the host's skin while the human is sleeping until the skin starts to ooze and the larvae then feed on the exudates.

INSECTS AND DISEASE

Most of the negative interactions discussed in this chapter thus far have dealt with the ways in which insects compete with us for foods, fibers, and structural materials. It is easy to understand how their attack causes economic loss because they either destroy our supplies or contaminate them so that they are no longer usable. Even the fleas, lice, bed bugs, mosquitoes, and related forms that attack us directly are seldom so numerous or severe that they do more than annoy us. However, there is another aspect of this negative insect–human relationship that far overshadows the competition for materials between insects and humans. It is the association that has evolved between disease-producing organisms and the insect pests of humans. To a mother in a tropical country, whose child is dying of an insect-borne disease, the importance of insects is more real and immediate than it is to an urban North American who is forced to pay an additional fraction of a cent per can of food because of losses inflicted by insects on a crop far removed from the consumer's location.

Most readers of this book are already aware that insects are known to be transmitters (vectors) of disease-causing organisms. The amazing part of this subject is that this well-known relationship was not discovered until the last decade of the nineteenth century. Thus there are people still living today who were born before we were aware that insects actually transmitted diseases. For instance, no one knew how people contracted malaria, yellow fever, or typhus—a situation we now find hard to comprehend.

The list of human diseases associated with insects is a long one (see Table 9.4). Among these are malaria, yellow fever, and typhus—primary diseases that have been the predominant cause of human mortality throughout recorded history. In the late 1970s World Health Organization estimated that there were 200 million people in the world with malaria.

Among the early highlights in the discovery of arthropod involvement in the transmission of disease is Patrick Manson's finding in 1878 that the larvae of *Wuchereria bancrofti*, the causative agent of filariasis in humans, could be found in the mosquito. Theobald Smith and F. L. Kilbourne demonstrated that the causative organism (a protozoan) for the livestock disease known as Texas cattle fever was transmitted from one cow to another by the cattle tick, *Boophilus annulatus* (Say). M. B. Waite of the U.S. Department of Agriculture demonstrated that honey bees carried the bacteria that cause fireblight from one apple or pear tree to another.

Aspects of Texas cattle fever and fireblight are examples of the complexities that often obscure what seems, superficially, to be a simple concept. When an arthropod acts as a vector of disease several elements are involved (see Table 9.5 and Fig. 9.48). The complicating factor in Texas cattle fever involved acquisition of the pathogen, whereas the method of transmission was the factor of concern in fireblight.

In order for a potential vector to become infective it must somehow acquire the pathogen. Yet research on *Boophilus annulatus* clearly showed that ticks that had never previously fed on an infected cow could introduce the fever in healthy cows. The ultimate answer was that, in fact, the ticks had acquired the pathogen, but not directly from a diseased cow. Instead, the offspring of infective female ticks became infective themselves because the pathogen invaded the eggs before they had been completely formed in the female's body. This is a case of "transovarial" transmission, a condition relatively more common with tickborne diseases than with those borne by insects.

The method of transmission of fireblight was shown to be linked to the honey bee's feeding on nectar of infected trees. The causative bacterium is present in the nectar of the infected plant. Thus, when the bee removes the nectar, it also picks up the bacterium, some of which may be dislodged when the bee visits another flower,

Table 9.4 Insect Transmittors of Organisms That Cause Disease in Humans[a]

Insect vector	Disease organism	Disease
Order Diptera		
Anopheles spp. mosquitoes	*Plasmodium* spp. (protozoan)	Malaria
Aedes spp. mosquitoes	Virus	Yellow fever
	Virus	Dengue
Culex spp. and *Aedes* spp. mosquitoes	Virus	Encephalitides
Mosquitoes, principally *Culex*, *Aedes*, *Anopheles*, and *Mansonia* spp.	Filaria (nematode)	Filariasis, including elephantiasis
Simulium spp. black flies	Filaria (nematode)	Onchocerciasis
Phlebotomus spp. sand flies	*Leishmania* spp. (protozoan)	Kala-azar
		Oriental sore
		Espundia
	Bartonella sp. (rickettsia)	Verruga
	Virus	Papataci fever
Tabanus spp. horse flies	*Bacillus anthracis* (bacterium)	Anthrax
Chrysops spp. deer flies	*Pasteurella tularensis* (bacterium)	Tularemia*
Glossina spp. tsetse flies	*Trypanosoma* spp. (protozoa)	African sleeping sickness
Order Siphonaptera		
Xenopsylla spp. and *Nosopsyllus* sp. fleas	*Yersinia pestis* (bacterium)	Bubonic plague*
	Rickettsia typhi (rickettsia)	Endemic typhus
Order Hemiptera		
Triatoma and *Rhodnius* spp. kissing bugs	*Trypanosoma cruzi*	Chagas' disease
Order Anoplura		
Pediculus humanis louse	*Pasteurella tularensis* (bacterium)	Tularemia*
	Borrelia recurrentis (spirochaete)	Relapsing fever
	Rickettsia prowazakii (Rickettsia)	Epidemic typhus
	Rickettsia quintana (Rickettsia)	Trench fever

[a]All of the diseases on this list, except for the two with asterisks, are exclusively transmitted by insects (under ordinary conditions).

thus passing on the pathogen because of mouthparts that are contaminated. This type of passage is called "mechanical" transmission and is the method commonly associated with many, but not all bacterial, fungal, and viral diseases.

Other diseases are spread by "biological" transmission, in which the pathogen acquired by the vector must undergo changes within the vector before it becomes infective. Thus an *Anopheles* mosquito, having just fed on a malarial person is incapable of transmitting malaria to another person for at least 4 days. This is the minimal time required for the sexual reproduction of the pathogen to take place within the

Table 9.5 The Elements Involved in Disease

Disease	Abnormal physiology (including growth and metamorphosis) of an organism
Pathogen	The "parasitic organism" that causes the host's physiology to be abnormal
Host	The organism in which a disease is produced by a pathogen
Vector	The organism that transmits the pathogen from an infected host to a previously noninfected host
Reservoir	A host organism that does not show any symptoms of disease when infected with a pathogen. It often maintains the pathogen over long periods

mosquito's body and the infective stage to be produced and become situated where it can be injected when the mosquito subsequently bites. The award of the Nobel Prize for medicine to Ronald Ross in 1902 was not simply for recognizing that mosquitoes transmitted malaria, but that only certain mosquitoes (Anopheline species only) were involved, that it could only be transmitted by the mosquitoes, not by dirty linens, "contaminated air" or other means, and that mosquitoes were not capable of immediate transmissability. A certain latent period exists between acquisition and the ability to infect. Note that several pathogens that are usually transmitted only by arthropods may be artificially transmitted by means of blood transfusion.

Arthropod-Borne Diseases of Humans

Malaria

A "three-factor" disease caused by a protozoan, malaria has historically been the greatest single killer of humanity. World Health Organization estimates are that one person dies somewhere in the world every 10 seconds from malaria. About 10 times as many cases exist where the patient does not die, but cause an estimated 15 days of disability per case. Malaria is estimated to have cost India $1 billion per year in the 10 years after World War II because of the productive time lost for the estimated 7,500,000 cases which occurred. The disease has occurred in all habitable continents. Although the disease is no longer endemic (native, indigenous) in the United States, it occurred throughout the United States until the 1930s, with an average 80,000 cases each year. It has been a factor in every war ever fought by the United States, including the revolutionary war.

The life history of the malaria parasite is complex (see Fig. 9.49). The key to halting the mosquito–human–mosquito–human cycle has historically proved to be through suppression of the mosquito population. Sri Lanka, a country with high rates of malaria was nearly malaria-free by 1963 as a result of massive mosquito control programs, based heavily on DDT. Control was so complete that the newly independent country essentially abandoned mosquito control, which resulted in such a resurgence of malaria that there were nearly a million cases in 1968. Another aspect of malaria control is through use of antimalarial drugs. However, new strains of drug-

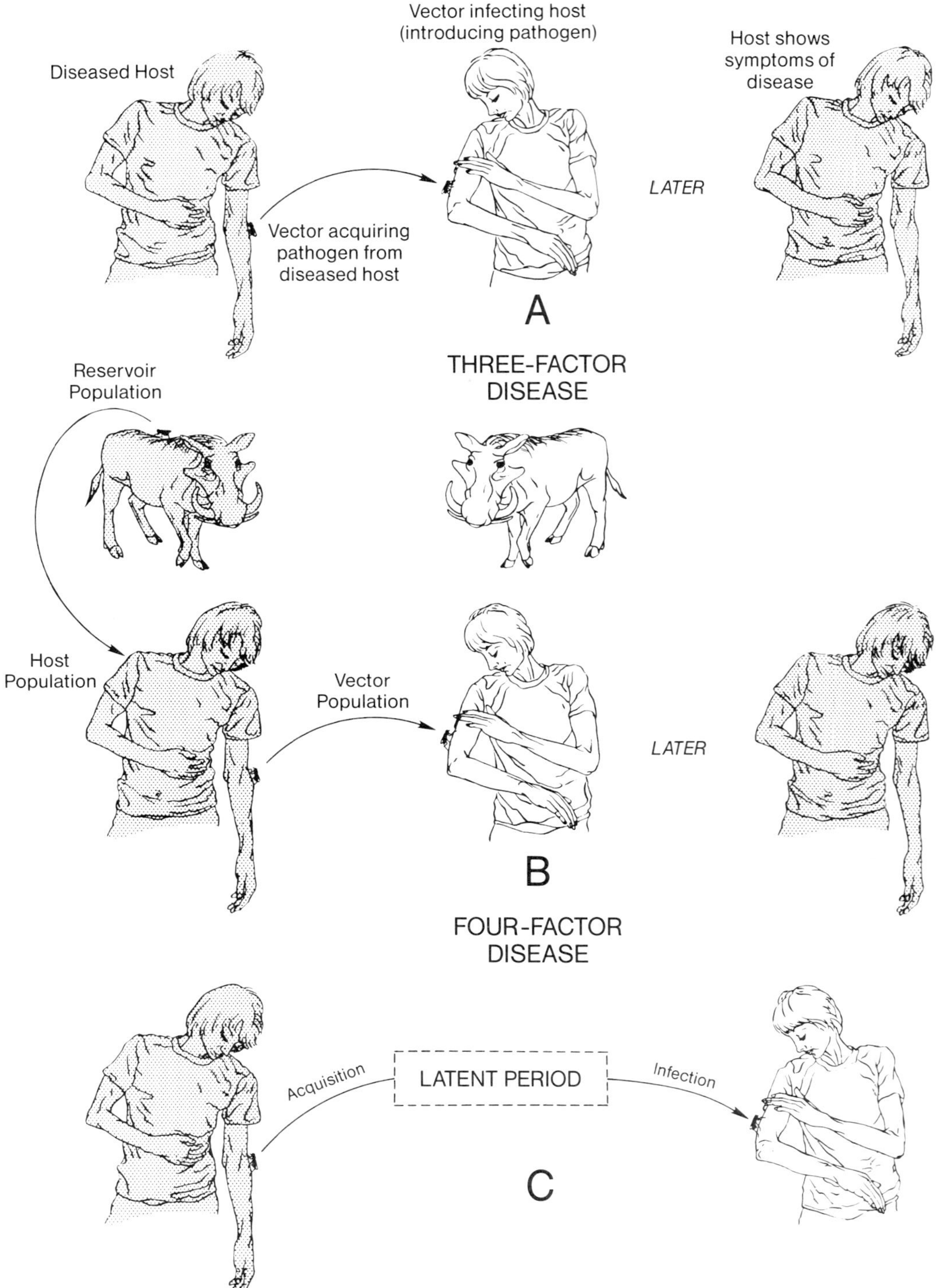

Vector infecting host (introducing pathogen)
Diseased Host
Host shows symptoms of disease
LATER
Vector acquiring pathogen from diseased host
A
THREE-FACTOR DISEASE
Reservoir Population
Host Population
Vector Population
LATER
B
FOUR-FACTOR DISEASE
Acquisition
LATENT PERIOD
Infection
C

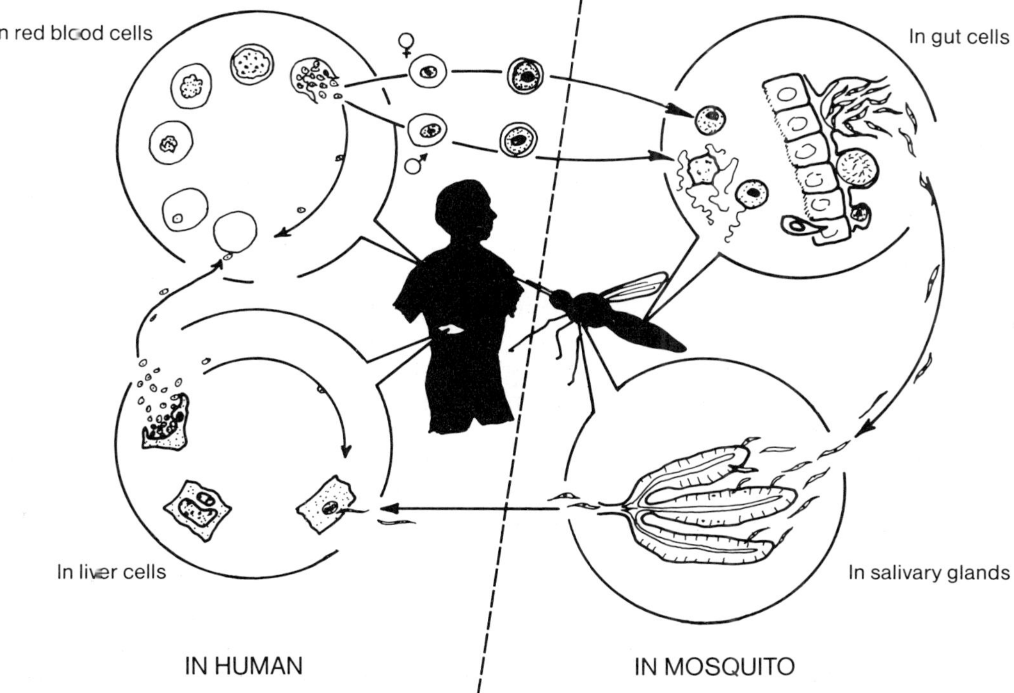

Fig. 9.49 Life cycle of human malaria. The pathogen, Plasmodium *spp., enters the human in the saliva of an infected* Anopheles *mosquito as it bites. The organism enters the human's liver tissue and multiplies. After one or more such cycles in the liver, the pathogens invade red blood cells. Here again, they multiply and cycle, destroying more red blood cells. Some of the pathogens form macro- and microgametocytes. When these two forms are picked up in the blood meal of an uninfected mosquito, they fuse, enter the wall of the mosquito's stomach and later release sporozoites. This form moves to the mosquito's salivary glands, penetrates, and is injected with the saliva the next time the mosquito feeds on a human.*

Fig. 9.48 The arthropod-borne disease complex. (A) In a simple arthropod-associated disease (e.g., malaria), the pathogen is picked up by a vector arthropod from a diseased host and transmitted to a previously uninfected host, which subsequently develops disease symptoms. This is a three-factor disease involving host, pathogen, and vector. (B) In more complex disease relationships (African sleeping sickness) the pathogen may be transmitted from a diseased host (human) to another human, but the pathogen may also exist in areas where there are no humans and other vertebrate species (wild animals, e.g., wart hog) act as reservoirs. Thus African sleeping sickness is a four-factor disease involving host, reservoir, pathogen, and vector. (C) Some aspects common to arthropod-borne diseases include Acquisition (when the vector picks up the pathogen) and Infection (when the vector passes the infective pathogen to a noninfected host.

resistant malaria have been reported from southeast Asia and South America and their incidence is spreading.

Filariasis

Among the filarial worm parasites of humans, *Wuchereria bancrofti* is probably the most widespread and serious. It, like the other filarial parasites, involves humans, filariae, and several species of mosquitoes as vectors. World Health Organization estimates that almost 300 million persons living in the tropical and subtropical areas of the world are infected with *W. bancrofti*, but because the disease is seldom fatal and only slightly debilitating in light cases, its prevention receives little attention.

Mature female filaria worms in the human body produce offspring, called "microfilaria" that circulate in the blood and are picked up when a mosquito feeds. In fact, these microfilariae appear in the peripheral circulatory system only for that part of the day when the potential vectors feed; they are concentrated in the visceral organs at other times. Once the microfilariae are picked up by a vector, they go on to the next phase in their development and reach the infective stage. Those microfilariae in the human's blood that are not picked up by a vector do not mature past the microfilaria stage. This is an evolutionary mechanism that prevents the rapid buildup of large parasite populations that would kill the host in a short time.

Infective filariae that are injected when a vector bites eventually end up in the lymphatic tissue. When these populations become numerous, as with repeated parasitization over a long period, the filariae block the lymph nodes, and, in combination with chemicals they exude, cause a grossly disfiguring condition known as "elephantiasis" (see Fig. 9.50), usually of the breast, scrotum, arms, or legs. The human filaria, *Onchocerca volvulus*, transmitted by black flies of the genus *Simulium* is even more devastating because it can cause blindness. This parasite is restricted to parts of Africa and Central America.

Another widespread filarial parasite, *Dirofilaria immitis* (Leidy), with a life history similar to *W. bancrofti* is becoming common among dogs in temperate North America. This parasite is commonly called the dog heartworm because the mature parasites congregate in the heart and pulmonary arteries. If large numbers (125 or more) of heartworms are present, they can greatly impede blood flow and heart valve function in the host, often debilitating the dog so much that slight exertion causes collapse.

African Sleeping Sickness

This is a disease caused by trypanosomes (flagellate protozoans), which, together with a similar trypanosome disease of domestic animals called nagana, has inhibited the development of almost one-quarter of the land area of Africa. In fact, these two trypanosomes have been referred to by environmentalists as the "saviors" of much of Africa from the ravages of exploitation. Domestic horses, camels, dogs, and mules do not survive exposure to nagana, and cattle or sheep who survive, do not thrive.

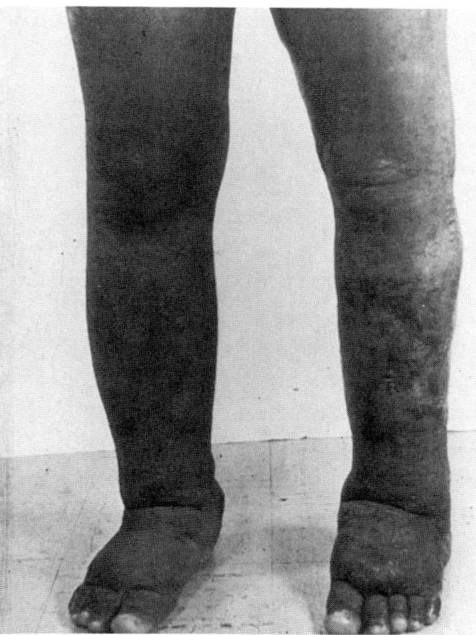

Fig. 9.50 Elephantiasis is caused by the mosquito-borne filarial parasite* Wuchereria bancrofti *and is a result of lymph blockage and chemicals released by the parasite. The greatly swollen appendages and thickened, coarse skin are symptoms from which the condition derives its name.

The disease in humans has become more widespread in Africa following colonialization and the breakdown of tribal barriers. It has been responsible for several catastrophic epidemics. For example, between 1906 and 1911 in the area of Uganda around Lake Victoria, two-thirds of the total population of 300,000 died from sleeping sickness.

Trypanosomes of two different species are transmitted by species of tsetse, genus *Glossina.* Immediately following the acquisition meal, trypanosomes may be transmitted but transmissability tapers off after a day. Then, about 18 days later, the tsetse becomes infective again, following the multiplication and migration of the pathogen within the vector's body. The trypanosomes are located in the blood of the host during the initial phase of the disease. Later they can be found in the cerebrospinal fluid, a diagnostic characteristic of sleeping sickness, which involves drowsiness and languor. Eventually the victim becomes comatose and dies, usually as a result of starvation.

Four factors are involved in African sleeping sickness: human host, trypanosome pathogen, tsetse vector, and wild vertebrate reservoir. Attempts at control of the disease have involved environmental manipulation to eliminate tsetse resting and pupation sites. Other programs have resulted in much wildlife extermination by hunters in attempts to eliminate the vertebrate reservoir. All attempts thus far have yielded only partial success because trypanosomes, like the *Plasmodium* (malaria) organism, have not yet proved amenable to the development of vaccines. Several new discoveries have increased the feasibility of antitrypanosome and antimalaria inoculation programs.

Typhus

This is a group of diseases caused by viruslike microorganisms known as rickettsiae and transmitted by fleas or lice from rodents.

The natural history of typhus rickettsiae is complex and involves numerous vectors and hosts, in addition to more than one species of rickettsia. Humans may become involved in the flea-borne domestic rodent typhus, termed murine typhus, or in the louse-borne epidemic form.

Typhus ranks with malaria, cholera, and plague as one of the four worst contagious diseases of mankind. It is interesting that all four are associated with insects and that neither malaria nor typhus are transmitted under any circumstances except by arthropod involvement. The role of typhus in human history is so important that victory in armed conflict between cities or nations was more often than not determined by the differing levels of disease between the two armies. Epidemic typhus was the most serious overall disease in World War I, killing 3,000,000 Russians alone.

Figure 9.51 illustrates the probable evolution of the two typhus rickettsiae that attack humans.

Humans contracting epidemic typhus exhibit different levels of reaction; children usually have moderate cases, but fatalities often occur in cases involving older persons. Victims demonstrate high fever, severe headaches, and confusion; a brick-red spotty eruption appears on the torso on the fifth or sixth day, and later spreads to the appendages.

Rickettsiae are also responsible for Rocky Mountain spotted fever, a tickborne disease closely related to the typhus organism.

Yellow Fever

Yellow fever is caused by a virus transmitted by mosquitoes, predominantly *Aedes aegypti* (L.). The impact of yellow fever on the development of the tropical and subtropical Americas was severe. For example, yellow fever killed 9000 persons in 1855 in the city of New Orleans. There was a total of 125,000 illnesses and 12,000 deaths throughout the southern United States in 1878. By 1905 the last outbreak of the disease in the United States killed 911 persons. Just prior to this, it had been determined that the vector of yellow fever in urban situations was *Aedes aegypti*, a relationship proved by Walter Reed and his coworkers of the U.S. Yellow Fever Commission in Cuba.

Suppression of yellow fever epidemics proved difficult. The difficulty was not how to stop yellow fever outbreaks, for *Aedes aegypti* control in urban settings proved highly satisfactory. The problem rather, was one of understanding where the initial cases came from. The yellow fever victim can only infect mosquitoes for a few days

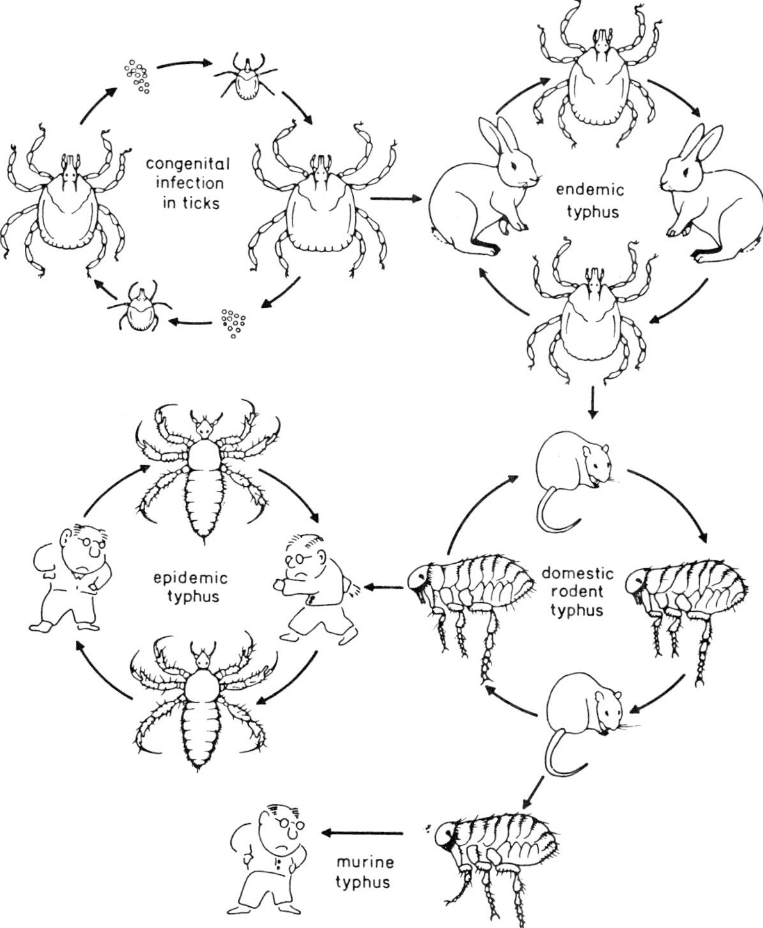

***Fig. 9.51** Probable evolution of the rickettsial typhus fevers of humans. Neither murine (flea-borne) nor epidemic (louse-borne) typhus is transmitted by the bite of the vector. Each involves instead, the vector's fecal contamination of a wound, bite, or scratched area on the human. The recent association between the typhus rickettsia and the louse is demonstrated by the relatively rapid death (8–12 days) of the "vector" after it acquires the pathogen. In fact, typhus appears to be as serious a disease of lice as it is of humans.*

before either dying or developing a life-long immunity. So, where did the outbreaks originate?

In the late 1920s A. Stokes determined that monkeys could be infected with the virus. It was also discovered that several species of mosquito, which ordinarily did not bite humans were the vectors of the monkey disease, called jungle yellow fever. As it turns out, two rather distinct disease cycles exist in nature, with only rare or accidental transmission from one cycle to the other. The mosquito vectors of the jungle disease are active only high in the jungle forest canopy, and therefore do not usually transmit the disease to humans. However, when woodcutters felled these trees the mosquitoes would sometimes bite the woodcutters, thereby transmitting yellow fever from the monkey population to the human population. When the woodcutters return to their villages *Aedes aegypti* would transmit the yellow fever virus to the rest of the population. The complex life cycle is illustrated in Fig. 9.52.

Plague

This disease is another one of monumental importance in the history of humankind. For example, in the first great plague, often termed the Justinian plague, in the mid-500s (A.D.), the disease swept out of Egypt and followed the great trade routes to Constantinople. At its height, the epidemic killed 5000–10,000 residents of Constantinople each day.

The second great epidemic was caused by crusaders who brought the disease back from the Near East into western Europe. It is estimated that 25,000,000 persons in western Europe perished. 111,000 citizens of Cairo, Egypt died in a few short months during the preparations for the Fifth Crusade (1198–1204). By 1956 less than 700 cases of human plague were reported in the entire world—a contrast to the half-million cases a year formerly occurring in India alone.

In 1894 the causative organism for plague was shown to be a bacterium, *Yersinia pestis*, and a year later the role of rats as hosts was established. In 1898 transmission of the plague bacillus by fleas was determined. Thus we have a case where a human is substituted as the host in a disease essentially of another type of animal. In 1914, the Indian Plague Commission proved that the rat flea, *Xenopsylla cheopis* (Rothshild) was the principal vector of the bubonic form of the disease, and that the plague bacilli

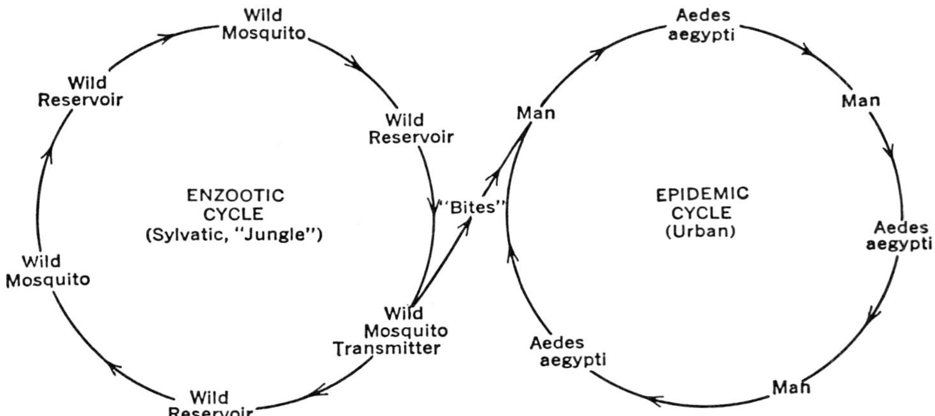

Fig. 9.52 The natural history of the yellow fever virus. Two distinct cycles exist; the jungle form as a disease of wild animals, primarily monkeys and vectored by tree top dwelling mosquitoes, and the urban or human cycle vectored by Aedes aegypti *(L.).*

multiply in the flea gut and block movement of materials. The hungry flea regurgitates during its repeated subsequent attempts to feed and thereby can inoculate several hosts (human or rodent) before the temporary obstruction of the gut breaks down. A second form of plague, the pneumonic type, is a disease of the lungs, transmitted directly by sputum from one human to another, but this usually follows an outbreak of the bubonic type. A third type, septicemic plague, involves the blood and may be picked up through the mucous membranes or through a break in the skin.

About 2 days after becoming infected, there is a sudden high fever, with accelerated respiration and rapid pulse. Some cases exhibit maniacal delirium, others extreme lethargy or coma. In about 75% of the cases a definite "bubo" or swelling occurs, most often in the inguinal area or the armpit, but sometimes in the neck. The primary bubo is a hard, tender swelling of the lymph nodes that may eventually erupt and disappear after draining. Secondary buboes may develop in other regions of the body. Unless treated with antibiotics, cases of pneumonic, septicemic, and bubonic plague are generally fatal. In the past, a 60–90% mortality was common for the bubonic form.

Plant Diseases

The insects that feed on plants have been exposed to the same disease vector potentials as have the insect vectors of human diseases. These plant feeders sometimes have intimate associations with plant pathogens where after the pathogen is picked up by the vector several alternatives may occur before the vector can become infective: the pathogen changes ("cyclodevelopmental" diseases), or the pathogen undergoes changes and also multiplies ("cyclopropagative" diseases), or simply multiplies within the vector ("propagative" diseases). The most common type of transmission, however, by plant disease vectors as compared to animal disease vectors is mechanical transmission. Most plant virus disease transmissions involve contamination of an insect's mouthparts with virus particles and re-insertion into another plant after a relatively short period, sometimes only seconds. Thus, most of the plant viruses require no latent period (also often called an incubation period) prior to becoming infective. In fact, in most instances the longer the delay between acquisition of the virus and the next feeding attempt, the less infective the aphid becomes. Such viruses are termed "nonpersistent" viruses, typified by the cucumber mosaic virus, transmitted by *Myzus persicae* (Sulz.), the green peach aphid. At the other end of the virus scale are the "persistent" viruses that have an incubation period, a particular species as vectors, and retention (persistence) of the virus through more than one stage in the vector's life history. The leaf roll virus in potatoes, again transmitted by *Myzus persicae* (Sulz.) is typical. Note that this is the same aphid listed earlier for the nonpersistent virus. This single aphid species has been shown capable of vectoring over 100 different virus diseases to plants in 30 different plant families, but this large number of disease associations is not common.

We will now describe some typical plant diseases vectored by insects. Table 9.6 lists the plant diseases according to the taxonomic status of the pathogen. We will

Table 9.6 Insect-Borne Diseases of Plants[a]

Diseases	Host Plant	Principal vector
Fungi		
Dutch elm disease *Ceratocystus ulmi*	Elms	Smaller Europena elm engraver *Scolytus multistriatus*
Chestnut blight *Endothia parasitica*	Chestnut	Various beetles
Blue stain *Ceratostomella ips*	Pines	*Ips* bark beetles
Oak wilt *Ceratocystus fagacearum*	Oaks	Nitidulid beetles
Brown rot *Sclerotinia fructicola*	Plum, peach, cherry	Plum curculio *Conotrachelus nenuphar*
Perennial canker *Gleosprium perennans*	Apples	Wooly apple aphid *Eriosoma lanigerum*
Blackleg *Phoma lingam*	Cabbage	Cabbage maggot *Hylemya brassicae*
Ergot *Claviceps purpurea*	Small grains, esp. rye	Flies and bees
Scab *Actinomyces scabies*	Potato	Potato flea beetle *Epitrix cucumeris*
Bacteria		
Bacterial wilt of curcurbits *Erwinia tracheophila*	Squash, melons	*Diabrotica* and *Acalymma* beetles
Stewart's disease *Erwinia stewartii*	Corn	Corn flea beetle *Chaetocnema pulicularia*
Fire blight *Erwinia amylovora*	Apples, pears	Bees, flies, beetles
Bacterial soft rot *Erwinia carotovora*	Cabbage, potatoes, others	*Hylemya* flies
Bacterial rot *Pseudomona melophthora*	Apples	Apple maggot *Rhagoletis pomonella*
Olive knot *Pseudomona savastanoi*	Olive	Olive fruit fly *Dacus oleae*
Rickettsia		
Pierce's disease	Grapes	Leafhoppers, especially *Carneocephala* and *Draeculacephala*
Phony peach disease	Peach	Same as for Pierce's disease
Mycoplasma		
Aster yellows	Many vegetables and flowers	Six-spotted leafhopper *Macrosteles fascifrons*
Peach yellows	Peach	Plum leafhopper *Macropsus trimaculata*
Phloem necrosis	Elm	The leafhopper, *Scaphoideus luteolus*
Peach-X disease	Peach, cherries, nectarines	Several leafhopper species
Pear decline	Pears	Pear psylla *Psylla pyricola*
Stubborn disease	Citrus	Several leafhopper species
Stunt	Corn	Several leafhopper species

Continued

Table 9.6, continued. Insect-Borne Diseases of Plants[a]

Diseases	Host Plant	Principal vector
Viruses		
Cucumber mosaic	Curcurbits	Several aphids and cucumber beetle
Yellow mosaic	Beans and other legumes	Aphids
Common mosaic	Beans	Aphids
Curly top	Sugar beets, beans, tomatoes	Beet leafhopper *Circulifer tenellus*
Yellow dwarf	Small grains, grasses	Many aphid species
Tristeza	Citrus	Aphids
Spindle tuber	Potato	Aphids
Leaf roll	Potato	Aphids
Spotted wilt	Tomato	Thrips of three genera
Leaf curl	Cotton	*Bemisia* spp. whiteflies

[a]Not a complete list.

discuss one disease each caused by a bacterium, fungus, and virus as examples of the associations that may exist.

Dutch Elm Disease

Elm trees, especially American elms, have always been the predominant shade and boulevard tree species throughout much of North America. In fact, during the first decades of the twentieth century we were especially grateful that elms had been so widely planted when a fungus (*Endothia parasitica*) essentially eliminated the American chestnut. The blight was devastating because it wiped out the 50% of the eastern hardwood timber stands comprised by the chestnut.

It is hard to imagine the impact on a community that has just lost its most valuable and abundant timber trees (chestnut) and then another fungus (*Ceratocystus ulmi*) comes along and wipes out all of the boulevard elms. Yet this is exactly what has happened to many areas in the eastern United States because of fungal diseases of trees (see Fig. 9.53).

The Dutch elm disease fungus was apparently introduced into North America in elm burl logs imported from Holland for use as veneer. In Europe the two primary vectors of the disease are *Scolytus scolytus* Fabr. and *Scolytus multistriatus* (Marsham). The bark beetle species *S. multistriatus* had already been introduced and had become established in North America about 20 years prior to the introduction of the fungus. Thus by the time the burl logs were shipped in 1931, all of the elements necessary for an epidemic of gigantic proportions were present: the fungus pathogen (*Ceratocystus ulmi*), an efficient bark beetle vector (*Scolytus multistriatus*), and a highly susceptible host population of elms.

Scolytus multistriatus, like many bark beetles, undergoes its larval development in the nutrient-rich cambium layer of trees, just beneath the bark on dying and newly killed elm trees. The larvae pupate in the region, and the adults emerge through holes

(A)

(B)

Fig. 9.53 (A) An elm-lined residential street in Rumsford, Maine, prior to Dutch elm disease. (B) The same street 6 years later, after Dutch elm disease struck, eliminating the towering elms.

in the bark. These adult beetles fly to the tops of healthy trees and bore into the thin bark of twig crotches to feed (Fig. 9.54) before seeking out newly dead or dying elms as oviposition sites. When appropriate material is found the beetles bore out nuptial chambers in the bark for mating, then the female tunnels along the cambium with the grain of the wood, making a straight egg gallery and depositing a series of eggs along each side. The larvae hatch out and bore away from the egg gallery, making a fan-like gallery pattern as illustrated in Fig. 9.55.

If an elm has been killed by Dutch elm disease and then is used as a breeding site

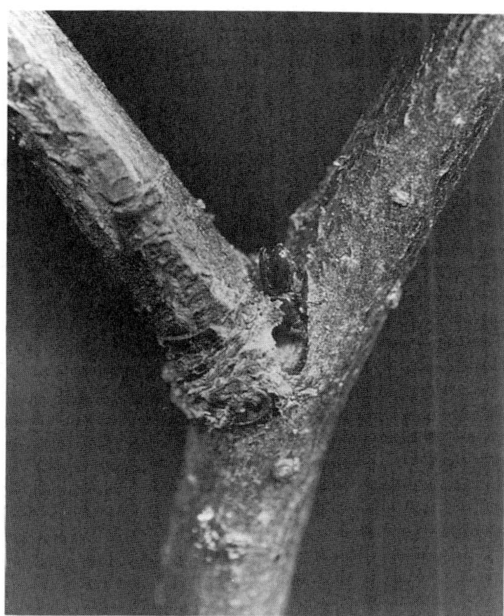

Fig. 9.54 Twig feeding in elms by adult Scolytus multistriatus. *Infection of healthy elms with Dutch elm disease occurs at these feeding sites when the beetle's body surface is contaminated by* Ceratocystus ulmi *spores picked up when the elms logs in which the beetles developed contain the fungus.*

Fig. 9.55 A gallery system produced by Scolytus multistriatus *beneath the bark of elm. The short straight gallery running with the grain of the wood is the egg gallery, bored out by the female after she has been mated. The expanding galleries fanning out from the egg gallery are produced by individual larvae as they feed and grow.*

by *S. multistriatus*, the new adults are often contaminated with fungal spores adhering to the outsides of their bodies as they emerge. Then, when they feed on twig crotches in actively growing elms, the fungal spores are deposited and infection of the tree occurs. The fungus eventually spreads throughout the tree's vascular system and causes production of gums which clog up the system, resulting in the tree's death.

Increased tree mortality provides additional habitats for bark beetle breeding, and therefore more opportunities for fungus transmission. A graphic presentation of the life history of *S. multistriatus* and its interactions with Dutch elm disease fungus is given in Fig. 9.56. Note that a native American bark beetle, *Hylurgopinus rufipes* (Eich.), can also transmit the fungus, but is not as significant a vector since emerging beetles tend to feed on less actively growing tree tissues than does *S. multistriatus*.

Fire Blight

The fire blight disease of apples and pears is caused by the bacterium *Erwinia amylovora* (Burrill).

These bacteria produce a toxic material while growing, which poisons the plant. Flowers and twigs are killed, and large branches and trunks may be girdled so that the whole tree dies. Early flower infections first result in a water-soaked appearance, followed by shriveling and drying up of terminal growth (see Fig. 9.57). The bacteria

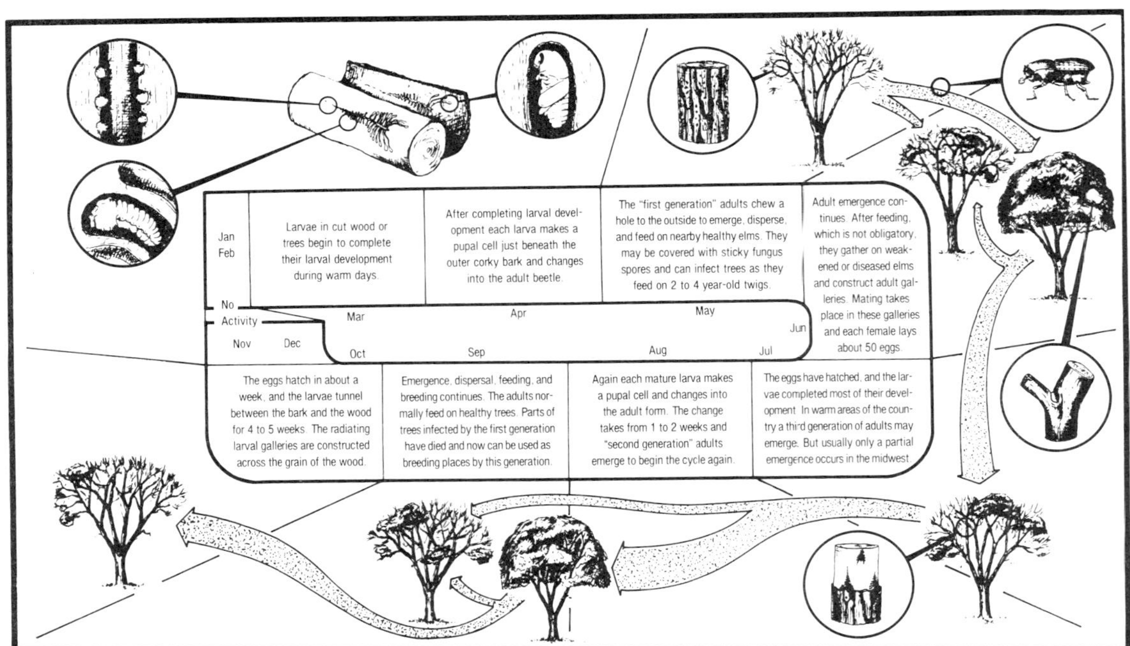

Fig. 9.56** **The seasonal cycle of* Scolytus multistriatus, *the smaller European elm engraver. This beetle is the primary vector of Dutch elm disease in North America.

Fig. 9.57 Early symptoms of fire blight. The tip wilts as blight extends down from the tip of the terminal.

overwinter in the edges of infected areas, called cankers, on larger stems. In the spring these bacteria again become active, multiply, and swell out through cracks in the bark as a sticky sweet ooze. Flies, beetles, and other insects attracted to the ooze pick up the bacteria and spread the infection when they contact flowers as was shown by Waite (1892) in the first report of insect transmission of a plant disease. The complete disease cycle of the fire blight organism, *Erwinia amylovora* (Burrill) is given in Fig. 9.58.

Fire blight attacks many species in the plant family Rosaceae in addition to pear and apple, including neighboring wild or ornamental plants. The importance of pollinators as vectors of the bacterium has been thoroughly documented and the bacterium has been shown to survive up to 20 days in bee hives, but it apparently cannot overwinter there. Although local spread of fire blight is primarily by insects, additional spread occurs through rain splash and even by wind blown strands of the dried bacterial ooze. However, natural movement of the bacteria over long distances does not occur except with the accidental aid of humans. Thus, the Atlantic ocean served as a barrier to fire blight bacteria until as recently as 1957, when *E. amylovora* was first found in England. In the few short years since it reached England the disease has spread through most of Europe.

Potato Leaf Roll Virus

Of the 22 different groups of plant viruses currently recognised by the International Committee on Taxonomy of Viruses, nine groups have viruses with aphids as vectors.

Why are aphids so effective as virus transmitters when so many other insects also

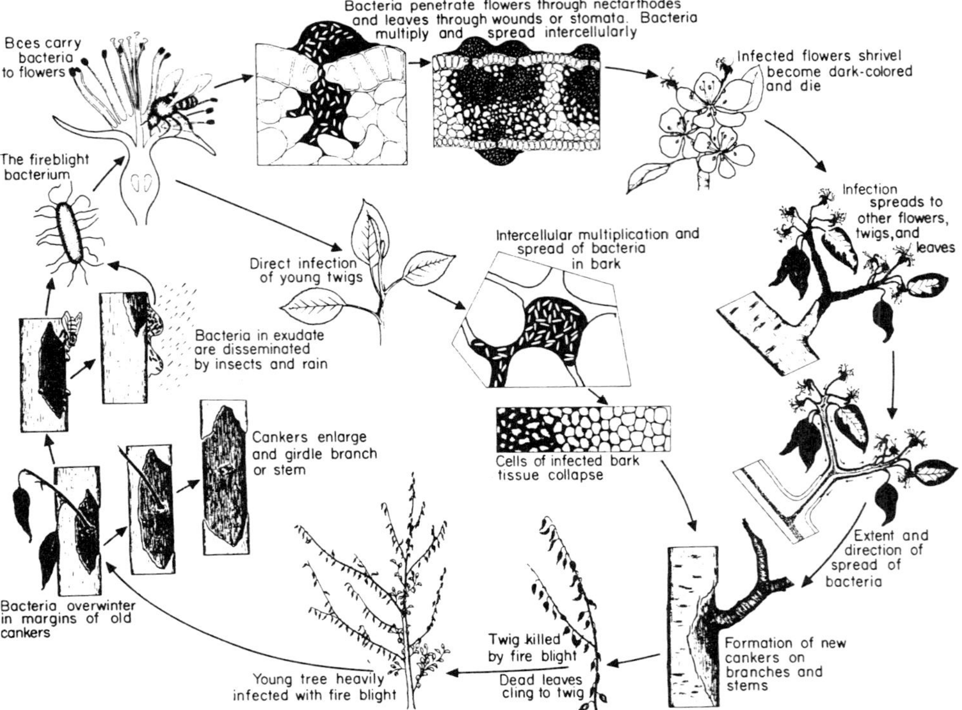

Fig. 9.58 The disease cycle of* Erwinia amylovora *in pear or apple. This was the first bacterium shown to cause disease in plants.

feed on plant sap? The answer may be because their mouthparts create much less cell injury when inserted into tissues than most other sap feeders. This is critical because virus particles cannot survive unless they get into a relatively normal cell. Piercing-sucking insects, like plant bugs and leaf-footed bugs inflict a greater degree of physical damage at the feeding site. In addition, the chemical nature of the salivary secretions greatly inhibits virus transmission. To illustrate, cells injected with virus particles during a feeding probe by *L. lineolaris* would be affected by both the physical lacerations caused by the feeding stylets (needlelike mouthparts) and the toxic saliva of the insect.

Viruses in general are composed of two basic units, a nucleic acid core encased within a protein coat. The core may be of either of the two nucleic acid types, DNA or RNA. Most aphid-vectored plant viruses are single-stranded RNA type, potato leaf roll virus (PLRV) is a double-stranded DNA type.

Virus particles or virions of PLRV isolated from *Myzus persicae* are hexagonally shaped and 2300 nm in diameter (10,000 nm = 1 micron (micrometer); 1000 microns = 1mm).Compared to other plant viruses, PLRV virions occur in very low concentration within the sap (phloem contents) of the potato plant. In spite of this, the plant undergoes some dramatic changes caused by disruption in the phloem's translocation capabilities, so that carbohydrates accumulate in the foliage, rather than moving down the vine and into the tubers. The leaf tissue thickens and the leaf edges roll up toward each other, hence the name of the disease. The tubers are also affected so that a ring of discolored tissue appears about 1 cm beneath the tuber's skin. This "net necrosis" is often visible in potatos purchased by consumers.

A total of nine aphid species have been reported to be capable of transmitting PLRV, but the only one of significance is *M. persicae*, the green peach aphid. Because PLRV is essentially a phloem disease, it is not transmitted by stylet contamination, that is, the virus is not picked up during brief probes of surface tissues as are so many of the "stylet-borne" or "nonpersistent" viruses transmitted by aphids. In fact, extended feeding of at least one hour is required for the aphid to acquire the virus, from the time the aphid first penetrates the cells of the plant surface and reaches the nutrient-rich phloem. Another unusual characteristic of PLRV is that the aphid can remain infective after acquiring PLRV for up to 21 days. This is in contrast to the nonpersistent viruses where the vector can remain infective for a few hours at most, and sometimes only for minutes or even only seconds. In fact, PLRV appears to multiply in the vector's hemolymph as demonstrated by repeated (15 times) blood transfers from virus-carrying aphid to virus-free aphid still resulted in the production of an aphid capable of transmitting PLRV. Such serial transmission is indirect, though strong evidence that the virus does multiply in its vector.

PLRV induces many changes in the host plant, some of which are more important than others. The changes in foliage are of little apparent consequence since the leaves and vines wither and die prior to harvest anyway. The foliar accumulation of carbohydrate, however, does affect tuber yield greatly: a light infection with PLRV reduces tuber yield 50% and heavy infection reduces yields up to 92%. Control of this virus in potatoes is a difficult problem, because it is often compounded by the presence of potato virus Y, potato virus A, potato virus S, potato virus M, potato aucuba mosaic virus, potato calico virus, and cucumber mosaic virus (CMV). All of these viruses are RNA types and are nonpersistent aphid-borne diseases requiring control strategies different than those used for PLRV.

SUMMARY

Almost every crop, including timber, is attacked by insect pests at every phase from seed to finished commodity. Most of the damage is caused by feeding, which includes chewing, boring, and sap sucking. Other damage may be caused by the egg-laying habits of female insects. The results are reduced crop growth, lower quality products, and abnormal plant development, as well as the creation of avenues of entrance for fungi and other pathogens that cause disease.

Crop pests include foliage chewers (grasshoppers, caterpillars, beetles, and sawflies), sapsuckers on stems, leaves, fruit, or buds (aphids, plant bugs, scales), tunnelers (caterpillars, beetles, fruit fly maggots), or subterranean root feeders (beetles, grubs, root maggots).

Because of their valuable aesthetic properties, ornamental perennials require more pest control than forest plantings and present difficulties because of size and location. Major pests include borers, bark beetles, defoliators, and scale insects. Conifers are more susceptible to injury by defoliation than are deciduous perennials.

Pests found in wood used in structures include termites and powderpost beetles that feed on the wood itself, and also carpenter ants that make their nests in structural timbers. Pests in homes and commercial establishments include the four common domestic species of roaches. Pantry pests include beetles and moths that attack dried foods. Clothes moths and carpet beetles attack fibers produced by animals, e.g., wool, furs, felt, and hides.

Insect pests of animals and humans include those that bite or sting as defense against disturbance, e.g., bees, ants, and wasps. Poisons injected by stingers may result in allergic reactions (anaphylactic shock) of the victim if the victim's immunologic defense system has been activated previously.

Annoying insects like the house fly irritate animals, thereby causing reduced production (milk, body weight gains). In humans discomfort results in lost work efficiency and recreational pleasure.

Some ectoparasitic insects (e.g., lice) spend their entire lives on the host animal—even the egg stage is glued to the animal's body covering. Active stages use highly modified tarsi for clinging. Infestation occurs primarily during contact between hosts at mating or from parent to offspring.

Intermittent parasites (e.g., bed bugs) are in contact with their hosts for short periods, usually to feed. While bed bug types live in their host's habitation, the non-insect ticks fall off after feeding and have to "catch" another host along trails frequented by appropriate animals.

Fleas are parasitic only as adults; immature stages usually occur in the host's habitation. Only adult female mosquitoes feed on blood, as do horse flies, deer flies, and black flies. The blood proteins are used for production of eggs. Eggs are laid on or near water and larvae are aquatic. Both sexes of the more advanced tsetse flies are blood feeders. Larvae remain within the adult female's body, nourished by "milk glands" until mature enough to pupate after being deposited.

The three groups of protelean insect parasites (mostly Diptera) of vertebrates are parasitic only in the larval stage. Those that attack the digestive–respiratory systems (e.g., horse bot) use mouth hooks to fasten to the system's inner surface, causing blood loss and irritation before being passed out with the feces to pupate. Warbles and bots form perforated skin cysts on mammals, often after extensive migration within the host's body. Both hide and carcass of livestock suffer damage. Whereas most blow fly species attack dead animals, the screwworm infests open wounds on live animals. Enlargement of the wound increases its attraction to ovipositing females. Such infestations often result in death of the host.

Many insects have been implicated as vectors of disease organisms in animals, humans, and plants. Most vectors acquire the pathogen by feeding on an infected host, but some vectors transmit after acquiring the pathogen directly from the female parent by transovarial passage during egg formation. In some vectors a latent period exists between acquisition and becoming infective. Reservoir hosts may act as a source of the pathogen without showing symptoms of disease.

Mechanical transmission occurs by contamination of the vector's body, usually mouth parts, and simple transfer at the next feeding. In biological transmission the pathogen undergoes changes within the vector's body, sometimes including propagation of the pathogen.

Malaria involves a human host, a mosquito vector, and a protozoan pathogen. The pathogen's sexual stage occurs in the mosquito. Filariasis is a disease caused by worms transmitted between humans by mosquitoes. Like malaria, microfilaria worms undergo changes in the mosquito once they have been picked up from the blood of an infective host. The African sleeping sickness life cycle involves humans, tsetse, protozoans, and animal reservoirs. Typhus is a rickettsial disease usually transmitted to humans by fleas (murine typhus) or lice (epidemic typhus). The yellow fever has two separate life cycles, one involving tree top mosquitoes and monkeys and another involving humans and the *Aedes aegypti* mosquito. The two cycles overlap when the human host takes the place of the monkey. Plague in the bubonic form is transmitted by fleas from the normal rat host to humans. Flea transmission occurs when the flea's gut gets plugged by bacteria. In its pneumonic form plague can be transmitted directly from human to human.

Many plant pathogens are classified according to whether they develop, multiply, and/or change in the vector. Many, especially the viruses, are often mechanically transmitted and may persist in the vector over long periods or, more commonly, for only a brief period (nonpersistent types).

Dutch elm disease is caused by a fungus most commonly transmitted from a newly dead elm to a healthy tree by bark beetles that have developed in the dead tree and become contaminated by fungal spores. Fire blight bacteria are mechanically transmitted by wind, rain, and insects. Insects pick up the overwintering bacteria at cankers or in the nectar of flowers and carry the bacteria to the next plant. Plant viruses are easily transmitted by aphids because of the slight damage inflicted by the aphid's syringelike mouthparts.

Whether insect competition for food, fiber, and structural materials is more significant than their direct attack on humans and the involvement in disease transmission is a moot point. The facts supporting the overall importance of insects as a significant factor in human existence cannot be denied.

SUGGESTED READINGS

Baker, W. L. 1972. Eastern Forest Insects. Misc. Publ. No. 1175 U.S. Department of Agriculture Forest Service, U.S. Govt. Printing Office, Washington, D.C.

Carter, W. 1973. "Insects in Relation to Plant Disease." 2nd ed. John Wiley & Sons, New York.

Ebeling, W. 1975. "Urban Entomology." Division of Agricultural Sciences, University of California, Davis, California.

Furniss, R. L., and Carolin, V. M. 1977. Western Forest Insects. Misc. Publ. No. 1339. U.S. Department of Agriculture, Forest Service. Govt. Printing Office, Washington, D.C.

Harris, K. F., and Maramorosch, K. (eds.) 1980. "Vectors of Plant Pathogens." Academic Press, New York.

Harwood, R. F., and James, M. T. 1979. "Entomology in Human and Animal Health." 7th ed. Macmillan Publ. Co., New York.

Johnson, W. T., and Lyon, H. H. 1976. "Insects that Feed on Trees and Shrubs." Cornell University Press, Ithaca, New York.

Metcalf, C. L., Flint, W. P., and Metcalf, R. L. 1962. "Destructive and Useful Insects." 4th ed. McGraw-Hill, New York.

Pierce, W. D. 1975. "The Deadly Triangle, a Brief History of Medical and Sanitary Entomology." Privately published. Los Angeles, California.

Sweetman, H. L. 1965. "Recognition of Structural Pests and Their Damage." W. C. Brown Co., Dubuque, Iowa.

Westcott, C. 1964. "The Gardener's Bug Book," 3rd ed. Doubleday and Co., Garden City, New York.

Zinsser, H. 1942. "Rats, Lice, and History." Little, Brown, Boston, Massachusetts.

10

Insect Control

From the material presented in the preceding chapter, it is obvious that insects often interfere with human plans, desires, and needs. Such interference may be severe enough to result in human death through disease or starvation. Because of these facts, people often conclude that all insects are "bad" and should be eliminated whenever possible. However, while it is true that certain insects or groups of insects cause highly significant problems for humanity, the proportion of these insect species is minuscule compared to the total number of insect species.

It is estimated that there are 2 to 3 million animal species existing in the world today. Of these, Mayr (1969) states that 1,071,000 have been described and/or classified (see Table 10.1). The total number of described insect species constitutes about 70% of all known animals. However only 2-5% of these are pest species.

Table 10.1 Number of Species in Major Animal Groups

Phylum	Animal	Number of species known
Protozoa	Single celled animals	28,350
Porifera	Sponges	4,800
Coelenterata	Jellyfish, corals, etc.	5,300
Platyhelminthes	Flatworms	12,000
Nemertinea	Ribbon worms	800
Aschelminthes	Roundworms, rotifers, etc.	12,500
Mollusca	Snails, clams, squid, octopus, etc.	107,250
Annelida	Segmented worms	8,500
Arthropoda		
Arachnida	Spiders, scorpions, ticks, mites	57,000
Crustacea	Crabs, shrimp, lobster, water fleas	20,000
Chilopoda	Centipedes	2,800
Diplopoda	Millipedes	7,200
Insecta		750,000
Echinodermata	Starfish, sea urchins, sea cucumbers	6,000
Miscellaneous		
Small phyla		7,800
Chordata		
Chondrichthyes	Sharks, rays	550
Osteichthyes	Bony fish	20,000
Amphibia	Frogs, salamanders	2,400
Reptilia	Snakes, lizards, turtles	6,300
Aves	Birds	8,600
Mammalia	Mammals	3,700
Miscellaneous chordata		1,375

We will begin our discussion of insect control by asking two questions: "What is a pest?" and "How does an insect become a pest?"

WHAT IS A PEST?

Basic answers to this puzzle are a little complex in that they involve an understanding of some ecological principles. Achievement of pest status for a given species is explained in terms of (1) where the various species of insects "fit" in an ecosystem, (2) the artificiality of some ecosystems, and (3) how the "balance of nature" may work to eliminate potential problems.

Where Insects Fit in an Ecosystem

The "ecosystem" is the totality of the living and nonliving components of a given area. Since these components are linked together in many ways (e.g. space, energy, interactions), an ecosystem is, in effect, a functional system. This system may be analyzed on the trophic relationships involved, on the evolution of the ecosystem, and on the feeding relationships existing within the ecosystem. The sizes of ecosystems vary. For example, the northern hardwood forest ecosystem covers many thousands of square miles, whereas the water-filled tree hole is another unique, but much smaller, ecosystem.

Energy is the basis for the interrelations of plants and animals with the abiotic (nonliving, nonorganismal) components in an ecological system. Since energy is required by every plant and animal to maintain its complex structure and to produce new individuals, an analysis of the energy relationships in an ecosystem is instructive.

The sun is the basic source of energy entering an ecosystem. Much (about 50%) of the entering solar energy gets reflected back; however, some of the remaining energy is absorbed, which raises the temperature in the environment. Aside from heat, only a small part (about 5% or less of that initially available) of the energy is retained by the system—that part that is captured by the physiological mechanisms of green plants known as photosynthesis. The plant binds the sun's energy into high energy chemicals (sugars, starches) to maintain itself and to build new tissue or offspring. About 1% of the total originally entering the system (after the plant has used what it needs to survive) is all that is available for the animals to use in their survival and reproduction.

Energy flow involves sources of nutrition (feeding), and the various steps through which the sun's energy passes are known as "trophic levels." The first trophic level is known as the producer level. Only at this level is the sun's energy obtained directly in a usable form. Next come the consumer trophic levels. In the first consumer trophic level, the organisms feed on the green plants that make up the producer level; these animals are known as herbivores. Animals in the third trophic level (or second consumer trophic level) obtain the sun's energy by feeding on the herbivores, which, in turn, obtained the energy from the plants. Animals that feed at this third level are carnivores.

As can be seen in Fig. 10.1 50–80% of the energy is lost at each level; therefore there can be only a few trophic levels. Not shown in the diagram is the decomposer level, made up primarily of bacteria, fungi, and protozoans that break down dead organisms into the less complex chemicals required by plants to capture additional solar energy.

Analysis of an ecosystem based on trophic levels at first appears easy. All one has to do is to categorize the organisms as producer, consumer, or decomposer. In the simplest systems the result is a *food chain* starting with a plant (alfalfa) being eaten by an insect (alfalfa weevil), which is preyed upon by damsel bugs (*Nabis rufiscollis*). In this food chain it is easy to visualize how energy from the sun eventually becomes available to the damsel bug and therefore to appreciate the basic energy relationships within this system. The problem is that such chains seldom exist in nature. The relationships are usually much more complex.

It is possible that the field of alfalfa can exist as a "pure" stand, that is, with few other producers (weeds) within the stand, making the first link in the food chain fairly realistic. However, the numbers of organisms acting as herbivores on alfalfa may be numerous. One sequence is alfalfa–cow–human, another is alfalfa–pea ap-

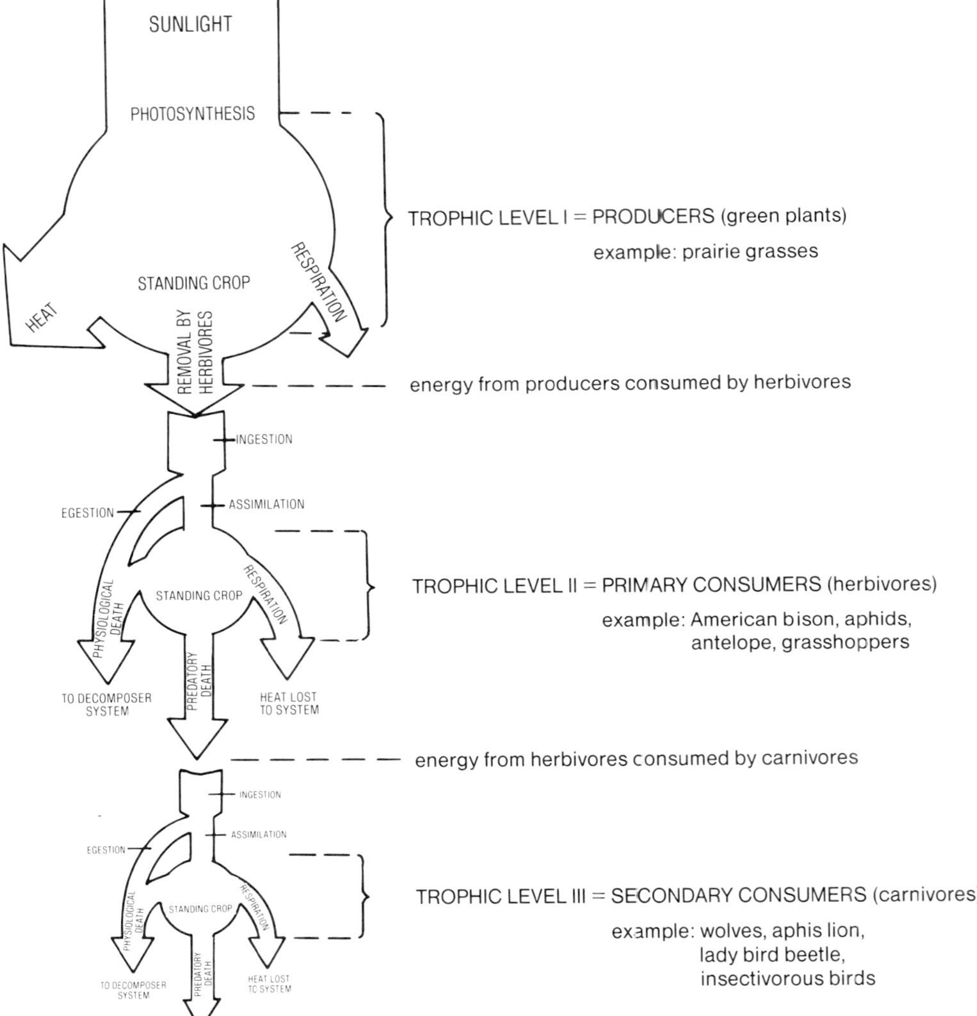

Fig. 10.1 Trophic levels—the system of sequential levels consisting of organisms with common feeding habits, linked and dependent on a continuous energy flow. Organisms in the same trophic level derive their food from plants by the same number of steps.

hid–aphid lion, and so on. The chains become interconnected when predators consume several kinds of prey or when the herbivores feed on several plant species. Thus, the food relationships existing in an alfalfa field become so enmeshed that the simple food chain becomes a complex food web (see Fig. 10.2).

Whether or not an insect species is a pest is determined in part by the insect's trophic role, as well as by its specific feeding habits. Herbivorous insects feeding on crop plants are often pests, whereas carnivorous insects in the same ecosystem may be considered beneficial if they consume pests.

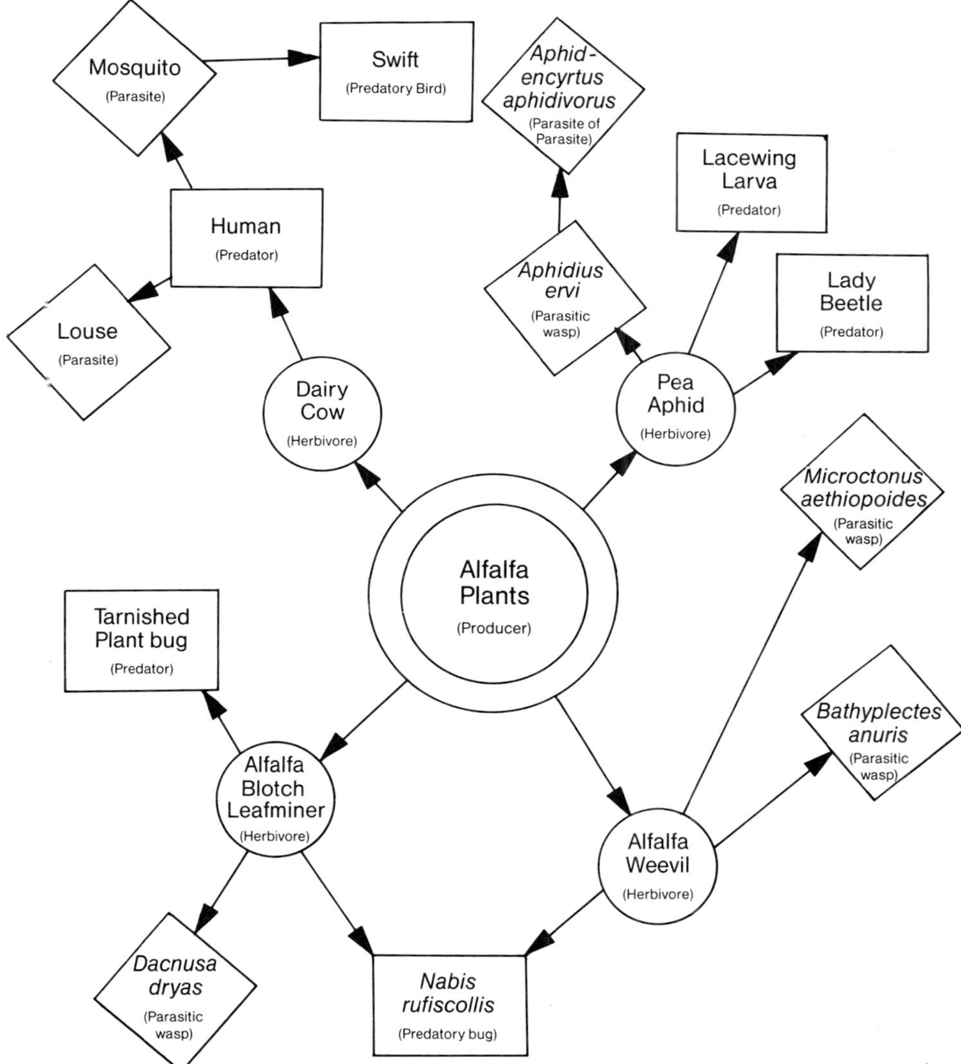

***Fig. 10.2** A portion of the food web that exists in the alfalfa ecosystem. The complexity of the complete food web in alfalfa is immense, since at least 600 animal species may be involved, including over 590 arthropods alone.*

Herbivorous insects feeding on plants that are not utilized by humans are considered neutral in terms of their impact on humans. Since most plants are not utilized directly by humans, the insects feeding on such neutral plants are themselves neutral. Also, those carnivorous insects, whether parasite or predator, that rely on such neutral insects would likewise be neutral. On the other hand, if the herbivores fed on plants that humans consider valuable then the herbivores may be pests, and their parasitic or predatory natural enemies may be considered beneficial.

Even the status of insects that are detritivores (feeding on nonliving or decaying organic matter) varies. If the organic material is debris, the insect is neutral. If it is a wedge of cheese the same insect would be considered a pest. We shall see in subsequent sections of this chapter that the pest status of a given insect depends on more than just what the insect feeds on or its manner of feeding, but also on the population level the insect reaches.

At this point it becomes obvious that the "pest status" of insects is a human construct—one that involves the mutual dependency of humans and various insects on natural resources. When our needs are in conflict we categorize the insect as a pest.

The Artificiality of Some Ecosystems

Since *Homo sapiens* is a part of nature, it may be argued that anything done by humans to the world is "natural." Thus, heavily urbanized areas, strip mines, and corn fields may be considered as natural since they have been developed by a part of nature—mankind. It is also possible to identify massive ecological disturbances that are not attributable to human activity (fires, earthquakes, violent storms), but such arguments become merely philosophical and tend to mask the concept of ecosystem artificiality and the level of change in ecosystems caused by human activity.

A city block in the business district of a large metropolitan area is a unique, and largely artificial, ecosystem. The soil is almost completely covered with surfacing that is impervious to water. Little plant and animal life exists other than humans and their associated organisms. Plant life is restricted for the most part to ornamentals used in street plantings or indoor plants, with a few weeds growing in the soil where the intentionally grown plants are maintained, plus some growing from cracks in sidewalks or streets. The animal life may be of the pet variety (dog, cat, bird, fish), the parasites of humans and pets, plus animals that have found it possible to live in urban settings (pigeon, rat, mouse, roach, and a few others). The point is that no one has difficulty appreciating the artificiality of the urban ecosystem. However, the corn field is in some ways as artificial as the city block.

In a corn field only a single plant species is tolerated or considered ideal. Other plants that occur there are classified as weeds and are removed by physical (cultivation) or chemical (herbicidal) means. Animals present in the field are predominantly based on the corn plant as the producer trophic level. If the corn herbivores (primarily insects) become abundant they are eliminated by chemical means. Soil fertility is analyzed and adjusted by means of chemical additions, usually in the form of synthetic fertilizers.

Not only is the plant species diversity limited, but the age and structural diversity of the corn population is restricted too. All the corn plants are the same age and at the same stage of development because the seed is commercially produced hybrid stock that is genetically uniform and planted at the same time. Such conditions of uniformity hold true with most annual crops as well as some perennials, like trees in forest plantations.

Ecosystems in areas that are not heavily disturbed by humans are either in a stable state or undergo a series of gradual changes until they reach a stable state. For example, a field that previously has been farmed will not remain an open and barren field for long after it has been abandoned. Grasses and annual broad-leaved plants invade and begin to grow the first year. These are followed by shrubs and then by trees that do well in open surroundings. These trees are often species of pines in North America. Pines do poorly in shaded conditions so that the initial stand of pines in an area is usually replaced as other tree species grow. These shade-tolerant trees can be oaks, hickories, beeches, hemlocks, or other species depending on locality (see Fig. 10.3) and often constitute the stable or *climax* state in terrestrial ecosystems. Such a group of plant species and their associated animal forms are termed "biotic communities." Major terrestrial climax communities of the world are termed grasslands, deserts, coniferous forests, or deciduous forests based on the climate in the area. The stability rather than the species involved is what constitutes the climax community, which is characterized by a higher level of species diversity than communities in the developmental stage.

The changes that occur in an area as it progresses toward the climax community are collectively termed "ecological succession." A characteristic of ecological succession is that later stages in the succession tend to accumulate more and more organic matter. The later stages are less productive, but tend also to be more stable. Therefore much of humanity's agriculture has been focused on using the maximum productivity so typical of early successional stages, precisely those stages that are least stable.

Age in Years	Vegetational stage	Dominant plants
Initial		Bare field to crabgrass
1	Grasslands	Crabgrass and horseweed
2		Crabgrass and aster
3 ↓ 20	Grass and Shrubs	Broomsedge and shrubs, then finally, young pines
25 ↓ 100	Pine Forest	Pines mature Broomsedge disappears Hardwood understory appears. Later the pines start to die out, are replaced by oaks and hickories from understory
150+	Oak-Hickory Forest	Oaks and hickories Climax

Fig. 10.3 Ecological succession as it occurs in the southeastern United States on an abandoned agricultural field. Note that the early developmental stages occur rapidly, but that the climax, a stable state, is not reached for many years.

Monocultures

Today's society seems to follow the maxim that bigger is better; at least many facets of our world fit this pattern. The business world is more and more heavily monopolistic with fewer corporations controlling larger segments of commerce. The same holds true for the food and fiber industry where larger acreages are devoted to fewer uses than in earlier decades. The family farm of the early twentieth century was devoted mostly to subsistence agriculture, that is, the owners lived on the land and grew most of what they needed on the farm, in addition to commodities intended for sale. Each farm had a wood lot, a few fruit trees, and a large home garden, plus a variety of animals (dairy cows, chickens, and pigs). Today, successful farms, at least in the midwestern United States, tend to grow only a few crops, notably corn and soybeans. Along with this trend toward decreased diversity is the increase in production of those few plants and animals being raised on farms. The result is termed "monoculture." This term, meaning single culture, could also apply to small acreages, however, the current usage links lack of crop diversity with the trend to increased farm size.

A consequence of reducing crop diversity is the resulting reduction in the overall diversity of consumer organisms. This is easy to visualize when a species of tree is removed from an area. If all of the oaks in an area were eliminated, then all of the organisms that required oaks would also be eliminated, as would be the organisms that are indirectly tied to oaks, such as predators that rely on oak herbivores. Thus, when the oaks are eliminated, a whole complex vanishes. On the other hand, when a "natural" area is brought under cultivation and planted to corn, the insects that live in the corn ecosystem benefit. Certainly, the ability of an insect to locate an adequate and nutritious food source is a requisite for survival, and one that eliminates many individuals. A high population of corn plants practically guarantees that corn herbivores will find a host. Thus, the importance of food as a *limiting factor* (a condition, either maximum or minimum, that approaches the extremes a species can tolerate) is reduced. Another aspect of this is that the insects that prey on or parasitize corn insects also benefit when the corn ecosystem is increased, provided all of the other elements necessary for survival are present. Often, a factor is absent from a monoculture, which may act as the limiting factor for the natural enemy of the insect. For example, many of the wasps whose larvae parasitize aphids, caterpillars, and other corn herbivores require nectar for survival as adults. Increased corn plantings, and therefore reduced field margins will practically eliminate the wild flowers that act as the source of nectar, thus also reducing the numbers of wasps that can survive.

A further consequence of monocultures is the ease with which the system is upset. This is especially true when wide areas have been planted to the few commercial hybrids usually available for any given crop. Since these are developed from only a few genetic lines of plant material, susceptibility to a "new" pest by one common hybrid would endanger a substantial area planted to that crop.

The appearance of a "new" insect or disease organism on a crop is usually the result of either an introduction of the pest into previously unexploited regions, or a change in the organism's ability to utilize a particular crop. Whereas insect invasions into new geographic areas have been documented many times, the "switch" of a known species to new crops has only rarely been documented. One example of such a switch involves the greenbug [*Schizaphis graminum* (Rondani)] and the feed grain sorghum, *Sorghum bicolor* (L.).

In the great plains of the American southwest, sorghum had been grown successfully (almost 14 million acres in 1967) because it not only tolerates the environmental conditions of the region, but it was not used as a food plant by the greenbug, a local aphid that usually attacks barley, wheat, oats, rice, and corn. In 1968, however, during a severe greenbug season, a new biotype of greenbug was reported on sorghum, which could successfully use sorghum as a food plant. Thus, large areas planted to sorghum expressly because it was resistant to a regional pest were now susceptible to the new biotype. This topic is covered in more detail later in the chapter under the subsection termed "Host Resistance."

Elimination of Potential Pests by "Balance of Nature"

As was shown in Chapter 1, most species of animals maintain a fairly stable population level over long periods of time (for insects, decades). However, within a season or between successive years population levels may fluctuate widely at a given location. The exact reasons for such fluctuations are attributed by scientists to different causes.

Chapter 1 presents one scheme whereby the environmental components that interact to affect population size are grouped according to physical and chemical factors, other organisms, food, and a place to live. One aspect to be considered here is the response that insects have to varying conditions of the environment. For example, insects respond to variables such as temperature, light, and moisture in two different ways. In the two-tailed response (Fig. 10.4), suboptimal conditions at both extremes cause such responses as reduced vitality and lower reproduction. In a one-tailed response, such as the response to oxygen levels by terrestrial animals, there is a minimum level required, but no maximum since there is never too much oxygen. Thus, environmental conditions affect insect numbers by producing maximum growth and development in optimal conditions, but suppressing growth and development under suboptimal conditions.

The classification in Chapter 1 only briefly distinguishes between factors whose severity of action on the population is independent of the population density from those whose severity of action depends on the population density. An example of density-independent action would be unusual weather such as a late spring frost, which might kill all individuals regardless of density. A counter argument is that if individuals in protected sites do survive, then the cold snap is indirectly density dependent, because its action is only effective against those individuals that have not been able to secure a sufficiently protective site (which must then be a density-dependent factor).

Density-dependent factors may be of two types: direct and inverse. A direct density-dependent factor has a greater effect in higher populations than in low populations. For example, virus diseases of insects kill a small percentage of low populations, in part because transmission of the disease through contact occurs rarely in sparse populations. The same disease moves quickly through dense populations be-

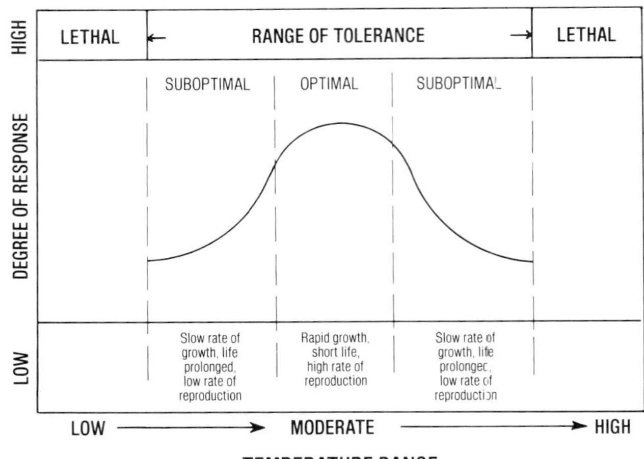

***Fig. 10.4** Diagram of an organism's response to an environmental variable. Note that this is a diagram of a "two-tailed" response with suboptimal conditions at both high and low ends of the range.*

cause of frequent encounters between diseased and healthy individuals. Inverse density-dependent factors have a greater effect in sparse populations than in dense ones. For example, in a low-density aposematic species (1) each predator (e.g., bird) requires a specific number of aposematic contacts before learning to avoid the aposematic pattern, and (2) very sparse aposematic populations could interrupt reinforcement of avoidance just through the low frequency of encounters between them and potential predators.

Most insect species never become very abundant, that is, enough to become pests, because as their numbers increase so do the numbers of their natural enemies (parasites, diseases, and predators). In addition, greater populations of a species at some point begin to affect adversely the food supply. Therefore, the tendency for a given species to become more abundant because of its inherent ability to produce numerous offspring is offset by the response of the environment, either through increased pressure from natural enemies, or because of food problems (contamination, reduced growth), or both. Some examples of pest complexes will verify this relationship.

An analysis was made of the insects that live on corn plants in Minnesota. Plants were examined over a 5-year period, and the insects found on the plants were collected and sorted to trophic level. The results were that 63 species of herbivorous insects were using corn as their food plant. However, only 17 of the 63 species ever become pests of corn in Minnesota. That means that most of those herbivores, although present, never become abundant enough to cause problems in corn production. Also, out of the 17 pest species, only nine are common enough each year to warrent consideration of suppression. The majority never reach the pest status, a concept we will discuss later.

As another example, scientists at Cornell University determined that 591 species of insects and related arthropods lived in alfalfa fields. Of these 591, only 311 were herbivores, but this large number included pollen feeders and detritivores. However, if only those herbivores that are economically important are considered, then only eight insects are left. Again, the majority of herbivorous insects never become a problem in production of alfalfa.

A study of the insect fauna found in Wisconsin apple orchards showed that over 100 of the 760 species found were herbivores on apples or apple trees. Only 43 were economically important and less than 10 were considered to be really serious pests. Here again is a large herbivore species class with very few pest species involved. But a closer look reveals that some of the pests we now regard as serious have not always been regarded as such.

HOW A SPECIES BECOMES A PEST

Pests and Thresholds

The concept of being a "pest" is an interesting one for several reasons. First, it concerns the human viewpoint of an organism's "place" in nature. Second, it is a quantitative concept involving population levels. Third, a given insect's pest status may change over time or under different circumstances. A pest, then, is an organism that reduces the availability of some resource valued by humans, such as a commodity, or one's health, or aesthetic pleasure. As such, it is an economic concept, too.

Few insects are pests individually because of their small size and the resulting lack of magnitude of their effects. For example, a single greenbug has practically no effect on its host, corn. However, in very large numbers the same species can kill a plant by the quantity of poisonous saliva injected into the plant tissues. Thus, the size of the pests' population is important, but only if the population and the damage it causes can be tied to the monetary value of the crop.

Entomologists use the term "economic threshold" to describe the lowest population density that will cause economic damage. Of course, any damage causes loss in value but economic damage as used in this definition is that point at which the losses become greater than the cost of applying some type of population suppression technique.

The long-term average density levels of a population without artificial manipulation of an insect tend to determine whether an insect is an economic problem or not. If the "general equilibrium population levels" are usually below the economic threshold then the insect usually is a nonpest. If it is usually above the level, then the insect is a rather constant pest. Of course, there are cases between these extremes, in which an insect sporadically reaches pest status or does so for a small, but regular, part of its seasonal life history (see Fig. 10.5).

The economic threshold for mosquitoes on humans in most of the continental United States has not been established. This is a result of our inability to quantify the levels of mosquitoes the human population can tolerate. At present, this level of tolerance appears to depend on exposure. Those who have been exposed to large

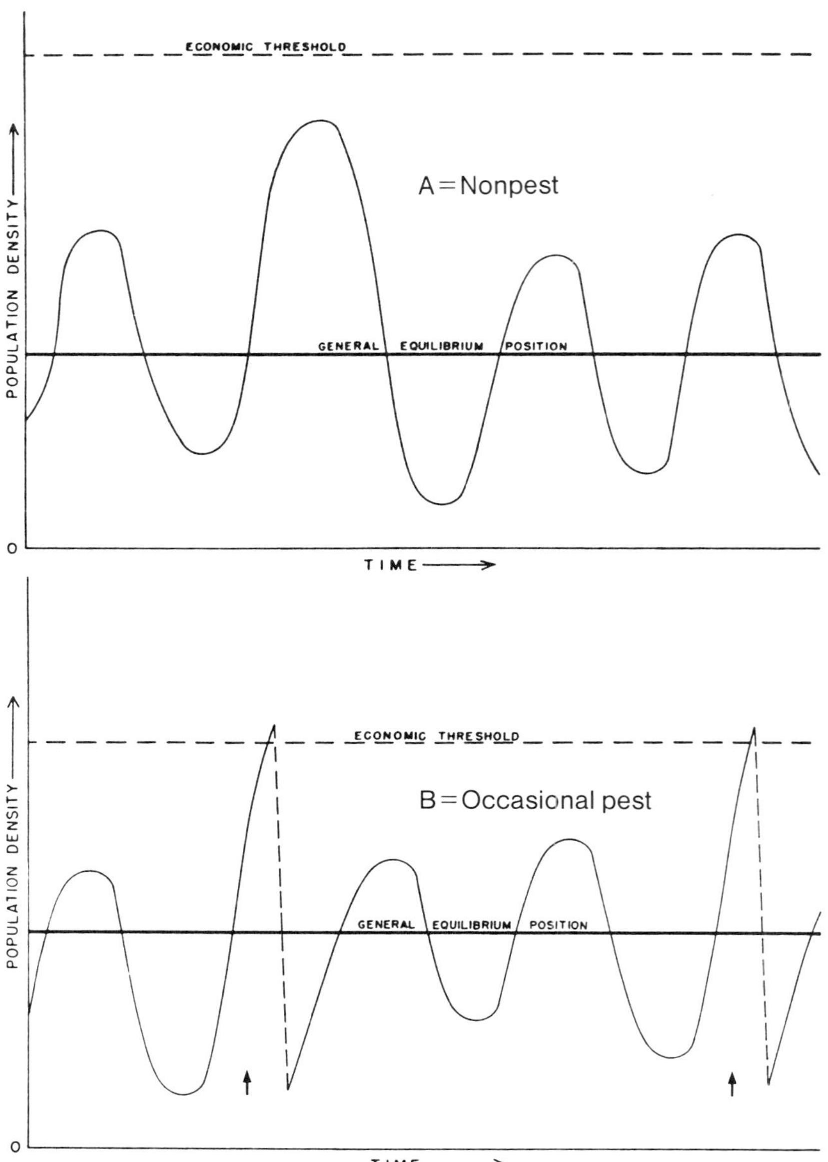

Fig. 10.5 Typical population levels of insects from the economic aspect. (A) Nonpest insect (Lygus on forage alfalfa) never reaches economic threshold. (B) An occasional pest (armyworm on corn) sporadically reaches the economic threshold. (C) Perennial pest (Colorado potato beetle on potatoes) often, though not always, reaches the economic threshold. (D) The perennial pest (codling moth on apple) is always present at densities that exceed the economic threshold. ↑ = pesticide application

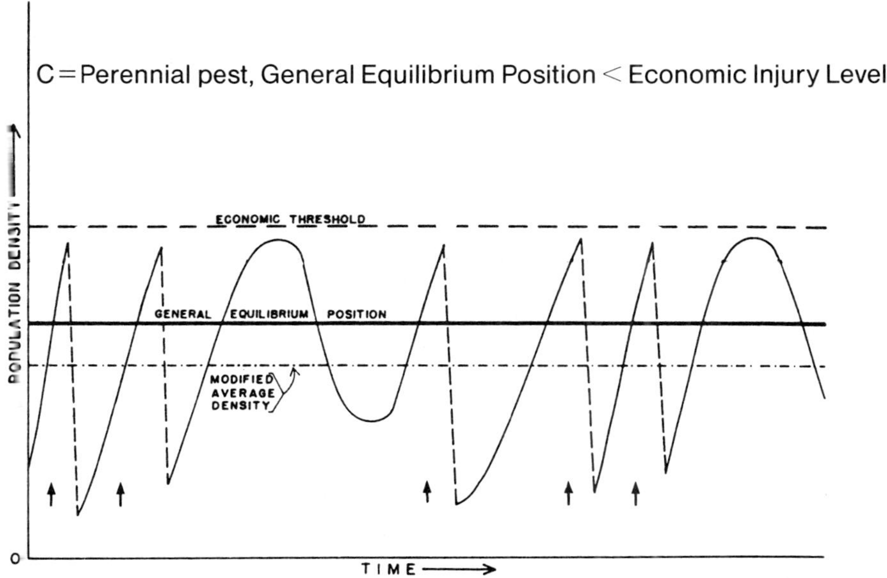

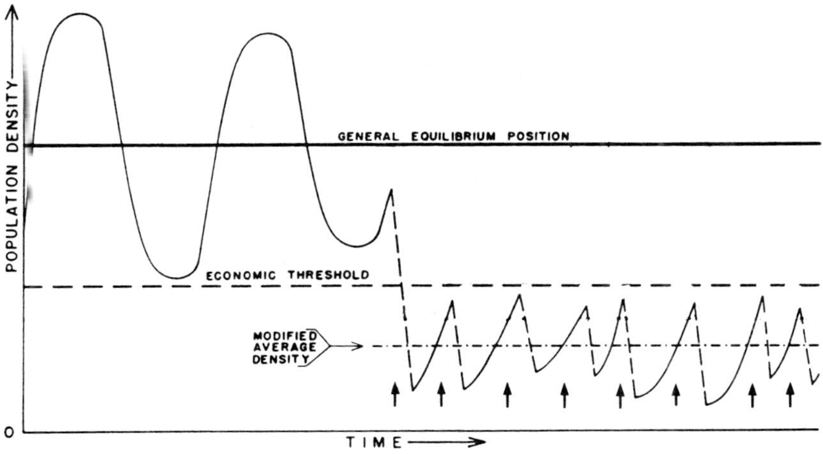

Fig. 10.5, continued.

populations of mosquitoes are more tolerant than those who have not. Basically, mosquitoes in the United States are a nuisance to humans. But, if a mosquito-borne disease (encephalitides, yellow fever, malaria, or dengue) is involved, the economic threshold of mosquitoes falls, essentially to zero. In other words, mosquitoes that might be vectors of a serious or fatal human disease are not tolerated in any numbers at all, while rather high populations of the same species are tolerated if no disease is known in the area.

The establishment of an economic threshold involves several factors: (1) the object of the attack, (2) the timing of the attack, and (3) the nature of the injury. For example, even slight chewing injury by an insect to a developing apple is enough to cause extreme loss in market value. Chewing injury to the foliage of plants being raised for cattle feed causes much less concern, since visual signs of injury do not affect an animal's feeding. Only when the chewing becomes severe enough to interfere with the plant's growth and storage capacity is the level reached.

The "action threshold" is the density at which control measures are applied to prevent an increasing population from reaching the economic threshold. There is always some lag time between the point at which a population is found to be approaching the economic threshold and when some type of control can be applied. In the meantime the insect population may continue to grow and/or multiply.

Consumer Demands and Economic Thresholds

The majority of modern society no longer lives in contact with agriculture and the problems of growing foods and other commodities (e.g., cotton, wood). People accustomed to urban life have learned to demand unblemished foods. As consumers we have been taught to look over merchandise before it is purchased to ensure that there are no flaws. Why then should we pay the same price for an apple with a worm in it as for a worm-free apple. If we reconsider though, a single worm really does very little real damage and with removal of it and a little fruit tissue the fruit can be consumed. However, the desire for blemish-free produce has caused growers to revise their economic thresholds downwards so that fewer and fewer insects can initiate an extensive and often expensive control procedure. These production costs are passed on to the consumer.

The demand for pure foods and the increased use of insect controls for cosmetic improvement of the foods has not always resulted in overall improvement of food quality. Growers do not necessarily discard infested farm produce, but use it in different forms which may mask insect infestation. For example, it is difficult to determine if worms were present in apples (or oranges, or tomatoes) once they have been squeezed for juice. And, it is impossible to recognize insect infestation once foods like corn, wheat, and potatoes are extensively processed into corn meal, wheat flour, and dried potatoes.

Consumer demands and the responsible government regulatory agencies, like the Food and Drug Administration have combined to "improve" the quality of foods, when possible. This is not always easy, because of the difficulty of eliminating minute insects from certain commodities. The solution has been to set tolerance levels of acceptance, in other words, levels of insect presence in foods that are acceptable. These are called defect action levels (DAL). For such delicate and difficult to clean fruits as raspberries, with their tight clusters of individual fruits, the DAL set by FDA is four insect larvae per 500 grams. This DAL does not even include the very small thrips, mites, and aphids that infest such fruits in large, though inconspicuous numbers. The DAL for spinach is an average of 50 thrips, mites, or aphids per 100 grams of spinach or an average of eight leaf miners per 100 grams. Of course, such insects are more readily concealed in chopped spinach than in whole leaf spinach.

INTEGRATED PEST MANAGEMENT

Analysis of a Pest Life System

In order to envision the ways in which numbers of a pest species can be suppressed below the economic threshold, it is necessary to analyze the environment of the pest. Instead of trying to study the complete ecosystem in which the pest is found, however, it has proved more fruitful to look at those elements of the ecosystem that are directly involved in the life of the subject species. In this case, the subject species is the pest insect and the organization of ecological information takes the form of a pest life system (see Fig. 10.6).

It is critical to remember in using the pest life system that (1) the object is to keep the pest from reaching the economic threshold of abundance and (2) since the pest life system is a functional one, each element's interactions with the others serves to

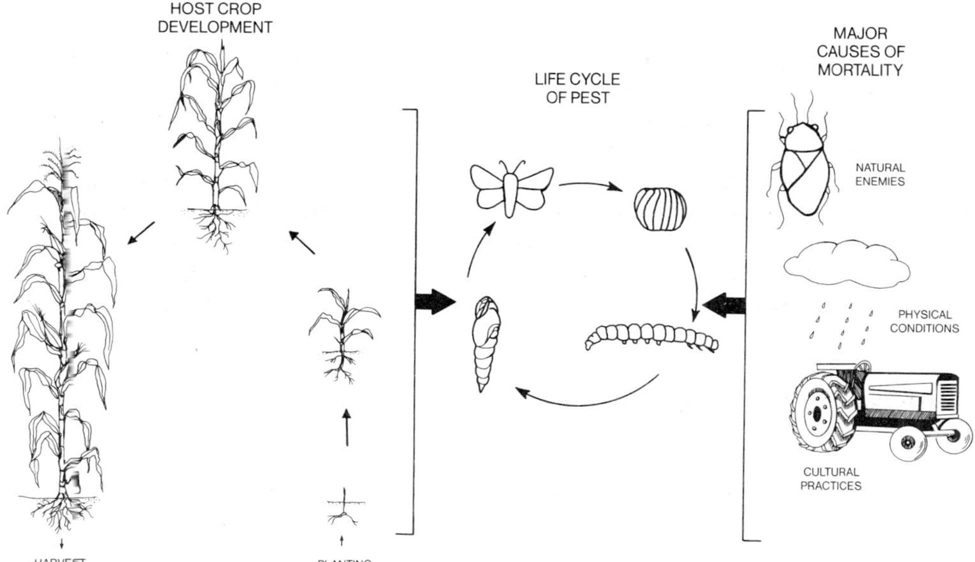

Fig. 10.6 Basic elements of a pest life system. Although this is an oversimplification, it is a starting point for a realistic analysis of the key factors acting on the insect.

emphasize that careless manipulation of one element may disturb the entire system. Taken together, these two aspects show the direction that modern pest suppression has taken—a direction based on careful monitoring of pest populations to determine trends, coupled with judicious use of a variety of suppression techniques that provide minimal disruption to the ecosystem. This adequately describes the control *strategy* (science of planning and directing large-scale operations) known as pest management. Before discussing this strategy further it is necessary to look more closely at both monitoring and ecological disruption.

Many years ago, entomologists recognized that when they tried to control one insect pest they often changed the ecology of the controlled area—the crop—to the point where another arthropod that had not previously been a problem began to be one. A good example is the codling moth caterpillar and mites on apples. In this case, the insecticide DDT was a new chemical (1940s) that proved to be remarkably effective against codling moth but nontoxic to mites. Most growers soon began using DDT as a control, and the levels of losses that had been previously attributed to codling moth were greatly reduced. However, applications of DDT also killed many other insects in the orchards, including the insect predators of the plant feeding mites. When the lacewings, lady beetles, and other insect predators were eliminated, the mite populations exploded into astronomical numbers. Thus, the mites were allowed to reach their economic threshold because the ecosystem had been disrupted.

It is also true that control of insect pests sometimes leads to outbreaks of other pest organisms. For example, the mechanical removal of suckers (water-shoots) on tobacco plants removes the favored site for development of the second generation tobacco bud worm population. In addition, it stimulates development of larger leaves on the main stem. One problem with this practice is that the disease tobacco mosaic virus is spread throughout the field during the removal of the suckers.

It is through the observation of such events that pest control scientists began to appreciate that the various methods of control for a given insect pest each had different effects on the rest of the crop ecosystem. The selection of control procedures must be evaluated with respect to its effect on the other pests (or potential pests) present in the crop. This approach, which depends on the use of multiple tactics in a compatible manner to maintain pest damage below the economic threshold, is termed "integrated pest management" (IPM). All key pests (those capable of causing major losses) including insects, weeds, and diseases are considered. Most important is the philosophical approach to the management of the entire pest complex in an economically and ecologically sound manner, thus optimizing control.

Monitoring Pest Populations

The only way that a complex of pests can be managed effectively is by monitoring pest levels at intervals throughout the year or cropping season. Since populations of insects usually occur in very large numbers, it is impossible to census every individual in an area. For example, 24 gypsy moth egg masses containing from 100 to 1000 eggs were found in one 20-minute inspection of tree trunks by one person in one-quarter acre area. Thus, a potential of 20,000 or so larvae could have hatched, in that quarter acre, just from the egg masses discovered. This does not include the egg masses that were missed. The solution to the problem of quantifying population levels is to select a sample (part) from the overall population and then to infer the characteristics of the whole population from the characteristics of the individuals in the sample.

Samples of populations are made in many different ways, depending on habitat, size, and the type of insect, including some of its biological characteristics. Common techniques involve use of an insect net to sweep insects from foliage (grass, alfalfa), or by separating live insects from debris (e.g. forest litter) by use of heat and light. Other techniques involve direct counts of insects per unit (leaf, plant, fruit, cow, etc.) or of damage.

Making Action Decisions

An example of the type of sampling information used by pest managers is shown in Fig. 10.7 for cotton. In this example, a paid insect scout has visited several fields and has reported both the numbers of specimens and percentage of damage for the key

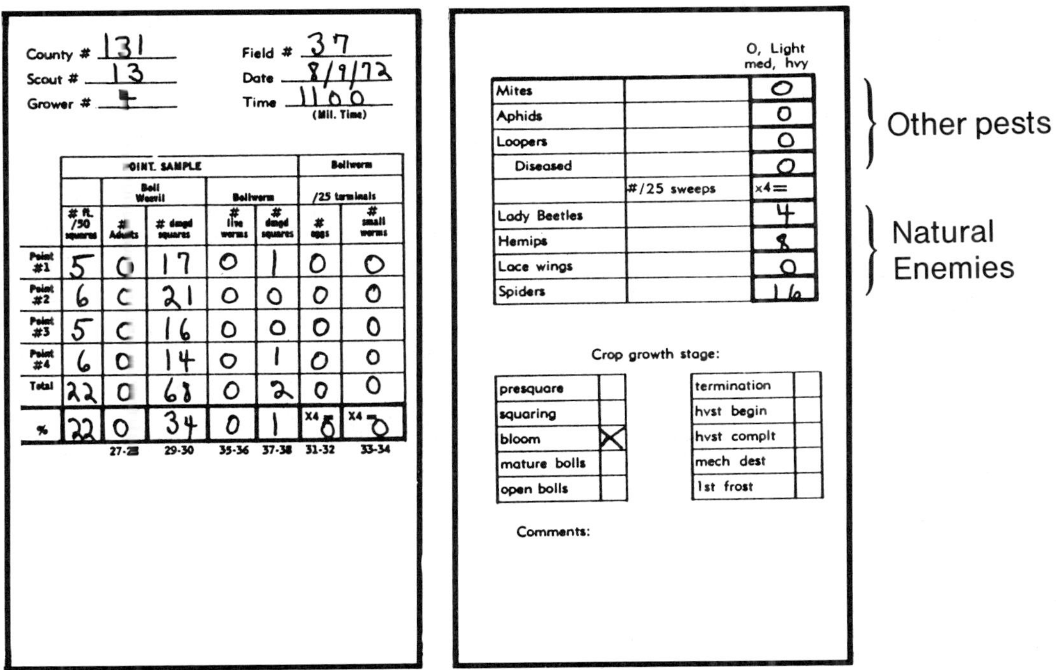

County # 131 Field # 37
Scout # 13 Date 8/9/72
Grower # 4 Time 1100 (Mil. Time)

	POINT. SAMPLE					Bollworm	
		Boll Weevil		Bollworm		/25 terminals	
	# ft. /50 squares	# Adults	# dmgd squares	# live worms	# dmgd squares	# eggs	# small worms
Point #1	5	0	17	0	1	0	0
Point #2	6	0	21	0	0	0	0
Point #3	5	0	16	0	0	0	0
Point #4	6	0	14	0	1	0	0
Total	22	0	68	0	2	0	0
%	22	0	34	0	1	X4 = 0	X4 = 0
		27-28	29-30	35-36	37-38	31-32	33-34

		0, Light med, hvy
Mites		0
Aphids		0
Loopers		0
Diseased		0
	#/25 sweeps	x4=
Lady Beetles		4
Hemips		8
Lace wings		0
Spiders		16

Crop growth stage:

presquare		termination	
squaring		hvst begin	
bloom	X	hvst complt	
mature bolls		mech dest	
open bolls		1st frost	

Comments:

Fig. 10.7 A scout work card for cotton. Data on population levels of pests and their natural enemies, plant damage, and crop growth stages are gathered at frequent intervals to track the pests so that they can be maintained below their economic thresholds.

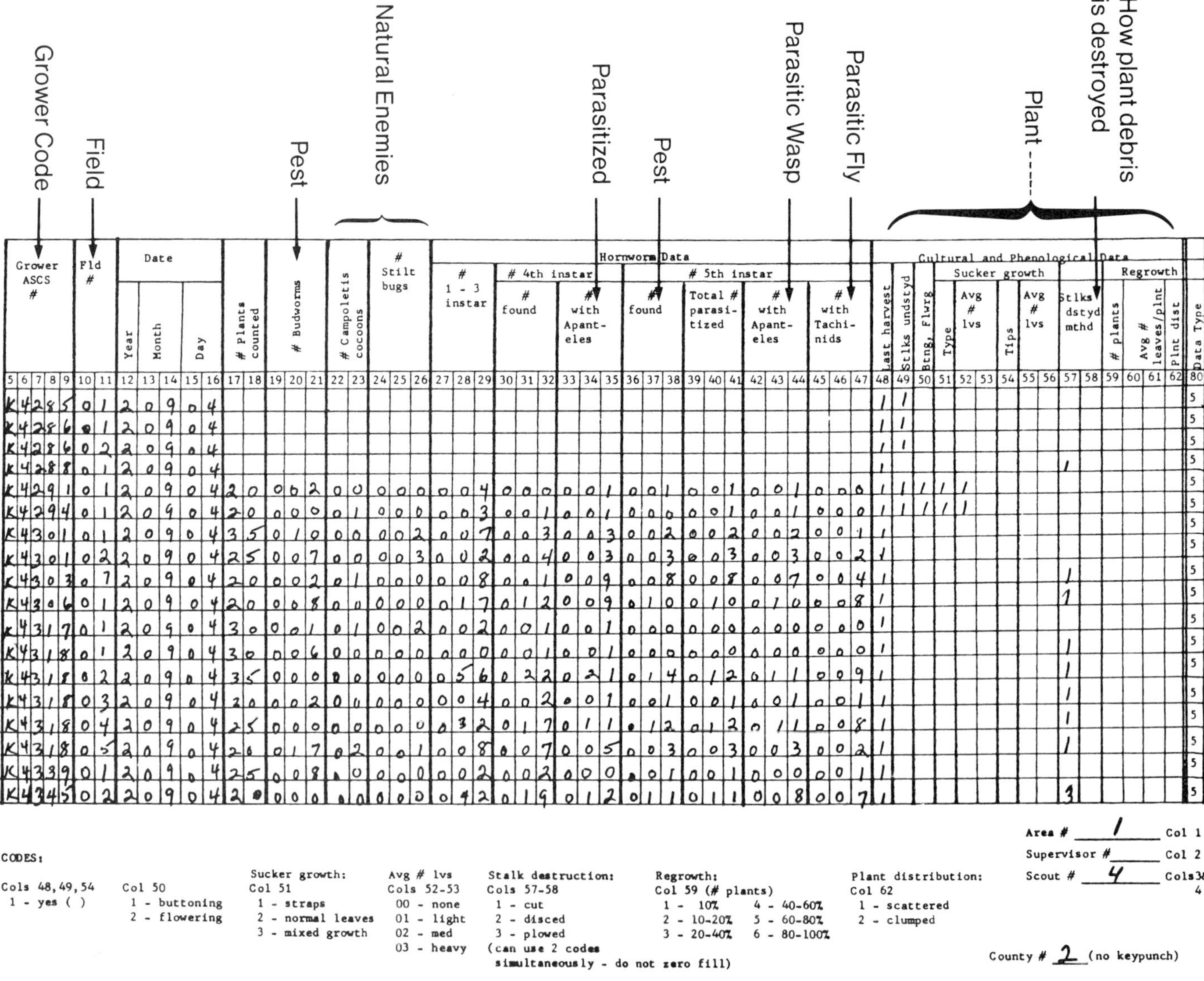

Grower ASCS # (5–9)	Fld # (10–11)	Year (12)	Month (13–14)	Day (15–16)	# Plants counted (17–18)	# Budworms (19–21)	# Campoletis cocoons (22–23)	# Stilt bugs (24–26)	Hornworm Data: # 1–3 instar (27–29)	# 4th instar: # found (30–32)	# 4th instar: # with Apanteles (33–35)	# 5th instar: # found (36–38)	# 5th instar: Total # parasitized (39–41)	# 5th instar: # with Apanteles (42–44)	# 5th instar: # with Tachinids (45–47)	Last harvest (48)	Stlks undstyd (49)	Btng, Flwrg (50)	Sucker growth: Type (51)	Sucker growth: Avg # lvs (52–53)	Tips (54)	Avg # lvs (55–56)	Stlks dstyd mthd (57–58)	Regrowth: # plants (59)	Avg # leaves/plnt (60–61)	Plnt dist (62)	Data Type (80)
K4285	01	2	09	04												1	1										5
K4286	01	2	09	04												1	1										5
K4286	02	2	09	04												1	1										5
K4288	01	2	09	04												1							1				5
K4291	01	2	09	04	20	002	00	000	004	000	001	001	001	001	000	1	1	1	1	1							5
K4294	01	2	09	04	20	000	01	000	003	001	001	000	001	001	000	1	1	1	1	1							5
K4301	01	2	09	04	35	010	00	002	007	003	003	002	002	002	001	1											5
K4301	02	2	09	04	25	007	00	003	002	004	003	003	003	003	002	1											5
K4303	07	2	09	04	20	002	01	000	008	001	009	008	008	007	004	1							1				5
K4306	01	2	09	04	20	008	00	000	017	012	009	010	010	010	008	1							1				5
K4317	01	2	09	04	30	001	01	002	002	001	001	000	000	000	000	1											5
K4318	01	2	09	04	30	006	00	000	000	001	001	000	000	000	000	1							1				5
K4318	02	2	09	04	35	000	00	000	056	022	021	014	012	011	009	1							1				5
K4318	03	2	09	04	20	002	00	000	004	002	001	001	001	001	001	1							1				5
K4318	04	2	09	04	25	000	00	000	032	017	011	012	012	011	008	1							1				5
K4318	05	2	09	04	26	017	02	001	008	007	005	003	003	003	002	1							1				5
K4339	01	2	09	04	25	008	00	000	002	002	000	001	001	000	001	1											5
K4345	02	2	09	04	20	000	00	000	042	019	012	011	011	008	007	1							3				5

Area # 1 Col 1
Supervisor # ______ Col 2
Scout # 4 Cols 3& 4

CODES:

Cols 48, 49, 54
1 - yes ( )

Col 50
1 - buttoning
2 - flowering

Sucker growth:
Col 51
1 - straps
2 - normal leaves
3 - mixed growth

Avg # lvs
Cols 52-53
00 - none
01 - light
02 - med
03 - heavy

Stalk destruction:
Cols 57-58
1 - cut
2 - disced
3 - plowed
(can use 2 codes simultaneously - do not zero fill)

Regrowth:
Col 59 (# plants)
1 - 10%
2 - 10-20%
3 - 20-40%
4 - 40-60%
5 - 60-80%
6 - 80-100%

Plant distribution:
Col 62
1 - scattered
2 - clumped

County # 2 (no keypunch)

Fig. 10.8 Tobacco pest management data compilation form shows data that are not entered directly are coded according to the system listed beneath the form.

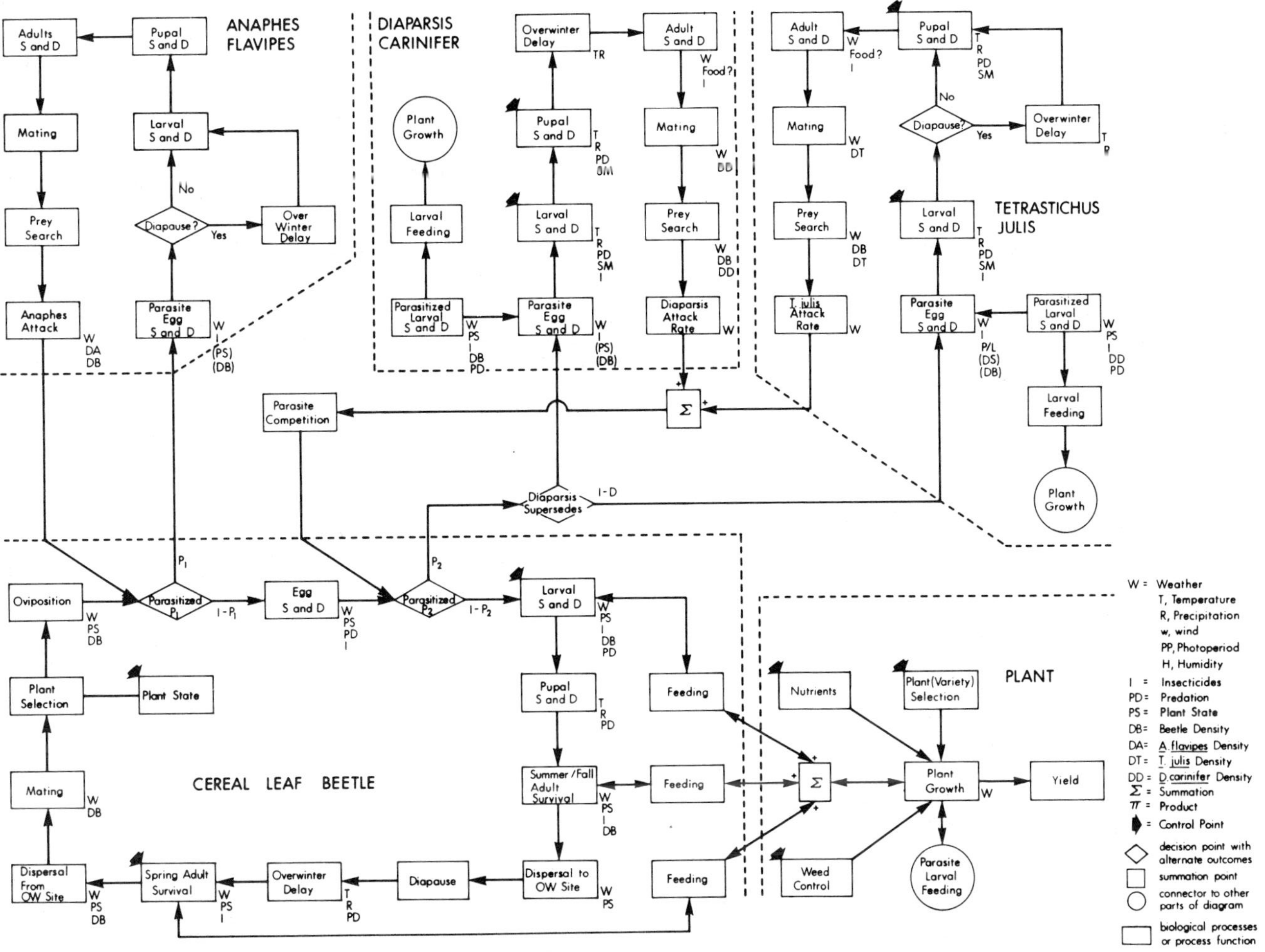

Fig. 10.9 Factors affecting the population levels of the cereal leaf beetle. Three major parasites' biologies and plant growth impinge on development of the pest population. Large arrows (◆) indicate points at which controls might be initiated.

pests, boll weevil and bollworm. Other information reported includes numbers of mites and aphids present, the plant density, and the crop's stage of development. Sampling data are summarized (Fig. 10.8) and are used to (1) determine population levels of pests; these levels are evaluated against (2) the economic threshold for the crop at its stage of growth. Two other factors are also introduced: (3) population levels of natural enemies (Figs. 10.9 and 10.10) and (4) predictions of short-term weather. Then, the four aspects are evaluated to predict the pattern of the pest population versus its natural enemies and crop maturation. If, at that point, it appears that the combination of weather and natural enemies will not hold the pest below its economic threshold, one or more control tactics are used (Fig. 10.11).

Because data for many components in the ecosystem must be considered in an IPM program, the easiest (and sometimes only) way to handle these masses of data is to computerize the whole operation. These data include abiotic factors (weather, soil, or water characteristics, topography); biological factors (insect, plant, mammal, fungus); amount of precipitation; levels of soil moisture; numbers of pest larvae; and growth rates of pests, crops, etc. IPM scientists have constructed mathematical models of real life processes (e.g., for numbers of eggs produced by a single insect under different conditions). These mathematical models may then be manipulated to determine probable outcomes that might occur under different conditions, a process called computer simulation.

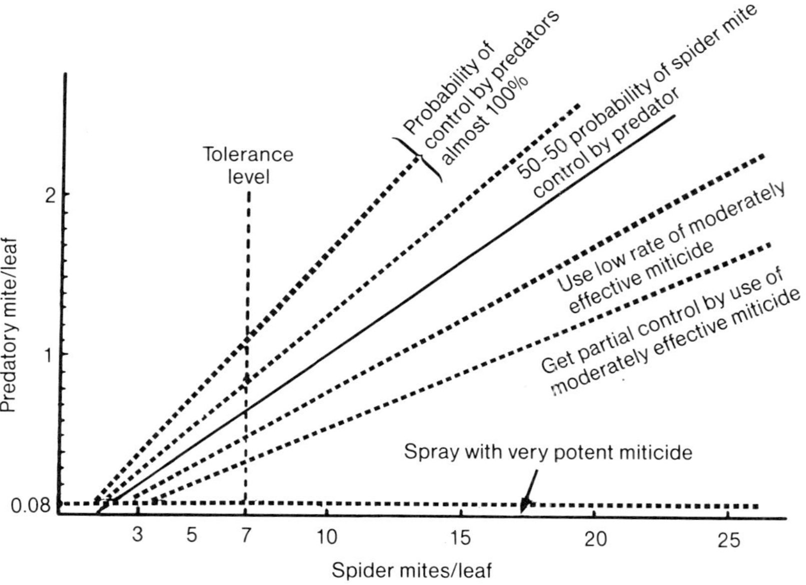

Fig. 10.10 ***The potential for pest suppression by natural enemies. The numbers of spider mites are counted and compared to the numbers of predatory mites. Specific control recommendation varies for each region of the graph; where the point falls for a given situation will determine what action should be taken, if any.***

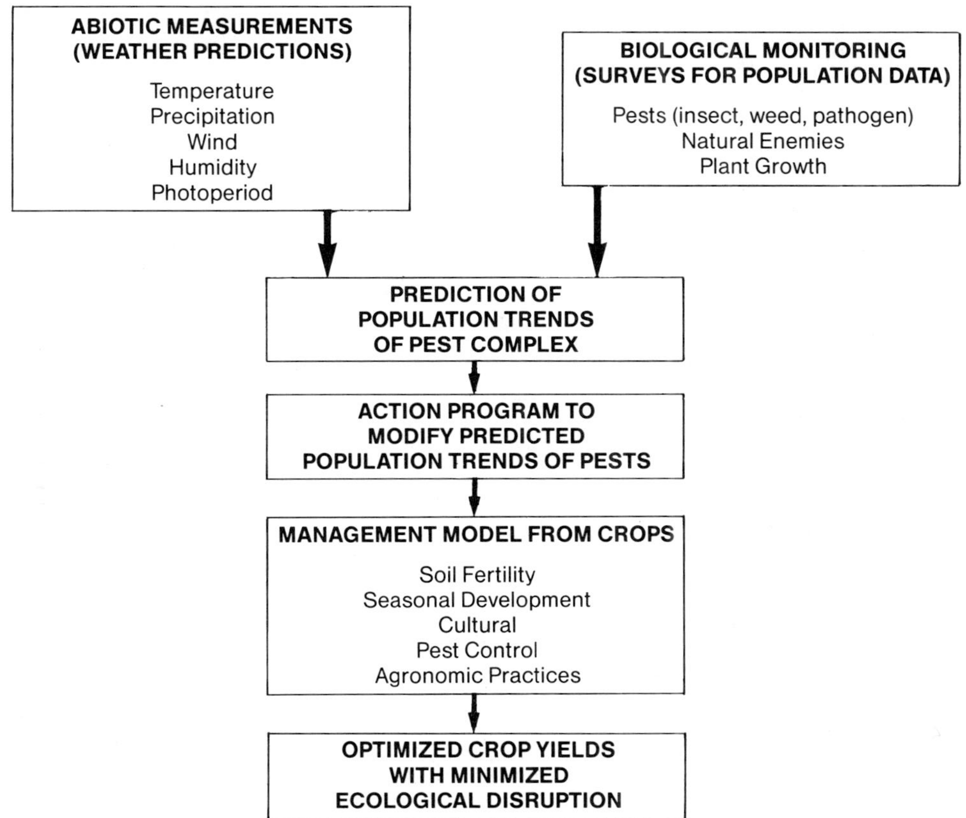

***Fig. 10.11** Elements in an IPM program to optimize yields of a crop are also considered in IPM programs designed to minimize the effect of insects important to public health. A key aspect is frequent communication between individuals responsible for each component part of the program.*

PEST SUPPRESSION TACTICS

The previous discussion in this chapter has dealt with the basic strategy in current use, that of maintaining pest populations below their economic injury levels through "integrated pest management." In several places throughout the discussion references have been made to "suppression technique," or to "manipulation of the environment," without explaining what these methods of control are. The remainder of this chapter spells out each of the tactics (skillful methods of action) available, with examples of their uses, and a comparison of the advantages and disadvantages of each type.

It is important to note that most of the efforts at insect control from the late 1930s into the 1970s were on a unilateral basis. That is, the strategy for control of a

APPLE SPRAYING SCHEDULE B
(Using Parathion and DDT)

Application	Materials to make 100 gallons of spray	To Control
DORMANT	DN compounds	Rosy aphid, bud moth

Note: This application is usually not necessary, except when bud moth or rosy aphid is a problem.

Application	Materials to make 100 gallons of spray	To Control
GREEN TIP to DELAYED DORMANT	Lime-sulfur 2 gallons, sulfur paste 8 to 10 pounds, or wettable sulfur 5 to 6 pounds	Scab

Note: Supplemental fungicide sprays or dusts may be needed prior to PETAL FALL spray to control scab, depending upon occurrence of rainy periods and amount of new growth. Mercury compounds (Puratized Apple Spray and Tag 331) may be used in supplemental fungicide sprays during the period of GREEN TIP to FIRST COVER for eradication of an early stage of apple scab infection. Mercury compounds are *eradicants only* and to be most effective should be used within *72 hours after a scab infection period* when previous protection against scab infection is questionable. Lime-sulfur may be used similarly through the DELAYED DORMANT period to eradicate an early stage of apple scab infection without danger of injury.

Application	Materials to make 100 gallons of spray	To Control
PRE-BLOSSOM PERIOD PRE-PINK	Sulfur paste 8 to 10 pounds, wettable sulfur 5 to 6 pounds, or ferbam 1½ pounds *plus*	Scab
	15% wettable parathion ½ pound	Rosy aphid

Note 1: Ferbam used *before* SECOND COVER on Golden Delicious and Jonathan *has caused serious fruit russeting,* especially when used with parathion.

Note 2: It is almost impossible to control rosy aphids after they have rolled the leaves. Make an effort to control them at this time. More than ½ pound of 15% wettable parathion per 100 gallons may cause injury on McIntosh, Snow, Early McIntosh, Kendall, Cortland and other varieties related to McIntosh.

Application	Materials to make 100 gallons of spray	To Control
PINK	Sulfur paste 8 to 10 pounds, wettable sulfur 5 to 6 pounds, or ferbam 1½ pounds *plus*	Scab
	15% wettable parathion ½ pound	Red-banded leaf roller, curculio, rosy aphid
IN BLOOM	When necessary, use sulfur paste 8 to 10 pounds wettable sulfur 5 to 6 pounds, or ferbam 1½ pounds for control of scab. If fire blight is a problem use 2-4-100 bordeaux or proprietary copper compounds at manufacturers' directions.	
PETAL FALL	Sulfur paste 8 to 10 pounds, wettable sulfur 5 to 6 pounds, or ferbam 1½ pounds *plus*	Scab
	15% wettable parathion ½ pound	Red-banded leaf roller, curculio

Fig. 10.12 A spray calendar typical of unilateral control strategies during the era of reliance on synthesized organic pesticides. Little attention is paid as to whether each cover spray is needed or if populations are below damaging levels.

Application	Materials to make 100 gallons of spray	To Control
FIRST COVER (7-10 days after Petal Fall)	Sulfur paste 8 to 10 pounds, wettable sulfur 5 to 6 pounds, or ferbam 1½ pounds *plus* 15% wettable parathion ½ pound	Scab Red-banded leaf roller, curculio
SECOND COVER (7-10 days after First Cover)	Ferbam 1 pound, or wettable sulfur 4 pounds *plus* 15% wettable parathion ½ pound *plus* 50% wettable DDT 1 pound	Scab Red-banded leaf roller, curculio, codling moth

Note: Sulfur paste used later than FIRST COVER may cause sulfur burn, especially under warm, humid conditions.

Application	Materials to make 100 gallons of spray	To Control
THIRD COVER (10-14 days after Second Cover)	Ferbam 1 pound, or wettable sulfur 4 pounds *plus* 15% wettable parathion ½ pound *plus* 50% wettable DDT 1 pound	Scab Red-banded leaf roller, curculio, codling moth
FOURTH COVER (Time to be announced)	Ferbam ¾ pound *plus* 50% wettable DDT 2 pounds or lead arsenate 3 pounds	Scab Apple maggot, codling moth

Note: The fungicide may be omitted in this spray application on scab resistant varieties, such as Jonathan, when scab is completely controlled. However, *if ferbam is omitted*, substitute in its place zinc sulfate 1 pound and lime 4 pounds (if lead arsenate is used instead of DDT).

Application	Materials to make 100 gallons of spray	To Control
FIFTH COVER (7-10 days after Fourth Cover)	Ferbam ¾ pound *plus* 50% wettable DDT 2 pounds, or lead arsenate 3 pounds	Scab Apple maggot, codling moth

Note. The fungicide may be omitted on scab resistant varieties, and if scab is completely controlled on susceptible varieties such as McIntosh and Red Delicious. However, *if ferbam is omitted*, substitute in its place zinc sulfate 1 pound and lime 4 pounds (if lead arsenate is used instead of DDT).

Application	Materials to make 100 gallons of spray	To Control
SIXTH COVER (Time to be announced)	50% wettable DDT 1½ pounds	Codling moth

Note: If mites are a problem, replace ½ pound of 50% wettable DDT with ½ pound 15% parathion or use commercial miticides at manufacturers' directions with 1½ pounds of 50% wettable DDT. If red-banded leaf roller is a problem or becomes a problem as late as 10 days before harvest, use DDD or TDE at manufacturers' directions.

Fig. 10.12, continued.

given pest was through reliance on a single tactic, usually insecticides. For example, pests of cotton, apples, and most other agricultural materials were controlled according to spray schedules. These schedules told the farmer when and what to spray, without regard to whether or not there were enough pests present to warrant an application (Fig. 10.12). As a result, a number of significant pest problems became apparent. Pest populations developed resistance to insecticides, treated populations flared up again (because their natural enemies were eliminated), and species that were previously of minor importance suddenly reached the pest status. It is because of these failures that entomologists developed the pest management strategy, based on minimum disruption of the ecosystem by encouragement of natural enemies, judicious pesticidal applications, and integration of other tactics as well.

The following list indicates pest suppression tactics in use today. They will be discussed in detail in the remaining pages of this chapter.

INVENTORY OF PEST SUPPRESSION METHODS

- Biological Methods
  - Natural Enemies
  - Resistant Host Varieties
- Chemical Methods
  - Insecticides
  - Insect Growth Regulators
  - Attractants
  - Repellents
  - Chemosterilants
- Autocidal Methods
  - Sterile insect release methods
  - Genetic manipulations
- Physicomechanical Methods
  - Exclusion
  - Energy—light, sound, heat, cold, moisture
  - Collection methods—trapping, suction
  - Destruction—crushing, grinding
- Cultural Methods
  - Crop rotation
  - Sanitation and crop residue destruction
  - Timing of planting and harvest
  - Maintenance of crop vigor—fertilizing, pruning
  - Soil and water manipulation—tillage, irrigation
- Regulatory Methods
  - Quarantines for exclusion and restriction
  - Suppression
  - Eradication

Biological Methods

The two principal tactics in this category differ in the way in which pest populations are suppressed through their use. The first is the now-classic use of natural enemies to reduce the pest population so that fewer pests are available to cause damage. The second is based on modifying the plant or animal that is the target of the pest's attack so that it is less susceptible to that attack, a technique centered around the concept of breeding a strain of host that resists the attack of the pest.

Natural Enemies

Organisms that are natural enemies of pest insects fit the classic definition of biological control, which is regulation of pest numbers by natural enemies. The ecological basis of this practice is that there exists some "balance" in the natural order of things. If a pest population begins to build up in an area, then the pest's own pests (its natural enemies) should also start to build up. The natural enemies should be on the increase because there are more food resources (the pest) available. The outcome of such an occurrence logically should be reduction of the pest by its natural enemies so that it no longer is at the economic threshold, and can thus be tolerated.

Predators. The kinds of natural enemies that can affect pest numbers are extremely varied. They include groups based on the type of activity, predation, parasitism, or pathogenicity.

Predators consume insects, some living entirely on an insect diet. The predator kills one prey, consumes it, and moves on to kill others. Since most predators are as large or larger than their prey (except group hunters, like some ants), they usually consume several prey during their life times. Predators among the vertebrates include the mosquito fish (*Gambusia* spp.) that have been introduced into bodies of water throughout the world to consume the aquatic stages of mosquitoes. The giant toad, *Bufo marinus* (L.), has been introduced to several Caribbean islands to control scarab beetle pests in sugar cane. Some groups of reptiles, birds, and mammals feed on insects, often exclusively (bats, lizards, woodpeckers, martins, swifts), but few instances of humans manipulating their populations exist.

Among the invertebrate predators of insects two groups are of primary importance: the generally predatory arachnids (spiders, Fig. 10.13) and the group of insects themselves. They range from the sedentary type like the ant lion that waits for prey to fall into its cone trap (Fig. 10.14), to the lacewing larvae and lady bird beetles (Fig. 10.15) that often "graze" on very small, but abundant pests (aphids, scales, mites, mealybugs) to the wasps that paralyze caterpillars or other prey and bring them to a nest to await consumption by the wasp's offspring (Fig. 10.16).

A classic example of the effectiveness of predators in controlling pests is that of the cottony cushion scale–Vedalia beetle story.

Cottony cushion scale (*Icerya purchasi* Maskell) is a sap-sucking insect that can live on many kinds of woody plants. Although it was known previously in other areas, it was first discovered in California in 1868 and by 1886 had spread to the newly

***Fig. 10.13** An orb-weaver spider waits for flying insects to hit its web and become entangled. The spider then paralyzes its prey with venom injected through its chelicerae. Other web-spinning spiders employ webs of different design to ensnare prey. Still others, such as the wolf and crab spiders, jump on or ambush prey without the aid of webs.*

developing citrus areas in Southern California. Scale populations on individual trees reached astronomic numbers, thereby reducing fruit quality and yield to a point where harvest was uneconomical. In effect, the citrus industry was being eliminated because of cottony cushion scale.

Efforts to control the scale with available techniques proved ineffective, so a different approach was sought. The scale was something of a rarity in Australia, the country of the scale's evolution. This condition was thought to be the result of natural enemy action. The problem was to determine which natural enemies had been left behind when the scale was accidentally introduced (Fig. 10.17).

Investigations in Australia, mostly by Albert Koebele of the U.S. Department of Agriculture, revealed two natural enemies of the scale. One was a parasitic fly, *Cryptochaetum iceryae* Will.; the other was a predacious lady bird beetle known as the Vedalia, *Rodolia cardinalis* (Muls.) (Fig. 10.18). Numbers of the two insects were collected and shipped to the United States. Because of delays in shipment and other

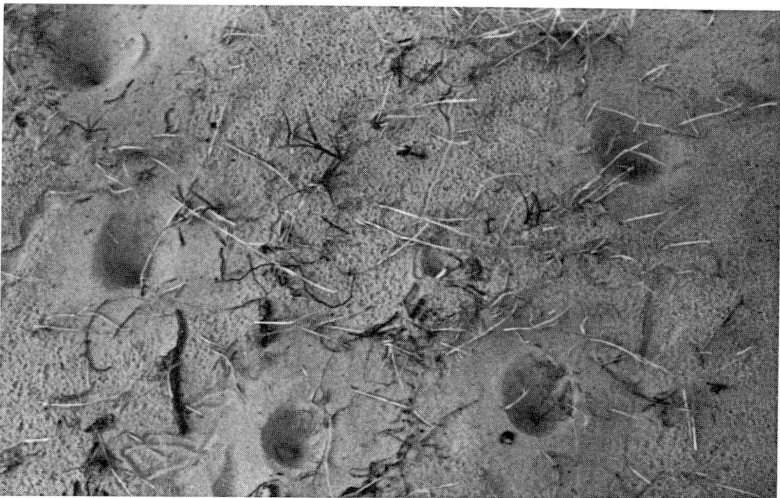

Fig. 10.14 The larval ant lion rests just beneath the cone of loose sand it has formed. Crawling insects entering the cone slide down the loose sides, sometimes encouraged by additional sand flipped at them by the ant lion. When the insect gets to the bottom of the cone it is seized by the ant lion's jaws and its body contents are sucked out. The adult ant lion looks much like a damsel fly.

Fig. 10.15 Lady bird beetle larvae feed on aphids in an aggregation. The aphids have a spectacular reproductive capacity, which, with the dispersive capacities of the winged forms, combines to allow survival in the face of such predation. From Tipton, Vernon J., editor, **Syllabus: Introductory Entomology**. *Copyright 1973 by Brigham Young University Press. Reprinted by permission.*

Fig. 10.16 A potter wasp provisions its nest with paralyzed insects. After sufficient "provisions" are laid up, the wasp will deposit an egg and the larval wasp will develop, feeding on perfectly preserved food (live, though paralyzed insects).

problems very few insects survived the trip (only 514 Vedalias, but almost 12,000 *Cryptochaetum* flies).

To test the effectiveness of these species, each was released under separate scale-infested caged citrus trees. Twelve days later examination of the *Cryptochaetum* colony revealed that most of the flies were dead; the fly evidently failed to reproduce and become established. The Vedalia colony, however, yielded an entirely different and remarkable result, for most of the scales had been consumed by the beetles. The beetles where then allowed to escape from the cage and in 6 months one release of 129 beetles had multiplied, rid an entire orange grove of scale and had spread over three-quarters of a mile from the original point of release.

Eventually both *Cryptochaetum* and *Rodolia* became established in California, the fly dominating the milder coastal areas and the beetle more important in the hotter inland areas. Since the original introductions, at a total cost of less than $5000, cottony cushion scale has not caused problems, except when its natural enemies have been inadvertently eliminated by pesticide application aimed at other pests.

Use of the Vedalia as a predatory biological control agent demonstrates several aspects of the principle of inoculative natural enemy releases. Such releases are intended to introduce a propagule (reproductive unit) of one species into the environment and then "let nature take its course," with the hope that the natural enemy becomes established at high enough population densities to suppress the intended target species, the pest. With the release of exotic animal agents the possible results are

1. Disappearance of all individuals after release, without any sign of reproduction
2. Repeated sightings of the released species in decreasing numbers, eventually

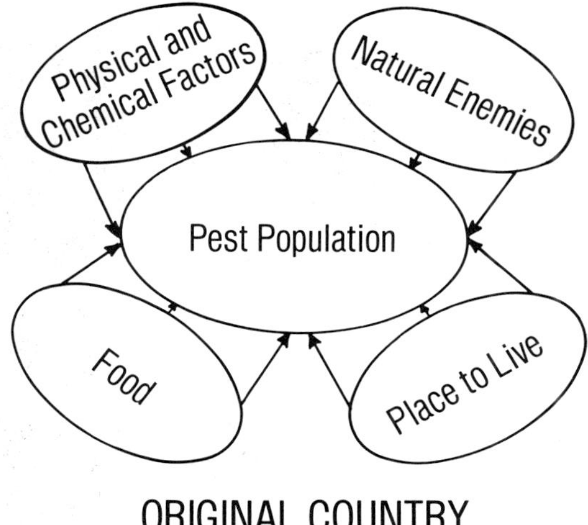

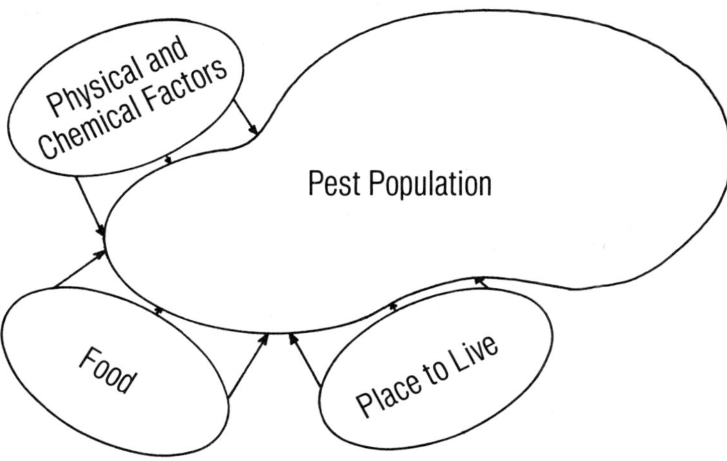

Fig. 10.17 Ecological imbalance as a result of accidental introduction of an insect pest into a new area. Without the action of natural enemies in suppressing population growth, the pest soon reaches epidemic proportions, provided the other requisites in the environment are suitable.

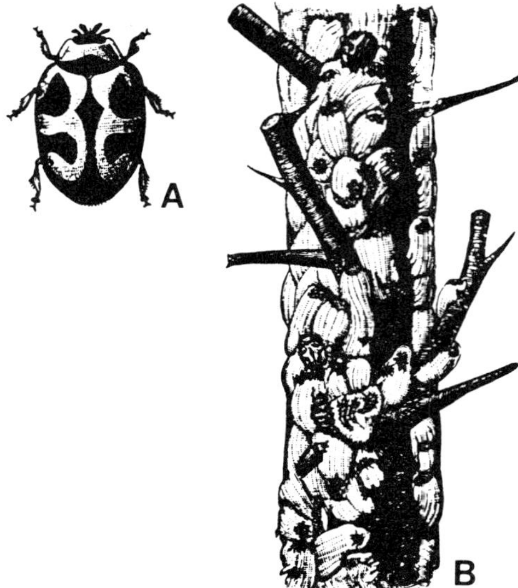

***Fig. 10.18** (A) The Vedalia beetle,* **Rodolia cardinalis** *(Muls.), (B) beetles and cottony cushion scales. Both the adult beetles and their larvae feed exclusively on the cottony cushion scale.*

resulting in disappearance. There is initial survival of individuals after release, but low reproduction (or none), leads to extinction

3. Establishment of the released species in the new habitat, but in such low densities that pest reduction is minimal
4. Establishment and "explosion" of the released species, as with the Vedalia.

Other aspects revealed are that the place to look for efficient and specific natural enemies is in the area where the pest originally evolved, for it is there that the enemies would also have evolved synchrony with their host's life cycle and sophisticated behavior in locating prey. In fact, "host specificity" (the range of species suitable for development of the natural enemy and acceptable to it) was considered to be an essential element in biocontrol successes for many years.

One aspect not adequately covered by the Vedalia case involves careful quarantining of the insect prior to its release in a new area. It is essential to check that potentially injurious insects are not introduced to a range of potentially suitable hosts that are themselves valuable (either plant or animal). A second aspect is that predatory and parasitic insects also have parasitoids, and introduction of a biocontrol species could be defeated by accidental introduction of that species' parasitoids. Another factor is that enough biocontrol agents must be collected and introduced to ensure an adequate chance of their survival in the new area. For example, if the shipments of Vedalia had been distributed at the rate of one beetle per acre instead of being confined to one tree, the chance of their ever finding mates when they were physiologically ready for reproduction would have been very slight. Large releases are often accomplished by mass rearing the biocontrol agents prior to release.

Mass rearing for predators is complicated by the fact that many predators act cannibalistically in the densities occurring in the limited space of rearing facilities. Also, acceptable prey or suitable substitute nutrient mixtures must be provided. In any case, factories for the production of millions of natural enemies that may be purchased for release in croplands are in operation in many locations.

Parasites. A parasite is defined as an organism that obtains its nutritional requirements from the body material of another organism (the host), lives in or on the body of the host and the host receives no benefits from the association, although it is not usually destroyed as a result of that association. A major difference between the association of insect parasites of other insects (termed parasitoids) and other parasitic associations regards the usual outcome. Parasitoids usually kill their hosts, whereas common parasites of other groups like tapeworms, lice, flukes, hookworms do not. Part of this difference is attributed to the relative similarity in body size of insect host and parasitoid. A caterpillar parasitized by a Tachina fly would be equivalent to a human parasitized by a maggot the size of a football. Most parasitoids are free-living in the adult stage, which permits a dispersal role; only the immature is parasitic and is usually very specific as to which species or genus of host it will attack. Metamorphosis is typically complete.

The largest group of valuable parasitoids is in the Order Hymenoptera (Fig. 10.19), where nine of the 16 superfamilies composing the order are all or partly parasitic. Six of the nine superfamilies make up a distinct, often formally acknowledged taxonomic group, the Parasitica. Parasitic flies also are important as natural enemies of pest insects. Of special significance are flies of the Family Tachinidae.

The difficulty in growing parasites is the necessity of providing a suitable host for

Fig. 10.19 A typical parasitic wasp of the Order Hymenoptera. This species is using its "stinger" (ovipositor) to inject eggs into a larval alfalfa weevil. The larval wasp will kill the host by consuming its body contents. Thus a pest is eliminated without adverse environmental effects.

parasitoid development. In this regard, parasitic wasps that attack scale insects and insect eggs have been most amenable to mass culture. The egg parasites are fed eggs gathered from moths that attack stored cereals. The scale parasites are fed scales grown on squash fruit or potato tubers. The success of techniques for growing millions of parasitic wasps allows entomologists to use these natural enemies in a way that overcomes two problems with traditional release methods. (1) The most desirable feature of the traditional biocontrol approach is that it tends to result in a permanent reduction in the general equilibrium level of the pest population, but the permanency of the control depends on the biocontrol agent's ability to survive in the new area year round. Many are unable to do so. (2) Sometimes the pest and natural enemy get "out of balance" due to weather factors, agronomic practices, or other pest control procedures.

When either of these problems occurs it becomes possible to use a technique called "inundative release," in which natural enemies are released in large enough numbers to cause an immediate and direct mortality of the pest population, but with no expectation of permanent regulation. It allows for use of effective natural enemies that may not be capable of colonizing the area because of overwintering. It also acts as a "surgical" treatment in that only given stages of particular groups of insects are destroyed as opposed to mass destruction of most insects by a pesticide application.

Russian governmental labs each rear up to 3,000,000 egg parasites per day. These parasitic wasps of the genus *Trichogramma* are released in corn fields at the rate of 40,000–60,000 per hectare three times during each of the two generations of corn borer, *Ostrinia nubilalis*. Costs per treatment are 23¢ per hectare per release. This compares favorably with the current $16-27/hectare for pesticidal control. In the United States over 3 billion parasites are reared per year.

Although inundative releases of parasitoids offer some exciting possibilities for use as biological controls, it has had much less impact thus far than inoculative releases of parasitoids. The parasitoid has been the focus of most of the efforts in biocontrol because parasitoids are often more host-specific than are predators. Several examples of inoculative parasitoid releases that have resulted in successful biocontrol are listed in Table 10.2.

Pathogens. Insects suffer from diseases caused by the same pathogenic organisms that attack other animals. A variety of protozoans, bacteria, fungi, viruses, and nematodes cause loss of appetite, listlessness, diarrhea, edema, and often death to their insect hosts. Since insects are often observed to have died under natural conditions as a result of disease, the potential use of insect pathogens as biocontrol agents has been a subject of interest. In fact, as early as 1664, references were made to the use of decoctions made from the bodies of diseased caterpillars sprinkled over infested plants. Obviously this was an early attempt to infect pests with disease, though prior to the proof of contagion.

The potential utility of pathogens is that the diseases caused by them are contagious. That is, an infected host becomes infective, or its exudates become infective, spreading the disease to other individuals of the host population. The difficulty with manipulating the pathogen and host insect populations to create disease outbreaks

Table 10.2 Successful Biocontrol Programs with Parasitoids

Pest	Area of success	Crop	Time	Parasitoid
Sugar-cane beetle borer *Rhabdoscelus obscurus* (Boisd.)	Hawaii	Sugar cane	1907–10	Tachinid fly
Spiny black fly *Aleurocanthus spiniferus* Quaint.	Japan	Citrus	1925	*Prosopaltella* wasp
Coconut moth *Levuana irridescens* B-B.	Fiji	Coconut	1925	Tachinid fly
Eucalyptus snout-beetle *Gonipterus scutellatus* Gyll.	So. Africa	Eucalyptus	1926	Mymarid wasp
Citrus black fly *Aleurocanthus woglumi* Ashby	Cuba	Citrus	1930	*Prosopaltella* and *Eretmocerus* wasps
Citriculus mealybug *Pseudococcus citriculus* Green	Israel	Citrus	1939–40	*Clausenia* wasp
Comstock mealybug *Pseudococcus comstocki* (Kuwana)	Eastern U.S.	Apple, pear, peach	1939–41	*Alloptropa*, *Pseudaphycus*, and *Clausenia* wasps
Red wax scale *Ceroplastes rubens* Maskell	Japan	Citrus, tea	1948	*Anicetus* wasp
Winter moth *Operophtera brumata* Linn.	Canada	Hardwoods	1955–60	Tachinid fly Ichneumonid wasp
Alfalfa weevil *Hypera postica* (Gyll.)	Eastern U.S.	Alfalfa	1960–75	*Tetrastichus*, *Bathyplectys*, *Microctonus*, and *Patasson* wasps
Walnut aphid *Chromaphis juglandicola* Kalt.	Western U.S.	Walnut	1959, 1968	*Trioxys* wasp

lies in the exact conditions, both physical and biological, required for spread of the disease, for infection to occur, and for production of viable material.

Bacterial diseases. Bacteria are present everywhere, in the digestive system of most insects as well as in the insect's immediate environment. Thus, bacteria become a factor in the death of insects when they penetrate the defenses of the host. These defenses are of two kinds: the outer chitinous cuticle covering the insect and the gut wall. If a break in the cuticle occurs through abrasion or wounding, the bacteria quickly penetrate into the body, multiply, and often kill the host. Likewise, passage through the gut wall may result in rapid infection and death. The true insect pathogenic bacteria possess toxins that can pass through the gut wall so these pathogens are effective only after the bacteria themselves or their toxins have been consumed. A typical example is *Bacillus thuringiensis* Berliner, one of the spore-forming bacteria. It is toxic to Lepidoptera, Hymenoptera, Diptera, and Coleoptera, and it can be cultured on artificial substrates. The spores retain their pathogenicity even in the dry state for up to 10 years, as does the toxin (Fig. 10.20). All of these characteristics contribute to the utility of *B. thuringiensis* because quantities can be produced inexpensively, and

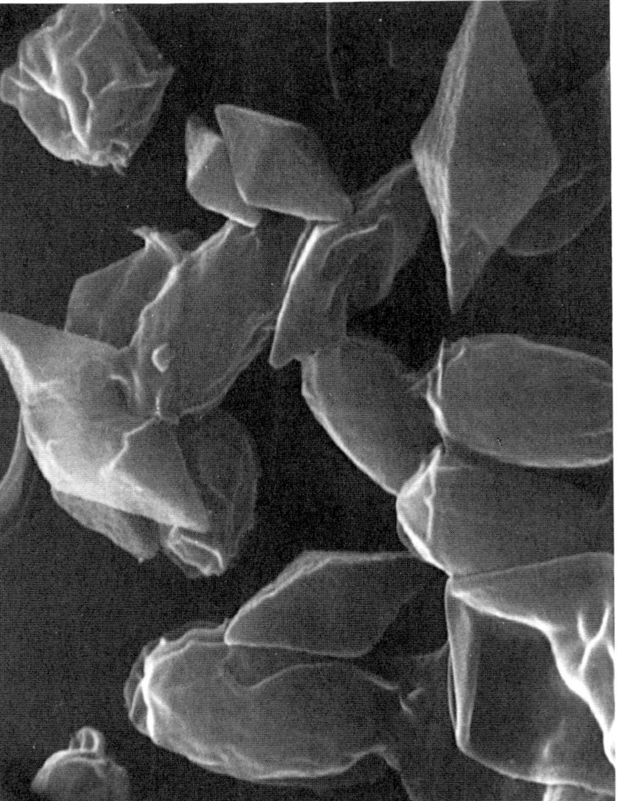

Fig. 10.20** Parasporal bodies of* **Bacillus thuringiensis,** ***one of the crystalliferous spore-forming bacteria. Through the process of sporulation (formation of a resting spore stage) the bacillus can withstand conditions too harsh for it in its active form. During sporulation the bacillus also forms the crystalline, proteinaceous "parasporal bodies" shown. The toxic nature of the parasporal body, in addition to several other toxins produced by BT, actually is responsible for death of susceptible insects.

suspensions of the spores or their parasporal crystals can be sprayed or dusted on foliage to protect them from specific pests. Related bacteria such as *B. popilliae* for Japanese beetles are effective pathogens, but since they have not yet been produced on artificial media, the expense in production prohibits broad utilization.

Fungal diseases. Because of the vastly different life cycles found among the many types of fungi, there is less known about fungal infections of insects than the other types of insect pathogens. Another factor contributing to this relative lack of knowledge is our low success in manipulating conditions so as to foster epidemic mycoses (fungal infections) among pests. This in spite of the often noticed insect mortalities caused by fungi in nature (Fig. 10.21). Recent work on aphids infected with *Entomophthora* fungi has shown that conditions of high humidity are not necessary to create fungus epidemics.

Fig. 10.21 **Entomophthora** ***fungus sporulating from cadavers of flies. In late stages of infection the dying fly clings to vegetation and the fungus soon permanently attaches it so that it remains on the vegetation after death.***

Many species must be ingested, as spores, with the host insect's food supply before they can infect. Others can enter through undamaged cuticle by dissolving the cuticle. Only a few entomophilic fungi have been reared on artificial media. Of these, *Beauvaria bassiana* (Fig. 10.22) is available commercially in Europe for application to agricultural commodities.

Fig. 10.22 ***The fungus*** **Beauvaria bassiana** ***attacking*** **Leptinotarsa decemlineata,** ***the Colorado potato beetle, one of its many hosts.***

Nematode infections. A wide variety of nematodes are associated with insects. Some nematodes that cause disease in vertebrates use insects as intermediate hosts and vectors, including the heartworm of canines and the filaria worms of humans. Another large group are parasitic on insects without requiring other hosts. These offer possibilities for use as biocontrol agents because (1) they have a rather selective host range and (2) at least one genus, *Neoaplectana*, has been cultured on artificial media. Of particular importance to their utility is the extended vitality of the infective stage. They can be kept for up to 3 years under refrigeration without loss of infectivity.

The juvenile *Neoaplectana carpocapsae* enter the host insect's haemocoel by actively burrowing through the gut wall (Fig. 10.23). They mature, mate, and produce eggs in the insect, which has since died. A second generation may be produced before juveniles leave the host body to await a new host. Mortality of the host insect is not caused by the nematode but by bacteria released by the nematode as it penetrates the gut wall. The bacteria multiply and cause a fatal infection of the haemocoel.

Virus diseases. The viruses that attack insects are very host specific. On the other hand, most of the insect pest groups that have been surveyed are attacked by at least one virus. Theoretically, the potential for pest suppression through creation of epidemics of viral disease is very appealing because only the host of the virus used would be eliminated, thus providing an effective tool with minimal ecological disruption.

Among the seven major categories of virus types thus far identified as pathogenic

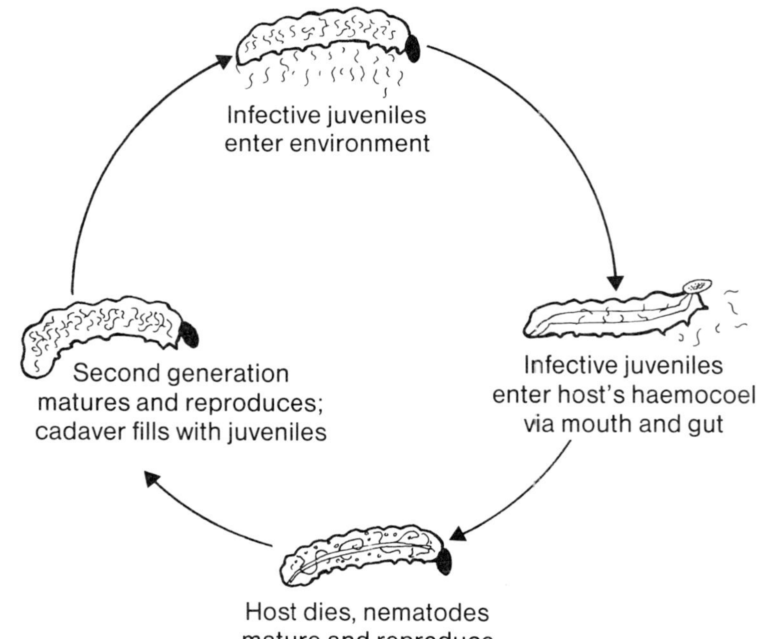

Fig. 10.23 Life history of* Neoaplectana *nematodes. In this case, bacteria released into the insect's haemocoel by ingested nematodes are the cause of the insect's mortality, not the action of the nematodes.

to insects, the baculoviruses offer the best potential as control agents. They are characterized as being enclosed within a proteinaceous inclusion body that is large enough to be visible under a light microscope. The two principal groups of insect pathogenic baculoviruses are the nuclear polyhedrosis viruses (NPV) and the granulosis viruses (GV), differentiated by the makeup of the inclusion body. The NPV inclusion body is relatively large, polyhedral in shape, and has many virions (virus particles) uniformly distributed throughout its crystalline protein matrix (Fig. 10.24). The granulosis virus inclusion body is much smaller than that of the NPV and contains only one virion. Baculoviruses can be stored dry for long periods. The inclusion bodies preserve infectivity of the virus from action by solvents or enzymes. Solar radiation, however, rapidly lowers infectivity of inclusion-type viruses.

Insects become infected by ingestion of material contaminated with viral matter.

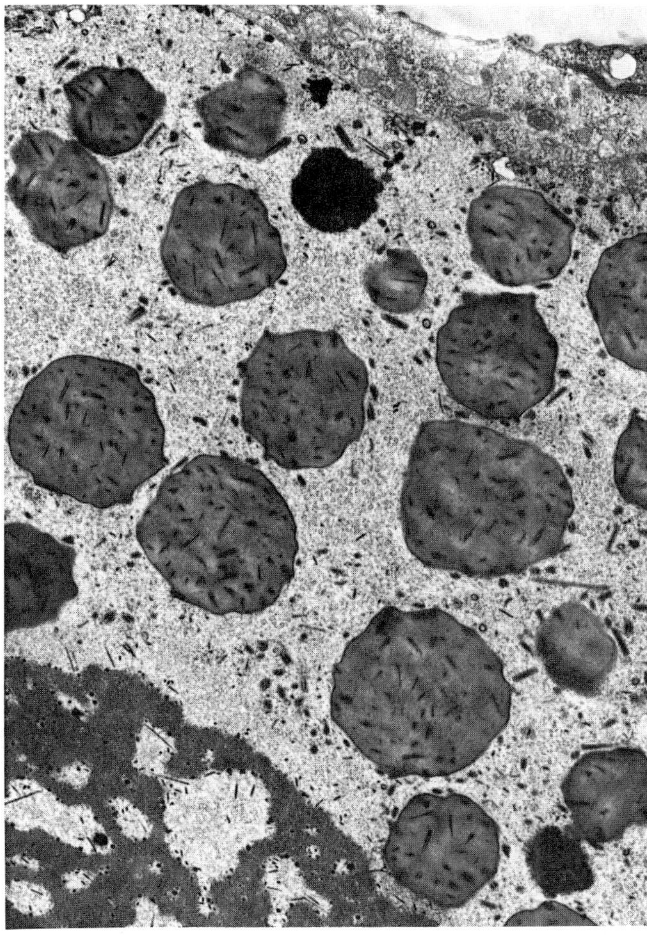

Fig. 10.24 Nuclear polyhedrosis virus in a caterpillar. The virions are visible within the inclusion bodies, which provide protection from chemical degradation in the environment.

Some evidence also suggests that *transovarial* transmission (from parent to forming egg) and contamination of wounds may also act to spread virus. An additional method is through the contaminated ovipositor of a parasite that had previously stung a virus-diseased host. After the virus is released from the inclusion body in the host's gut, it infects the midgut and then replicates and soon fills the body of the infected insect with inclusion bodies. Prior to its death, the insect often moves high up on the host plant's foliage and becomes attached (Fig. 10.25). After death the infected cadaver's contents become a liquefied mass of inclusion bodies. When the integument ruptures, the contents splash down over the foliage, thus facilitating infection of subsequent hosts that feed on the contaminated foliage.

Insect viruses are most effective when high densities of hosts make rapid spread through the population possible. Unfortunately, natural epidemics of virus in pest insects usually occur only after the pest has already inflicted heavy damage. The goal in using viruses as biocontrol agents then is to create these artificial epidemics before the number and size of the pest reaches the economic threshold. Such epidemics are feasible with foliage-feeding beetles and caterpillars, but less so with internal feeders because of the difficulty in getting the virus into their habitat. For example, the codling moth feeds within the developing apple and is not exposed to the virus.

In order to use viruses effectively, three criteria must be met. (1) The virus must be in a form that can be stored over extended periods. (2) It must be made available in large enough quantities to become commercially feasible. (3) It must be safe to the person using it as well as to nontarget organisms in the environment. While the first criterion is met by the nature of the inclusion bodies, the other two present major difficulties.

***Fig. 10.25** A caterpillar killed by virus hangs from its host plant. At this stage the integument forms a sack that holds the liquefied body contents, laden with inclusion bodies. Infected caterpillars often climb to the top of their hosts before death, and rupture of the integument releases the contents which splash over and contaminate more foliage, effectively aiding spread of the disease.*

Because viruses only replicate (multiply) in living cells, not on artificial media as *Neoaplectana* or *Bacillus thuringiensis* do, only limited quantities of any single virus is available. To overcome this obstacle, three approaches are used. Field collections of naturally infected larvae can provide some material, as can induced epidemics in lab colonies of the pest species, provided the species can be maintained under artificial conditions. The most potential, however, lies in the use of tissue culture techniques. Colonies of host insect cells are grown on sterile nutrient substrates and then infected with virus. After maximum viral replication has occurred, the inclusion bodies are extracted for use.

Safety in the use of insect viral diseases is a major concern of the scientific community, since the materials to be spread in the environment offer potential hazards, both to humans and to the other organisms exposed. Current knowledge of baculoviruses suggests that they are unable to infect vertebrate cells. Therefore, the probability of any *oncogenic* (tumor causing) or *teratogenic* (embryonic malformation causing) effects are very low. However, vertebrates, including humans, could have allergic or toxic reactions as a result of virus and inclusion body entry through the respiratory or gastrointestinal systems or through the eyes. The U.S. Environmental Protection Agency, in conjunction with other concerned groups have developed sound guidelines for testing the safety of baculoviruses. At least seven baculoviruses of Lepidoptera and one of a sawfly (Hymenoptera) have been tested according to the recommended forms and none have yielded either toxic or allergic responses. Most of us have been exposed to baculoviruses in our environment and use of these baculoviruses as insect controls may increase our exposure, but not introduce us to unknown elements.

A final aspect of the use of host-specific biocontrol agents merits discussion. It concerns the relation between the damage potential of a pest and the cost of developing a biological control program for that pest. The cost of inoculative releases of natural enemies is relatively low, especially if the releases are successful and equilibrium levels of a pest are lowered permanently as a result. Considerably more cost is involved in mass rearing of predators and parasitoids for inundative releases or of growing pathogens on artificial media. The costs of these programs narrows the number of pests that can be controlled. And, using the current technology, only the few most severe, pests can be considered for control using viral diseases. Breakthroughs in areas of virus culturing techniques or in modifying the virulence of such diseases and their dissemination could greatly expand the range of pests that are candidates for such biocontrol.

Host Resistance

Resistance in a host is that quality that enables it to avoid, tolerate, or recover from the attack of pests that would cause greater damage to other members of the same species under the same conditions. Thus, a resistant crop variety would produce a crop in spite of the presence of a pest species. Several problems exist in the employment of this type of pest suppression, however. Among these are identification of resistance, incorporation of the resistance with other desirable host qualities into an acceptable host, understanding the mechanisms of resistance, and the length of time required to produce resistant varieties. The benefits to be derived from such programs include low cost and environmental safety.

Resistance is determined by comparing a test variety and some standard, usually a commonly available, variety. For example, experimental "cultivars" (synthetic varieties produced by plant breeders) of apple are often compared with the common red delicious apple for a wide range of characteristics, including resistance to insect, mite, or disease attack. To determine resistance, studies are made in which the varieties being compared are actually exposed to the same pest pressure. They may be tested within the same orchard or greenhouse, or they may be artificially infested with identical populations of pests. Evaluations such as percentage of the experimental cultivar infested with the pest is compared with percentage infestation of the common variety in the same experimental area are made. Because comparative studies are done under the same test conditions, differences in infestation reflect true resistance in the host rather than annual or geographical variations in pest abundance.

Ecological Resistance. *Phenological Asynchrony.* Many plants are susceptible to attack by a pest only at a specific stage of growth. If planting dates are manipulated so that the susceptible stage of the host does not occur when the pest is active, then injury should be avoided. For example, the late season planting of corn followed in the midwest United States avoids most of the attack by first generation European corn borer. Such a practice may backfire, however, in areas where two generations of borer exist.

Genetic Resistance. *Nonpreference.* This characteristic enables hosts to avoid attack. Most insects locate their food supply by chemical or visual cues from a host. Hosts that have no such cues or have the cues masked are subjected to lower levels of attack, that is, they survive because of the modified behavior of the pest.

In order to identify the feeding stimulants that cause a pest to choose a particular host feeding-choice experiments are done. Pests are offered two or more sites for feeding to determine which they prefer. It has been found that if extracts from a susceptible variety of plant are applied to leaves of a nonpreferred variety, the pest may choose the nonpreferred variety. This confirms the presence of a phagostimulant or ingestion-stimulating chemical in the susceptible variety, which is either lacking or masked chemically or physically in the resistant foliage.

Antibiosis. Even when pests locate hosts through the perception of appropriate cues and consume host material, they may encounter a type of resistance known as "antibiosis" (against life). This occurs because the chemical composition of the host material ingested is modified in some way. The host may possess toxic materials that poison the pest; the host may lack one or more essential nutrients or the nutrients are not in adequate balance for survival of the pest; the host tissue may contain antimetabolites.

Symptoms of antibiosis are expressed in a variety of forms, including death at any stage prior to reproduction (unsuccessful overwintering, failure to complete development, etc.). It can also be expressed in abnormal growth rates or food conversion, or malformed or stunted adults that may have reduced fertility and fecundity.

2,4-Dihydroxy-7-methoxy-1,4-benzoxazine-3-one (DIMBOA) is a chemical whose presence in corn seedlings protects them from the attack of caterpillars because of its toxic qualities. Most corn varieties become susceptible to caterpillars like the European corn borer as the plant grows and the levels of DIMBOA are reduced. Once the nature of DIMBOA's activity was determined and its identity known, subsequent varietal improvements in dent corn resulted in vastly more resistant varieties with much higher levels of DIMBOA in the mature plants and therefore much more resistance to corn borers (Fig. 10.26).

Tolerance. The ability to suffer the attack of pests and still survive is a type of resistance termed "tolerance." The host tolerates pest attack without loss of yield because of the structural characteristics of the host.

Corn, sugar cane, and squash plants are bred for stem rigidity against chewing insects. This may take the form of closer packing of the very hard vascular tissues, heavier cell walls, or more layers of strengthening cell types. Hairiness or presence of hooked hairs (trichomes) may provide resistance to small sap-sucking insects like aphids and leafhoppers by entangling or restricting movement of the pests (Fig. 10.27). Plants have been bred that produce prop roots on the corn faster than they are chewed up by corn rootworms (Fig. 10.28).

The genetic basis of resistance in plants has been shown to be of two types. In oligogenic resistance, the factor controlling resistance is under the control of one or two major genes. In polygenic resistance, several to many genes contribute to the complex interaction resulting in resistance. Oligogenic resistance is much easier to incorporate into a variety than is polygenic resistance.

Implementation of a host resistance program involves many steps. The first step is to identify the sources of material available, based on the assumption that resistance is probably present within the gene pool of the crop. This may require collecting trips to gather varieties of the crop from the area in which it evolved, since the crop has had the longest time to evolve resistance to its pests in its area of origin. The varieties, wild strains, and close relatives are of special value. Another source for germ plasm, the pool of genetic variation possessed by a species, is in germ plasm banks, such as the four maintained by the U.S. government for the very purpose of preserving these varieties in an increasingly uniform world.

The next step is to subject these varieties to the attack of pests for which resistance is sought. This can be done through field evaluations using natural infestations or controlled infestations either in field or greenhouse. Evaluations are usually conducted with many individuals of each variety (mass screening), since variation also exists within varieties as well as between them. Evaluation of resistance is made by estimating damage levels or by counting insect survivors. Success in finding resistance among existing varieties is proportional to the number and diversity of varieties tested and the efficiency of the screening technique.

After resistance has been determined and its nature identified, the third step is to incorporate the resistance into an acceptable crop variety that can be economically competitive with commercial varieties. This critical step is the real weak link in

Fig. 10.26 (A) Damage by European corn borer to susceptible dent corn. (B) The resistant variety shows little evidence of feeding by the borer, although the plant was hand infested with an egg mass of the insect. Such protection conferred by the presence of toxic DIMBOA in the plant tissues demonstrates the resistant quality of antibiosis.

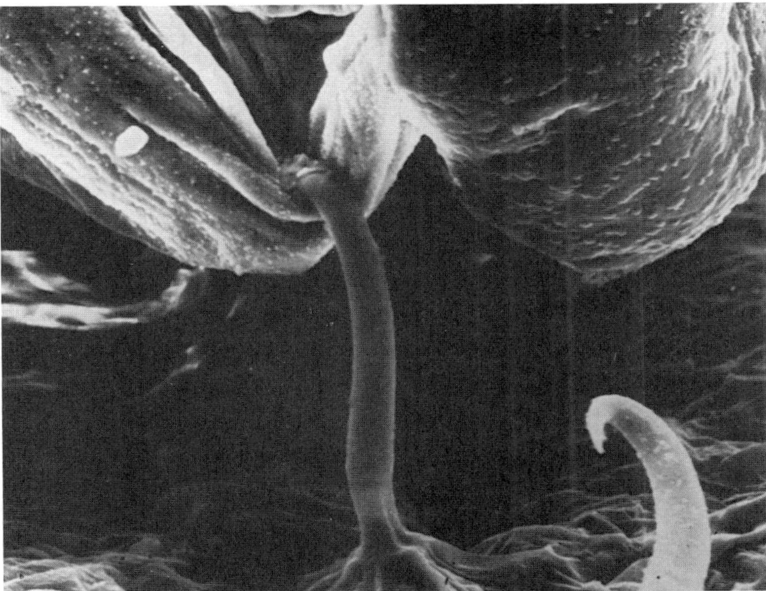

Fig. 10.27 A hooked leaf hair (trichome) on a bean leaf is embedded in the body wall of a leafhopper nymph. Nymphs thus entangled often die.

Fig. 10.28 Comparison of corn rootworm damage in a susceptible variety (left row leaning over) and a resistant variety. The resistant variety produces the roots that prop up the plant faster than the prop roots are destroyed by the corn rootworm larvae.

developing resistant varieties; the new resistant varieties are often not as acceptable to the consumer as existing varieties. For example, the rice varieties that were developed to be resistant to several of the major insect pests and diseases produce well, but the harvested rice does not store well and its cooking qualities and taste are less satisfactory than other varieties.

Since 10 to 25 years may be required for the complete development of a highly resistant crop variety, alternate strategies using the technique have been followed. The first is to pursue resistance to several pests simultaneously so that varieties resistant to several pests are available. A second strategy is that of producing several moderately resistant varieties instead of one that is highly resistant. The advantages of the former are that moderate resistance occurs more frequently in germ plasm than does near immunity. Also, a greater diversity through release of several moderately resistant varieties allows adapted varieties to be produced that more nearly match local agronomic and ecological conditions. In addition, varieties may be changed at shorter intervals when lower levels of resistance are sought. Most important, however, is that several moderately resistant varieties will each take over a smaller proportion of the market than would a single highly resistant one that would tend to monopolize a wide geographic area and open the area to new biotypes of pests that may evolve.

Pest organisms also evolve under the pressure of modified hosts so that host varieties that have been developed and released because of their resistance to a given pest may be susceptible to new strains of the pest. Thus, breeding programs offer only temporary solutions to pests. Plant pathogens usually overcome the resistance bred into varieties in less time than do insects. Recent advances in genetic engineering suggest that genetic resistance can be incorporated into acceptable cultivars. Transfer of genetic material will facilitate rapid development of resistant varieties, thereby reducing the time required by the traditional approach.

Chemical Methods

To most people, the concept of insect control is synonymous with the use of insecticides. As has been demonstrated previously, a wide range of techniques other than chemicals is available for suppression of pests. In this section, we will discuss not only the *pesticidal* (*pest killing*) chemicals so commonly considered, but other chemical uses, such as repellents, attractants, sterilants, and growth regulators.

Insecticides

"Insecticides" are pesticides that are specifically toxic to insects. Other types of pesticides include herbicides (for plants), nematicides (for round worms), molluscicides (for snails and slugs), rodenticides (for rats and other rodents), and algicides (for algae).

More than 400 insecticides are registered for use in the United States and they have been grouped into various classifications in an attempt to better understand them. One of the older systems groups them according to how the pesticide enters the body: fumigants through the respiratory system, stomach poisons through the digestive system and contact poisons directly by penetration of the cuticle. The most common grouping today is based on the chemical nature of the pesticide or its source (Table 10.3).

Several of the groups mentioned in Table 10.3 warrant discussion either because of their current importance or because of their roles in development of modern pesticide technology. Those listed but not discussed are less important today than in the past.

Types of Insecticides. *Synthetic inorganics*, notably the arsenic based compounds, were the first pesticides relied on for insect control because of their relative effectiveness against the boll weevil, a new pest in the early 1900s for which there was no other satisfactory control, and chewing insect pests of trees. The disadvantages are that piercing-sucking insects are not controlled because arsenicals must be ingested to be effective, and compounds that contain high levels of water-soluble arsenic are taken in by the roots of plants and tend to burn the foliage.

Botanicals are derived from plant materials. The three most common examples, pyrethrin, rotenone, and nicotine, are highly toxic to insects. Both pyrethrins and rotenone are also very safe to mammals. This combination of high insect toxicity with low toxicity to mammals creates ideal products for use in human habitations and in home gardens. A limitation to their wider use is the high cost of producing them. The plants from which pyrethrin (African chrysanthemums) and rotenone (tropical legumes of the genera *Lonchocarpus* and *Derris*) are derived require considerable hand processing. Pyrethrin is extremely effective as a space spray (in an enclosed area) for flying insects and is the most common active ingredient in fly sprays. It usually contains piperonyl butoxide as a "synergist" (chemical that increases the activity of other chemicals).

Organochlorines include several groups of insecticides that are of great historic interest. As a group, the organochlorines are very persistent, with residual activity of

Table 10.3 Classification of Insecticides Based on Chemical Nature and Source

INORGANIC: "Mineral" origin; were never a part of, nor a product of, living organisms.
- *Natural inorganic* mined from the earth. Examples: sulfur, cryolite.
- *Synthetic inorganics* from inorganic chemicals. Examples: lead arsenate, sodium floride.

ORGANIC: Always contain the element carbon; have at some time been a part of (or are a product of) living organisms.
- *Natural organics* occur in nature.
  - *Botanicals* derived from plants. Examples: nicotine, pyrethrins, rotenone, essential oils, vegetable oils.
  - *Others* petroleum oils[a]
- *Synthetic organics* synthesized in a laboratory or factory.
  - *Organochlorine compounds* (hydrocarbons). Examples: DDT, methoxychlor, dieldrin, chlordane, Kelthane.
  - *Organophosphates* Examples: malathion, Diazinon, parathion, dimethoate, Meta-Systox-R, naled, phorate.
  - *Carbamates.* Examples: carbaryl, Baygon, carbofuran.
  - *Synthetic pyrethroids.* Examples: resmethrin, permethrin, fenvalerate.

[a]Note that of the groups mentioned above, this group kills insects in an entirely different way, by blocking their respiratory openings, thus physically suffocating them. Principal use has been with surface oils on mosquito-infested bodies of water. Oils are also used on plants to kill insect eggs and scale insects.

several weeks to many years (Table 10.4). Their mode of action is still not understood, but involves abnormal nerve transmission.

The first organochlorine insecticide, DDT, had such an effect on pest control that it helped to revolutionize both agriculture and public health. The two characteristics of DDT that made it such a spectacular insecticide were its high toxicity to many kinds of insects and its persistence. The pesticides available at that time—arsenicals

Table 10.4 Comparison of Toxicities and Persistence of Selected Insecticides from the Major Chemical Groups[a]

Chemical	Environmental persistence (half life in soil, in months)	Insect	LD_{50}[a] (μg/g)	RAT LD_{50}[a] (mg/kg)	
				Oral	Dermal
Arsenicals					
		Hornworm	85	125–185	
Lead arsenate	1–4	Colorado potato beetle	140–240		
Botanicals					
Nicotine sulfate	<1	Aphid	48	50	140
		Locust	500		
		Hornworm	4650		
Rotenone	<1	Silkworm	3	60	>1000
		Locust	4700–7000		
Pyrethrins	<1	Louse	42	200	1800
		Fly	31–38		
		Mosquito	0.5–1.0		
Organochlorines					
DDT	36–120	Roach	10	87	1931
		Fly	6–8		
		Mosquito	5–8		
Chlordane	4–12	Fly	1.6	283	580
		Roach	1		
Organophosphates					
TEPP	<1	Bee	1.2	0.2	2
		Locust	4.4		
Parathion	1–4	Roach	1.0	3	4
		Locust	0.7		
		Fly	1.0		
Malathion	<1	Western spruce budworm	29	885	4000
		Douglas fir tussock moth	1000		
Diazinon	4–12	Western spruce budworm	19	66	379
		Douglas fir tussock moth	15		
Carbamates					
Carbaryl	1–4	Western spruce budworm	20	307	2000
		Douglas fir tussock moth	25		
Carbofuran	4–12	Western spruce budworm	6.0	8	10,200
Synthetic Pyrethroids					
Resmethrin	?	Western spruce budworm	0.26	1500	3040
Permethrin	?	Face fly	0.369	2000	

[a]The LD_{50} figures are the dose of toxicant required to produce 50% mortality of a population. The figures for insects are topical applications (direct treatment of the body surface). Those for the rat are through the mouth (oral) or skin (dermal).

and botanicals—either had to be applied at heavy rates (3–15 lb/acre) or were so expensive that their use was prohibitive on all but a few crops. In addition, the botanicals had a residual activity of only several minutes to a week. In comparison, DDT is not soluble in water so it will not wash off in rain, and it can be applied to leaves or to the walls of houses in malarial zones. Its residual activity could still kill mosquitoes 6 months to a year later. DDT provided excellent control when applied at the rate of 1 lb/acre, which, in combination with a price of 22¢ per pound allowed growers to protect even low value crops from insect damage, something that was not possible with earlier pesticides. However, the persistence of DDT and its solubility in lipids caused incorporation and accumulation in animal fat and thus in the environment.

Perhaps the greatest legacy from DDT was the interest it stimulated in the synthesis of new compounds that could match or exceed the toxicity of DDT without having its detrimental qualities. The early work led to other even more persistent organochlorines, like dieldrin, aldrin, and chlordane, all of which have also been banned. Further research has produced insecticides with more desirable qualities than these mostly abandoned compounds.

Organophosphates were originally developed as byproducts of research on nerve gases as chemical warfare agents. Some of the earliest types, and several still used today, are almost as toxic to humans as they are to insects (see Table 10.4) because they attack the cholinesterase involved in nerve impulse transmission by both vertebrates and insects. Organophosphates are used in spite of their very high toxicity to humans because they have such short residual life. Thus, insecticides can be applied to nearly ripe fruit just a few days before harvest and will have killed the pests and then broken down into harmless residues before pickers are required to handle the fruit. Also, since some organophosphates are water soluble, they can be picked up by the plant and translocated throughout the tissues in high enough concentrations to kill insects feeding on the tissues, a quality known as systemic activity.

Further work on organophosphates has yielded such materials as malathion, a potent insecticide with a low mammalian toxicity that is available for use by the general public. Other organophosphates, like ronnel, provide systemic activity to livestock and thereby protect them from cattle grubs and other pests. Diazinon is a safe, relatively persistent organophosphate, making it a suitable product for home use on gardens and lawns.

Carbamates are the synthetic analogs of the methylcarbamate ester of physostigmene. This compound is the toxic constituent of the Calabar seed used in West African religious rituals. Carbamates act as potent esterase inhibitors, but are highly unstable, as demonstrated by the short duration of their toxic symptoms and cholinesterase inhibition. The chemical atropine sulfate can be used as an antidote in humans. As a group, carbamates have short residual activity. Several compounds in this group are used against mites, molluscs, fungi, and weeds, as well as those with insecticidal activity.

The synthetic *pyrethroids* have only recently been developed, having been discovered during attempts to synthesize the toxic molecules of naturally occurring pyrethrins, in an effort to circumvent the high cost of producing natural pyrethrins and the uncertain availability from foreign suppliers. In the process of synthesizing new pyrethroids, an attempt was also made to improve the instability of pyrethroids to sunlight. It was successful, and it is now feasible to apply residual films of contact pyrethroids.

Because of the complex procedures required in their manufacture, synthetic pyrethroid compounds are very expensive, compared to standard organophosphates on a weight for weight basis. However, pyrethroids have been found to be as much as 10 times more effective in the field than the most potent of the three earlier groups of synthetic organic insecticides. Therefore much less material is required per acre to give the same level of insect control, and actual expense is less than more traditional insecticides. Pyrethroids are effective against moth and butterfly caterpillars; they also kill the eggs and adults. Many beetle, fly and bug species are also killed by synthetic pyrethroids.

As a group, synthetic pyrethroids are nerve poisons, as opposed to organophosphates and carbamates, which are acetylcholinesterase inhibitors. Like most of the organochlorines, the synthetic pyrethroids are almost insoluble in water, but soluble in lipids. Thus, once applied to a surface they will not wash off with water. Also, they may be used as mothproofers since impregnated cloth will retain the chemical, even after many washings. Several act as flushing agents, that is, they irritate insects and cause them to move about, thus contacting more of the chemicals that had been applied. Tetramethrin has low toxicity, but good knockdown (immobilization), whereas bioresmethrin is very toxic, but lacks quick knockdown. A combination of the two provides both rapid immobilization and high death rates. Because of low mammalian toxicities (see Table 10.4) some synthetic pyrethroids will be especially useful for direct application to livestock against ectoparasitic insects and biting flies. Past practices for biting flies have consisted of daily applications of natural pyrethrins, a procedure limited to animals of especially high value becuase of the labor involved, but the residual toxicity of the synthetics will allow for treatment of livestock that are not tended often.

Unless problems develop with the use of synthetic pyrethroids, they seem to be the answer for pesticidal control of insects in the near future. Other chemicals, notably those natural or synthesized products that cause abnormal behavior rather than mortality, will also play a major role. They will be covered later in this section.

Chemical and Biological Characteristics of Pesticides. *Solubility.* Most insecticides are solids at room temperatures. Those pesticides not very soluble in water, when prepared using the appropriate formulation technology, cannot be removed by rainwash. Therefore, thorough coverage of surfaces, e.g., leaves, or an animal's body, will provide protection until the pesticide breaks down. On the other hand, water solubility increases the systemic activity of insecticides within the bodies of animals or plants.

Persistence. The length of time a pesticide remains toxic after being applied is its persistence, or residual activity. Even though the breakdown products of a pesticide may remain in a location for a very long time, unless they remain toxic, they are not termed persistent (see Table 10.4). With moth-proofing, persistence is a desired characteristic, or with systemics in plants, where suppression of pests during a critical growth period is essential. However, when a systemic-treated plant is about to be harvested for consumption by humans, persistence of the insecticide would be a problem. The difficulty lies in balancing the degree of persistence of a particular insecticide with both the length of time of residual activity and its degradation to nontoxic products just prior to its use. It is possible to increase an insecticide's residual activity by adjusting its formulation.

Biomagnification. The lipid-soluble organochlorines, noted for their persistence, are also difficult for most organisms to metabolize once ingested. Thus, an organochlorine after ingestion ends up in the fatty tissue in concentrations of several parts of the compound per million (ppm) parts of fatty tissue even in an environment with low (2 ppm or less) levels. If the organism is consumed by a predator, the concentration of the persistent pesticide in the predator may be even higher (Table 10.5). This pattern occurring in only two or three trophic levels results in very high levels in a secondary or tertiary predator. When the levels reach the point where they produce toxic symptoms, the predator dies, or fails to reproduce. This has resulted in deaths of many birds (eagles, ospreys, grebes, and woodcocks) and fish (lake trout). Most humans carry a considerable pesticide load in their fatty tissues and this is one of the reasons DDT and related compounds were taken off the market.

Formulation. As pure compounds, most insecticides are quite toxic to humans and in a physical form that is unsuitable for use. In addition, applications of actual insecticide often cover many acres, and at rates of less than one pound of insecticide per acre. Regardless of whether the pure chemical is a solid, liquid, or gas at room temperatures, the problem then is one of insecticide formulation—how to process insecticides into a final physical form that will improve their safety, storability, applicability, and effectiveness.

Because such small amounts of pure insecticide are required for large areas, the traditional approach to applying insecticides is to dilute the small amount of active ingredient with some inert material. The three basic forms of *diluent* are solid, gas, and liquid. The solids used are usually inert clays, talc, or plant materials. The gas is usually air, and the liquids are either water or some petroleum solvent.

One of the earliest formulations used was a dust. First the insecticide is dissolved in an organic solvent, then the resulting solution is adsorbed onto clay or talc particles. The dust formulation is relatively safe to the applicator because it is already in a

Table 10.5 Biomagnification of Persistent Pesticides

Location	Species	Pesticide	Level of pesticide	Concentration factor	Trophic level
Clear Lake, Calif.	Plankton	DDD	12.5 ppm	250	Herbivore and producers
Clear Lake, Calif.	Frogs	DDD	100.0	2000	Carnivore
Clear Lake, Calif.	Sunfish	DDD	600.0	12,000	Carnivore
Clear Lake, Calif.	Grebes	DDD	1600.0	32,000	Secondary carnivore
Long Island, NY	Plankton	DDT	0.04	1	Herbivores and producers
Long Island, NY	Shrimp	DDT	0.16	4	Herbivores
Long Island, NY	Minnows	DDT	2.00	50	Carnivore
Long Island, NY	Ring-billed gull	DDT	75.0	1875	Secondary carnivore
Wisconsin	Herring gull eggs	DDT	200.0	40,000	Carnivore

dilute form, and it may be applied without very elaborate equipment. One problem is the ease with which slight winds cause the dust to drift. Granular formulations, made in about the same way as dusts, use larger particles of inert clays or plant materials, resulting in materials of about the same texture as coffee grounds. Granulars are not subject to drift, but their use is limited to soil applications. Most currently used soil and lawn insecticides come as granulars and can be applied using fertilizer spreaders. In fact, premixed insecticide-fertilizer granulars are available.

A few insecticides are water soluble and can be dissolved to produce a dilute spray suitable for use on plants. However, most insecticides are not directly soluble in water. Solutions in petroleum products, e.g., deodorized kerosene, are used by structural pest control operators and homeowners on rugs, floors, and furniture. A concentrated solution of insecticide in organic solvent may be diluted with water if an emulsifying agent, a chemical that stabilizes the suspension of oil droplets in water, is added. Emulsifiers are detergentlike and will keep the minute droplets suspended in water for 24 hours or longer, which is an advantage when using a spray tank. The product thus formulated is called an "emulsifiable concentrate" (EC). Because it is concentrated, bulk of the EC product is minimal and shipping costs are kept down. This same concentration causes problems though, because of the hazards inherent in handling concentrated toxic materials, especially since many pesticides can be absorbed through the human skin.

Wettable powders (WP) and flowables (F) essentially are insecticidal dusts that contain a wetting agent that permits the powder to be mixed easily with water. The commercial product usually has a high concentration of insecticide. Since the insecticide is bound to the powder, WPs and Fs are safer to handle than ECs, with less hazard to the operator. For the same reason, WP and F sprays are less likely to cause phytotoxicity than are ECs, a condition sometimes occurring when the EC breaks down. A problem with WPs is that the powder tends to settle to the bottom of the spray tank. Most commercially available sprayers have an agitator in the tank to keep the dilute spray thoroughly mixed.

Mist blowers disperse fine droplets of insecticide using high-speed fans that propel air. Mist blowers provide for more efficiency in applying pesticides because only the supply of moderately concentrated insecticide is carried in the machine and no large water tank or massive machinery are required. However, because the droplet size in mist blowers is smaller than in hydraulic sprayers there is more problem with drift from the target area (Table 10.6). Fogger formulations result in even smaller droplet size. Aerosols are very small droplets suspended in air after the propellant evaporates, as it does immediately upon release.

Slow release formulations may take the form of insecticide-impregnated plastic strips that are hung in enclosed areas to control flying insects. The same principle is used in flea collars where the vaporizing insecticide kills insects that approach the neck area of the pet. Some insecticides are micro-encapsulated in about the same way that ink is in no-carbon copy paper. The capsules are intended to provide slower release of the insecticide or to protect nontarget organisms.

In summary, the same insecticide is often available in a wide variety of formulations to fit the user's needs. Hazards to applicator, type of application equipment available, specific nature of the pest to be controlled, and other factors determine which formulation is to be selected for use.

Table 10.6 Influence of Spray Droplet Size on Drift of Spray Away from Its Target[a]

Diameter of droplet (μm)	Type of droplet in nature	Time for droplet to fall 10 feet	Horizontal distance droplet would travel in 3 mph wind before it would fall 10 feet
0.5		6750 minutes	308 miles
5.0	Fog	66 minutes	3 miles
100.0	Mist	10 seconds	409 feet
500.0	Light rain	1.5 seconds	7 feet
1000.0	Moderate rain	1.0 seconds	4.7 feet

[a]The surface–volume relationship of friction (a surface function) to weight (a volume function) keeps very small droplets airborne for long times, reducing control over where the droplet ends up.

Limitations in the Use of Insecticides. One of the most important problems in using insecticides is the development of insecticide resistance. Insect populations become resistant to pesticides with repeated pesticide applications. As far as is known, there is a minute fraction of the entire population of many insect species that is naturally tolerant to particular pesticides. This preexisting tolerance allows those few individuals possessing it to survive an application that decimates almost all others. Those who survive pass the resistance on to the next generation resulting in a greater percentage of resistant individuals. This subject is covered in considerable detail in Chapter 3. A further complication is the matter of cross-resistance. This occurs because the major groups of insecticides each have a particular mode of action, that is, the way in which the insecticides in that group poison the insect. When the population becomes resistant to a specific insecticide, it may also become resistant to all the other insecticides which function in the same way.

A second problem is the resurgence of a pest after it has been treated if its natural enemies are killed by the pesticide, too. Work is being conducted to study and develop insecticide-resistant populations of natural enemies. Current studies are predominantly in the tree-fruit and cotton ecosystems. The appearance of "new" pests as a result of pesticide use has already been described too.

Both the phenomenon of the new pest and resurgence may be attributed to the lack of specificity common to most of the insecticides currently available. In fact, the cost of developing new insecticides is so staggering—$25-50 million for each new compound in 1986—that chemical companies will not try to develop a compound to kill a single species unless that species is one of the 10 or 15 most important pest species. Some specificity is inherent in particular insecticides, such as systemics for piercing–sucking insects or special formulations of certain microencapsulated insecticides for chewing insects. Other ways to protect nontarget organisms include timing of application and leaving certain areas unsprayed to serve as refuges for natural enemies.

Insecticides are very useful in knocking down large populations. For example, a population of several million cutworms per acre can be reduced to a few hundred per acre with a single application but three to five additional applications may be required to reach eggs, pupae, or genetically diverse forms. Other methods of control work well at low pest densities, but not so well with the large numbers so effectively controlled with pesticides.

The dangers of pesticidal contamination of the environment and of toxicity to those using pesticides is very real and is mentioned here along with the other problems of pesticidal use. These two subjects are covered in more detail later in this chapter under governmental regulations.

Insect Growth Regulators (IGR)

Several rather new compounds have been developed that do not kill the insect directly, but interfere with the normal growth and development of the insect. IGRs act in several ways. The most common interferes with the insect's ability to reach the adult stage because it possesses a degree of juvenile hormone activity. Thus the larval–pupal or pupal–adult molt is not normal, the molting individual does not escape from its old cast skin, or the resulting individual is abnormal (see Fig. 10.29). One IGR, Altosid, is especially effective against mosquitoes; the larval stage receives the treatment and thus the population is reduced before it becomes a problem. Because larval development in floodwater mosquitoes is very synchronous—flooding initiates hatching of the eggs—this unstable material can be applied precisely during the last larval instar, the stage most sensitive to IGRs.

An IGR with different activity is Dimilin, a chemical that interferes with chitin formation during the process of molting. Since chitin is a key material in the insect integument, interference with chitin formation causes death at the molt.

IGRs offer some exciting alternatives to the use of standard pesticides because IGRs seem to affect only the arthropods, and are effective only if they are present during an insect's sensitive period. Because of this latter aspect they also seem to provide some degree of protection to natural enemies. Drawbacks to expanded use of

Fig. 10.29 Unsuccessful emergence of alfalfa weevils treated with an experimental insect growth regulator. Although the treated larvae were able to pupate, the resulting adults were unable to free themselves from the pupal cuticle and died. From J. W. Neal Jr. and A. A. Hower with acknowledgment of USDA.

IGRs are many. They are typically unstable, a problem that has been overcome somewhat by incorporation with other materials to produce a slow-release formulation. A major barrier to further development of IGRs is the mass of legal requirements in registering an IGR for use. Since IGRs are categorized as pesticides, they must conform to the same legal requirements for registration as a pesticide. The maze of legal requirements are presented in the section on governmental regulations.

Insect Attractants

Because of their small size and relative lack of visual acuity (Chapter 6), insects live a much more chemically mediated life than do humans. Their ability to discriminate between concentrations of chemicals and between closely related chemicals is much more highly developed than it is in humans. In fact, human reliance on vision is so strong that it is difficult to imagine living in a world whose signals are primarily chemical, whether taste or smell.

The variety of chemical sources that attract insects is myriad. Foods, mates, oviposition sites, nesting sites, and prey habitat are among the many chemical-emitting sources that affect one or more species of insects. Chemicals emitted by one individual of a species that cause a behavior in another individual of that species are termed "pheromones." Collectively the pheromones, allomones, and kairomones are called semiochemicals (see Chapter 6 on insect behavior).

The first insect pheromone identified was the sex pheromone of the silkworm *Bombyx mori.* As with most Lepidoptera, the female moth produces and emits a pheromone when she is physiologially receptive for mating. The male, often with its relatively more elaborate antennal receptors, homes in on the source of the pheromone, the female, and mating occurs. Although the silkworm's pheromone (bombycol) was not identified until 1959 the attraction of male moths to females because of odor has been known for many years.

Since the first isolation of bombycol a whole series of sex attractants have been identified (Table 10.7). These have proved to be remarkably specific, that is, each species of insect typically has its own unique sex attractant chemical. Sometimes the attractant is a mixture of chemicals, and in these cases the particular mixture is unique for each species, or even for particular populations. Not all of the pheromones identified are sex attractants. The other types, with examples, are listed in Table 10.8.

Classes of attractants other than pheromones are also being investigated for their potential use in pest suppression. In biocontrol, the arrestant kairomones emanating from caterpillar frass (feces) have been identified and are being used to cause the pest's parasitoids to search for their hosts more diligently in low density situations. Food lures for Mediterranean fruit fly, Japanese beetle, and yellowjackets are being used commercially in control programs.

Incorporation of attractants into control programs involves three different strategies. (1) Most attractants are used in conjunction with traps so that seasonal appearance of a pest can be monitored (Fig. 10.30). Trap monitoring is used to determine when to apply pesticide applications and thereby to minimize the number of applications required to achieve the desired effect. (2) Several experimental attempts

Table 10.7 Examples of Commercially Available Insect Sex Pheromones That Have Been Synthesized and May Be Incorporated into Control Programs

Common name	Trade name	Species attracted
Disparlure	Disparmone Pherocon GM	Gypsy moth
Grandlure	Grandamone Pherocon BW	Boll weevil
Gossyplure	Pherocon PBW	Pink bollworm
Looplure	Cabblemone Pherocon CL	Cabbage looper
Muscalure	Muscamone	House fly
Codlelure	Codlemone	Codling moth
Virelure	None	Tobacco budworm
None	Z-11	Red-banded leaf roller European corn borer Oblique-banded leaf roller Smartweed borer

have been made to control pests by removing the males of the species by trapping them with pheromone traps. Successful results of such experiments depend on isolation of the experimental area and the host range of the pest studied. (3) The last method employs pheromones as mating confusants. In this strategy, the male's ability to locate its mate through pheromonal attraction is overridden by saturating the entire area with pheromone. Thus, the male has no gradient of pheromone that is essential in locating the female source of pheromone emission.

Repellents

Although the original definition of these chemicals, which cause insects to make deliberate movements away from the source of the chemical, is still valid, its meaning in insect control is being expanded. Today, insect repellents include all chemicals that protect animals or plants from insects by making them unattractive, unpalatable, or

Table 10.8 Types of Pheromones

Action	Function	Example
Sex attractant	Mate location	Gypsy moth, codling moth, honey bee
Alarm	Defense, escape	Aphids, honey bee, ants
Trail marker	Trail marking	Ants, primitive bees, termites
Arrestant	Localize searching for mates	Bark beetles, carpet beetles
Aggregation	Swarm integrity	Honey bees[a]

[a]Several specialized and rather unique pheromonal relationships have been identified among social insects. These have not been included in this table.

***Fig. 10.30** The inner surfaces of this trap contain a source of pheromone and are coated with sticky material. Attracted moths enter and become entangled in the sticky coating. Traps in apple orchards are inspected weekly to determine emergence of codling moth, red-banded leafroller, and several other moths that attack this crop.*

offensive. Some act as antifeedants or excitorepellents rather than causing the insect to avoid the source of the chemical.

Antifeedants or feeding deterrents are substances which, when tasted, result in cessation of feeding. This cessation of feeding may be either temporary or permanent, depending on the action and potency of the material. A great many chemicals extracted from plants have been reported to act as antifeedants for phytophagous insects. For example, meliantrol, which has been extracted from the leaves and bark of the Indian neem tree, is an antifeedant for desert locust (*Schistocerca gregaria*). This material is so strong that 1 ng/cm^2 on otherwise acceptable food sources will cause the locust to cease feeding. (One nanogram (ng) is one billionth of a gram.) A material extracted from certain composite plant species in the genus *Vernonia* is a powerful antifeedant for both southern armyworm and fall armyworm. Neither of these general feeders will consume species of *Vernonia* containing the antifeedant, but readily feed on *Vernonia* species lacking the compound.

Antifeedant chemicals appear to have the most potential for pest suppression if the plant species can be bred to contain antifeedants in high enough quantities to deter feeding, much like the DIMBOA example given earlier in this chapter. There are limitations to their use as materials applied to animals or plants for protection. Since antifeedants sprayed on plant surfaces seem to protect only the areas covered by the spray, all new growth or missed surfaces would be subject to attack. The antifeedant-protected areas would just force the pest to concentrate on unprotected areas. For animals, antifeedants would be ineffective because vectors of animal diseases could probably still transmit the disease in the process of initial feeding probes. In addition, the annoyance caused by the presence of pests flying around the host animal or human would still be offensive even though actual feeding were reduced.

Excitorepellents are chemicals that so irritate insects contacting them that they abandon the area. The uncoordinated behavior is due to acute disturbances of the nervous system. Pyrethrin is one pesticide with excitorepellent properties. It is employed to flush hidden insects, such as the nocturnal cockroaches, out of their hiding places so that contact can be made with other toxic residual materials. Creosote is a repellent used as a protective barrier strip to stop chinch bug invasion in corn fields.

Feeding deterrents may have their best application as protectants for nonliving organic material. The best repellents deter the feeding of the clothes moth or carpet beetle in wool and the termite or powderpost beetle in wood. However, the same problem of coverage exists with repellents for wood. The exposed surface still must be completely protected or the insects will use untreated areas as avenues of entry and then expand into the unprotected inner material.

For all of their varied uses, the classic type of repellents have proved to be of greatest value in protecting humans from attack by blood feeding insects. Those materials currently available are chemicals that slowly volatilize and tend to keep such insects as mosquitoes, the biting flies (stable, horse, deer, and black flies, greenheads, and punkies), ticks, and mites from feeding. Again, rather complete coverage is required in order for most of the repellents to be effective. Any untreated area e.g., the back of the neck, may be subject to attack even though the rest of the body is protected. Also, most repellents must be reapplied at frequent intervals. Their potency is reduced due to vaporization, absorption, dilution with perspiration or just being rubbed off. Some recently developed materials have considerable long-distance effect and may protect the whole person if only the hat band or some similar article is treated.

Considerable effort has been expended on research concerning repellents primarily because of the need to protect armed forces operating in insect-infested areas. In the South Pacific during the Second World War troop debilitation because of insect-borne disease was often higher than losses from combat. As a result of chemical research, repellents like diethyl toluamide, dimethyl phthalate, and a few others are now available to protect sportsmen, those involved in outdoor occupations, and the armed forces from the insects, mites, and ticks that make exposure so agonizing and dangerous.

Chemosterilants

Several hundred of these chemicals that interfere with normal reproduction have been discovered in the past two decades. That such chemicals might offer novel approaches to insect suppression was demonstrated by the success achieved in the screwworm eradication program of the U.S. Department of Agriculture. Although this program achieved insect sterility through radiation rather than with chemicals, the theoretical potential for chemosterilants became apparent.

The screwworm fly eradication program is explained more completely in the next section, but the points to be considered here concern the basis of its operation—release of insects that had been reared and then sterilized prior to release. The only satisfactory way of sterilizing the flies when the program was initiated was through exposure to radiation. If effective chemosterilants could be used, natural populations of insects could be treated, thus eliminating the need to rear them under artificial lab conditions. Many species that have never been successfully reared in the lab could be controlled. But, release of some type of insects, even if they are sterile, would create more problems than are solved. For example, release of large numbers of sterile disease vectors as a control technique would be ridiculous because the sterile insects could act as vectors just as well as the normal insects did. Thus, the short-term problem could outweigh the long-term benefits of such a program. Chemical sterilization of natural populations does, however, have theoretical merit.

Several groups of chemosterilants have been discovered. Of these, the alkylating agents have shown widest promise. They mimic the effects of radiation and result in lethal mutations of offspring or in their incomplete development. Alkylating compounds such as tepa and apholate have been incorporated into baits; both sexes of insects feeding on such baits are sterilized. In addition to sterilization of the target species, these compounds also dramatically reduced their competitive ability and length of life. Unfortunately, research on chemosterilants has also shown that the compounds identified to date are highly mutagenic to vertebrates. Thus, even use in bait formulations constitutes an unacceptable risk for practical application of chemosterilants as a viable suppression technique.

Autocidal Methods

Until 1955 insect pest suppression was based on increasing the environmental resistance (making it a less suitable environment) to the many offspring produced by pests and thereby to lessen their chances of surviving to maturity. Techniques such as pesticidal application, biocontrol with natural enemies, and use of resistant hosts have been employed to increase the effects of environmental adversity. The intention is to reduce the number causing damagé in the present generation and to lower the numbers contributing offspring to the next generation. The difficulty with this approach is that some individuals are left and do reproduce. The mate-finding methods (pheromones, swarming flights, etc.) employed by insects are so efficient that mating can often occur even in very sparse populations and a high percentage of the resulting offspring are often able to reach maturity in an environment free from competition and without too many natural enemies. Thus, in just a few generations the pest population is back at the pretreatment level or even higher if natural enemies do not recover.

In the early 1950s, research on the effects of gamma radiation showed that insects could be made sterile without significantly changing their behavior, if they were exposed to the proper dosage of radiation. Thus, it became feasible to alter the reproductive potential of insects in a naturally occurring population if they could be made to mate with a sterile rather than fertile partner, for fewer or no offspring may result from such matings.

The use of sterile matings was proposed as a control technique for the screwworm fly, *Cochliomyia hominivorax* (Coquerel). The end result was to be the complete eradication of this pest from the southeastern United States.

The screwworm is a fly maggot capable of devouring animal tissue once the animal's skin is broken. Adult flies are attracted to barbed-wire cuts, insect bites, and other open sores in adult animals in addition to the unhealed umbilical sites in newborn animals. The female deposits eggs at the site of the wound and the larvae bore into the wound increasing its size and causing the wound to ooze liquids that make the wound even more attractive to ovipositing female screwworm flies. The mature larvae leave the wound to pupate in the soil. Untreated livestock and wild animals often die as a result of attack, so that livestock losses in the southeastern United States were estimated to run about $20 million each year.

An insect with such economic importance as the screwworm has received considerable attention by entomologists, including those employed by the U.S. Department of Agriculture. The results of research by USDA scientists and others revealed several interesting facets of screwworm biology that are pertinent to the development of autocidal control technology. Among them is that very low populations exist in the field during the coldest parts of the year, and that the fly could be reared in large numbers on nonliving material. The key discovery was that the female screwworm mates only once in her entire life, whereas the male screwworm mates several times. A normal female mated to a sterilized male produces no offspring, therefore, flooding an area with sterilized males could greatly reduce and eventually eliminate the species from a discrete area, provided reinvasion from other areas could be prevented. The theoretical eradication of a natural population of screwworms by overflooding with sterilized mates is presented in Table 10.9. This technique, known as the Sterile Insect Release Method (SIRM), involves multiple releases of male screwworm adults that have been reared under artificial conditions and then sterilized while they were pupae by exposure to 5 kV of gamma radiation emitted by a cobalt-60 source. If the first generation to be treated is inundated with twice as many sterile males as exist in nature, then two of every three matings will be nonproductive, provided the sterilized males compete well with males in the normal population. The result of releases in the first generation is a 67% reduction and continued releases over four generations theoretically eliminates the insect from the treated area because the last few fertile members of the natural population are sought out by the sterile forms and neutralized.

Table 10.9 Theoretical Eradication of a Population of Insects Provided Females of the Species Mate Only Once

Assumed natural population of virgin females in the area	Number of sterile males released each generation	Ratio of sterile to fertile males competing for each virgin female	Percentage of females mated to sterile males	Theoretical population of fertile females each subsequent generation
1,000,000	2,000,000	2:1	66.7	333,333
333,333	2,000,000	6:1	85.7	47,619
47,619	2,000,000	42:1	97.7	1,107
1,107	2,000,000	1,807:1	99.95	Less than 1

The first successful trial of the SIRM on screwworms was on the island of Curacao in 1953. After 7 weeks of releases, 100% of the egg masses collected were sterile. Because of this success, screwworm eradication from the southeastern United States was attempted. Flies were reared in a "factory" on a medium of meat and blood, produced at the rate of 50 million flies per week, packaged in cartons holding about 440 flies each, sterilized, and then released from airplanes at the rate of 200 per square mile over approximately 85,000 acres. Results were spectacular, and the fly was eliminated from Florida over the 1958–1959 winter, but only at great effort and expense (over $10 million), with an estimated savings of $140 million overall.

A subsequent proposal to eliminate the screwworm from all of North America was made, based on a SIRM program that would eradicate the fly first from Texas and then by shifting the release area further and further south, to rid the continent of screwworms. Reinvasion from the south was to be prevented by a barrier zone in which continually released sterile insects would neutralize invaders.

Implementation of the SIRM program for screwworm eradication showed vastly reduced populations. Confirmed cases of livestock infested with screwworm fell from 49,484 in 1962 to 446 in 1965, so that by 1966 the U.S. Department of Agriculture officially considered screwworm eradicated from the United States. Those few cases reported were considered to be the result of fly invasion from Mexico. However, the bottom fell out of the release program in 1972 when 92,198 confirmed cases were reported in the United States. Although still unconfirmed, it is now believed that the female screwworm flies began to mate discriminately with normal flies rather than with the lab-reared, sterilized flies. The reared flies also proved to be less competitive and vigorous. In addition, ranchers had relaxed the practices that had previously minimized fly attack. Recently efforts have been made to develop effective chemical attractants to supplement the control program for screwworm flies.

To date, genetic methods other than SIRM are still in the experimental stage. All the techniques being studied depend on the mass production of special genotypes and subsequent release into the natural population of the species to be suppressed. In recognition that total eradication of a species is very difficult, some of the newer genetic controls aim at suppressing the pest population, not completely eliminating it. An additional consideration of major importance is the size of the natural population at the onset of a genetic suppression program, since the capacity for mass rearing an insect limits the numbers that can be released. Therefore, combining techniques has been suggested. For example, tobacco hornworms are so numerous in their area of infestation that a genetic release program would have only a minor effect. A realistic approach would be one in which the natural population is treated first with insecticides, which are particularly effective against large populations, followed by a genetic release program such as SIRM, which is particularly effective in low populations. A high proportion of sterile males would be available to seek out and neutralize the few normal females left in the area after the insecticidal treatment.

Proposed genetic controls of many other types are being investigated for possible use. The two with the highest probability of success involve conditional lethals and translocations. Conditional lethals are genetic alleles that are fatal under a particular condition, as, for example, high temperature. This technique might allow for introduction of mutant individuals into a population in early spring when temperatures are favorable. By the time that high midsummer temperatures prevail, a large portion of the population could then be carrying the temperature-sensitive conditional allele, and they would die. Translocations are exchanges of chromosomal material between nonhomologous chromosomes, which are not fatal if both parents are homozygous for the translocation. However, matings between parents that are not homozygous for the translocation result in significant decreases in gamete viability. Thus, the population is dramatically suppressed, with reductions of 35–100% possible.

Genetic control offers other possibilities besides those mentioned previously. Among these is the displacement of a pest strain with a form that is innocuous. Some strains of known vector species have much lower capacity for disease transmission than others. Thus, replacement of a vector strain with one that is not a transmitter solves the disease problem without leaving an ecological void, the empty niche of the vector, to be filled by reinvasion.

Physical Control

Physical properties of the environment are sometimes very effectively used to destroy insects or to prevent them from causing damage. Barriers, for example, window screens, keep flying insects from entering buildings. Dusty furrows, especially with crankcase oil at the bottom, may stop the movement of armyworm caterpillars or chinch bugs into corn fields from neighboring areas. Sticky bands around tree trunks act as barriers to prevent such crawling insects as wingless adult cankerworm moth and gypsy moth caterpillars from moving up the trunk. These barriers may not always work, however, since gypsy moth caterpillars are blown between trees as they spin down on silken strands. Cankerworm females are sometimes so abundant that they cross the sticky bands by crawling over previously entangled females or after dust and other debris has covered the band's surface.

Extremes of temperature, both high and low, are used to kill insects. Low temperatures are less effective on many insects from the temperate region since they have evolved tolerances to the naturally occurring low temperatures of winter, in the range of −20°F. However, many of the stored-product pests evolved in the tropics and die if chilled to even 35°F. Therefore, products stored in grain elevators located in the upper midwestern states (e.g. Minnesota) are protected by circulation of air through the elevators during the rigorous winter months when temperatures are often below 0°F. High temperatures in the range of 130°F for three or more hours are effective in killing almost all insects. Such high temperatures may be employed in protecting stored foods, clothing, or other materials that can withstand such exposure. High temperatures are also employed to rid bulbs of insect pests contained within them. The trick is careful control of temperature, so that the level needed to kill the insect is reached and held but not exceeded, since only a few degrees difference will kill the bulb, too.

Radiant energy is often employed in insect control. Sound waves in the ultrasonic range can kill insects, but sound "shadows" that result when sound waves are blocked often allow insects to survive exposure to lethal doses because they were concealed. Light is used, usually in conjunction with a trapping device, to control insects, not because of any mortality due to the light, but because many insects are drawn to some lights. Many light traps employ "black" lights that emit ultraviolet radiation, rather than incandescent lights, since many insects are especially responsive to those wavelengths. Insects that are attracted are drawn in by suction from a fan or simply fall into a container and are killed by drowning, electrocution, or mashing, or are simply held until the trap is emptied. Elements of a typical light trap are presented in Fig. 10.31.

Light traps, like other kinds of insect traps have some inherent problems that must be solved before their use as controls can be broadly accepted. A primary problem in their general use is that the attraction to light varies with the species, the sex, and the physiological state of the individual involved. We don't know how far away insects perceive and are attracted to lights. We suspect that intensity may be attractive at one level and repellent at others. Therefore, light traps (and pheromone traps, too) are used primarily to determine when and where given species appear so that timing of other suppression techniques can be pinpointed to maximize their effect.

Cultural Methods

Cultural control of insect pests involves manipulation of the host plant or animal by humans in ways that reduce or eliminate the pest's attack. Manipulations may include destruction of host residues, crop rotation, mixed cropping, special planting dates, or management of water availability. Regardless of which particular cultural

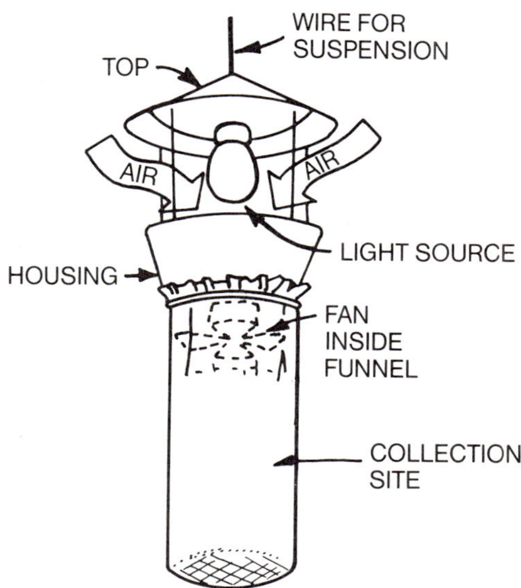

***Fig. 10.31** Insects are attracted to the light source of an insect light trap, then drawn in by the fan's suction to the collection site where they may be killed by insecticide, by the fan itself, by falling into a liquid where they drown, or by electrocution.*

control is employed, they are all based on the concept that each insect usually has a particularly vulnerable stage in its seasonal history that can be discovered if the insect's biology is known in sufficient detail. For corn rootworms the oviposition site is critical. For European corn borers it is the overwintering larva that is most vulnerable. With others, the adult might be most easily destroyed. These common sense controls often are not particularly expensive, but may be very effective, especially when conducted on an area-wide basis.

Crop Rotation

While insects like Japanese beetle, grasshoppers, white grubs, harlequin bugs, and wireworms may be categorized as general feeders, most insects have fairly restricted diets. This is true for many of the pests on our major crops. Asparagus beetles live only on asparagus, cabbage aphids on plants of the cabbage family (Brassicae), and the squash vine borer on squashes, cucumbers, and melons.

Those dietary specializations that make insects so efficient in locating and utilizing resources can also be turned against them through crop rotation. For example, adult Northern corn rootworm *Diabrotica longicornis* (Say), produce one generation each year, depositing their eggs in late summer and fall at the bases of corn stalks. Since this is the only site where rootworm eggs are laid the obvious control is simply to plant a crop other than corn in a field where corn has been grown the previous season. After a year's production of another crop the field can be replanted to corn with freedom from rootworm attack.

The frustrating point in the elimination of rootworm by rotation of corn with another crop is that growers often fail to follow the recommendation for economic reasons. They find it more profitable to grow corn season after season using pesticides rather than alternating with a crop of lower value. A further consideration to farmers is that crops other than corn often require different types of machinery, which can involve great cost.

Variations in Planting and Harvesting Dates

The seasonal development of many temperate zone insects and plants is tied closely to the progression of the seasons through photoperiodicity or temperature-driven processes. Therefore, most of the individuals of a brood hatch at one time in a given location. The alfalfa blotch leafminer, *Agromyza frontella* (Rondani), overwinters as a pupa in the soil. The earliest flies emerge on the same day that pin cherry, *Prunus pensylvanica* blooms in western Massachusetts, both processes being essentially heat-driven events. The adult population tapers off a few weeks later after mating and egg laying are complete (Fig. 10.32). The larvae begin to develop between the surfaces of alfalfa leaflets and leave to pupate in the soil. Three to four generations per year are typical.

Standard practice in New England for harvesting alfalfa calls for waiting until 10% of the plants are in flower. This allows for optimal fiber bulk and protein, which

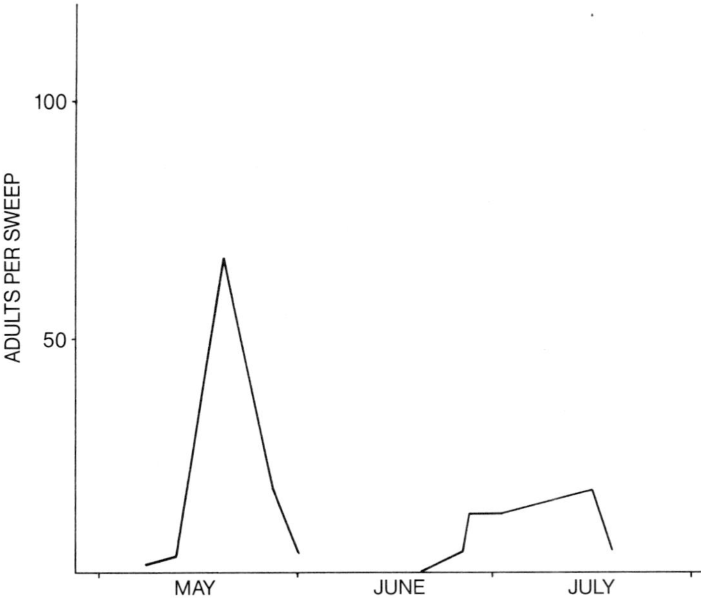

***Fig. 10.32** Population pattern of adult alfalfa blotch leafminer,* **Agromyza frontella** *(Rondani), in spring and early summer as shown by monitoring the numbers of adult flies taken by sweeping an insect net through alfalfa. Data are from unpublished work by A. J. Alicandro and T. M. Peters for western Massachusetts in 1980.*

drops as stems mature. However, by the time 10% bloom is reached almost all of the alfalfa blotch leafminers are in the soil where they are safe from the violence caused by the harvesting process.

Early harvesting of the first cutting of alfalfa, which is cut three to four times a season in Massachusetts, greatly reduces the numbers of leafminers in the second generation because they are not sufficiently mature to complete pupation even if they manage to get out of the alfalfa before its drying and baling is complete. Therefore, it is recommended that growers harvest when leaf miners begin to be seen. This occurs about 10 days after pin cherry blooms. Harvest at this time will greatly reduce the next generation because leaf miners are killed in the drying process, and those adults of the previous generation that are still alive have no place to deposit eggs since it takes several days of regrowth before alfalfa leaves are available for feeding or oviposition. Our data show that early harvesting yielded 0.24 adult leafminers emerging per 0.23 ft^2 of soil, whereas 44.0 adult leafminers emerged per 0.24 ft^2 of soil in alfalfa cut at the standard 10% bloom stage of growth.

Whereas only three cuttings of alfalfa are usual in northern areas, some regions of the southwestern United States can produce alfalfa on a continuous basis with eight or more cuttings each year. The crop is lush, and many insects reside in this ecosystem. However, the drastic change that an alfalfa field undergoes at harvest produces a concomitant change in the insect fauna. As the alfalfa is cut and left to dry in the field, those insects capable of flying to adjacent sheltered areas may survive, but those that cannot fly perish in the hot sun. Pests reinvade quickly, but not their natural enemies. Harvesting of alfalfa, leaving strips of halfgrown alfalfa as a refuge for natural enemies, was tested. As a result of experiments like these, continuous production of alfalfa is now accomplished by strip harvesting to reduce losses caused by insects. Fields are harvested in strips so that when one strip is ready for harvesting, there is a strip of half-grown alfalfa on either side of it (Fig. 10.33). Insects driven from the harvested area have a refuge just a few yards away. The effect of this system on survival of natural enemies in alfalfa is shown in Table 10.10.

The larva of the Hessian fly, *Phytophaga destructor* (Say), feeds on juices exuding from wounds caused by biting the tissues within the leaf sheath of the wheat plant. This causes abnormal plant growth and results in broken stalks rather than harvestable heads from infested plants. The minute, frail, non-feeding adult fly emerges from its puparium in the fall after early fall rains soak the puparia. In average temperatures of 66°F adults will emerge in 7 days (11 days at 60°F). The flies are rather

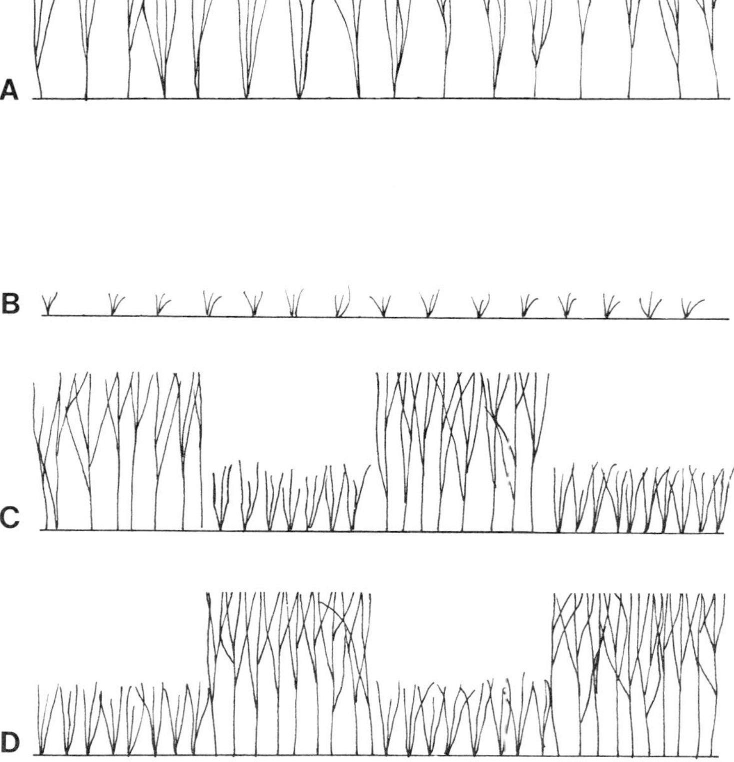

***Fig. 10.33** The strip farming technique used in alfalfa production conserves natural enemies by providing adequate adjacent refuges. (A) Mature alfalfa ready for harvest. (B) After harvest the soil is exposed and insects leave or die. (C and D) When mature strips are harvested, insects seek refuge in partially mature strips.*

Table 10.10 Conservation of Natural Enemies by Strip-Harvesting Alfalfa

	Average number per acre		
Natural enemies	Regular farming	Strip-farming	Increase
Lady beetle			
adults	46,000	205,000	159,000
larvae	11,000	232,000	221,000
Green lacewing			
larvae	195,000	206,000	11,000
Parasitic wasps	70,000	287,000	217,000
Big-eyed bugs	199,000	401,000	202,000
Predatory spiders	105,000	1,094,000	989,000
Totals	626,000	2,435,000	1,809,000
Total natural enemies per square foot	14	56	42

synchronous in emergence and die in 4 days or less. Only the wheat that has emerged by the time flies are present is subject to attack, so if planting is delayed so that the shoots do not break ground during the few days while flies are ovipositing, the crop escapes attack.

Other Cropping Techniques

Fallowing, that is, leaving land barren or unused for a season is an agronomic practice used to conserve moisture in arid regions and it also gives some relief from insects, weeds, and diseases. If the land is plowed so that no growth occurs, insects hatching have no foods and perish. The same technique is used on pastures to rid them of some kinds of ticks. *Mixed cropping* has some appeal in cotton areas where the plant bugs that can cause considerable damage to cotton prefer the alfalfa strips planted among the cotton. The damage done by the plant bugs to alfalfa is minimal. *Interplanting* is a technique espoused by some gardeners under the title of companionate planting. Such combinations as beans and potatoes are said to reduce pest populations, but data are inconclusive. Further testing of combinations is necessary.

Sanitation. Clean cultivation of fields in fall by plowing is effective against a variety of insects using crop residues, weedy field edges, and plant trash as overwintering sites. Some species move to field margins and survive on weeds or volunteer plants as the crop on which they usually feed is harvested. Others are left in the fields with the crop residue remaining after harvest. Plowing buries the larvae-infested crop debris and prevents them from infesting the following year's crops.

Insects in fallen fruit may be the prime source of individuals in the subsequent generation. Destruction or removal of drops in orchards will reduce the populations of insects such as apple maggot and plum curculio. Control of the house fly can be accomplished by removing a significant percentage of its breeding sites (human and animal excrement preferred, but garbage and other decaying organic material acceptable). In this case, a 50% reduction of breeding area will result in a 50% reduction in the size of future generations.

Sanitation versus No-Till Agriculture. Several insect pests benefit by clean culture of agricultural fields. For example, the pale western cutworm, *Agrotis orthogonia* Morrison, prefers to oviposit on barren soil. However, the majority of insects benefit by conditions in which weeds or dead plant parts not yet broken down by decomposition bacteria, fungi, and arthropods have accumulated. Not only may some of the material serve as an alternate food supply (e.g. nectar for adults from weed flowers), but the accumulations of plant trash provides hiding places as well as moderating the temperature and moisture stresses of barren soil. With this in mind, the new agricultural practice known as "no-till" merits some discussion.

Weeds, the most constant and serious competitors of crop plants the world over, have been controlled by humans since the beginning of agriculture by *tillage* (disturbance of the soil as a means of destroying weeds by mechanical injury, other pests by exposure as well, and as a preparation of the soil for planting or for moisture conservation).

No-till agriculture depends on selective herbicides for weed control in the crop rather than conventional tillage practices (see Fig. 10.34). Planting is accomplished by drilling seeds into a narrow slit cut through the soil and plant trash that accumulates on the soil's surface as a consequence of not plowing. Synthetic fertilizers are used rather than crop rotation with legumes to build nitrogen in the soil.

The no-till technique, or modifications of the practice called reduced-till have opened up vast acreages not previously amenable to row-crop production. Thus, a technique that was still in the experimental stage in 1966 was adopted on 6.5 million acres by 1975 and is expected to become the conventional approach to production on 65% of the land used for feed grains, wheat, and soybeans in the United States by the year 2000. What implications does adoption of no-till technology have for insect pest suppression on these crops?

Damage by insect pests in no-till systems is more severe than on the same crop under conventional tillage practices. This is due in part to the lack of control that conventional tillage offers by destruction of overwintering and oviposition sites, as well as by exposure or mechanical destruction. The result is that the corn rootworm population in corn is about four times as large in no-till as compared to conventional till. Black cutworm, *Agrotis ipsilon* (Hufnagel), attacked 15% of the no-till corn seedlings, but only 1% in conventionally tilled corn grown in Ohio.

No-till techniques are likely to change the pest species complex somewhat, as it has already done with the weed complex, so that different problems will have to be faced. For example, mice and slugs, neither of which are problems typical in conventional-till crops have proven to be pests of some significance in no-till systems. Control of the pest insect complex will continue to rely heavily on pesticides, as it does now. However, the reduction in habitat disturbance that no-till provides is conducive to establishment of an extensive biocontrol agent complex that could hold many pests at levels below their economic thresholds.

(A) (B)

***Fig. 10.34** (A) Conventional tillage agriculture versus (B) reduced tillage agriculture. In conventional tillage, weeds are controlled by physical means, plowing and cultivating. In reduced tillage, weed control is primarily accomplished with chemical herbicides (note dead weeds in B). Reduced tillage minimizes soil loss by erosion, but enhances pest insect survival in the mulch that accumulates.*

Governmental Regulations

Examination of the significant events listed in Table 10.11 show a marked shift in emphasis by governmental agencies over time. The early thrust of legislation concerning insects was aimed at preventing the entry and establishment of pests not indigenous to the region through four basic strategies: exclusion, containment, suppression, and eradication. Later efforts were more concerned with pesticides and their uses, while maintaining the vigor of the earlier, more traditional efforts.

"Quarantines" (that is, restrictions on movement of infested or potentially infested plant or animal materials for the purpose of controlling the spread of a pest) are invoked by governmental agencies as a method of excluding exotic pests from possible invasion of previously uninfested areas. Such quarantines are most effective if they reinforce natural ecological barriers like the ocean, mountains, and desert that permit California to maintain such effective exclusion of pests. A quarantine along the political boundaries of a state like Iowa would be far less effective in restricting movement of insects. It would, however, restrict spread to what the insect was capable of under its own powers. Quarantines provide for regulation of pest move-

Table 10.11 A History of Significant Events in Regulatory Entomology

1873	First regulation directed against a foreign insect. Germany regulated entry of American grape phylloxera.
1881	California passed plant pest quarantine legislation. This was the first government agency in the United States to attempt to retard importation and spread of insect pests and plant diseases.
1892	First federal plant pest quarantine bill proposed in Congress but failed to pass.
1896	Ohio State Horticultural Society passed resolution asking federal government to prevent entry of foreign insects with imported plants and plant products. This was the first concerted effort to obtain federal legislation for this purpose.
1897	First federal plant pest quarantine bill drafted in Washington following introduction of San Jose scale into eastern states. Bill failed to pass.
1905	Federal Insect Pest Act of 1905 enacted, enabling federal government for first time to regulate importation and interstate movement of living insect pests.
1906	USDA starts Texas fever tick eradication program. About 750,000 square miles including 985 counties in southern States were placed under USDA–state quarantine to eradicate Texas cattle fever.
1910	Passage of Federal Insecticide Act sets standards for insecticides and fungicides, prohibits inclusion of phytotoxic materials, forbids false claims on labels.
1912	National Quarantine Law approved. United States last important agricultural country in world to pass such legislation. Prior to this time, United States said to have become a dumping ground for refuse nursery stock.
1917	Plant quarantine inspection service established on Mexican border. It provided a more effective means of safeguarding nation against introduction or spread of foreign pest insects.
1922	Isle of Wight disease of honey bees prevented entry into United States.
1929	Mediterranean fruit fly found for first time in United States in Florida. USDA Secretary requested Congress to appropriate $4½ million for eradication of pest. The insect was eradicated from the United States (16,000 square miles). This was the first time any insect pest had been eradicated from any area.
1935	Preflight inspection and clearance of aircraft departing from Hawaii to the mainland of the United States inaugurated as plant quarantine method.
1938	Passage of federal Food, Drug and Cosmetic Act ensures safety of food for human consumption by establishing tolerances for lead and arsenic in food and required that white powdery insecticides be colored to prevent confusion with flour.
1940	First specially built and equipped inspection station building constructed at Hoboken, N.J., for handling imported plants, bulbs, and seeds from foreign countries.
1945	Provision made for supervision of treatment of agricultural commodities in foreign countries to qualify plant products for importation.
1947	Passage of Federal Insecticide, Fungicide, and Rodenticide ACT (FIFRA) adds rodenticides and herbicides to the pesticides covered by the 1910 law, regulates marketing and labeling, and requires registration of pesticides entering interstate commerce with USDA. Company must prove product (1) is effective for its intended uses, (2) will not injure humans, wildlife, crops, livestock if used as directed. Federal inspectors could seize and impound nonconforming products.
1954	Miller Amendment to federal Food, Drug and Cosmetic Act of 1938. Set pesticide tolerances (residues allowable) in raw agricultural commodities.
1958	Food Additives Amendment to Food, Drug and Cosmetic Act. Extends restrictions previously applied only to pesticides to all food additives. The Delaney clause forbids tolerance of a proven carcinogen in foods at any level. Set pesticide tolerances for processed foods.
1972	Federal Environmental Pesticide Control Act (FEPCA) and amendment of FIFRA (discussed in text).

ment by complete exclusion of suspect materials or admission only of treated or inspected materials. The difficulty in quarantine conformity lies in the faults to which quarantines are prey:

1. The tendency of some state regulatory groups to inspect rigidly when they have adequate local supplies of seed, sets, and stocks and then to let down all barriers when local supplies fail.

2. Seeds, plants or commodities inspected during the growing season, but ignored the rest of the year.

3. Pressures applied by special interest groups to induce quarantine officials to erect outright economic barriers disguised as plant quarantines.

4. Pressures applied by special interest groups to weaken or remove local restrictions to benefit their economic ambitions.

5. Retention of antiquated regulations.

6. Laxity in enforcement of regulations.

7. Lack of funding to accomplish the required tasks.

U.S. governmental agencies are also charged with the tasks of controlling insects already in the country through containment, suppression, and eradication.

Containment. The government attempts to contain introduced pests that have not attained their full geographic infestation potential for two reasons. First, it is cheaper to restrict movement of a pest than it is to suffer the depredations of its attack on a crop. Second, containment of a new pest within a restricted geographic region enhances the possibility that eradication might be accomplished if a breakthrough in control technology occurs. Examples of containment programs with some degree of success are the gypsy moth and the Japanese beetle programs.

Area-wide suppression programs superseding state boundaries are conducted by federal agencies for such pests as grasshoppers and forest pests. Cooperative suppression programs between countries exist for locusts and fruit flies.

Eradication programs of some newly introduced pests have been rather spectacularly successful. The Mediterranean fruit fly, *Ceratitis capitata* (Wiedemann) has been eradicated at least five times from the continental United States. Parlatoria date scale, *Parlatoria blanchardi* (Targioni-Tossetti) was eradicated from the United States over a period of 12 years. The eradication program for screwworm fly, *Cochliomyia hominivorax* (Coquerel) with its successes and failures was described earlier in this chapter.

In addition to the regulatory roles that governmental agencies provide, many also maintain laboratories to provide research and development to support their regulatory functions.

The federal government has assumed a growing responsibility for regulation of the chemicals with which pests are controlled. Early pesticidal problems were more concerned with protecting the grower from fraudulent sales of ineffective materials. The trend has been reversed now, so that the consumer, not just the grower, is also protected. The Environmental Protection Agency (EPA) was formed in 1970 to centralize the enforcement of legislation designed to protect the consumer and the environment from pesticidal contamination, as well as contamination from other sources.

The most recent major legislation enacted is the latest major revision of FIFRA, known as FEPCA (the Federal Environmental Pesticide Control Act). This law closed almost all of the loopholes known to exist at the time the law was passed. The new elements in FEPCA not previously covered in federal legislation are as follows:

1. Infractions against FEPCA regulations are punishable with fines and/or imprisonment. A method of enforcement is provided.

2. Pesticide use inconsistent with the label is prohibited.
3. Pesticides are categorized as general-use or restricted-use.
4. Pesticides in the restricted-use category may only be applied by certified applicators.
5. All pesticides, including those sold within the state of manufacture, must be registered.

The prior restrictions of FIFRA still hold true, in addition to these new regulations. It is interesting to note that a citizen in 1978 could not obtain penicillin without a prescription, but could purchase enough of the deadliest pesticides known to wipe out an entire city without any questions being asked or proof of ability to use these materials properly.

Federal laws require the following information to appear on the label of a pesticide:

Name and address of the chemical company, its EPA registration number, trade name of the material, and type of pesticide, list of all active ingredients, kind of formulation, hazard statements, directions for use, net contents, storage and disposal precautions.

Comparison of Pest Suppression Tactics

ADVANTAGES	DISADVANTAGES
Regulatory Controls	
Keep pests from becoming established in new areas. Contain introduced pests in restricted areas so that eradication programs might be feasible.	Thorough inspection impossible with quantities of materials now being shipped. Do not work well unless natural ecological barriers exist. Failures often very obvious.
Cultural Control	
Relatively simple and inexpensive. No contamination of environment. Modification of practices normally used in farming operations.	Application usually separate from the damage. Timing may be crucial—does not give complete economic control. Difficult to evaluate degree of control.
Physical Control (Heat, cold, electricity, radiation)	
No toxic chemicals.	Mostly possible only in confined areas. Expensive to carry out. Hard to ensure uniform treatment. Traps require maintenance.
Mechanical Control (Exclusion: tanglefoot, screens, etc.)	
No toxic chemicals.	Screens work only on indoor pests. Screens not practical for very small species.
Resistant Varieties	
Semipermanent. Inexpensive over the long run. Do not destroy other organisms or contaminate the environment. May affect entire population of the pest; applicable on low value crops where other techniques are too expensive.	Need public subsidy. Mainly limited to annuals. Initially expensive to develop; slow development time permits some damage depending on type of resistance. When breeding for one desirable characteristic other good qualities may be lost.

Biocontrol using Parasites and Predators

Frequently permanent; may need repeated reintroduction in some areas. Inexpensive. Does not disturb other organisms or contaminate the environment. May result in overall reduction in pest population.

Need public subsidy. Slow response to pest population fluctuations. Allow limited to heavy damage. Many exotics fail to become established after release. Those established are frequently ineffective because of low populations. Not completely compatible with chemical applications.

Biocontrol with Bacteria, Nematodes, and Viruses

Mostly highly specific. No environmental hazard. Hopefully some will give quick kill. No phytotoxicity, little evidence of resistance buildup.

Mass production outside the host's body has not been accomplished, except for *B. thuringiensis.* Registration for use is necessary. Not completely compatible with chemical applications. Strict timing is sometimes essential. Commercial product will likely continue to be very costly.

Chemical Pesticides

Fast and positive results can give almost complete protection. Clean products please consumers. Perfected by industrial research; no public subsidy. Often the only effective control available.

Residue problems. May kill nontarget organisms. Post-application flare-ups possible. Normally insignificant species may become important. Do not give area-wide population reduction. Resistance can develop and contamination of environment is possible.

Repellents

No environmental hazard.

Most only operate on contact, although some new ones may operate at a distance. Mostly on animals or people to repel blood feeders. Must be applied frequently.

Attractants

Usually highly specific. No hazard to environment. Highly efficient monitoring system

Expensive to identify and manufacture. Traps require frequent maintenance. Effect varies with abundance of the pest. Not applicable to many pests.

Endocrines

Little or no mammalian toxicity. High specificity required for successful use.

Expensive to develop. Need more work in this area; still in their infancy.

Chemosterilants

Do not have to rear and release to get control. Highly specific.

Presently available materials are too toxic or otherwise unsuitable. Resistance can develop.

Sterile Male Technique

Theoretically possible to eradicate a species. Noncontaminant of environment. Does not harm other animals; highly specific.

Many factors required or desirable to increase chance of success: Females mate only once, must be able to mass-rear, released insect must not be harmful, must be low natural population at some time, must be restricted to a discrete geographic area. Resistance (behavioral) may develop. Requires much equipment expertise and public subsidy.

Pest Suppression Technology in the Future

The trend toward IPM will be extended to all situations, for management of pests, regardless of the specific type, still involves the same factors—monitoring and utilization of a mix of tools—but only when necessary, based on analysis of pest population trends. Since IPM requires lots of labor, the whole field of insect suppression may be expected to become more labor intensive.

Tactics in pest suppression will continue the trend toward specificity of action. This trend is being fostered by the appreciation that the use of broad spectrum pesticides can have unpredictable and catastrophic outcomes. Pesticides will continue to be a significant tool, but their further refinement to increase specificity of action should produce the ability to "surgically" remove a given species without significant disruption to the rest of the ecosystem. Few new pesticides will be developed, unless this work is taken on by governmental agencies, because of the skyrocketing costs of registering such chemicals prior to their sale.

Applied genetics will continue to grow in importance. Strains of domestic plants and animals will continue to be bred for greater combined resistance to insect and disease attack. This trend will accelerate over the next 20 years with genetic engineering capable of producing a genetic fusion of many desirable characteristics into our plant and animal stocks. This trend will be especially valuable in developing countries. Similar work on natural enemies will make them hardier and more effective.

Genetics, in combination with other tactics, will also be used to attack some of our major pests by creating disruptions in the natural populations. Deleterious genetic shock will be introduced, only to take effect one or more generations later, after it has spread through the wild population.

From the insect's viewpoint, life will become more difficult as the level of sophistication in humanity's weaponry continues to increase. However, the house fly, gypsy moth, and potato leafhopper, along with the rest of the insect world, share a combination of traits that has never failed them in their struggle to survive in spite of changes in the environment. Their genetic plasticity, in combination with their mobility and short generation time, allows for evolution of genetic combinations to suit practically every situation imaginable. It is probable that they will continue to plague humankind in spite of the novel ways we develop to suppress them.

SUMMARY

Ecosystems are all of the living and nonliving components of a given area. The basis for ecosystem analysis is energy, which flows from the producer trophic level through one or more consumer trophic levels, with energy lost at each level. Food webs consider the producers of an area, the various herbivores that consume them, and the many predators, parasites, and diseases of the herbivores. Secondary consumers (predators, parasites) may be consumed by higher level consumers. The food of an insect may determine whether it is or can become a pest.

Most ecosystems modified by humans are "artificial" to some extent. Agricultural cropland typically has only one type of plant present, with little diversity in plant age or genetic composition. Such monocultures are mandated by the economics of management and equipment availability. However, monocultures encourage pest exploitation.

Only 2–5% of the known insect species are pests. Potential pests are often kept in check by suboptimal conditions of the physical environment and by density-dependent natural enemies. Thus, few of the insects that feed on crop plants cause economic losses because of their low populations.

Insects become pests when they reach economic thresholds. Such levels change depending on crop, stage of development, and potential for disease transmission. Because pest suppression techniques take time, they should be applied when pest populations reach the action threshold. Consumer demands for unblemished products are forcing growers to lower economic thresholds to produce cosmetically acceptable foods.

Intelligent management of pests depends on a thorough understanding of each pest's life system, that is, the elements that control their population densities, as well as the possible effects that suppression techniques against any one pest will have on the other elements of the ecosystem. In order to manage pests, crop ecosystems are monitored regularly by sampling organisms. Sampling data are analyzed according to crop stage and weather predictions to determine whether pest populations will reach economic thresholds. If they do, then appropriate suppression tactics are applied.

Prior to the 1970s insect control was unilateral, with reliance on a single method (often insecticide use), and considered only the pest involved. Development of insecticide resistance, posttreatment flare ups, and "creation" of new pests by insecticides forced entomologists to become applied ecologists, using a variety of suppression techniques to control pests when control is required, but with minimal disturbance to the ecosystem.

Biological methods include the use of the natural enemies of insect pests to suppress them. "Biological control" with predators and parasites has proved effective, especially with parasitic wasps, because of their host specificity. Since many of our serious pest insects have been introduced, most of the successful programs have been based on introduction of natural enemies from the country where the pest originated. Recent successes in mass rearing of insect parasites and predators have allowed for inundative control programs.

Some insect pathogens (e.g., bacteria) have proved effective, especially if they can be mass produced on artificial media and stored without losing potency. Fungi and nematodes offer some promise, but there are few successes so far. Viruses show great promise, because inclusion bodies protect virions outside the host cell. Also, viruses have extreme host specificity. Viruses have not been mass produced. Problems with registration for use must be overcome before the potential use of virus diseases can be fully exploited.

Resistant host varieties exhibit two types of resistance: ecological (phenological asynchrony, induced) and genetic (nonpreference, antibiosis, tolerance). The classic evaluation of resistance in host germplasm and incorporation into acceptable cultivars takes many years. Genetic engineering will significantly reduce the time required to develop resistant hosts.

Insecticides are the most widely used chemical method of suppressing populations because of their rapid action. Organochlorines, like DDT, are seldom used now because of persistence and biomagnification. Organophosphates and carbamates show little persistence. The newest class, synthetic pyrethroids, are extremely toxic, but expensive. Insecticide formulations with solid, gaseous, or liquid diluents offer a variety of forms and application techniques that modify hazard and specificity of action. Insecticide use may lead to development of resistance in pest populations.

Insect growth regulators affect maturation (e.g., Altosid) or molting (e.g., Dimilin). Attractants (e.g., sex pheromones) are used in monitoring, but show some promise in control by trapping out one sex or by inhibiting mate location in an attractant-saturated environment. Recent research on repellents indicates that the total coverage necessary with commercially available compounds may not be required with newer compounds that have effect at greater distances.

The chemosterilants so far identified have proved too dangerous to be used. If new ones are developed, they could be used to sterilize natural populations, thereby eliminating the need to rear and sterilize insects with radiation prior to their release in sterile insect release programs like the screwworm. Such autocidal techniques are most effective with low populations and could be used after insecticidal treatment (most effective with high populations). Conditional lethals and translocations offer some promise for other genetic control programs.

Physical control employs barriers, temperature extremes, and light traps. Cultural methods have been effective in reducing pests through crop rotation, planting-harvesting dates, and sanitation. New reduced-tillage agriculture is creating different or increased pest problems.

Governmental regulations for quarantines help keep foreign pests from invading, and slow the spread of those already in the country. Some introduced pests have been eradicated. FIFRA and FEPCA are enforced by the Environmental Protection Agency to improve control of pesticide use and public safety.

As a strategy, IPM will increase in importance. New genetic tactics will provide a greater variety and faster development of resistant plants and animals. However, the genetic plasticity of insects will enable them to adapt to the environmental modifications we develop to suppress them.

SUGGESTED READINGS

Huffaker, C. B. (ed.) 1980. "New Technology of Pest Control." J. Wiley & Sons, New York.

Huffaker, C. B., and Messenger, P. S. (eds.) 1976. "Theory and Practice of Biological Control." Academic Press, New York.

Maxwell, F. G., and Jennings, P. R. 1980. "Breeding Plants Resistant to Insects." John Wiley & Sons, New York.

McEwen, F. L., and Stephenson, G. R. 1979. "The Use and Significance of Pesticides in the Environment." John Wiley & Sons, New York.

Metcalf, R. W., and Luckman, W. H. (eds.) 1975. "Introduction to Insect Pest Management." John Wiley & Sons, New York.

Odum, E. P. 1971. "Fundamentals of Ecology." W. B. Saunders, Philadelphia, Pennsylvania.

Pal, R., and Whitten, M. J. (eds.) 1974. "The Use of Genetics in Insect Control." Elsevier, New York.

Ware, G. W. 1975. "Pesticides, an Autotutorial Approach." Freeman & Co., San Francisco, California.

11

The Diversity of Insects

One of the most interesting aspects of entomology is the vast variety of species that live in our world. However, so many species exist that reference to any particular one requires some scheme that will distinguish it from all others. The scheme for grouping insects into categories is based on similarities between species within each group.

CLASSIFICATION

In order to place individual units, like species, into groups, the first step is to decide the group's characteristics. Classification is the process of setting up such groups. These groups can be based on many kinds of similarities, e.g., color, shape, size, type of food eaten, habitat, how it respires, etc. In biology these classifications provide the highest degree of predictability when the groupings are based on phylogenetic relationships. "Phylogeny," which is the evolution of a genetically related group of organisms (i.e., species) as distinguished from the development of the individual, provides a highly predictable basis for groupings because these similarities are inherited.

Phylogeny as a Basis for Classification

Two difficulties may occur using the phylogenetic basis for most insect classifications. These are (1) similarities between different species may be due to phenomena other than descent from common ancestry, and (2) such descent from a common ancestor cannot be tested directly.

In addition to resemblance as a result of common ancestry, resemblance may also arise by convergent evolution, by mimicry, and even by chance. Each of these phenomena compounds the problems of insect classification by forcing the classifier to discriminate between the different types of similarity. In convergent evolution, two unrelated species may end up with similarities as a result of adaptations to similar ecological problems, e.g. streamlining of aquatic bugs and beetles, which aids in their life styles in water. Mimicry, an adaptive resemblance not involving common descent, was discussed earlier. Similarities due to chance probably do not exist, but only reflect our inability to determine why some species which are apparently unrelated, based on other characteristics, share some common characteristics. Therefore, phylogenetic classifications must separate out the types of resemblance not due to common descent so that the phylogenetic pattern is clarified.

Since animals usually mate only with members of the same species there are no positive data to yield objective direct testing of relationships except at the species level. Therefore, we must infer phylogenetic relations by indirect evidences of common ancestry. Such inferences are made on criteria from a wide variety of characteristics including structural (external, internal, chromosomal), physiological (proteins, isozymes, secretions), ecological (habitat, food, seasonal patterns, parasites), behavioral (communication, courtship), and geographic (distribution patterns) characteristics. The most widely used of this whole array is structure because it is the only characteristic available for use with the specimens that are stored in our museum collections. Also, most classifications based on structure have also been verified by other criteria. Because nonstructural criteria often require special collecting techniques in order to have suitable materials for comparative studies, these nonstructural criteria are not usually used unless the classification derived from structural characteristics is unsatisfactory. When unsatisfactory, a classification allows for little predictability concerning the ecology or habits of a species, which reflects a weakness within the currently used classification.

The basic unit in a classification is the species. The species is grouped with other very closely related species (one or more) into a genus. Closely related genera are grouped into a family, families into an order, orders into a class, and classes into a phylum. Thus, a classification represents a relationship between a number (sometimes a very large number) of species. Since the groups are ordered into a series of progressively more inclusive groups, they are called a hierarchical classification. The hierarchy is usually expressed in the following way, using the house fly as an example:

Phylum	Arthropoda
Class	Insecta
Order	Diptera
Family	Muscidae
Genus	*Musca*
Species	*domestica* Linn.

Such a classification provides a nested set of names which not only expresses some degree of relationship but also provides the key names under which literature concerning a particular insect species may be found.

Naming of Species

The scientific name of a species consists of its generic name and its species name. This two-part system of names is called the binominal nomenclature, and is somewhat analogous to the system of names for humans in which relationships are reflected by surnames. When a specimen is identified as *Musca domestica* Linnaeus, this indicates that the specimen has essentially been compared with a previously identified specimen or with a carefully worded (and usually illustrated) description of *Musca domestica* Linnaeus, which was named and described by a scientist at some earlier time. The name of the person who originally described the species is a part of the species' scientific name. Thus, *Musca domestica* Linnaeus was originally described by the Swedish scientist of the eighteenth century, Carolus Linnaeus. When a new species is being described, a single specimen is chosen and designated as the *type* of the new species. Such types are of great scientific importance, since they carry the name of the new species and are the ultimate source for comparison by future workers. They are usually deposited in large repositories such as the British Museum (Natural History), or the Smithsonian Institution (National Museum of Natural History) for maximum security and availability.

Sometimes the author's name following the species will appear in parentheses, e.g., *Aedes aegypti* (Linnaeus). This indicates that the species *aegypti* was described by Linnaeus, but he originally placed it in another genus. Some later worker decided that the species *aegypti* Linnaeus really should be included with other mosquitoes into the genus *Aedes* because of their mutual similarities. Since the opinions of scientists do not always agree on interpretations of phylogenetic relationship and also since new information regarding the species is continually being added, classifications are under continual revision.

Many insects have common names, some of which are used in this book. The trouble with using common names for a species is that common names vary from one place to another, especially between countries. For example, the "ladybug" in America is called Mareinkafer in Germany, Lorita in Honduras, Nyckelpiga in Sweden, and Leppakerttu in Finland. Therefore, the scientific community uses binominal nomenclature to refer to a given species because the system provides worldwide uniformity.

The species, whether we refer to it by common or scientific name, is both a genetic unit, forming a reproductive community, and an ecological entity. It is also an evolutionary unit, in which the populations comprising the species are tied together by a common gene pool (all of the genes in a population). The classic definition of a species is that it is a group of interbreeding populations that are reproductively isolated from other such groups. Since populations comprising a species share a common gene pool, they demonstrate some exchange of individuals between populations, thus gene flow between populations exists. However, when the gene pool is no longer a common one between isolated populations, gene flow is interrupted. Then "speciation," the establishment of intrinsic barriers to gene flow by the development of reproductive isolating mechanisms, begins to occur. The isolated populations begin to accumulate differences from each other as each continues its evolutionary track of the ecosystem in which it lives. Reproductive isolation may involve spatial isolation of the populations, including the occupation of different habitats, seasonal or temporal isolation, behavioral differences especially with regard to courtship and mating, and mechanical isolation in which the genitalia are structurally incompatible. Even if mating occurs between individuals of two different species the usual result is death of the gametes before fertilization, or the hybrid offspring die prior to birth or are sterile.

TAXONOMY

Taxonomists, who are scientists specializing in this area, construct classifications to reflect the relationships among and between groups of insects. Thus each classification is the result of a scientist's judgment as to the number of groups and the hierarchical level to which each group is assigned. Biologists differ as to which classification should be followed and what the specific basis should be. Research in this area is called "taxonomy," the theoretical study of classifications and the basis or bases for classification. At present there are three major theories as to the appropriate basis for classification. The "Phenetic" school bases classification on overall similarity without regard to phylogenetic history, whereas both the "Phylogenetic" and "Evolutionary" schools are based on phylogeny. They differ in that the "Phylogenetic" followers base classification on the recency of descent from a common ancestor and the "Evolutionary" group bases their classifications not only on recency of descent, but also on the degree of divergence between groups. Over the past two decades the trend has been toward quantitative comparisons of characteristics.

Although considerable research is underway on the diversity of insects, most of the work is in the area of theoretical taxonomy as just defined; in the process of speciation with its genetic, behavioral, and ecological aspects; or in descriptions and reevaluations of species groups, genera, or (rarely) families. Even under this intensive taxonomic investigation, the major groupings of insects have proved to be relatively stable, with only occasional changes as a result of new insights into relationships.

MAJOR GROUPS OF INSECTS

Major groups of insects are called Orders. The orders may themselves be grouped into Subclass Apterygota (the primitively wingless orders) and Subclass Pterygota (have wings now, or did have but lost their wings).

We will now discuss in greater detail the most numerous and economically important orders, covering their life history, phylogeny, and economic impact, with some detail on particular species of economic importance.

Order Orthoptera (Plate 11.1: Grasshoppers, Crickets, Locusts, and Katydids)

This order of relatively large-sized insects has two primary features that characterize the groups within the order and all active stages. These features are (1) most forms are jumping insects and (2) most feed on vegetation. Exceptions are some leaf-rolling forms, which consume aphids, and mole crickets, which not only do not jump (their hind legs are not enlarged), but may consume insect grubs along with their diet of plant roots, or be strictly carnivorous. But, as a group, the important Orthoptera are plant feeders and are jumpers. The jumping is facilitated by the expanded femoral segment of the hind leg, but walking is its usual method of covering short distances. Jumping is resorted to only when the insect is startled.

The Orthoptera are admirably equipped to feed on plants because of their powerful chewing mouthparts. These mouthparts are so unspecialized in nature that they are often studied in entomology laboratories as the basic type of chewing mouthparts.

The two pairs of wings in Orthoptera differ from each other both in form and texture. The forewings, called tegmina (sing. tegmen) are relatively narrow and parchmentlike, with visible venation. The hindwings are folded lengthwise into a number of pleats which allow the very wide hind wing to be covered by the forewing when not in use. The wings are flexed back along the body at rest. The hindwing is membranous, but sometimes brilliantly colored. Some species are flightless, have short wings, or are entirely wingless.

Many of the Orthoptera demonstrate behavioral patterns tied to the production and reception of sound. The special receptors for sound (tympana) are located on the sides of the first abdominal segment or on the front legs. The sound-producing mechanism operates on the same principle as running a stick along a picket fence. The wing vibrates as a series of pegs or ridges on the hind leg or other wing are rubbed along it, producing the sound. The songs or noises produced by each species are relatively distinct. Songs are usually produced by the males of the species and serve to attract the female. Some crickets can produce more than one kind of song, each serving a distinct purpose, e.g., courtship, aggression, or alarm.

Plate 11. 1

ORDER: ORTHOPTERA

(derivation: orthos=straight,
ptera=wings)
Common Names: Grasshoppers, Crickets, Locusts, Katydids

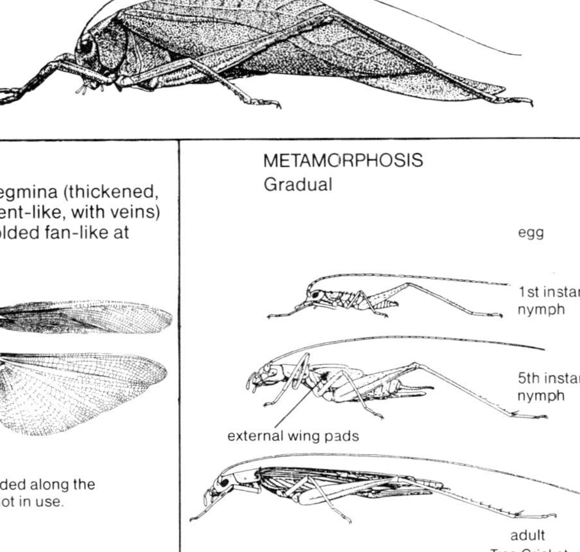

MOUTH PARTS
Chewing

WINGS
1st pair=tegmina (thickened, parchment-like, with veins)
2nd pair folded fan-like at rest

wings are folded along the body when not in use.

METAMORPHOSIS
Gradual

OTHER CHARACTERISTICS

1. Acoustic

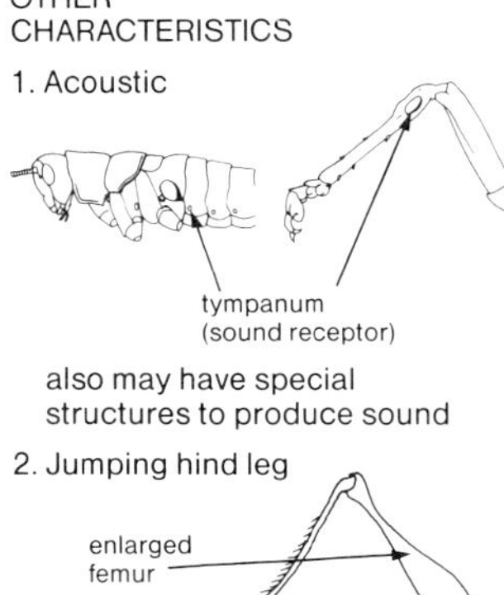

also may have special structures to produce sound

2. Jumping hind leg
3. Plant feeder (usually)
4. May exhibit gregarious swarming behavior and morphology

BIOLOGY

The female usually uses her strongly developed ovipositor to place her eggs into the substrate (plant parts, soil). The young emerge and develop to adults which find mates through acoustic behavior.

ECONOMIC IMPACT

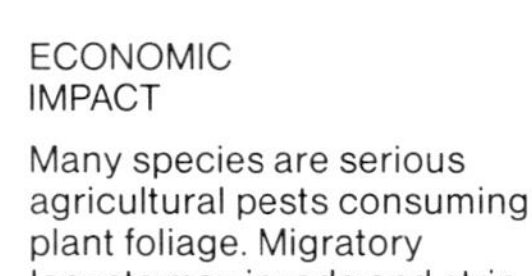

Many species are serious agricultural pests consuming plant foliage. Migratory locusts may invade and strip large areas of vegetation.

OTHER COMMON FORMS

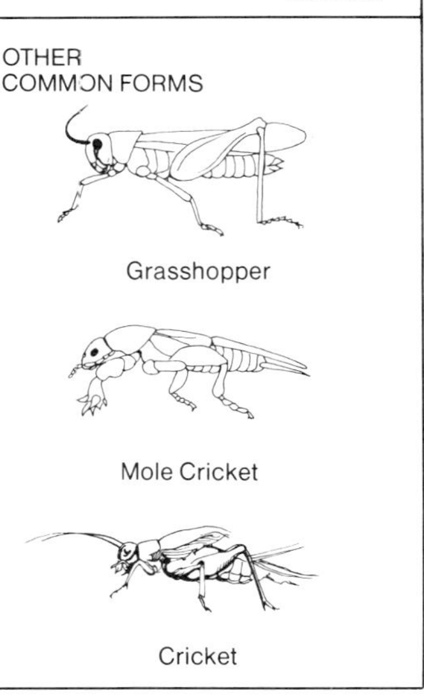

Many species are cryptic, both in coloration and texture, that is, they match foliage and other plant parts or the substrate. A common combination of cryptic coloration with the flash effect of brilliantly colored hind wings enables banded-winged grasshoppers and other Orthoptera to elude predators.

Morphologically, the Orthoptera may be characterized not only by the jumping hind legs and type of wings, but by a dorsal collarlike first thoracic plate, which extends down on each side and around to the bottom of that segment. The collar is saddle-shaped in many grasshoppers. Females deposit packets (oothecae) of eggs in the soil or individually in plant tissue. Many of the long-horned (long-antennae) grasshoppers use the long sword or spearlike ovipositor to pierce plant tissue. The short-horned grasshoppers dig holes in the soil with stubby ovipositors.

Orthoptera exhibit a classic type of gradual metamorphosis, that is, the immature form looks like the adult, except that it lacks functional wings. Young and adult basically feed on the same foods and live in the same locations.

Phylogeny and Synopsis

The early ancestors were similar to the Order Plecoptera, but had long running legs. The legs were attached more to the side of the body than under it, thus they were already evolving the jumping habit. The wings were also more to the side than lying flat atop the body. The general relationships between the various groups of Orthoptera are presented in Fig. 11.1. The groups are divided into two major suborders, the

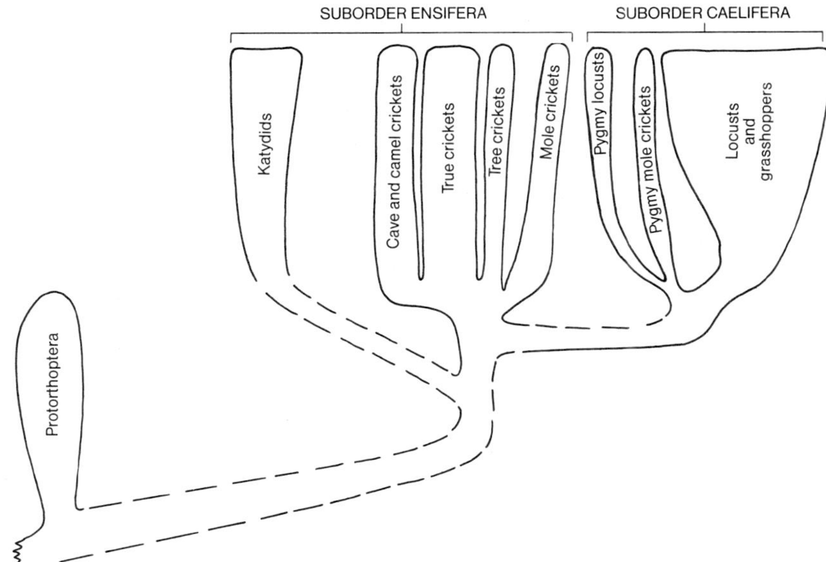

Fig. 11.1 Phylogenetic relationships among the living Orthoptera. The top line represents the groups living today. Points lower in the figure represent progressively earlier time periods. The Protorthoptera were the ancestral forms that led to the Orthoptera.

Suborder Ensifera being the older of the two. Ensifera have a sword-shaped ovipositor and antennae of 30 or more segments, usually longer than the body. Tympana are on the front legs, and stridulation is produced by the file and scraper on the wings. The earlier forms were evidently predaceous, with long prehensile spines and long pointed jaws. Their diet has evolved from a carnivorous one at first to omnivorous, then to phytophagous. Groups within this suborder are the katydids, the wingless cave and camel crickets, the mole crickets, the true crickets, and their near relatives, the tree crickets. This group has invaded both aboveground phytophagous habitats, including trees (katydids and crickets) and caves (cave and camel crickets), and subterranean niches (mole crickets)

The Suborder Caelifera has short antennae, and the tympana, if present, are on the first abdominal segment. The ovipositor is stubby. If they stridulate, they do so by rubbing the hind legs over the wings or by snapping the wings while in flight. Except for a few very small and minor groups, the Caelifera comprise essentially the grasshoppers and locusts. They are phytophagous and lay their eggs in soil by means of a digging, rather than stabbing, ovipositor. This suborder is much newer, geologically, than the Ensifera, having sprung from the ancestors of the modern cricket.

Economic Impact

As a group, the Orthoptera compete for plant materials with mankind. Among the crickets, many species cause serious agricultural losses by attacking such crops as peas, beans, tomatoes, cotton, and grains, both in the seedling stage and later when the part used by man is maturing or harvestable. In the United States, the Mormon cricket, a short-winged, long-horned grasshopper, marches into various agricultural crops from its rangeland breeding areas. Mormon crickets will feed on at least 250 different plants.

Grasshoppers are serious pests of many crops and have tended to be more serious as land is first brought under cultivation, later becoming less important as habitat is modified. This pattern occurred in the early agricultural history of western and midwestern North America in the last century. It has also been a consistent pattern in the virgin lands program of the USSR since World War II. The problem is no different when grasslands are used for grazing of domestic animals. Grazing, especially on communally held lands, frequently leads to overgrazing, with concurrent increase in exposure of the soil to the sun. Warmer soil temperatures favor development of many grasshopper and locust species, which in turn, contribute to more destruction of remaining vegetation.

It is not difficult to see why Orthoptera have been so successful in attacking many of our crops. Considering that a number of these crops are domestic grasses (e.g., wheat, oats, rice, and corn), is it any wonder that so many of the polyphagous Orthoptera, which evolved close associations with native grassland plants, do well when man cultivates domestic grass species in their habitat?

The notorious locusts that have devastated large geographic areas and caused famines because of destruction to agricultural crops belong to Orthoptera. Vast numbers of these insects still occur, but international cooperation in some areas, combined with telecommunications and area-wide governmental control programs, have isolated an increasingly urban western world from the realities of plagues of locusts that still occur elsewhere in the world.

Phase Polymorphism. Locusts are species of Orthoptera that can exist in more than one form, depending on the environment in which they live (see Key Economic Group: Locusts). Those in isolated conditions where contact with other individuals of

KEY ECONOMIC GROUP: LOCUSTS

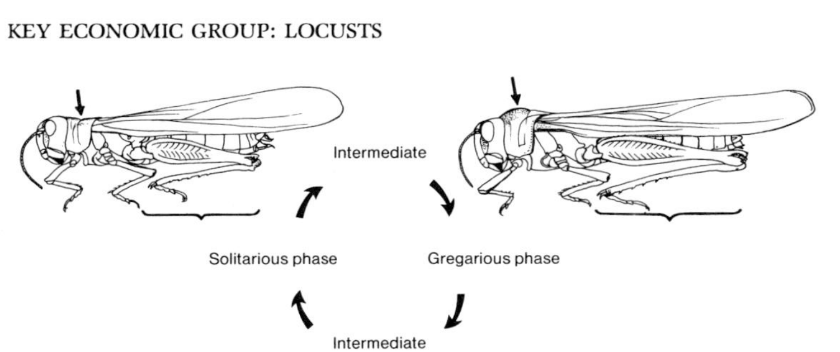

Distinguishing Characteristics: Much like the nonmigratory species of grasshoppers in overall appearance. Expresses phase polymorphism: an extremely solitary, relatively sedentary form and a gregarious one that is restless, active, and has different body proportions. Note differences in head size, hind leg length, and shape of the collar behind the head.

Occurrence: Found on most continents of the world, each area having one or more species peculiar to that geographic region. Erroneously considered to be limited to the dry desert areas of Australia, North America, and Africa, but many other species live in cooler dry areas and are restricted in hot regions to areas of high moisture.

Economic Impact: Known to feed on many types of vegetation. Although found more in connection with grasslands and grassy areas, the migratory phase is notorious because almost all vegetation in the paths of the swarms may be stripped. Single swarms can cover up to 1000 square kilometers and contain as many as 10 billion individuals. Swarms on the move cover as much as 3000 kilometers in a single generation. Locusts have affected human history by causing famine and emigration from large areas of the world.

Control: The huge areas covered require population suppression efforts to be international in scope to be effective. Unsettled political conditions and irregularity of locust swarm occurrence lowers locust control priorities. Effective control requires continual monitoring of breeding areas to keep populations below the levels required for gregarious phase to appear through early use of chemical or other control techniques.

the same species is rare, mature into "solitarious" adults. These forms tend not to aggregate and are not very restless. They produce many small eggs and some adults have increased longevity in this solitarious form. The same species matures into a rather different adult if it is reared in an environment densely populated with its own kind, and where there is continual contact with others. The resulting "gregarious" adults form and maintain aggregations, even when flying. They produce fewer, but larger eggs than their solitarious counterparts. The color of the adult as well as its body proportions are so different in these two forms that earlier scientists placed the solitarious and gregarious forms of a single locust species into two distinct species. Experiments have shown that, in addition to a range of forms and colors, behavioral and physiological differences exist within a single species. The solitarious phase tends to have larger hind legs and a smaller head than the gregarious phase, as well as absolute differences in body size. In nature, all transitional stages between the two extremes can be found.

Order Hemiptera (Plate 11.2: Aphids, Scales, Leafhoppers, Spittlebugs, and "True" Bugs)

A characteristic that is present throughout the order and the key to its success is the piercing-sucking type of mouthparts they possess. Evolution of the ability to pierce through the protective covering of an organism and tap into its living fluids have allowed many hemipterans to develop a continuous contact with their food supply, without necessarily damaging the host organism as much as a chewing insect does. For instance, a scale insect can settle onto the stem of a perennial plant, tap into the food supply coursing through the stem, and remain in that location for the rest of its life with a secure source of energy.

Most groups of Hemiptera are winged, either with a uniformly textured forewing (Suborder Homoptera) or with the forewing thickened at the base and membranous at the tip (Suborder Heteroptera). The larger forms are strong fliers, but the majority of smaller forms are very much at the mercy of even weak air currents. Even so, true bugs, aphids and leafhoppers commonly make seasonal shifts in geographic location on a continental scale.

With few exceptions, Hemiptera exhibit a gradual metamorphosis, with external development of wings. Some types, for example, female scales and wingless aphids, develop without any real changes in form. Others, e.g. whiteflies and male scales, have a quiescent nonfeeding stage comparable to the pupa. In these forms the appearance of large wing pads at a molt indicates considerable internal development prior to their appearance.

Although the Homoptera are mostly terrestrial and phytophagous, two groups of Heteroptera have become established in the aquatic habitat, where most are carnivorous. Among the terrestrial Heteroptera have evolved some animal parasites, insect predators, and many groups of plant feeders. These terrestrial Heteroptera commonly possess repugnatory glands that produce an unpleasant odor. The groups with repugnatorial glands and those that can "bite" readily are often aposematically colored, frequently with a black and red pattern.

Plate 11. 2

ORDER: HEMIPTERA

(derivation: hemi=half, ptera=wings)
Common Names: The true bugs, Aphids (plant lice), Cicadas, Leafhoppers, Scales, Mealybugs, Whiteflies

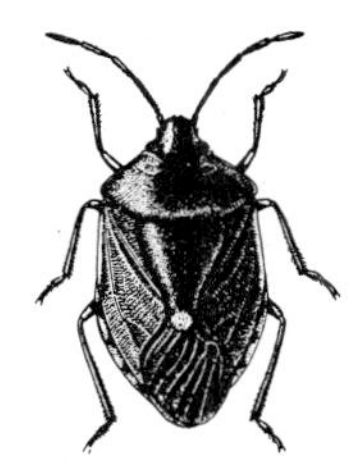

Stinkbug

MOUTH PARTS
Piercing-sucking

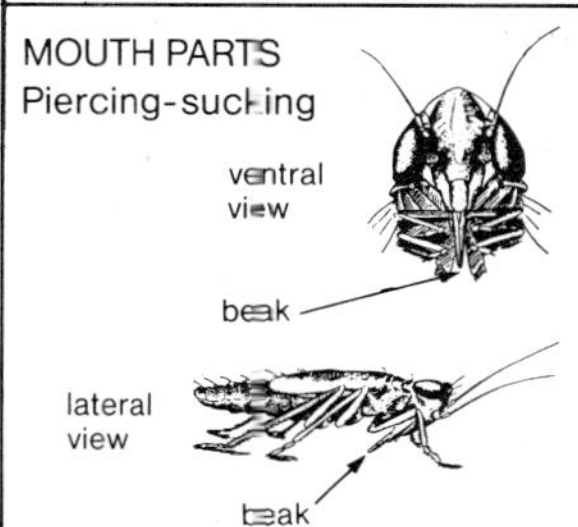

WINGS

When present, usually 2 pairs (1 in ♂ scales); forewing thick at base and membranous at tip in Suborder Heteroptera (e.g. stinkbug pictured above) or uniform in texture as in Suborder Homoptera (e.g. this leafhopper) Hind wing entirely membranous and smaller than forewing.

METAMORPHOSIS
Gradual

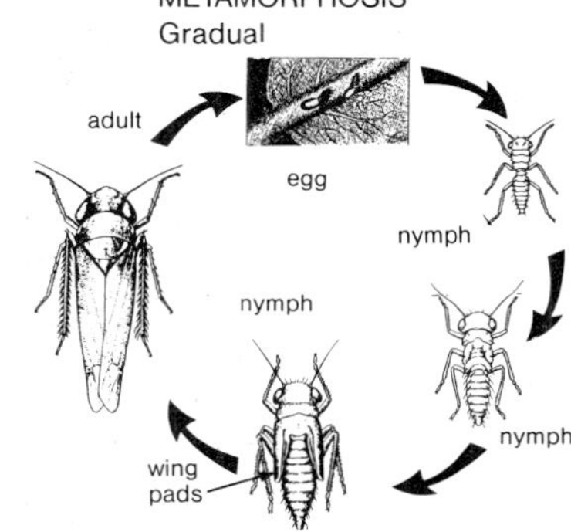

OTHER CHARACTERISTICS

1. All Suborder Homoptera are plant feeders
2. One large group of families in Suborder Heteroptera are associated with water.
3. Some, like aphids, have very complex life histories (e.g. metagenesis, and see key economic group pages).
4. Many have sound-producing organs.

BIOLOGY

Diversity of unusual life styles is often linked to the unique method of feeding by the Order. Problems of feeding on very dilute foods are frequently offset by the permanent association possible with long lived (e.g. perennial) hosts.

OTHER COMMON FORMS

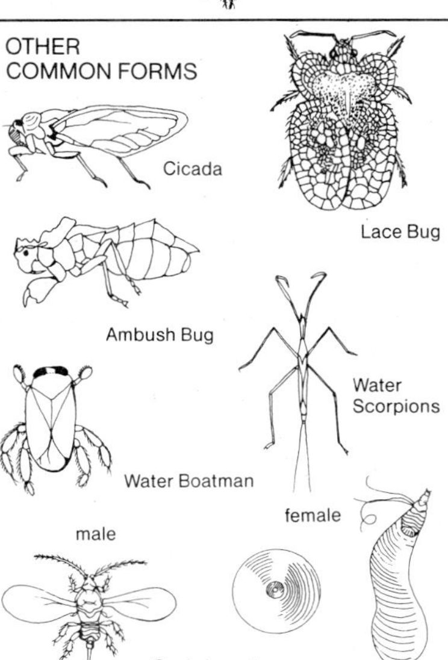

ECONOMIC IMPACT

Many Homoptera are serious plant pests. Severity is due to type of feeding which results in discoloration similar to that produced by problems in soil fertility, plant diseases, and other conditions; with piercing-sucking mouthparts the Homoptera become "living hypodermics" and are the most common transmitters of viruses and other diseases of plants. Many homoptera are easily overlooked because of their commonly small size, often secretive appearances and habits, so that much damage may be caused before detection.

Some Heteroptera are important plant pests, but seldom transmit plant diseases. The bed bug and several other forms are blood suckers on vertebrates, a number in these groups have been incriminated as transmitters of human diseases.

Members of both suborders may produce sounds. Many Heteroptera stridulate (rub a "scraper" over a series of "teeth"), whereas the male cicadas have tymbals (sound-producing organs that work like the lid of a large tin can when it is popped in and out).

The secretion of wax, a characteristic of the insect body wall, is highly evolved in several Homoptera. Some forms, for example, the hard scales, secrete a waxy covering over their entire bodies. Others, like mealybugs and some aphids, produce long waxy filaments that are useful in defense.

Phylogeny

The order is divided into two suborders that are generally quite different from each other, both in structure and in life style. The forewing texture in the Suborder Homoptera is similar throughout (thus the name derivation *homo* = uniform, *ptera* = wings). It is entirely a terrestrial plant–sap feeding group. The leafhoppers and their relatives (treehoppers, cicadas, and spittlebugs) are abundant, active, and highly mobile forms. The aphids, scales, mealybugs, and whiteflies all have one or more stages that are sedentary, so that unusual dispersal mechanisms have evolved. The newly hatched nymph of the scale disperses by both active crawling and by passive free-fall aerial means. Mealybugs have no winged stages so that dispersal is accomplished by means of crawling and by mechanical movement of plant parts, whether by man or wind, etc.

Suborder Heteroptera has forewings thickened at the base and membranous at the tip. They probably evolved away from the homopterous group as a result of the insect-blood feeding habit which developed in the Heteroptera. Three of the four major groups of Heteroptera are primarily predaceous (see Fig. 11.2). Two separate heteropterous groups evolved close ties with the aquatic environment: The water-strider group is a surface-dwelling type that exists on prey caught in the surface film or along the edges of water. The backswimmer group are subsurface insects primarily, but with some mechanism that allows most of them to use atmospheric oxygen supplies. Members of this group make up a large proportion of the free-swimming insect life in ponds.

The bed bug branch of the Heteroptera contains a number of families that are insect predators (assassin bugs, ambush bugs, damsel bugs, pirate bugs), a few vertebrate blood feeders (bed bugs, some assassin bugs), and an important plant-feeding family (the plant bugs).

The stink bug branch has mostly returned to an ancestral plant-feeding existence. However, the intimate ties between insect and plant host as exhibited in the Homoptera seem to be lacking in the more mobile stink bug, seed bug, squash bug types.

Economic Impact

The economic scale is tipped heavily against the members of this order. Although a few predatory Hemiptera are useful natural enemies of pest insects and scale insects have been the source of dye chemicals and shellac, at least one major pest species

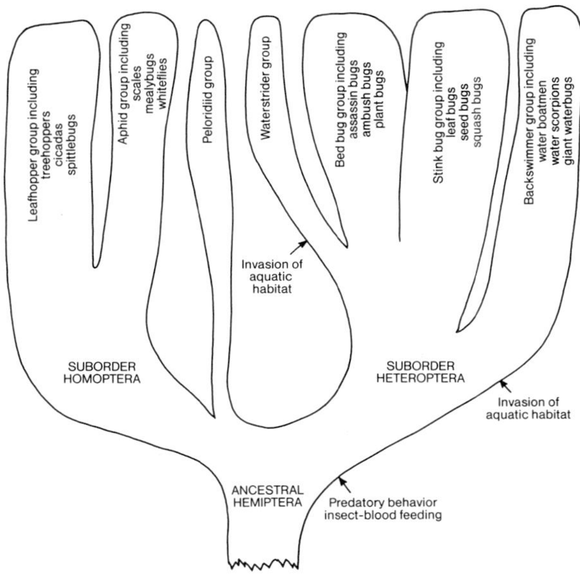

Fig. 11.2 Major relationships among the Hemiptera. Note that the Peloridiid group is a form intermediate between Homoptera and Heteroptera. Peloridiids occur in the southern hemisphere, living on mosses.

affects practically every crop plant cultivated by modern man. The annuals are attacked by aphids and leafhoppers; the perennials by scales and mealybugs, among others.

The sedentary habit, type of feeding, and small size of most Homoptera frequently mean that severe plant damage may occur before an infestation is detected and control measures are applied. Discoloration, stunting, and abnormal growth caused by the feeding of Hemiptera is easily confused with other plant problems. The injection of salivary chemicals by some aphids stimulates gall formation or can be toxic from the potato leafhopper, greenbug (an aphid), or tarnished plant bug.

Aphids, mealybugs, whiteflies, and scales all produce honeydew, a liquefied feces high in carbohydrates. Honeydew is sticky and fouls consumable plant materials, in addition to acting as a substrate for sooty mold, a black fungus.

Among the plant-feeding insects, the Homoptera is the most frequent group associated with transmission of plant disease. Aphids and leafhoppers are especially notorious as vectors of plant viruses, and other intracellular disease organisms. The Heteroptera are not common vectors of plant diseases, possibly because of the toxic nature of the salivary secretions in many of these species. The tarnished plant bug can destroy an entire flower bud or growing tip with a single feeding.

Some of the Homoptera cause considerable damage to plants with their ovipositors. Cicadas split the bark of woody twigs and insert their eggs. When emergence of adults is massive, widespread damage to orchards results.

Among the vertebrate blood-feeding Hemiptera, several assassin bugs feed on humans, but most are intermittent parasites of rodents. Some of the species are vectors of Chagas' disease, which is sometimes fatal to humans and other mammals.

KEY ECONOMIC GROUP: APHIDS

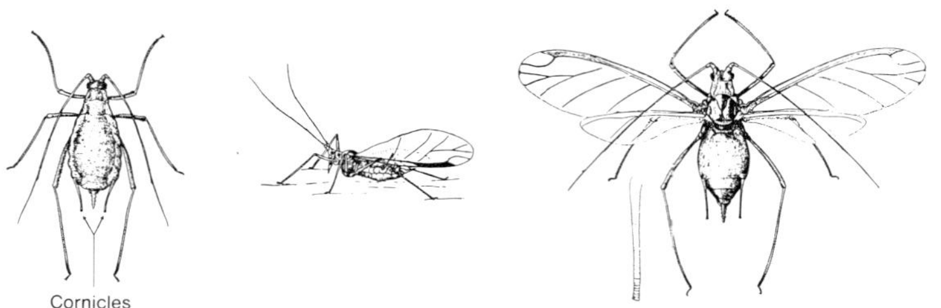

Distinguishing Characteristics: Differ from other Hemiptera by the presence of two distinct cornicles, tubelike structures on the abdomen that secrete a defensive fluid when disturbed. Feed on plant sap with their piercing-sucking mouthparts. Some species have very complex life cycles. Reproduction may be so telescoped that developing embryos in a parthenogenetic female may already be forming their own embryos.

Occurrence: Over 1300 species in North America. Habitats range from desert plants to temperate forests.

Host Relations: Most species live on only one kind of plant. Some species alternate between a variety of annuals during the summer and a perennial the rest of the year.

Economic Impact: Direct removal of plant sap causes cell injury or death. Injury is complicated by changes in plant growth such as stunting, cessation of root growth, discoloration, abnormal thickening or twisting. Aphids secrete large amounts of honeydew, which makes commercial plants unpleasant to handle and supports growth of sooty mold, a black fungus. This makes plants unsightly and commercially unacceptable. Aphids transmit more plant viruses than any other group through mechanically contaminated mouthparts, which allow viruses to be injected as they feed.

Control: Because of high reproductive capacity, especially in low temperatures, relatively sedentary life of wingless forms, and piercing-sucking feeding habit control is difficult. Chemicals must reach sedentary forms while quiet or through systemic insecticidal activity.

The bed bug family consists of a number of species that feed on man. Because bed bugs feed repeatedly on humans throughout their lives, they are highly suspect as disease vectors. However, the bed bug has not been demonstrated to be a vector of human disease in nature, even though it can experimentally carry a number of insect-borne disease organisms.

Order Coleoptera (Plate 11.3: Beetles and Weevils)

These insects typically are characterized by complete metamorphosis, chewing mouthparts, and two pairs of wings, with the front pair modified as veinless, heavy

Plate 11. 3

ORDER: COLEOPTERA

(derivation: coleo=sheath,
ptera=wings)
Common Names: Beetles, Weevils

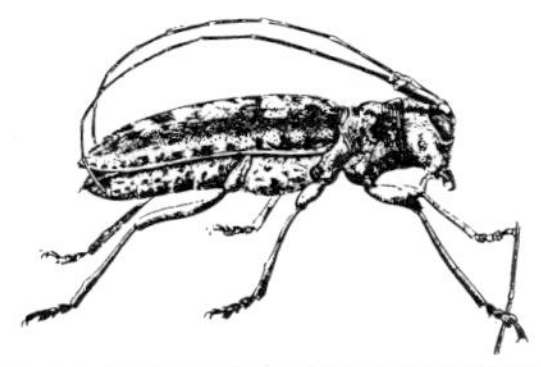

"Long-horned"
Wood-boring Beetle

MOUTH PARTS
Chewing

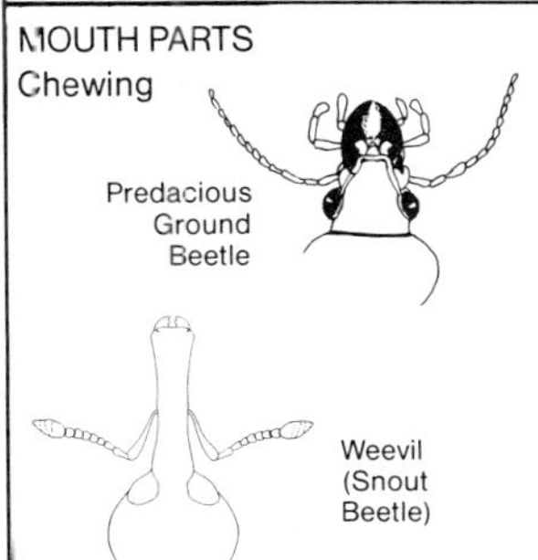

WINGS

2 pair; forewings usually thick, veinless, horny, usually covering most of abdomen, sometimes truncated. Hind wings, if present, membranous and may fold under forewings when at rest.

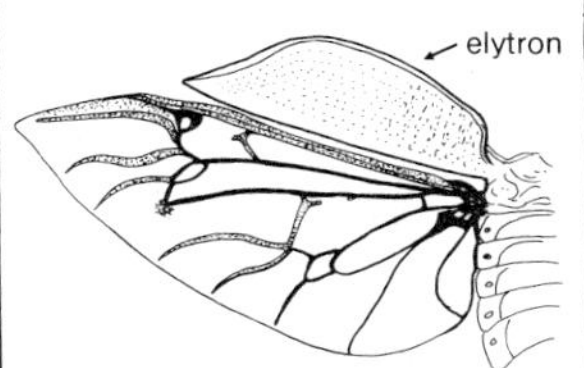

METAMORPHOSIS
Complete

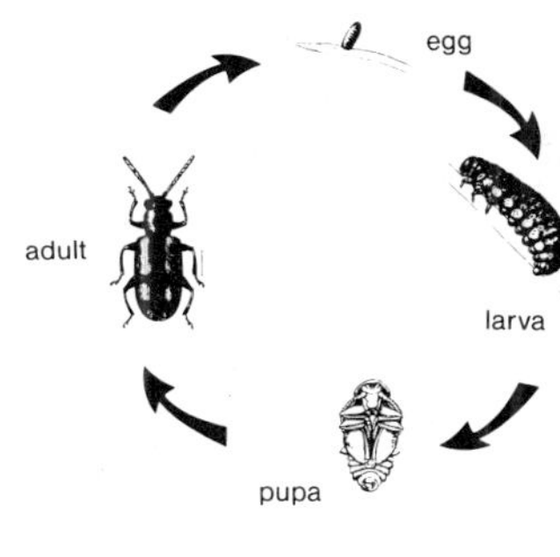

OTHER
CHARACTERISTICS

Structural integrity; heavily constructed exoskeleton with massive reinforcement of forewings permit exploitation of environments where physical protection of wings is at a premium, i.e. subterranean environment, boring in wood and other hard or abrasive substrates.

Subterranean tunneling by scarab beetles to create a site for deposition of eggs.

Beetle galleries exposed by removal of bark and wood.

ECONOMIC
IMPACT

Many serious pests of crop plants and of trees. Foliage feeding by leaf beetles and some scarabs. Subterranean damage by white grubs, wireworms and rootworms. Seeds and stored products attacked by weevils and flour beetles. Growing woody materials are attacked by bark beetles, weevils, and borers. Structural timbers attacked by powderpost beetles. Animal products, like fur and woolens attacked by carpet beetles.

Several groups e.g., lady birds, ground beetles, and checkered beetles are predatory and significant as biological control agents of pest groups.

wing covers (known as elytra) and the hind pair as folding membranous flight wings. Yet these characteristics reveal no obvious reason for the remarkable success of the Coleoptera within the animal world. Based on the diversity of species, not only is the Coleoptera the most abundant group of animals, but about one out of every four living species of animals is a species of beetle.

One of the secrets to the success of Coleoptera is the "structural integrity" of the adult beetle. Typically, it is heavily constructed and, with the hard wing covers, presents a nearly impervious suit of armor to the environment. The wing covers often have an interlocking mechanism where both the wing margins come together down the center of the back and the wing margin touches the side of the abdomen.

Protection of the folded membranous flight wings by the wing covers has allowed beetles to exploit subsurface habitats more than any other group of adult insect. Not only are the flight wings folded out of the way when not in use, but they are protected while the adult is boring into wood, soil, or some other substrate. The streamlining of the beetle's body by the wing cover mechanism also allows for low resistance to drag in water.

The spiracles open beneath the wing covers. Aquatic beetles capture a bubble of air between the wings and abdomen as they submerge. Then the submerged beetle can withdraw oxygen from the bubble, since it is in contact with the abdominal spiracles. Terrestrial beetles also derive some advantage from an air chamber beneath the elytra, especially those in very dry environments, like stored flour. The air chamber, with an intermediate humidity level, reduces the differential between the high humidity of the air in the insect's tracheae and the low humidity in the overall environment, thus reducing evaporative water loss.

Coleoptera have evolved into almost all of the niches that other insects fill. There are leaf-feeding larvae (leaf beetles) ecologically equivalent to moth or sawfly caterpillars; borer larvae (metallic wood borers and long-horned wood borers) equivalent to moth and horn-tail borers; aquatic predators (diving beetles) equivalent to water boatmen and backswimmers; fungus feeders (shining fungus beetles) equivalent to fungus gnats; dung inhabitants (scarabs) equivalent to face, house, and horn flies; and carrion feeders (burying beetles, carpet beetles) equivalent to blow fly maggots. The list could extend to many other beetle groups. In addition, the beetles are the best adapted to subterranean life and invasion of wood, especially as adult insects. In addition, beetles are probably the most sophisticated and numerous of the insects that produce light. Although many beetle groups also produce characteristic sounds, we have little knowledge of how these sounds are used in their lives.

Phylogeny

The primitive beetle ancestors were derived from megalopterous types that also gave rise to the primitive Hymenoptera and a third branch that includes all of the other insect orders that undergo complete metamorphosis. Two important suborders of beetles have evolved from the primitive beetle ancestors (see Fig. 11.3).

In some ways the suborder Adephaga is primitive, because its thoracic structure has not evolved the rigidity of the more advanced suborder Polyphaga. The Adephaga,

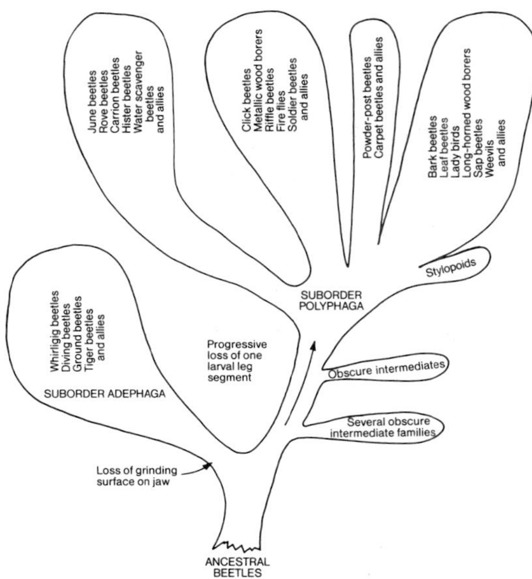

Fig. 11.3 Relationships among the Coleoptera.

however, have made some major evolutionary modifications in evolving as efficient predators. Among these are a prognathous head (mouth parts directed forward instead of down) and the loss of the mola (grinding surface) on the mandible so that it functions only in biting, but not grinding. Adephaga have evolved on land (ground and tiger beetles), on the water's surface (whirligig beetles), and beneath the surface (diving beetles and their allies).

Between the Adephaga and the Suborder Polyphaga are two groups of small, rare beetles that are important in illustrating the evolutionary trend toward progressive reduction in the number of leg segments in the larvae. The mandibles of Polyphaga are more like ancestral forms in that they have the grinding mola.

The Polyphaga is divided into five major groups. The stylopoids are unusual internal parasites of other insects, frequently Homoptera and Hymenoptera. Stylopoids have much-reduced forewings and are so unusual—their life histories and appearance—that they have often been considered as a separate order (Order Strepsiptera). Generalizations about the other groups are almost meaningless because of the wide variety of ecological niches each has filled through evolution.

Economic Impact

It is not surprising that since Coleoptera is the largest group of organisms in the world, its economic effect is widespread. The activities of most of the beetles are far enough removed from human activities that interactions are indirect at best. Indeed, most families of beetles are plainly neutral as far as their direct influence on human existence.

KEY ECONOMIC SPECIES: BOLL WEEVIL

Larva and Pupa

Adult

Distinguishing Characteristics: Only the weevil group has the chewing mouthparts at the tip of a ventral snout. Also have elbowed antennae and fat, legless larvae. Adults overwinter in plant debris and trash away from the cotton field. Survival depends on amount of cover and winter temperatures. Weevils enter fields at plant appearance and attack blossom buds and early bolls. Eggs are deposited in feeding holes and the legless larvae destroy the buds or bolls. After transformation from the pupa, the adult chews its way out. There may be as many as eight generations per year, with considerable overlap.

Occurrence: Not native to the United States, it is thought to have originated in Central America where one race still breeds on wild cotton. First reported in 1892, it rapidly spread from point of entry in southern Texas throughout the Gulf and Atlantic coastal states.

Economic Impact: Estimates of insect losses in cotton average 19% of potential crop yields, of which 8% is attributable to the weevil. Average losses in the 1950s ran $200,000,000, making boll weevil one of the most costly insect pests in the United States.

Control: Since the 1920s insecticides have been used to control boll weevils. Aerial application was developed primarily as an antiweevil technique. However, pesticide-resistant weevil populations were detected in the mid-1950s and were widespread by 1960. Integrated techniques of population monitoring, suppression of populations through precisely timed pesticide application, crop residue destruction, and preservation of natural enemies are current management practices. New integrated systems using boll weevil sex pheromone promise better precision in weevil control.

Among the benefits derived by humans from beetle activities are those associated with beetle predation on other organisms. Notable among the predatory groups are the ladybird beetles. These "ladybugs" are used in applied biological control more than other predatory beetle groups because their prey is often from the plant pest group (aphids, scales, mealybugs, mites) and a modest level of prey specificity is evident. For example, control of cottony cushion scale by the Vedalia beetle (a ladybird beetle) is one of the classic successes in applied biological control. Other groups, like the ground and tiger beetles, are sometimes important, although less specific in prey selection than some of the ladybird beetles.

Beetles have also been used in biological control of weeds; for example, control of St. John's wort in California by *Chrysolina* leaf beetles and destruction of the dung habitat of flies in Australia by dung beetles. The breakup and penetration of dung by the dung beetles results in rapid drying of the dung, thus reducing its suitability for fly maggots.

The negative interactions between man and beetles primarily involves our mutual needs for particular materials. Powderpost beetles attack wooden flooring, furniture, and tool handles. Others, like the flour beetles and grain weevils attack stored vegetable materials while their counterparts, the carpet and hide beetles, demolish preserved animal materials including furs, skins, woolens, felt materials, and leathers.

Dead trees are attacked by a variety of wood boring beetle families before they can be harvested and processed. However, of greater importance is the activity of such insects as the bark beetle in weakening and killing trees through girdling of the trunk and repeated attacks which open the tree to infection. This is how Dutch elm disease fungus is spread through its association with elm bark beetles.

Most crop plants have at least one serious beetle pest that attacks some part of the plant. As is the pattern in many insect groups, different species attack seeds, stem, leaf, root, flower, and fruit. If a beetle has not evolved that is restricted to a specific plant, chances are that one of the generalized feeders such as white grubs, wireworms, or scarab adults can attack it.

Order Lepidoptera (Plate 11.4: Moths, Butterflies, and Skippers)

Adults are characterized by two structural attributes: mouthparts in the form of a coiled siphoning tube that can be uncoiled for use in sucking nectar or other plant juices, and membranous wings covered with scales. Although the mouthparts may be vestigial in some moth groups, the scales are invariably present when the adult is winged. Adult Lepidoptera have prominent compound eyes, antennae, and usually a set of palps associated with the mouthparts. The body and legs, as well as the wings are frequently clothed in scales, or are hairy. This body covering acts as an insulator so the insect can retain some of the heat generated by the flight muscles.

Lepidopterous larvae are typically caterpillars, which feed on plants or plant material. In addition to the three pairs of thoracic legs, caterpillars have up to five pairs of fleshy prolegs on the abdomen, each proleg tipped with few to many stiff hooks, called crochets. The larvae may be distinguished from sawfly caterpillars by the crochets on the prolegs, the number of prolegs (more than five pairs in Hymenoptera, five or fewer in Lepidoptera) and the larger number of eyes (only one pair in Hymenoptera).

Larvae of Lepidoptera are capable of spinning silk. Some, like tent caterpillars, webworms, and bagworms, construct silken structures in which the larvae dwell. Many groups spin cocoons in which they spend their pupal stage.

Plate 11. 4

ORDER: LEPIDOPTERA

(derivation: lepido=scale,
ptera=wings)
Common Names: Butterflies, Moths, Skippers

Adult butterfly stage of Alfalfa Caterpillar

MOUTH PARTS

Siphoning; the coiled tube straightens out for insertion into a nectar or water source.

Many adults have no functional mouth parts.

WINGS

2 pair usually present, at least partly covered with overlapping scales. Wing texture membranous; often brightly colored

METAMORPHOSIS

Complete

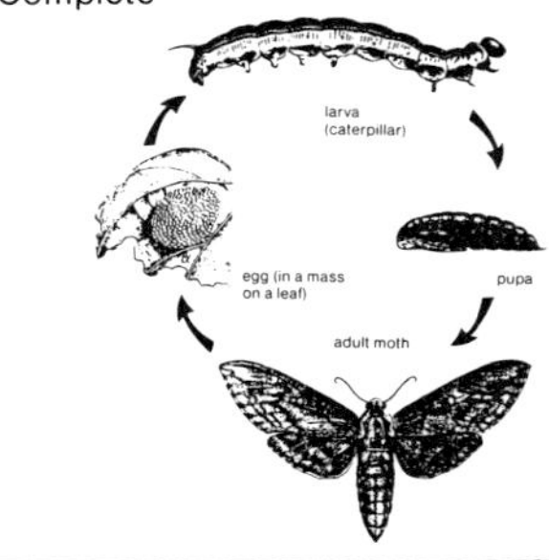

OTHER CHARACTERISTICS

1. Larval stage is usually phytophagous
2. Adults feed on liquids, if at all.
3. Larvae have 5 or fewer pairs of fleshy, abdominal prolegs, with hooks (crochets).
4. Larvae are capable of spinning silk; many construct cocoons, or other silken structures (webs, tents, mats)
5. Adults of some families possess abdominal tympana (ear drums).
6. Many adults produce sex pheromones.

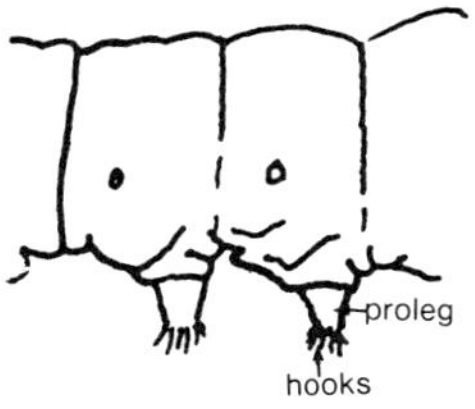

proleg
hooks

ECONOMIC IMPACT

Although most pest species of Lepidoptera feed on foliage, many crop plants have one or more pest species that attack the buds, fruits, stems or other plant parts. Economic losses may result from plant breakage, reduced photosynthetic surfaces, and injury to growing tips. Stored cereal products are severely affected by meal moths. Animal products such as woolens, feathers, hair, and felt are attacked by clothes moth species. Key pests of tobacco, cotton, corn, apples, and several forest trees are Lepidoptera.

Adults of most species are short-lived, although some, like monarchs, live for months and may migrate over hundreds of miles. Pheromones in the form of sex attractants have been identified for many Lepidoptera.

Coloration has been utilized in more ways by Lepidoptera than by any other group of insects. This may be linked to the relatively small repertoire of alternative defense mechanisms available to Lepidoptera. Although larvae in some groups have evolved repugnant taste, stinging hairs, or a dense covering of hairs and spines as defense against predation, many have developed cryptic colors and behaviors to elude natural enemies. A great number of leaf-feeding caterpillars match the color and texture of host plant parts. Others, like cutworms, have adopted nocturnal activity periods as a way to hide from enemies while feeding. Although many uses of color may be found, adult coloration generally falls into two major groups, basically following behavioral patterns. Butterflies are typically clothed in bright colors which are apparent while they are actively flying in the light part of the day. Moths are most often cryptically colored, at least in the position they remain in during daylight hours. Some of the moths with bright colors have such color confined to the hind wing and cover the brightness with a cryptic forewing when at rest.

Phylogeny

Primitive lepidopterous forms show many similarities to the caddisflies, especially in some characteristics of their wing venation. These primitive forms also still retain functional chewing mandibles in the adult and might best be designated, therefore, as the "jawed moths." Their larvae are detritus feeders, with long antennae and numerous prolegs. Two other small, but distinct, forms constitute intermediate groups between the primitive and the familiar moths and butterflies. In one group the adults are active during the day, with legless leafmining larval forms. The second intermediate group is very heterogeneous, and includes the swifts, so called because of their rapid flight. These forms are often root feeders or wood borers as larvae.

The advanced Lepidoptera includes almost all of the very familiar groups as well as most of the economically important ones. These advanced forms are characterized by the evolution of a copulatory aperture in the female, separate from the egg-laying aperture. Additionally, a trend toward different venation in the fore and hind wings is realized in the advanced groups, resulting in reduced venation in the hind wing as compared to the forewing.

The advanced Lepidoptera are divided into three groups: the microlepidoptera, the macromoths, and the butterfly-skipper group (see Fig. 11.4). Adult microlepidoptera are small and inconspicuous, with narrow, fringed wings. These include the grass moths, clothes moths, fruit moths, bagworms, and other economically important groups. Adult macromoths are typically rather fast fliers, without clubbed antennae, and many spin silken webs or cocoons as immature forms. The greatest numbers occur in the cutworm–armyworm group and the looper–measuringworm group. The most spectacular moths are to be found among the sphinx and giant silkworm groups.

Butterflies evolved from the large-moth branch of Lepidoptera. They are charac-

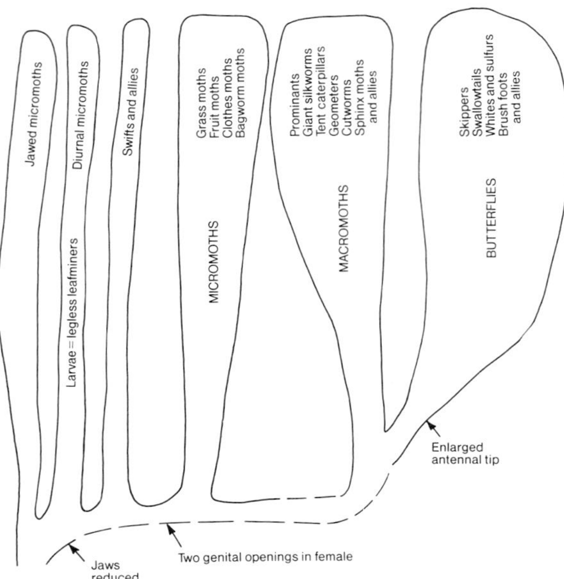

Fig. 11.4 Relationships and evolutionary trends in the Lepidoptera.

terized as being diurnal adult forms (active during the day), with clubbed antennae, and slow, erratic, fluttering flight. Butterflies commonly close their wings together vertically at rest, whereas moths usually fold them flat along the body.

Economic Impact

Except for the silkworm and the industry involved in silk production, some pollinating species and the sheer esthetics of some adult forms, the Lepidoptera are pestiferous whenever they are associated with humans. Adult Lepidoptera are innocuous but the larvae feed almost exclusively on plants and plant products, therefore there is competition for food and fiber from plants and forest products.

Damage by lepidopterous caterpillars is primarily injury from chewing, most commonly leaves, thereby reducing the photosynthetic surface of the plant. When the leaves are the plant part consumed by humans, the food may be so contaminated by droppings that human consumption is halted. Defoliation also reduces production of other plant parts, e.g., fruit, although the losses incurred are indirect. The fruit itself may be attacked directly by caterpillars, as the apple or walnut is by the codling moth. Growing tips of many plants, like tobacco, may be heavily injured by bud moths. Even after the plant materials are stored, they may be attacked as cereal crops are by the grain and flour moths. Not only is a portion of the stored material consumed, but the part left is heavily contaminated by insect fragments, feces, and cocoons, thereby significantly lowering the marketability of the commodity. Even some animal products, such as woolens, felt, and feathers, are attacked by clothes moths.

The Lepidoptera include key pests for a majority of North American crops. These are the European corn borer, corn earworm, and fall armyworm for corn, the bollworm and pink bollworm for cotton, and the hornworm and budworm for tobacco. The key pests in apples are the codling moth and the red-banded leaf roller, and in cabbage, the cabbageworm. In forests, key pests include the gypsy moth, spruce budworm, and tussock moths.

KEY ECONOMIC SPECIES: *Heliothis zea*

Larva

Adult and Pupa

Common Names: Corn earworm, Tomato fruitworm, Cotton bollworm, False tobacco budworm

Distinguishing Characteristics: Larva varies from green or pink with alternating light and dark longitudinal stripes, to almost uniform black. Pupa overwinters in the soil, the adult emerging in the spring to fly at dusk. Single eggs are deposited preferentially on the silks of corn, but also on tomato leaves and stems, growing tips of tobacco, or leaves and buds of cotton. Larvae destroy all or part of the plant product by chewing. Larvae are cannibalistic, so that solitary individuals are usually found.

Occurrence: Distributed throughout U.S., it is less severe in the extreme temperate areas where only one or two generations may be produced in a year. Extreme infestations are a result of early season dispersal flights from milder climates in which the insect can overwinter.

Economic Impact: From 10 to 15% of corn grown for canning is destroyed by *H. zea*. Additionally, about 1% infestation may be sufficient to halt sale to consumers in retail outlets. The practice of chopping off the tips of the ear (where the corn worm resides) is a common market packaging technique.

In tomatoes a single larva may damage as many as 12 fruits resulting in average losses of 12.9% in untreated fields over a 10-year period. On cotton, the larvae travels from one boll to another as it matures, thereby destroying all the bolls on one branch. In tobacco, it damages the leaves during early development. In alfalfa, *H. zea* does not reach levels that cause economic injury.

Order Diptera (Plate 11.5: Flies, Midges, Gnats, Mosquitoes, Bots, Warbles, and Keds)

The primary characteristic of this order is the presence of only one pair of functional flying wings; the pair on the last thoracic segment is reduced to a pair of halteres, alternating gyroscopic balancing organs. As a group, the Diptera are the best fliers among the insects and have a highly mobile, relatively large head, with large eyes and conspicuous, sometimes very long, antennae. These sense organs are vital to rapid, highly maneuverable flight. A number of species swarm as adults. The swarms are frequently involved with mating. In some dixid midges as few as six or eight males may constitute a swarm, whereas in the African kungu fly a swarm consists of a cloud of flies 1500 feet high.

Dipterous larvae have specialized in efficient utilization of decaying organic matter. Most advanced forms breed in semiliquid decaying organic matter and extract protein from it. Some primitive forms are aquatic or require very moist environments, e.g. decomposing plant material, "over-ripe" mushrooms, and boggy soils. Other primitive types live in streams, ponds, and water-filled tree holes, hoofprints, or even tin cans. Those that live in temporary water environments have extremely rapid development, a characteristic that has been strongly selected for in these forms.

Many of the primitive forms (black flies, sand flies, no-see-ums (punkies), and mosquitoes) have taken up bloodfeeding as adults as a method of acquiring sufficient protein to produce eggs. However, since high protein demand is keyed to egg production, only the females are bloodfeeders. Males do not require external sources of protein to produce sperm.

Some of the next more advanced group of flies also feed on vertebrate blood, e.g., horse flies, deer flies, and some snipe flies. Active and rapacious carnivores known as robber flies pursue their insect prey on the wing, pierce the body and suck out the contents. Bee flies are closely related to robber flies, however, they obtain their protein as predatory or, in some cases, parasitic larvae; the adults feed on nectar.

Some bloodfeeders are found in the next more advanced group. Both sexes of the tsetse fly, stable fly, and horn fly feed on blood. However, most of the flies in this group either carry over enough protein within their bodies from larval life to produce eggs, or feed on available free liquids, including exudates from wounds. These flies are predominantly an evolutionary experiment in the larval utilization of abundant, though otherwise poorly exploited, organic matter. Larvae can be found in rotten wood, dung, carrion, and fermenting and rotten fruit. Some live in the water, frequently as herbivores. Others have gone from carrion feeding in the larval stage to living in wounds to consuming the vertebrate host. In this group the most interesting type of reproduction is in the tsetse. It produces offspring singly. The single egg is held within the female's body, and the newly hatched larva is supplied with nutrients by glands within the mother's body. The larva does not leave the female's body until it is mature and ready to pupate.

Several of the most specialized groups have become larval parasites of mammals and are known as bots or warbles. They live in the nasal passages, sinuses, windpipe, or digestive tract, or like the ox warble, settle in the muscles beneath the skin after a

Plate 11.5

ORDER: DIPTERA

(derivation: di=two, ptera=wings)
Common Names: Flies, Mosquitoes, Gnats

House Fly

MOUTH PARTS

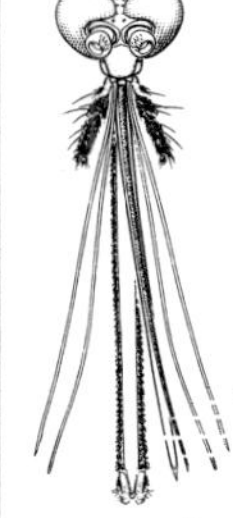

Variable, many have piercing-sucking (mosquito) or sponging (housefly), some with reduced and non-functional mouthparts

piercing-sucking (adult ♀ mosquito)

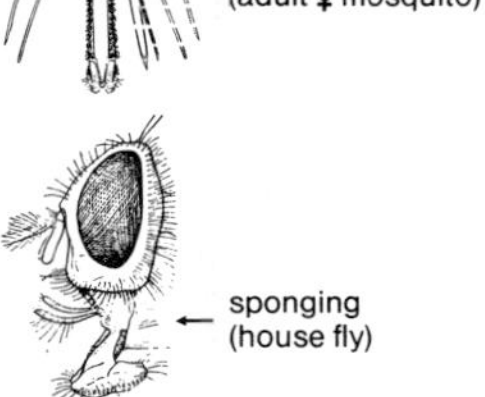

sponging (house fly)

WINGS

When present, 1 pair membranous, reduced venation, 2nd pair reduced to a pair of sensory clubs (halteres) that act as gyroscopic stabilizers.

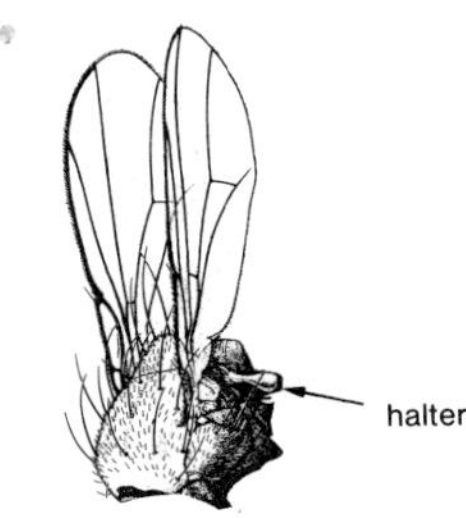

METAMORPHOSIS

Complete

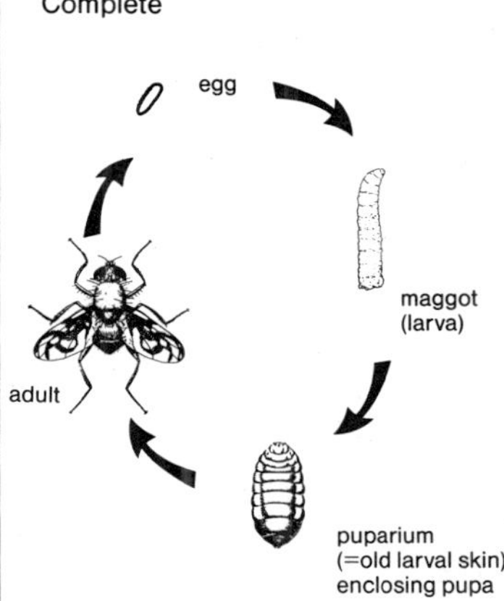

LIFE HISTORY

Eggs—Dropped or placed in the larval habitat, sometimes injected by female's ovipositor. Hatch soon after deposition.

Larvae—Most are aquatic or in semi-liquid environment. Legless, frequently without head capsule (=maggot).

Pupae—Primitive groups may have active aquatic form. Advanced forms pupate within barrel-like last larval skin (=puparium).

Adults—Primitive forms slow fliers and fragile; advanced forms more robust and strong fliers.

ECONOMIC IMPACT

Many pests of humans and animals; some are plant pests. Frequently involved in transmission of disease. Although some groups are beneficial (e.g., Tachina fly parasites of insects) probably the most harmful order overall.

OTHER COMMON FORMS

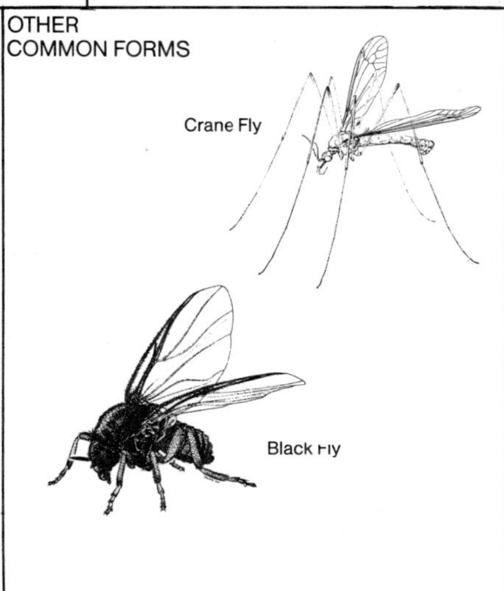

long migration through the body. The family Tachinidae are internal parasites of insects. They have become important as biological control agents in several instances.

The last group of flies produce their young much like the tsetse, as described above. The adult flies are parasites on bats, some other mammals, and on birds. Some, like the sheep ked, illustrated in Plate 11.5, are wingless and because of their flattened, leathery body and widely separated legs, are easily confused with true ticks.

Phylogeny and Synopsis

The overall relationships among the Diptera are presented in Fig. 11.5. The close relatives of Diptera are most probably the fleas (Order Siphonaptera). Both of these groups originated from an ancestral type generally like the Mecoptera (the scorpionflies). In the Diptera the venation became greatly reduced and the second pair of wings evolved into balancing halteres.

The Diptera are usually organized by taxonomists into three suborders. The Suborder Nematocera is typified by mosquitoes, fragile long-legged and long-winged adults with long threadlike antennae. The larvae usually have a head capsule and are aquatic. The Brachycera (e.g., horse flies) have shorter antennae with fewer segments and are more robust fliers. Larvae of many are predatory or parasitic, and many have evolved away from the aquatic habitat. The Suborder Cyclorrhapha is by far the largest group. They have radiated (evolutionarily) into a wide variety of ecological niches, mostly larval niches. The adults are usually robust (e.g., blow flies) and excellent fliers, and most feed on free liquids.

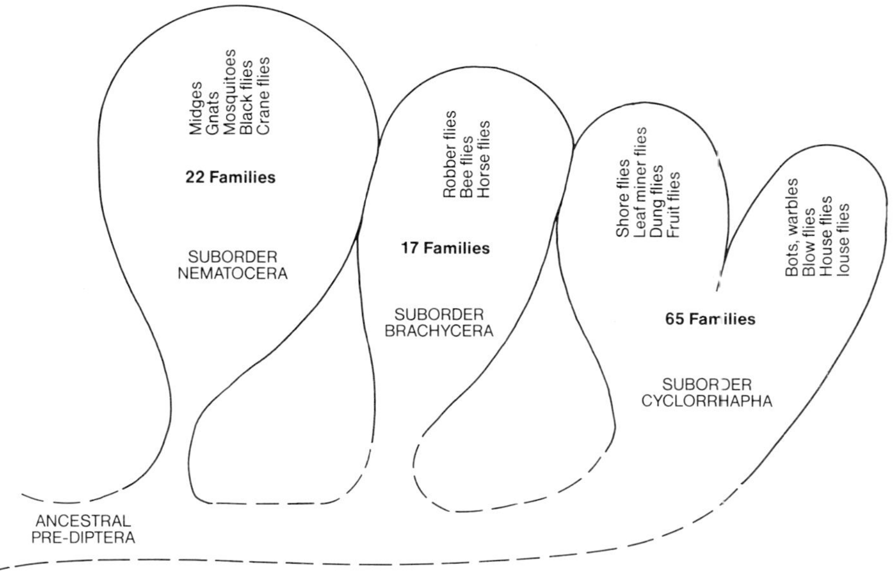

***Fig. 11.5** Relationships among the Diptera. The division of the Suborder Cyclorrhapha is based on the presence (right) or absence (left) of a flap at the base of the wing.*

Economic Impact

The most impressive contributions of the Diptera to man's welfare are as natural enemies of pest insects. Notable among these are the hover flies, whose larvae are blind and legless, though still effective predators of aphids. Other dipterous natural enemies are the parasitic tachina flies, which attack European corn borers (among many other Lepidoptera) and the parasitic–predatory maggots of the family Sarcophagidae that are so effective against the gypsy moth. As we have seen earlier, certain species of flies infest dead and/or dying animals, among which are a few that feed on dead and decaying flesh within a wound of a living animal, thus improving the conditions of the wound by leaving only the healthy tissue behind. These also "inject" the wound with chemical secretions that keep the wound relatively free from infection and encourage healing. The general activity of many species of flies in consuming, and thus removing, decaying organic matter is ecologically helpful in eliminating unsightly and odorous material from the environment. This removal also releases the materials tied up in organic material for recycling.

KEY ECONOMIC GROUP: MOSQUITOES

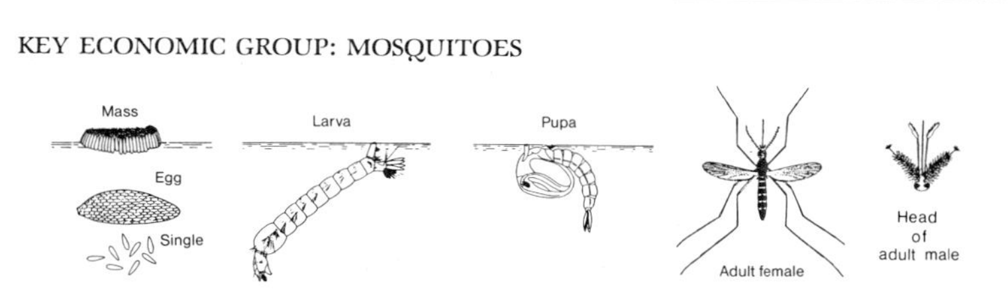

Distinguishing Characteristics: These Diptera have at least some scales on their wings and a long proboscis. Adult females have sparsely haired antennae, whereas males have bushy antennae. Larvae are filter feeders or occasionally predatory; pupae are active swimmers. Adults often mate after female encounters male in male swarm.

Occurrence: All types of aquatic habitats in egg, larval, and pupal stages, e.g., rock pools, tree holes, snowmelt and rain pools, swamps, artificial containers, salt marshes. Adults most common near aquatic habitat, but often occur considerable distances away.

Host Relations: Each species tends to feed mainly on one or a few groups of hosts in nature. They are known to locate their host by attraction to dark or contrasting color, temperature, and humidity gradients, CO_2, and various host-emitted odors.

Economic Impact: Known to attack all types of terrestrial vertebrates. Transmit the following diseases of humans: malaria, yellow fever, dengue, some encephalitids, filariasis. Public annoyance and nuisance at outdoor functions and in agricultural production areas. Incalculable losses in man-years of labor due to malaria and other diseases transmitted by mosquitoes or to secondary infections in victims of these diseases. Milk and meat production is lowered due to annoyance.

KEY ECONOMIC SPECIES: SCREWWORM FLY

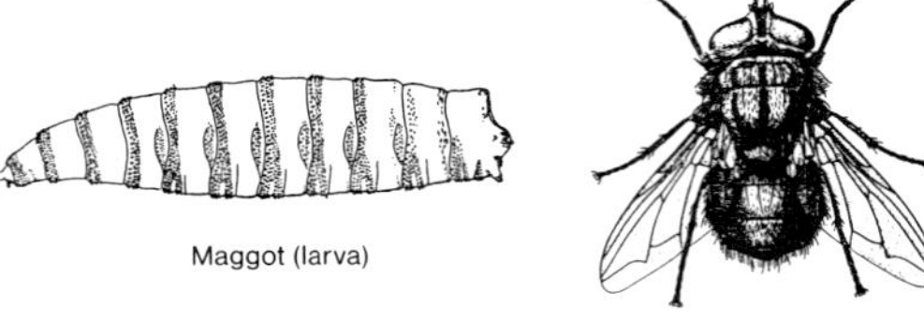

Distinguishing Characteristics: The maggot has slightly raised ridges covered with small spines on each segment resembling slightly the threads on a screw. Adult is metallic bluish-green fly, larger than the house fly, with reddish-yellow face and three black stripes on the thorax. Female is incapable of breaking the skin of the host so eggs are laid near wounds. Hatching maggots enter the wound and in a day or two begin to tear at the adjacent healthy tissue with their mouth hooks. Maggots secrete a toxin that prevents the wound from healing and causes a discharge and foul smell, which attracts additional screwworm flies. Mature larvae leave the wound and fall to the ground where they pupate. A generation may be completed in 3 weeks.

Occurrence: Southern parts of the United States, south to Argentina. Frequently brought to northern states in infested livestock, but these infestations are eliminated annually by cold winter temperatures. Subject of a massive eradication program by USDA, it has been eliminated from the southeast United States.

Economic Impact: Attacks cattle, horses, mules, sheep, hogs, goats, man, dogs, and other domestic animals and wildlife. An infested wound, if untreated, becomes increasingly attractive to further oviposition by screwworm flies. Infested animals hide and refuse to eat. Death ensues unless the animal is found and treated. Newborn farm animals are frequently attacked at the navel.

Control: Control via sterile male release program is enhanced because (1) females mate only once, (2) male can be sterilized without detectable loss in competitive ability compared to normal males, (3) low winter population is restricted to southernmost parts of continental United States, (4) species can be mass-reared—up to one billion per week have been reared and sterilized.

The Diptera are most important as pests of humans and animals. In temperate regions those important primarily as bloodfeeders are mosquitoes, black flies, horse flies and deer flies, horn flies, and stable flies. However, in the tropics many species frequently act as vectors (transmittors) of organisms causing such diseases as malaria, dengue, yellow fever, and filariasis borne by various mosquitoes; tularemia by a horse fly species; onchocerciasis by black flies; leishmaniasis (kala-azar and oriental sore) by sand flies; nagana and African sleeping sickness by tsetse flies.

Bot and warble fly larvae are internal parasites. They debilitate livestock, both through physical interference with flow through the animal's digestive system if large numbers of larvae are present, and by removal of small amounts of nutrients by each larvae. These parasites also cause loss of blood, nausea, discomfort, and pain because they attach to the intestinal walls with their clawlike mouthparts. The screwworm, a wound inhabitor, causes death in untreated livestock when large numbers of larvae attack a wound, enlarging it by feeding, and making it more attractive for additional egg-laying by screwworm females.

Numerous other Diptera are nuisances in the existence of humans and livestock. Many of these flies, though some are nonbiting, cause irritation by clustering around the eyes and mouth, where they feed on mucous secretions. Several, like the face fly, have been incriminated in the transmission of eye diseases. Some flies become nuisances in temperate regions because of their overwintering habits. Face flies and cluster flies seek out attics, church belfries, and crawlspaces in structures as autumn approaches. Not only do some of these flies show up in the living areas within houses on warm early spring days, but a substantial percentage die in the attics and walls during winter. When abundant enough, these can become a food source for carpet beetles and other insect carrion feeders which in turn can become a nuisance.

Several groups of Diptera are pests of growing plants. Not only is ripening fruit attacked, but other materials are subject to attack by bulb flies, onion and cabbage maggots, and the carrot rust fly, among the many dipterous plant pests. Perhaps the most serious pest of wheat is the larva of a primitive dipteran, the Hessian fly. Most other major crops have at least one serious pest among the Diptera. Additionally, several diseases of plants are transmitted by Diptera.

Order Hymenoptera (Plate 11.6: Ants, Bees, Wasps, Hornets, and Sawflies)

This is the third largest order of insects, with approximately 100,000 species identified. Not only is it a large group, but it has diversified into a number of ways of living that have made it overall a significant order of insects. The group is characterized as having essentially chewing mouthparts, which may be modified somewhat for lapping, as in the honey bee. The two pairs of wings are membranous and are hooked together into single flight surfaces by modified hairs on the leading edge of the hind wing that hook into a groove on the trailing edge of the forewing. Wing venation is somewhat reduced, with fewer veins than found in more primitive insects, but with many cells of clear membranous areas formed by the branching of the veins. The wings can be folded back along the body when they are not in use. Although the presence of wings is typical of the order, some species never have wings and others, like the ants, only have special individuals (reproductives) that are winged.

All Hymenoptera exhibit complete metamorphosis; the larva differs in form with the group. In the primitive foliage-feeding groups the larva is a caterpillar, whereas in the social insects it is much like a weevil grub, with a head capsule, but frequently without legs. The parasitic Hymenoptera sometimes have larvae without many distinguishing structural characteristics, being simply wormlike, legless, with reduced mouthparts, and a weakly developed head capsule.

Plate 11.6

ORDER: HYMENOPTERA

(derivation: hymeno=God of marriage, refers to union of fore and hind wings by hooks; ptera=wings)
Common Names: Ants, Bees, Wasps, Hornets, Sawflies

Yellow Jacket

MOUTH PARTS

Chewing, frequently modified for lapping

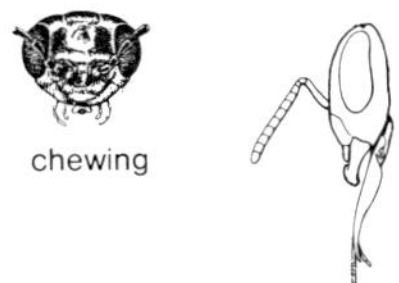

WINGS

2 pairs, membranous, joined together by series of hooks that originate in the hind wing and hook into a groove on trailing edge of fore wing.
Venation frequently forms cells (enclosed areas).

METAMORPHOSIS

Complete

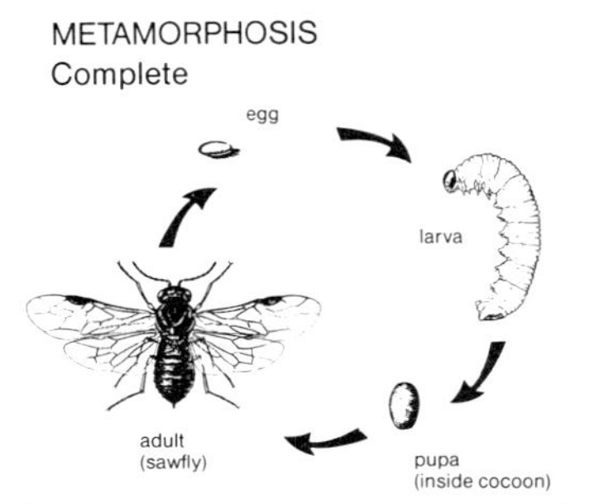

OTHER CHARACTERISTICS

1. Ovipositor of female like a sabre saw in primitive groups (A); a hypodermic needle-like stinger in advanced groups (B)

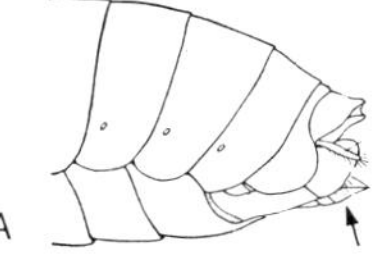

2. Variety of life styles: parasitism, predation, social and solitary existence; foods include pollen and nectar. Exhibits haplodiploidy.

ECONOMIC IMPACT

Primitive forms (sawflies, horntails) are plant pests, as defoliators or borers. Advanced groups are important as natural enemies of pest insects, parasitic wasps are especially valuable. Most significant Order in the pollination of plants, and therefore, in production of many crops. Producers of natural products e.g., honey, wax. Some species e.g., leafcutter ants, are pests.

OTHER COMMON FORMS

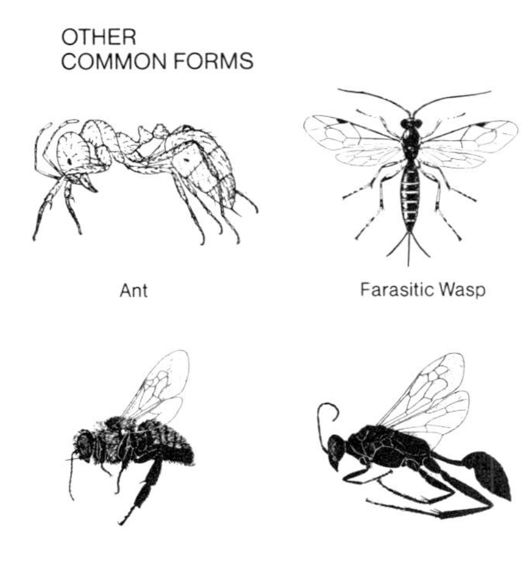

Evolution of the Hymenoptera has been greatly influenced by the ovipositor of the female. The primitive sawflies have a sabersaw-like ovipositor which is used to cut slits in plant tissue into which eggs are thrust. The slender stabbing stinger of the more advanced Hymenoptera is a modification of the same structure. The role of the ovipositor in hymenopterous evolution is covered in the section on Phylogeny and Synopsis.

Life-style among the Hymenoptera is extremely varied. It ranges from the solitary existence of many thread-waisted wasps to the complex social order to be found among the ants and social bees. Larvae may be parasitic, feeding on the body contents of some other insect. Larvae of colonial forms may be kept in groups or cells, where they are fed by adults with larval food varying from honey and pollen to partially digested animal materials, depending on the species involved. Adult foods also vary over the same wide range.

Larvae frequently form cells or cocoons in which they pupate. The cocoons of sawflies are very dense, tightly woven structures that look much like a kidney bean in color and size. Other Hymenoptera spin delicate white cocoons, and still others form no cocoon at all.

Adult Hymenoptera are usually very good fliers, with concomitant excellence of vision. Most are active during the day, but some are nocturnal. The adults frequently display sophisticated behavioral patterns, especially in the social groups. Pheromones have been linked to more than a dozen different behaviors in the honey bee alone. The nest provisioning and social Hymenoptera are among the insects that demonstrate the sense of memory. Most of these forms must be able to return to the nest or colony after foraging trips. Since the foraging is often by flight, these insects have an ability to relocate the nest by means of landmarks.

Phylogeny and Synopsis

Ancestral forms had a sawlike ovipositor, much like today's sawfly, and relatively large, almost net-veined wings similar to today's lacewing (Order Neuroptera). The insects evolved into larval plant-feeding forms and are represented by the sawflies, typical of the Suborder Symphyta, with their caterpillar-bodied larval forms (see Plate 11.6). From this life-style a group evolved into a stem-boring larval form, like the horntails and closely related stem sawflies. At this point in hymenopterous evolution it has been theorized that individuals of one group overtook other insects that were also boring in plant stems. This pattern was successful and those horntaillike Hymenoptera encountering other insects in burrows had high survival rates, eventually leading to a parasitic existence.

In the advanced Suborder Apocrita the most primitive forms are all insect parasites (braconid and ichneumonid wasps), with the constricted abdomen and stabbing ovipositor typical throughout the suborder. From these, with a major improvement in ovipositor mechanism, evolved the specialized parasites, some of which have reinvaded the plant world as gall wasps, others have evolved into highly specialized parasites of other insects.

The addition of paralyzing chemicals to the stabbing ovipositor occurred in the next more advanced group to evolve. By paralyzing prey, the wasp can complete oviposition without danger from a struggling host. This same group also contains the ants, all of which are social, and the wasp–hornet–yellow jacket forms, many of which are social species.

Nest provisioning by wasps is the next evolutionary trend, which is facilitated by the paralyzing sting. This frees the wasp to increase in body size relative to its larval food supply, because many small paralyzed prey can be supplied for a large-sized larva. It is a short step, evolutionarily, from the provisioning of a nest with food prior to depositing the egg to continuing the provisioning process after the larva has hatched, that is, to caring for the offspring.

Bees provision their nests with plant material rather than paralyzed prey. The bees, while closely related to provisioning wasps, are distinguished by the type of hairs they possess, especially on the thorax. The hairs are branched, looking much like feathers, and provide a large surface area to which pollen grains adhere on contact. Only a few bees are truly social, with the honey bee the most familiar representative of a group containing bumble bees, leafcutter and cuckoo bees, among others. The overall pattern of evolution in the major groups of Hymenoptera, with the significant characteristics is presented in Fig. 11.6.

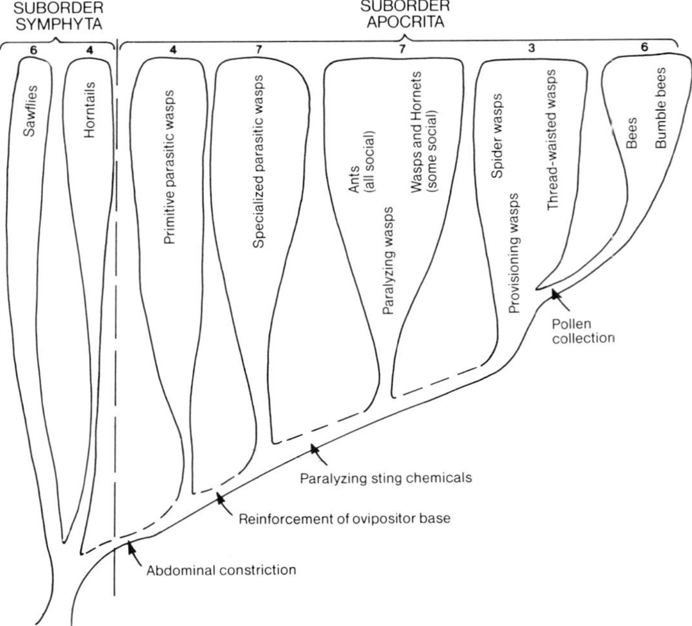

***Fig. 11.6** Evolution in the Hymenoptera. The key to the advanced Suborder (Apocrita) is development of a "stabbing" ovipositor and a constricted abdomen. Together, these allowed for precise egg laying by parasites. Evolution of paralyzing chemical production in the "sting" mechanism allowed for preservation of food (paralyzed prey). True colonial behavior appears among the groups in several of the advanced lines. The number of families in each group is indicated.*

Economic Impact

As a whole, the Hymenoptera comprise one of the most important orders of insects. Many are important as beneficial insects, but some groups are pests. Among the pests are the phytophagous sawflies, ranking with the moths and bark beetles as the three most important forest pests. The stem sawflies are especially important in several grain crops, e.g., wheat, rye, and barley. Horntail larvae bore in the wood of both deciduous and coniferous trees, although most species only attack dead or weakened trees and are not especially serious if wood is utilized rapidly or kiln dried. Among the specialized parasitic groups, several are known to have evolved back into the use of plant materials and are serious pests, attacking seeds of clover and wheat among others.

KEY ECONOMIC SPECIES: HONEY BEE*

Distinguishing Characteristics: A true social insect with reproductive division of labor (only the queen lays eggs), overlapping generations, and cooperation among individuals in care of the offspring. Among the bees, the honey bee is unusual in that the entire colony overwinters. In other social Hymenoptera only the fertilized queen lives through the winter. New colonies are founded by swarming, in which the queen leaves with half of the worker population. Colony sizes of 60,000–75,000 individuals are common during the growing season. Wax, produced by glands on their bodies, is used to make cells in which young are reared and food is stored. Activities in the hive are controlled by chemical communication, including suppression of egg laying, alarm reactions, and sex attraction.

Occurrence: World-wide

Economic Impact: Significant in two ways: (1) pollination of crops and (2) production of honey and wax. Many fruits and vegetables are pollinated entirely or primarily by insects. Introduction of rented hives into the field at flowering increases pollination, resulting in production gains of 150–500%. Honey is used by food processors in large amounts, often as the sweetener in baked goods. Yields of 50–150 pounds of honey per colony make beekeeping an economically viable industry. Wax is used as a base in cosmetics and in candle production.

*Illustration reprinted by permission from E. O. Wilson (1971). The Insect Societies, Harvard University Press, Cambridge, Massachusetts.

Ants are household and agricultural pests. Not only may those in crops feed directly on the plants, but other species protect aphids by driving natural enemies away, thereby lowering crop productivity. Ants are among the Hymenoptera that may inflict serious injury to animals by stinging. Although the stings of many Hymenoptera (e.g., bees) are painful and may be inflicted in great numbers, there is a more serious aspect. The introduction of sting chemicals, which have some protein components, eventually sets up an immune reaction in the human's body. After this, the human is sensitized to the sting of that particular insect, and the next sting may cause anaphylactic shock, resulting in swelling, respiratory difficulty, and even death. In fact, more human deaths in North America result from the sting of insects than from the bites of poisonous reptiles.

The positive value of many Hymenoptera lies in the widespread occurrence of parasitism and predation within the Suborder Apocrita. Figure 11.6 shows that all but one of the major groups are predominantly parasitic or predacious, usually on plant-feeding insects. Therefore hymenopterous parasites are usually sought as biological control agents. In these cases, parasites have been found tc be more effective than predators, since parasites tend to be more specific in selecting a host.

The value of Hymenoptera as pollinators ranks as high in importance as their activities as natural enemies. Only a small group among the Hymenoptera normally act as pollinators (see Fig. 11.6), but their value to agriculture makes them a vital group.

MINOR GROUPS OF INSECTS

Plates 11.7 through 11.18 summarize the important aspects of the moderately important orders. The least significant orders are listed in Table 11.1, which also gives their basic characteristics. A key for identifying specimens is given in Chapter 12.

Plate 11.7

ORDER: COLLEMBOLA*

(derivation: colla=glue
embolon=bar)
Common Name: Springtail

* Some scientists exclude these animals from the Class Insecta

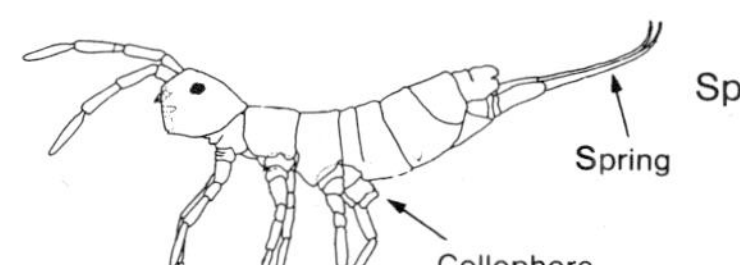

MOUTH PARTS

Chewing–embedded in head (cheeks grow down over the mouthparts)

WINGS

None

METAMORPHOSIS

None

Egg
↓
Immature
↓
Immature
↓
Immature
↓
Immature
↓
Adult–reproduction
↓
Adult–reproduction

OTHER CHARACTERISTICS

Most are minute and without a tubular repiratory system.

Most have a terminal "tail" that is fastened to a catch on the underside of the abdomen When the tail is released while pressure is exerted, the insect is thrown into the air, sometimes several centimeters. Hence the name, springtail. Another anterior abdominal appendage (the collophore) is thought to act in water regulation.

BIOLOGY

These subterranean or surface dwellers are numerous and well adapted soil dwellers. Primary requirements are moist conditions and abundant organic matter. Sometimes abundant in snow fields in late winter (snowfleas) Others live on the water surface in ponds and stream backwaters. Eggs may be laid in groups. Up to 5 generations/year.

OTHER COMMON FORMS

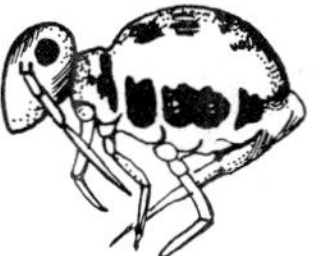

note lack of spring

ECONOMIC IMPACT

Significant in nutrient recycling because they consume organic debris. One is a serious pest of alfalfa and several cause problems in mushroom production.

Plate 11. 8

ORDER: THYSANURA

(derivation: thysanos=fringe,
ura=tail)
Common Names: Silverfish, Firebrat

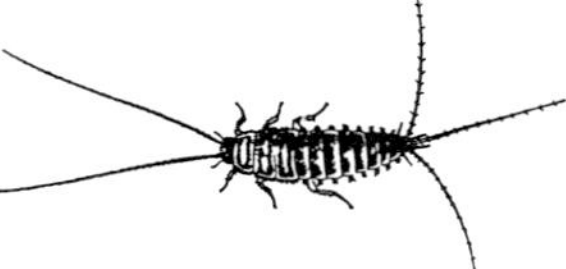

MOUTH PARTS
Chewing

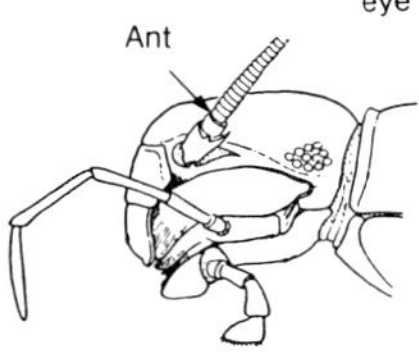

WINGS
None

METAMORPHOSIS
None

Egg
↓
Immature
↓
Immature
↓
Immature
↓
Immature
↓
Adult—reproduction
↓
Adult—reproduction

OTHER CHARACTERISTICS

Abdominal appendages (= styli) present on some segments (see illustration above).

The terminal tails are: a pair of cerci, and a median filament.

LIFE HISTORY

Eggs are laid singly in cracks; immatures look just like adults, feed on same foods, and live in same location; reach sexual maturity after several molts and keep molting and growing throughout life, even after they begin to reproduce. Adults of winged insects do not molt.

OTHER COMMON FORMS

The firebrat is much like the silverfish, but prefers higher temperatures, e.g., around steam pipes, furnaces, boilers, and is more brownish than silver in color.

ECONOMIC IMPACT

Domestic silverfish feeds on starchy materials in books, especially bindings, labels, also eats wallpaper paste. A frequent, but seldom significant pest.

Plate 11.9

ORDER: EPHEMEROPTERA

(derivation: ephemera=short-lived
ptera=wings)
Common Names: Mayflies, Shadflies

MOUTH PARTS

Vestigial (adults don't feed)

WINGS

2 pair, with many veins; forewings large and triangular; hindwings small, round, may be vestigial or absent; held above body at rest; cannot be folded down along body; many species with distinctly pleat-like corrugations of the wing surface. Such wings are used to produce essentially vertical flights.

METAMORPHOSIS

Gradual

Egg
↓
Nymph (many instars)
↓
Nymph (external wing pads in later instars) — Aquatic

↓
Subadult
↓
Adult — Aerial (terrestrial)

OTHER CHARACTERISTICS

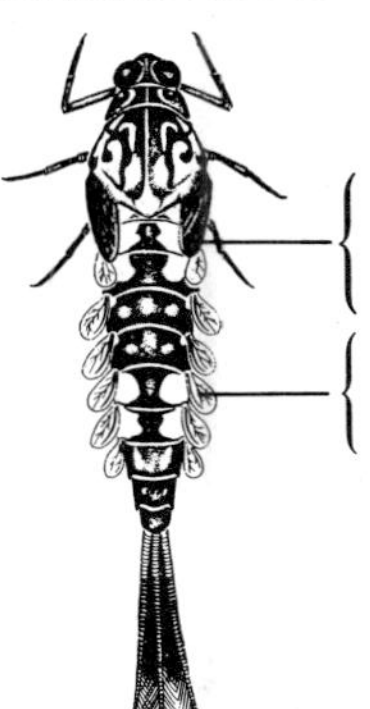

Mayfly subadults are the only insects that molt after the wings become functional. They are usually duskier and hairier than the subsequent adult. Subadults and adults live only a day or two; neither stage feeds.
Aerial stages have 2-3 long terminal appendages.

LIFE HISTORY

The eggs are laid into the water; the aquatic nymphs are predominantly herbivorous they remain in the nymphal stage for several months to a year or more. The dusky subadult (the fisherman's "dun") emerges from the split skin at or below the water surface, flies away, and molts several hours later into a smooth, brightly colored adult (the fisherman's "spinner"). The adult reproduces and dies in a day or so. Emergence of many species is synchronous.

ECONOMIC IMPACT

An important source of freshwater fish food. The mayfly nymph is the herbivorous browser of the freshwater environment. The most frequent insect models for fly fishing lures.

Plate 11.10

ORDER: ODONATA

(derivation: odous=a tooth)
Common Names: Dragonflies, Damselflies

Dragonfly

MOUTH PARTS
Chewing

WINGS
2 pair; net-like, with many veins; elongate; can't be folded along body; at rest, held horizontally (dragonfly) or vertically (damselfly); excellent fliers. A pigmented spot on leading edge of each wing near tip.

METAMORPHOSIS
Gradual

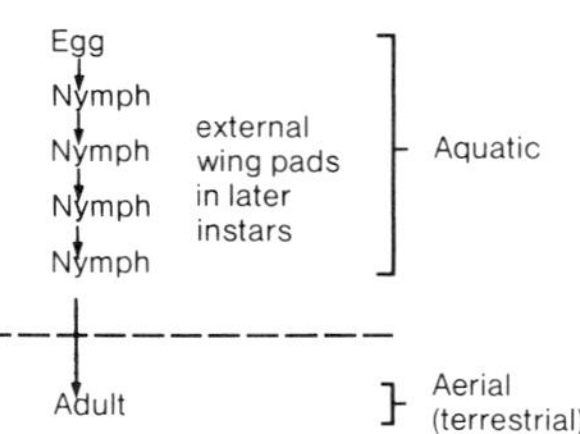

OTHER CHARACTERISTICS

1. Nymph with grasping "mask" (labium).

extended

retracted

2. Adults with large compound eyes.
3. Long abdomen.
4. Dragonfly nymph removes oxygen from water by means of gills in rectum (digestive system).

LIFE HISTORY

The dragonfly is the total predator, both as a nymph, with its extensible mask, and as a fleet-winged flying adult. They are frequently territorial and patrol stretches of pond or stream. Adults mate in flight and assume contorted positions in order to transfer sperm from genital structures on male's anterior abdomen (posterior in other Orders) to female.

ECONOMIC IMPACT

Generally considered as beneficial. Both nymphs and adults consume insects. Have been introduced to control mosquito larvae in the water and adults in the air.

OTHER COMMON FORMS

Damselfly nymph has external leaflike gills at end of abdomen. Swims by using gills as a tail. Female acult damselfly inserts eggs in aquatic plants.

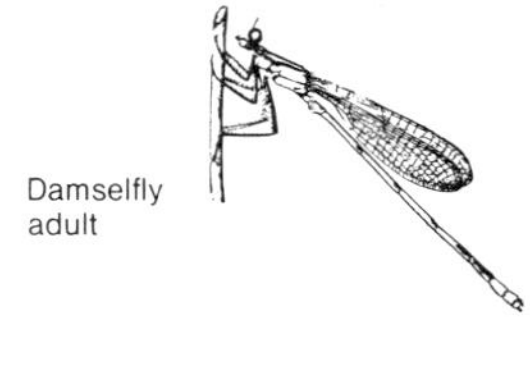

Damselfly adult

Damselfly nymph

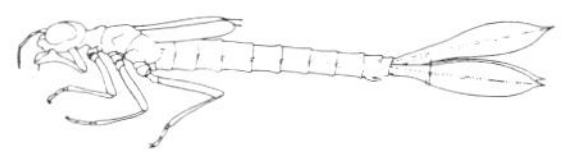

Plate 11.11

ORDER: DICTYOPTERA

(derivation: dictyo=net,
ptera=wings)
Common Names: Roach, Mantis

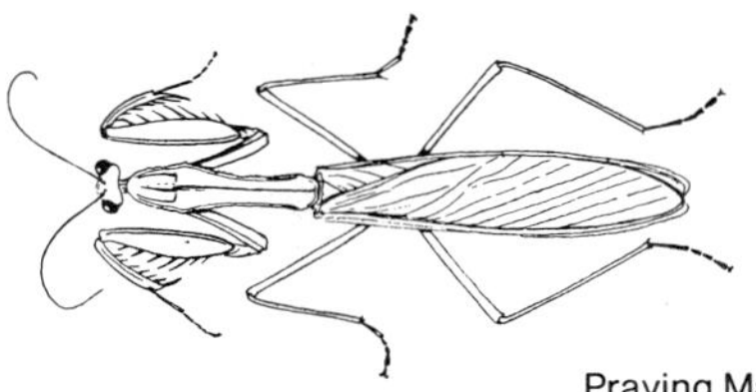

Praying Mantis

MOUTH PARTS
Chewing

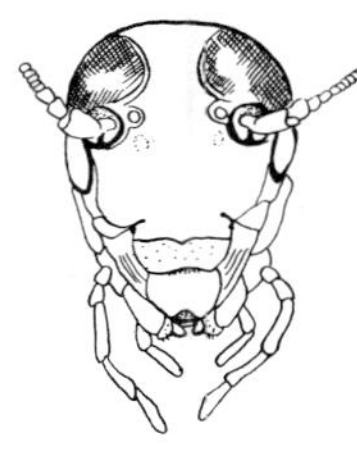

WINGS
Two pair; 1st pair=tegmina (sing.-tegmen), thickened parchment-like, with veins; 2nd pair folded fan-like at rest.

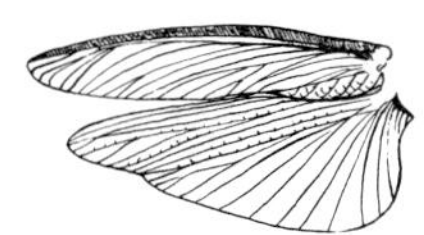

METAMORPHOSIS
Gradual

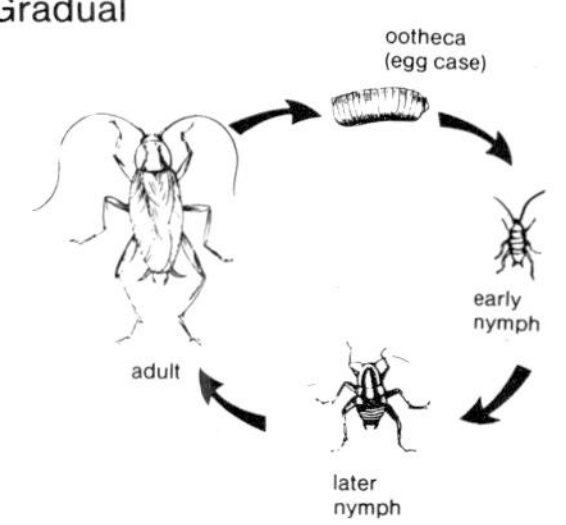

OTHER CHARACTERISTICS

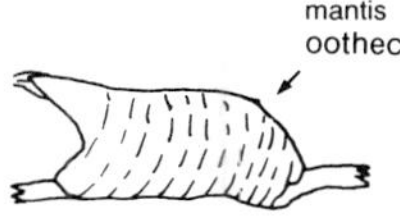

1. Eggs produced in packets (oothecae)
2. Many roaches carry the ootheca until the eggs hatch.
3. All mantids are predacious.
4. Roaches are general feeders.

BIOLOGY

Dictyoptera are primarily tropical, consisting of two very distinct groups both in feeding habits and appearance. Mantids are predatory in all active stages and have an elongate prothorax with grasping front legs. Roaches are omnivorous, secretive, often rapid runners with flat bodies. The egg packets produced by Dictyoptera are retained in the body by the more advanced roaches, thereby deriving increased protection. Most roaches are leaf litter dwellers. In nature, most temperate zone Dictyoptera probably have single annual generations.

ECONOMIC IMPACT

Although most of the more than 2000 known types of cockroach don't interact with man, a few species have become very serious pests. The foul odor, associations with human wastes, and nocturnal habits have made domestic roaches the most serious insect pests in our urban world. Because roaches frequently contaminate food supplies and may crawl over sleeping people, mechanical transmission of human disease is a threat. Roaches are widely used as laboratory animals for scientific study.

Mantids are of some value as predators on pest species, but are rather nonspecific in prey preference.

Plate 11.12

ORDER: ISOPTERA

(derivation: iso=equal,
ptera=wings)
Common Name: Termite

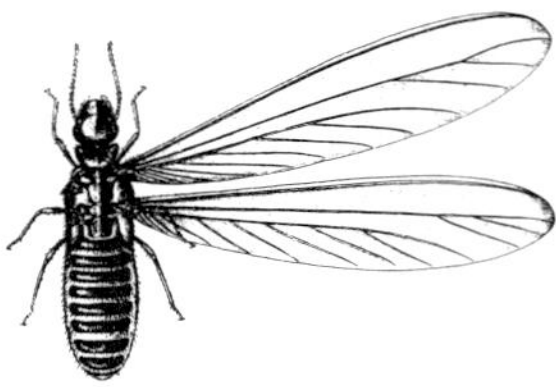

Reproductive
(wings of one side are removed)

MOUTH PARTS

Chewing

Chewing (worker)

In some forms mouth parts are modified for defense.

Crushing (soldier)

WINGS

2 pair, when present; membranous; venation very similar in fore and hind wing. The elongate wings are broken off by the reproductives after the nuptial flight, leaving short stubs. Removal is facilitated by a suture (line of weakness) near the wing base.

METAMORPHOSIS

Gradual

Stages

Egg
↓
Nymph
↓
Adult

OTHER CHARACTERISTICS

1. Colonial (all species).
2. Caste system (social polymorphism)—the presence of functionally different forms in the same colony.
3. Cellulose feeders:
    1. wood (by primitive types)
    2. wood and fungus (more advanced types grow fungi on partially digested cellulose, then consume it all
4. Symbiosis with cellulose-digesting protozoa and bacteria. Found in digestive tract of termites. Passed to offspring by anal-oral transfer.

BIOLOGY

Nuptial flights of ♂ and ♀ winged reproductives emerge from established colonies. After alighting and breaking off their wings, individuals wander about until encountering a reproductive of the opposite sex. The royal pair locates suitable wood (primitive termites), or alternately a stone (advanced forms) and begins tunneling. Initial offspring are first fed regurgitated foods, but later aid in colony growth and maintenance. Primitive subterranean termites nest in the wood or soil. Advanced forms create sophisticated structures (termitaria) with ventilation systems, cemeteries, and various chambers for reproductives and fungal farms. Such termitaria may be mounds up to 20 ft. high and oriented for optimum solar exposure.

ECONOMIC IMPACT

In nature, termites are beneficial because their use of cellulose rapidly returns wood materials to basic reusable components. Any wood product, especially that in contact with the soil may be attacked. Unless detected early, total destruction usually occurs. Control with special construction or chemical treatment protects woody materials, but only at considerable cost.

Plate 11.13

ORDER: PLECOPTERA

(derivation: plecos=plaited, ptera=wings)
Common Names: Stoneflies

MOUTH PARTS

Chewing, but often very soft and reduced

WINGS

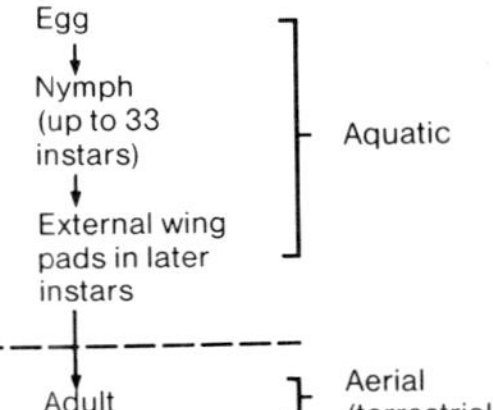

2 pairs of membranous wings, the hind wings are folded into plaits and are much wider than the forewings.
The wings are folded (flat over the back) along the body when not in use.

METAMORPHOSIS

Egg
↓
Nymph (up to 33 instars)
↓
External wing pads in later instars — Aquatic

Adult — Aerial (terrestrial)

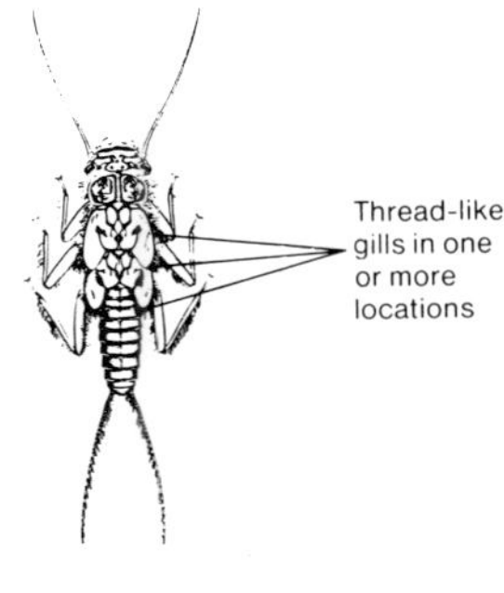

Nymph

LIFE HISTORY

Masses of eggs are deposited by the female into the water. The nymphs (predatory or leaf shredders) respire by means of gills and live 1 or more years before emerging as adults. Adults are usually found near the cold, nonpolluted stream habitats of the nymphal stage. Most nymphs prefer gravel or rocky bottomed streams. Several groups emerge as adults in late winter. They do not fly, but their early appearance avoids the presence of many predators.

ECONOMIC IMPACT

Second in importance only to the Ephemeroptera as food for freshwater fish in streams.

Plate 11.14

ORDER: MALLOPHAGA

(derivation: mallo=wool,
phaga=eat)
Common Names: Biting Lice, Bird Lice

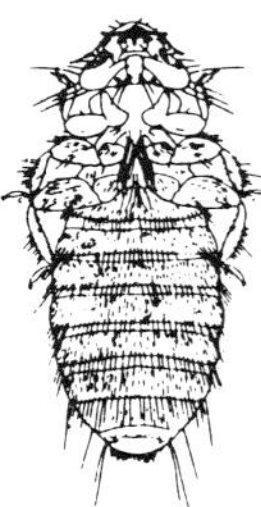

MOUTH PARTS
Chewing, but much reduced

WINGS
None

METAMORPHOSIS
Reduced Gradual

Egg
↓
Nymph
↓
Nymph
↓
Nymph
↓
Adult

OTHER CHARACTERISTICS

1. Primarily bird parasites; a few on mammals.
2. Never leave host's body.
3. Most have 2 claws, one group on mammals with claws like Order Anoplura.
4. Head wider than thorax (compare with sucking lice).
5. Most species are confined to a single species of host.
6. Body flattened and tough.

LIFE HISTORY

Eggs (nits)—attached to feathers or hairs.

Nymphs—like adults, but smaller and lighter.

Adults—the two-clawed forms move very rapidly through feathers. Singleclawed forms are more sluggish.

ECONOMIC IMPACT

Infested domestic fowl have lowered egg production; constant scratching frequently leads to secondary infections and loss of plumage or hair. Predisposition to disease is severe and death may result.

Plate 11.15

ORDER: ANOPLURA

(derivation: anopl=unarmed, ura=tail)
Common Names: Lice, Cooties

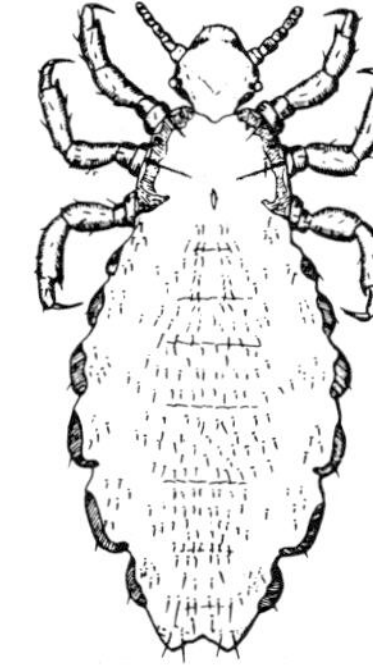

Human Body Louse

MOUTH PARTS

Piercing-sucking (retract into head when not in use)

WINGS

None

METAMORPHOSIS

Reduced Gradual

Egg
↓
Nymph
↓
Nymph
↓
Nymph
↓
Nymph
↓
Adult

OTHER CHARACTERISTICS

1. All are blood feeders
2. All are on mammals; most are quite host specific (only live on one species of host).
3. Never leave host's body (even the eggs are glued to the hair of the host), except the human body louse.
4. Claws fitted for clinging to hairs (see above).
5. Body hardened and flattened dorso-ventrally
6. Head narrower than thorax (cf Mallophaga)

LIFE HISTORY

Eggs (nits) —glued to hair of host

Nymphs — like adults, but smaller and usually paler in color

Adults — are sluggish and feed frequently.

ECONOMIC IMPACT

Lice inflict serious losses to livestock through irritation. Not only do they cause fatigue in infested humans, but the potential for epidemics of epidemic typhus and trench fever (both louse-borne) exists. Louse infestations prevail under human stress conditions, e.g. wartime or natural catastrophe.

OTHER COMMON FORMS

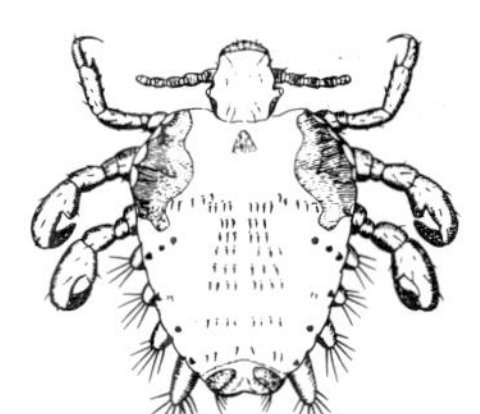

Pubic Louse

Plate 11.16

ORDER: NEUROPTERA

(derivation: neuro=nerve,
ptera=wings)
Common Names: Lacewings, Doodlebugs, Aphislions, Antlions

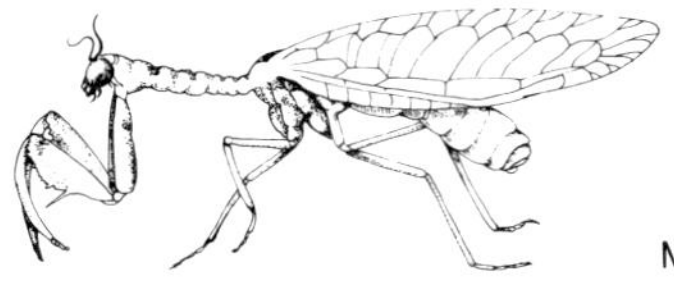

Mantisfly adult

MOUTH PARTS
Chewing

WINGS
2 pairs, membranous, with many many longitudinal and crossveins. Wings folded back (rooflife) over body when not in use.

METAMORPHOSIS
Complete

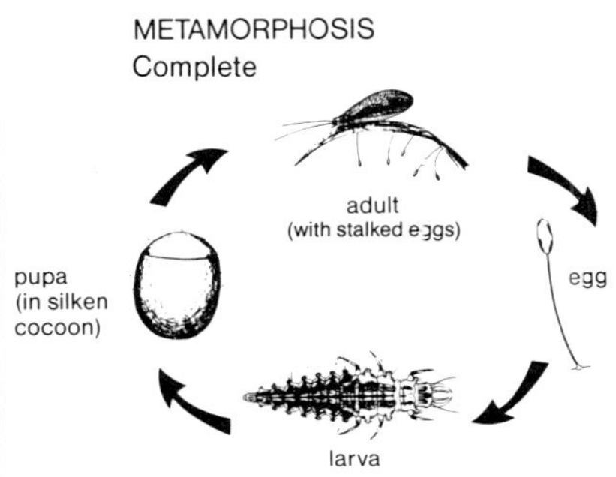

OTHER CHARACTERISTICS

1. Mostly predacious, especially the larval stage.
2. Biting larval mouthparts, but used for piercing and sucking out body fluids; frequently sickle-shaped

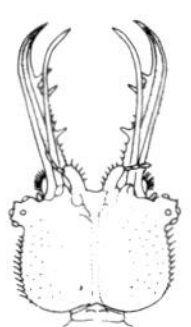

ventral view

3. Terrestrial (except for Spongillaflies living in freshwater sponges)
4. Weak fliers
5. Larvae produce silk in Malpighian tubules; spin silk from anus.

LIFE HISTORY (of Aphislion)

Eggs—Stalked (not all Neuroptera produce stalked eggs); little discrimination by adult in selection of site.

Larvae—Active predators locate prey by contact, not vision.

Pupae—Larva spins a pearl-like silken cocoon on underside of leaf.

Adult—Some feed on soft-bodied insects. Many are attracted to light.

ECONOMIC IMPACT

Beneficial as predators of many small, soft-bodied arthropod pests of plants, e.g. aphids, mites, scales, whiteflies. Mass rearing of some lacewing species for release is being investigated as a biological control procedure.

OTHER COMMON FORMS

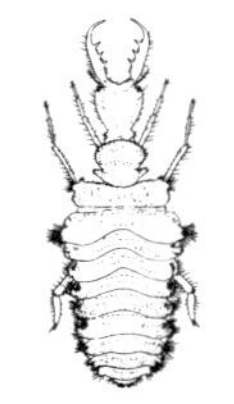

Doodlebag (Antlion larva), a sedentary predator

Antlior adult

Plate 11.17

ORDER: TRICHOPTERA

(derivation: tricho=hair,
ptera=wings)
Common Name: Caddisflies

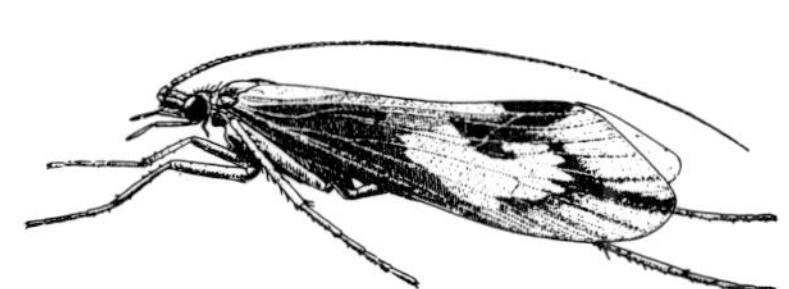

MOUTH PARTS

Chewing, but greatly reduced except for 2 pairs of large palps

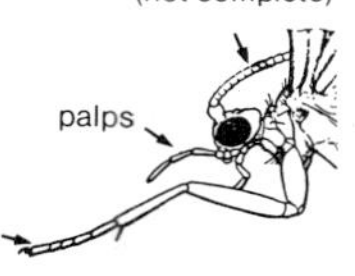

WINGS

2 pairs, large and membranous, densely covered with hairs (but not scales). Wing colors usually mottled somber browns. Held roof-like at rest.

METAMORPHOSIS

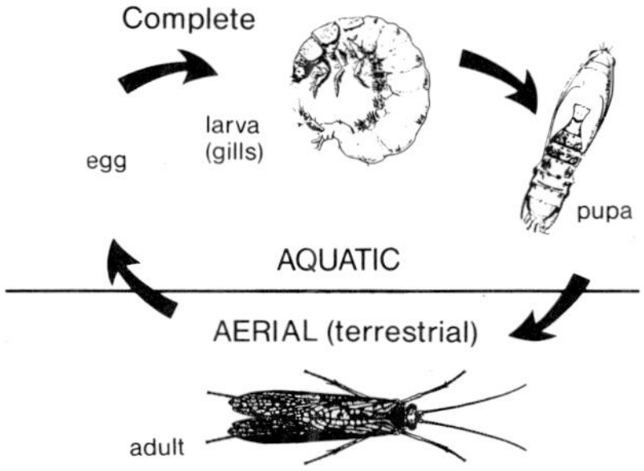

Complete

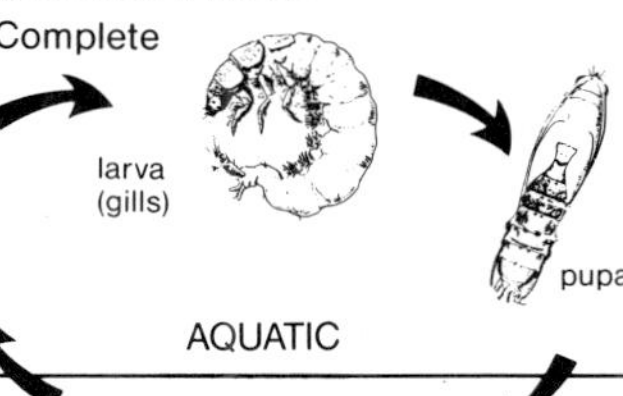

OTHER CHARACTERISTICS

1. Adults look like moths
2. Aquatic larval and pupal stages
3. Larvae produce silk
   - A. Some construct silken nets
   - B. Some construct cases, attaching sticks, gravel, etc. to exterior. Cases are characteristic in shape for each Caddisfly group.

LIFE HISTORY

Eggs—deposited in masses or strings in or over water

Larvae—bottom dwellers, graze on microorganisms, algae; some predatory, free-living, or build and attach silken nets in streams.

Pupae—spin cocoons, even within a case; cuts way out with jaws when mature; pops to surface

Adults—fly in evening; may lap liquids

ECONOMIC IMPACT

Important source of food for fish and other aquatic animals

COMMON FORMS OF LARVAL CASES

spiral

twig-case

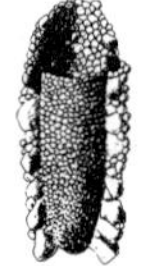

attached to substrate

Plate 11.18

ORDER: SIPHONAPTERA

(derivation: siphon=tube,
aptera=wingless)
Common Name: Fleas

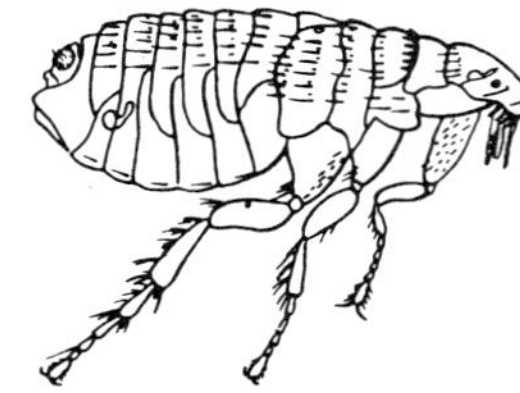

MOUTH PARTS

Piercing-sucking
(see above illustration)

WINGS

None

METAMORPHOSIS

Complete

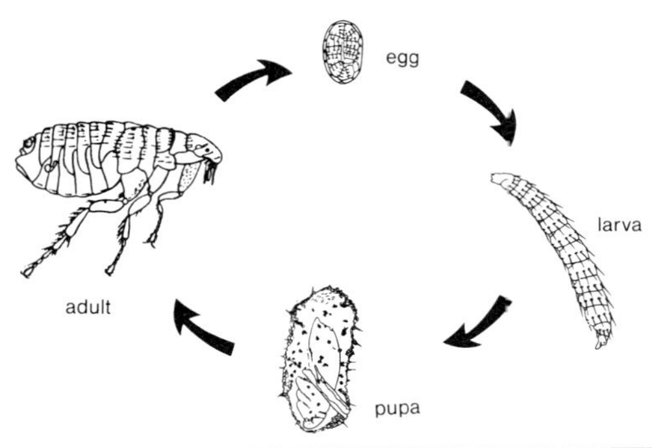

OTHER CHARACTERISTICS

1. Adult of both sexes are bloodsucking parasites of mammals and some birds.
2. Many are intermittent parasites, only visiting the host for a blood meal.
3. Some are important transmitters of specific vertebrate diseases.
4. Larvae are non-parasitic; usually live in nest of host.
5. Adults are good jumpers and runners.
6. Adults are laterally compressed, hard and leathery. Streamlined (antennae fit in grooves; posteriorly projecting spines).

LIFE HISTORY

Eggs—Produced by female in host's nest or on its body; usually end up in host's nest.

Larvae—Wormlike and legless non-parasitic scavengers; feed on organic debris in host's nest, including nutrient-rich feces of adult flea.

Pupae—Larva spins silken cocoon, pupates inside.

Adult—Some stay attached to host, e.g., sticktight flea; many find host only for blood meal, are attracted to heat and a moving body.

ECONOMIC IMPACT

Fleabites cause discomfort. Disease transmission e.g., plague and endemic typhus in humans, tapeworms in rats and dogs. At least the initial phases of the plague which decimated medieval Europe were spread from rats to humans by fleas.

Table 11.1 Additional Orders of Insects

Order	Derivation of Order name	Common Name (if any) + Appearance	Metamorphosis	Mouthparts	Wing characteristics	Additional aspects
Protura[a]	pro = first ura = tail		None	Sucking	None	Eyes lacking; no antennae (forelegs are tactile organs); anamorphosis
Diplura[a]	diplo = two ura = tail		None	Chewing	None	Conspicuous cerci, some pincer-like, many segmented in others; eyes absent
Dermaptera	derma = skin ptera = wings	Earwig	Gradual	Chewing	Forewing short, veinless; hind wing membranous, folded under forewing at rest	Forceps-like cerci; some parental care of offspring
Grylloblattodea	gryllus = cricket blatta = roach		Gradual	Chewing	None	Cold-adapted; live at high elevations and along snow fields
Embioptera	embio = lively ptera = wings	Webspinners	Gradual	Chewing	When present, two pair, long, membranous, with few weak veins	Gregarious; live in chambers of silk spun from front tarsi
Zoraptera	zoros = pure aptera = wingless		Gradual	Chewing	When present, two pair, long, membranous, deciduous (break off)	Under bark and in decaying wood; gregarious

(*continued*)

Table 11.1, continued. Additional Orders of Insects

Phasmida	phasma = apparition	Walking sticks 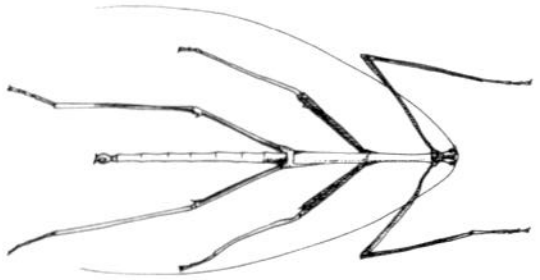	Gradual	Chewing	Forewing narrow; thick hindwing, fan-like	Solitary; feign death when disturbed
Psocoptera	psoco = rub small ptera = wings	Barklice, booklice 	Gradual	Chewing	When present, two pair, membranous, held roof-like at rest; few veins	Some become pests in humid storage of cereal grains
Thysanoptera	thysano = fringe ptera = wings	Thrips 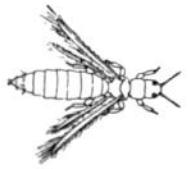	Advanced Gradual (two quiescent pre-adult stages)	Rasping-sucking	When present, two pair, thin, fringed with long hairs	Cigar-shaped, minute insects; without tarsal claws; exhibit haplodiploidy
Megaloptera	megalo = giant ptera = wings	Dobsonflies, alderflies 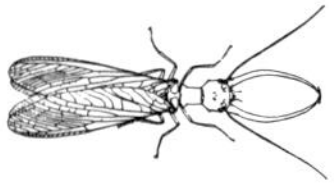	Complete	Chewing	Two pair, membranous, net-veined	Larvae are aquatic and predacious, with abdominal gills
Raphidiodea	raphis = needle	Snakeflies 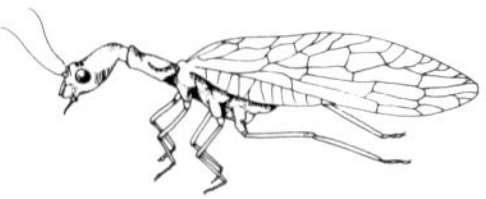	Complete	Chewing	Two pair, membranous, net veined	First thoracic segment long and neck-like; terrestrial predatory larvae
Mecoptera	meco = long ptera = wings	Scorpionflies 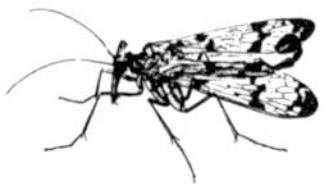	Complete	Chewing, at tip of a ventral extension of head	When present, two pair, long, membranous, with numerous veins, frequently spotted	Adult males have genitalia appearing like a scorpion's stinger

[a]Some scientists exclude these animals from the Class Insecta.

SUMMARY

Because insects are classified according to phylogeny, groups show inherited similarities. Progressively greater similarities are evident in each step in the nested hierarchy from Class to Order, Family, Genus, and Species. The scientific name of an insect consists of genus name, species name, and author. Scientific names are used to refer to species throughout the world because common names vary with location, culture, and language. The species is the only objective category because it is an interbreeding unit. Categories above the species level are subjective and grouped differently according to the taxonomist's best judgment.

SUGGESTED READINGS

Borror, D.J., DeLong, D. M. and Triplehorn, C. A. 1981. "An Introduction to the Study of Insects." W.B. Saunders, Philadelphia, Pennsylvania.

Mayr, E. 1969. "Principles of Systematic Zoology." McGraw-Hill, New York.

Merritt, R. W., and Cummins, K. W. (eds.) 1978. "An Introduction to the Study of Aquatic Insects in North America." Kendall/Hunt, Dubuque, Iowa.

Ross, H. H. 1974. "Biological Systematics." Addison-Wesley Publ. Co., Reading, Massachusetts.

Waterhouse, D. F. *et al.* 1970. "Insects of Australia." Melbourne University Press, Canberra, Australia.

12

Collecting, Mounting, and Identifying Insects

Many entomologists agree that one of the best ways to learn about insects is to make an insect collection. In fact, most entomology courses at the college level require an insect collection as a part of the course. Fortunately, field trips for the purpose of building an insect collection are so much fun that even the most apathetic student will begin to show some enthusiasm when a praying mantis, or other large insect is found.

The collection is a multipurpose learning tool. It teaches us where different insects can be found as well as something about their life histories. We can begin to become aware of insect diversity and their many adaptations as we collect and then examine our specimens. The purpose of this chapter is to aid you in collecting a wide variety of insects and then to introduce you to the proper techniques for mounting or otherwise preserving your specimens. As a final step, an identification guide to common insect groups will aid you in determining which major groups are represented in your collection and where your future efforts should be directed in order to fill in voids and thus improve your collection.

COLLECTING

Since insects occur in many kinds of habitats, it may take different kinds of collecting equipment to catch them. The image of an insect collector most common to people is one hotly pursuing a butterfly across a meadow with a gauzy insect net. Indeed, this is a productive collecting technique, but only one of the many types that have been developed. The student that relies on net collections alone dramatically limits the variety of insects that are caught. Many other kinds can be collected if beating, baiting, trapping, or emergence techniques are also used.

The following is a list of basic equipment that will help you make an excellent collection.

1. Nets. Three basic kinds may be used. First is the gauzy butterfly net. Second is a sweep net; it is sturdier than the butterfly net. Third is an aquatic net, often of very heavy construction. A wire hand-held strainer is often substituted for use in the water.
2. Killing jars. Pour an inch or so of wet plaster of paris into the bottom. After it is dry, soak with ethyl acetate. The fumes given off by the ethyl acetate will kill insects placed in the jar.
3. Vials partly filled with ethyl alcohol (80%).
4. A white enamel or plastic pan, over 1 inch deep.
5. Forceps.
6. Camelhair brush (small).
7. Some white cloth or plastic, about the size of a bedsheet.
8. A sheet metal funnel, 6–8 inches in diameter, with a frame to hold it upright and 4–5 inches off of a table's surface.
9. Pinning block.
10. Spreading board.

Insects occur in almost every terrestrial or freshwater habitat and the air surrounding these areas. Although collecting in most habitats will yield some insects, a few types of habitats are almost always more rewarding than others. For the new collector these are the rich and exciting areas in which to collect. Lush low vegetation like meadows, green roadsides, and alfalfa fields, especially if some flowers are in bloom, yield hundreds of specimens in a few minutes. Ponds, either the water's surface, the water's edge, or beneath the surface, are rich in a large variety of insects. Look under the bark of downed logs or roll the log over and look in the depression for termites, ants, beetles, and springtails. Another favorite collecting site is at bright lights on warm summer evenings, especially if the light is at considerable distance from other competing lights.

Insects are most abundant and easy to collect at the peak of the growing season or shortly after the rainy season in drier areas. Early in the growing season many insects will be in immature stages. In areas with moderate temperatures, late morning until noon is ideal, as is late afternoon. In deserts, wait until sunset to start collecting and continue up to midnight. Warm evenings almost anywhere are rewarding.

Techniques for collecting large numbers of insects often involve a rather blind gathering in of specimens.

1. *Extracting.* Place leaf litter, sod, fungi, duff, or nest material into a funnel and put an incandescent 100 watt bulb about 1 inch above its surface. The heat and light from the bulb drives the insects down through the material and into an alcohol-filled bottle under the funnel (see Fig. 12.1).

2. *Baiting.* Carrion, newly felled logs, animal dung, and rotting fruit all attract insects. Since many of these insects are not readily encountered except at their unusual breeding sites you may have to either create such settings or check naturally occurring sites. Mixtures of molasses and beer smeared on the bark of trees sometimes attracts many moths, wasps, beetles, and flies.

3. *Beating.* A large percentage of the insects at rest on a plant will drop to the ground if they are violently disturbed. This characteristic is exploited by spreading a white sheet (for visibility) under a bush and striking the bush a few times with a heavy stick. Many of the insects that fall will remain motionless for a few moments. If you wait a minute you will spot them as they crawl away.

4. *Netting.* Instead of relying on your vision to pick out a likely specimen and then catching it in your net (as in Fig. 12.2), resort to a net technique called sweeping. Often done with a heavy net, the collector sweeps through herbaceous vegetation, roadside grasses or meadows, taking in many of the unobserved insects resting on the leaves, stems or flowers through which the net passes (Fig. 12.3). If some care is taken to avoid a netbag full of plant parts, the whole contents can be placed in the killing jar and then stored in another container until a later period for examination. The same basic technique works in the water, except that the net's contents should then be placed into a white pan that has about an inch of water in it. A little patience is usually rewarded as water scorpions, mosquito larvae, damselfly nymphs, among others, work their way out of the vegetation and debris in the water. They are plainly visible against the white background and can be removed easily. Moving water yields

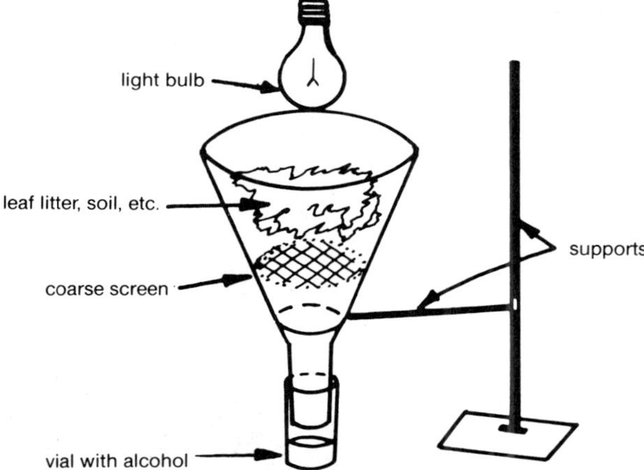

Fig. 12.1 The Berlese funnel for extracting insects from leaf litter, sod, and similar materials.

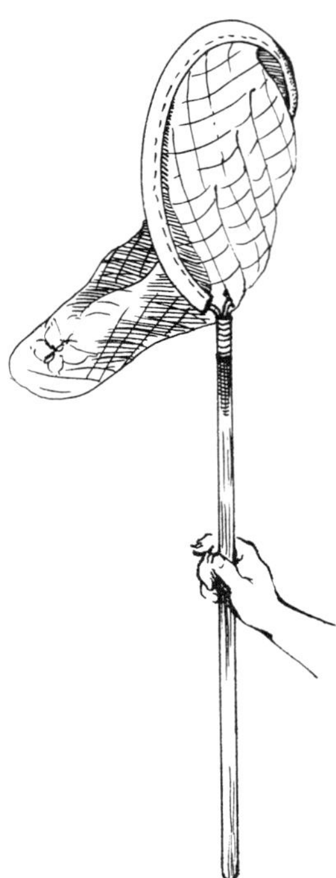

Fig. 12.2 After a specimen has been seen and caught in a net, it can be prevented from escaping by turning the net over on itself.

Fig. 12.3 Passing the heavy sweep net through vegetation catches many specimens that otherwise escape notice.

***Fig. 12.4** Aquatic insects drift with the current from the disturbed area into the waiting net.*

large numbers of specimens. While disturbing the bottom of the stream, hold the net downstream and insects under the rocks or on plants will be swept into the net (Fig. 12.4). Even greater quantities can be obtained if a 4-foot length of window screening, attached to two dowels, is held perpendicular to the current as bottom materials upstream are disturbed.

Several orders of insects could be overlooked by collectors unless they examine specific habitats. For instance, lice and fleas are numerous on wild mammals and birds, but exist practically nowhere else. Therefore, you must determine the habitats characteristically occupied by the members of the orders your collection lacks and then go to that location for a specific group. Another secret to a good collection is to keep at it. Instead of collecting insects only at predetermined times, as on collecting trips, keep your eyes open all the time. Carry a small vial in your pocket so that you will always have someplace to put your finds. It is amazing how often the unusual insect is encountered when you least expect it. Another trick you might try to fill in the one order or couple of families that are missing is to examine window sills along the inside base of the screen, light fixtures, and even spider webs.

PRESERVING

Like any other living organism, a newly killed insect begins to deteriorate. The primary changes involve loss of body fluids through dehydration. If the insect is relatively large, the soft body contents decay and the specimen will start to smell. Soft-bodied insects like caterpillars, grubs, or maggots shrivel up as they dehydrate. Only those insects with hard exoskeletons appear superficially to be unchanged after death. These, however, become very brittle and are easily destroyed. Because of the natural deterioration that occurs, insects must be preserved in special ways so that they can be identified and also to retain reasonably accurate resemblance to the insect as it appeared when alive.

Three basic procedures are used in preserving insects, depending on the size or rigidity of the individual specimen.

1. Hard-bodied forms that are not minute are usually mounted on insect pins. A

specimen may then be handled by the pin without destroying the insect so that even years after the soft body contents have dried up and the insect has become exceedingly fragile, the structural characteristics used to identify the specimen are still visible. The pin is inserted in the thorax, just to the right of the midline so that at least one-half of the specimen is completely open to observation (Fig. 12.5). Placement of the pin, standardized for some insects, should ensure that the insect is perpendicular to it and at a standard height (Fig. 12.6b). Standard heights are achieved by using a pinning block (Fig. 12.6a); after the specimen is properly positioned, its height can be adjusted by pushing the pin through the hole in the tallest level of the pinning block. The other levels of the pinning block are for positioning labels.

Butterflies, moths, and skippers are also preserved on pins, but the wings are spread when the insect is newly killed so that they will stay in position after the specimen dries. Figure 12.5 shows the proper wing positioning and Fig. 12.7 shows the insect on the spreading board. After the pinned specimen has been adjusted in the center groove of the spreading board so that the top surface is flush with the wing bases, each wing is positioned on the board's surface by inserting a pin just behind a main wing vein. This allows movement of the wing without tearing it. Then the wing is held in position with strips of paper so that it will dry flat.

Beetles, wasps, flies, and other hard-bodied insects that are too small to be pinned are usually "pointed." Triangles cut from heavy paper are positioned on a pin on the pinning block. The tip of the triangle is bent with forceps, glue is applied to the bent tip, and the insect is placed on the glue. Again, position the insect so that it is glued only on the right side of its thorax (Fig. 12.8), leaving one entire side exposed for examination under a microscope. Clear fingernail polish or white glue is recommended.

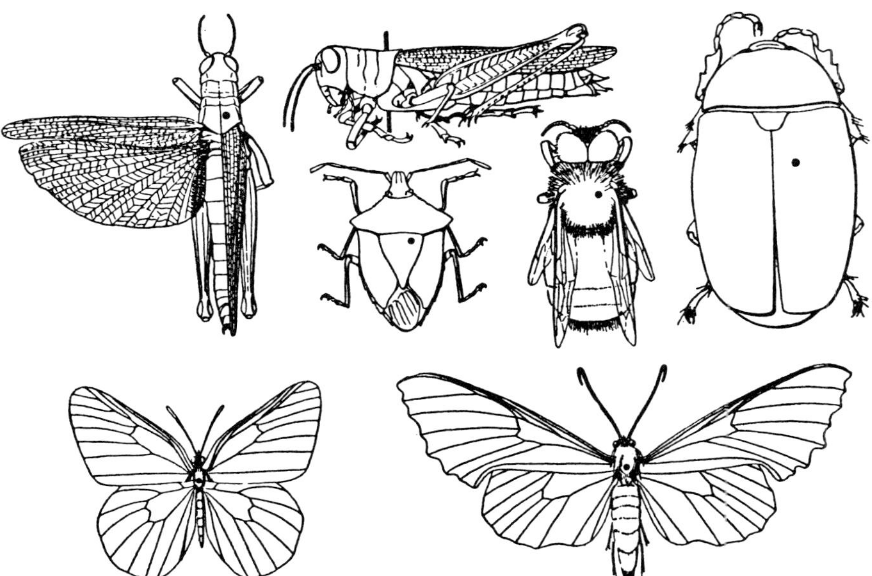

Fig. 12.5 The black dot indicates where the pin should be located for some common groups.

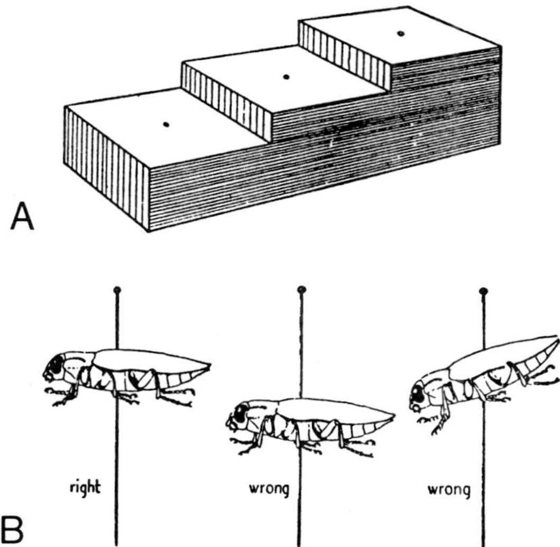

Fig. 12.6 ***(A) Use of a pinning block will ensure that specimens and labels are at uniform heights, thus increasing the orderly appearance of the collection. (B) In positioning the pin, be sure that the specimen will be in a horizontal position when it is subsequently placed in the collection.***

2. Preservation in alcohol is used with soft-bodied insects because they shrivel up if they are mounted on pins. A good mixture to use is 80% ethyl alcohol with a little glycerol in it. The glycerol will keep the specimens from completely drying out if the alcohol evaporates. One of the main difficulties with alcoholic preservation is that most containers, regardless of the type of closure, allow the alcohol to evaporate. This

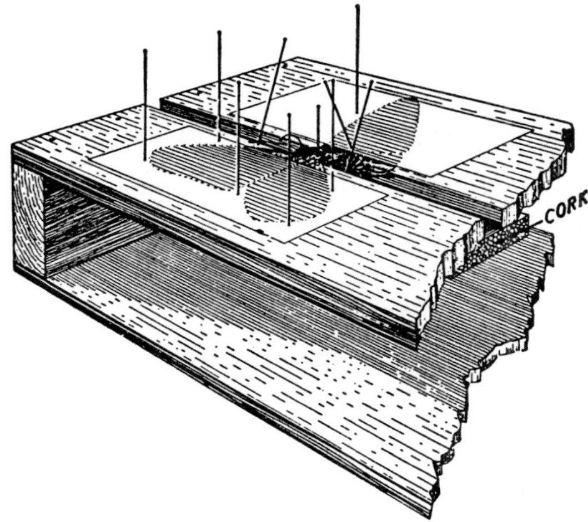

Fig. 12.7 ***The spreading board allows the collector to dry Lepidoptera with their wings displayed.***

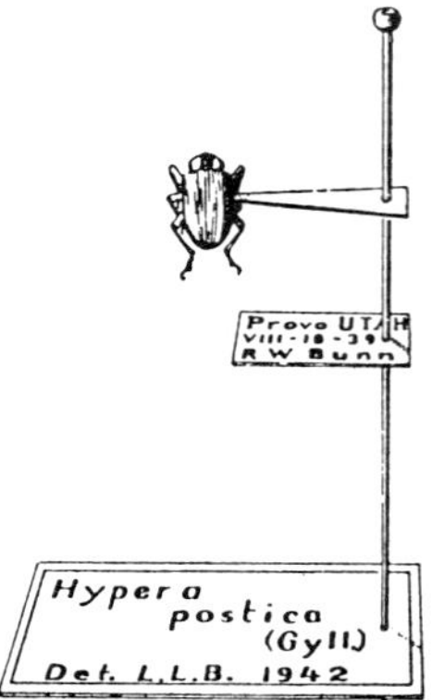

Fig. 12.8 ***Pointing a small insect on a triangle of paper attached to a pin will allow examination of most of its important structures. The result is a properly mounted and labeled specimen ready for the next step—identification***

necessitates periodic checking and refilling to keep specimens in good shape. If the specimen completely dries up, it can be soaked overnight in warm soapy water, then placed back in alcohol. Rubbing alcohol can be used if ethyl alcohol is not available. Another drawback of alcoholic preservation is that the specimen has to be removed from the vial and placed in a dish before it can be closely examined under a microscope.

3. A third method of preserving insects is reserved for fairly small and fragile specimens. These can be permanently mounted on microscope slides which then allows them to be viewed with a compound microscope. Preservation of specimens by slide mounts is not recommended for the beginning collector.

Softening Dried Insects

The student's zeal for collecting is seldom matched by a similar zeal for mounting the insects after they have been collected. Inevitably, some of the specimens will dry out before they can be mounted. However, there is a way that these specimens can be treated so that they are not broken as they are being mounted. In a tight container, like a coffee can, place several layers of paper toweling. Moisten the toweling with water and add a little phenol to inhibit mold formation. Then cut a disk of plastic window screen to fit over the toweling. Place the insects on top of the screen. In a day or so the specimens should be "relaxed" enough for mounting.

Labeling

The preparation of the insect specimen is not complete until it has a label identifying where it was collected and who collected it (Fig. 12.8). Early in the formation of a collection the specimens will probably be limited to one or a few locations, tempting the collector to keep the collection localities in his or her head. However, as the collection grows, it is not possible to remember where each specimen came from. Therefore, proper technique is to have at least one label with each specimen listing where it was collected, the collector's name, and date of collection. Often the location is given as the city or county and state. The date should be written using Roman numerals for the month and arabic for the day and year. Thus, October 7, 1986 is written X/7/86. This system is used because 10/7/86 could be interpreted as either October 7, 1986 or as July 10, 1986. The labels for specimens in alcohol-filled vials should be written in India ink and placed in the vial, not taped to the outside or pinned to the cork. Labels giving additional information like habitat, time of day, or what the specimen was doing when it was collected, are useful. Eventually, an identification label will be attached after the identity of the specimen has been determined.

STORAGE OF THE INSECT COLLECTION

Pinned specimens must be stored in boxes with very tight-fitting tops. This is necessary because a number of insects normally feed on dead insect remains and these pests (carpet beetles, booklice) can entirely demolish a good collection in just a few months. To counter this problem, the collection should be continually fumigated with paradichlorobenzene (moth crystals), which will kill the museum pests. However, students frequently forget that the fumes from moth crystals will only be effective in a tight container. In addition, care must be taken to enclose the moth crystals in some manner other than just placing them on the bottom of the insect box where they can destroy the collection as the collection box is moved.

Insect collection boxes have relatively soft pinning surfaces on the bottom so that the pins can be pushed in and hold the specimen securely. Such a surface can be cut from a sheet of cork, cardboard, or some kinds of fiber board. If plastic is to be used, test it before using because moth crystals can dissolve some types of plastic.

Students responsible for making an insect collection as a part of their course requirements normally will be supplied with the proper equipment, chemicals, and supplies by their instructor. However, such materials are also available from a number of biological supply companies, which service educational institutions. Names and addresses are readily available from the biological science departments of high schools, colleges, and universities.

IDENTIFICATION

Once the insect has been caught, killed, preserved, and labeled, the next step is to identify the specimen. When we look at objects that we are well acquainted with, we take in a great number of characteristics at a single glance. Thus, each of us can recognize different brands of cars, kinds of flowers, or vegetables, or whatever else has become familiar to us. In fact, the more experience we have the better we learn to discriminate between similar yet different objects. The same holds true in identifying insects. Eventually, enough experience can be gained to recognize all of the orders of insects as well as many of the common insect families, genera, or species. However, a technique is available for identifying insects even when the groups are not familiar to the student. This technique of identifying specimens uses an artificial device called a taxonomic key.

Taxonomic keys are constructed so that the user is presented with pairs (called couplets) of statements. The user compares each statement with the specimen and chooses which one fits the specimen best. Upon making the choice, the user reads across to the right-hand margin of the key where a number indicates which pair of choices (couplet) must be made next, and so on, until the identification has been completed. The end result of the keying process indicates the group to which the specimen belongs based on two important points: (1) the series of choices made were correct, and (2) the group to which the specimen belongs has indeed been included in the key. After identifying a specimen compare it with the illustrations and descriptions included for it in Chapter 11. If the specimen and descriptions do not fit, run it through the key again.

The following key will allow the user to identify most specimens of the insect orders students commonly collect. In addition, the insects need not be adults to be keyed. While pupae and eggs have not been included, the active immature stages have been, so that most of the insects can be identified.

TAXONOMIC KEY FOR IDENTIFICATION OF INSECTS TO ORDER

1a. Quiescent, immobile form; often encased within a cocoon or "shell" 2

1b. Active form; appendages distinct; or wormlike, but with a segmented body; wings or wing pads may be present 3

2a. Wings and wing pads absent; segmented body without legs or antennae; under a wax scale on plant (scale insects) ..... Hemiptera

2b. Appendages present (may include wing pads), glued to body, or free, but appear to be encased; or body within a case or cocoon ..... Insect pupae (rear if possible)

3a. Wormlike, with segmented body; wing pads not visible; compound eyes lacking; legs and antennae may be lacking (insect larvae) 35

3b. Not wormlike; body with distinct head, thorax and abdomen; compound eyes usually present; wings or wing pads usually present; antennae and 3 pairs of thoracic legs usually present 4

4a. Collected only from surface of a live warm-blooded vertebrate, or from its nest 5

4b. Collected from other situations 7

5a. Body laterally compressed (Fig. 12.9); piercing-sucking mouthparts . . . . Siphonaptera

Fig. 12.9

5b. Body dorsoventrally compressed (Fig. 12.10) 6

Fig. 12.10

6a. Piercing–sucking mouthparts; head narrower than width of thorax (only on mammals) . . . . Anoplura

6b. Chewing mouthparts; head wider than thorax (mammals or birds) . . . . Mallophaga

7a. Wings absent or represented by small wingpads on top or sides of 2nd and 3rd thoracic segments, the pads not reaching to the tip of the abdomen (insect nymphs and wingless adults) 8

7b. Functional wings present, usually reaching tip of the abdomen, or the top of insect's abdomen with a central line dividing the two sides; rarely the flight wings are folded under short wing covers 20

8a. Mouthparts chewing 10

8b. Mouthparts not of the chewing type 9

9a. Typical piercing–sucking mouthparts; antennae present, but palps of mouthparts absent; tip of leg bearing claws . . . . Hemiptera

9b. Mouthparts form a blunt cone; palps of mouthparts present; legs without claws on tarsi (minute insects) . . . . Thysanoptera

10a. Aquatic 11

10b. Terrestrial 13

11a. Lower lip forms an extensible elongate hinged "mask" fitted for grasping prey; with 2–3 leaflike terminal gills or gills lacking . . . . Odonata

11b. Mouthparts without such a "mask" 12

12a. Gills leaflike, attached at sides of several abdominal segments; 3 (sometimes 2) terminal filaments on abdomen . . . . Ephemeroptera

12b. Gills filamentous, frequently arranged in tufts at the bases of thoracic legs; 2 terminal filaments on abdomen . . . . Plecoptera

13a. Hind legs fitted for jumping, with expanded femur; posterior of abdomen with cerci more or less obvious, but not pincerlike . . . . Orthoptera

13b. Hind legs not expanded for jumping (if so, then wing pads are not present) 14

14a. Abdominal appendages lacking 15

14b. Abdominal appendages present, may be single to many segmented feelerlike cerci (Fig. 12.11) and/or terminal filaments or ventral appendages present 16

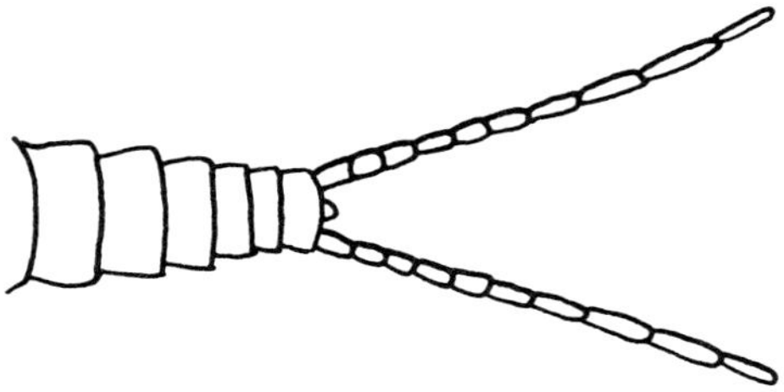

Fig. 12.11

15a. A highly constricted section between thorax and abdomen; a dorsally expanded "node" on one or two segments in the constricted area; antennae elbowed (ants) . . . . Hymenoptera

15b. No "wasp waist" constriction between abdomen and thorax, but three distinct body regions . . . . Psocoptera

16a. Abdomen with 3 long terminal appendages; at least abdominal segments 7–9 with pairs of peglike appendages . . . . Thysanura

16b. Less than 3 terminal abdominal appendages, if none at all, then thorax and abdomen are not distinctly divided into regions 17

17a. Fewer than 7 abdominal segments; elongate fleshy ventral lobe on first abdominal segment; forked terminal abdominal appendage usually present . . . . Collembola

17b. More than 7 abdominal segments; abdominal appendages are a pair of single to many segmented lateral or dorsal cerci 18

18a. Cerci one segmented and pincerlike (Fig. 12.12) . . . . Dermaptera

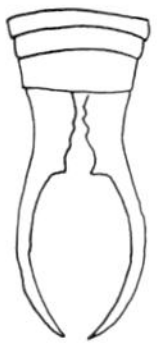

Fig. 12.12

18b. Cerci not pincerlike 19

19a. Cerci of 1 to 3 segments; colonial forms occurring in large numbers; soft bodied; not especially flattened; antennae like a string of beads . . . . Isoptera

19b. Cerci longer; not colonial, but may be gregarious; elongate first thoracic segment with grasping forelegs, or flat insects with very long antennae . . . . Dictyoptera

20a. A single pair of thoracic wings, the second pair reduced to club-like halteres (Fig. 12.13) . . . . Diptera

20b. Two pairs of thoracic wings, though they may be joined together 21

21a. Mouthparts of the chewing type 27

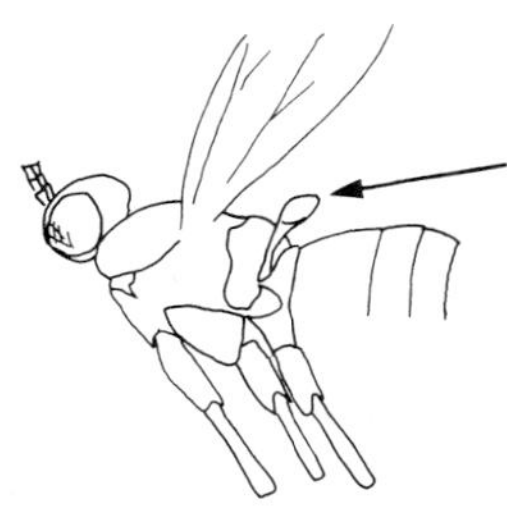

Fig. 12.13

21b. Mouthparts other than chewing; they may be vestigial 22
22a. Membranous wings covered with flat, shingle-like scales, often in bright colors . . . . Lepidoptera
22b. Wings not covered with scales 23
23a. Wings hairy, either with long fringe on edges or surface densely packed with short hairs 24
23b. Wings without hairs or with only a few isolated hairs 25
24a. Wings membranous; more than two longitudinal veins in each forewing, wing hairs short, but densely packed . . . . Trichoptera
24b. Wings very narrow, with only 1 or 2 longitudinal veins; fringe formed by hairs on leading and trailing wing margins as wide as wing itself (minute insects) . . . . Thysanoptera
25a. Mouthparts piercing–sucking, a segmented beak without palps . . Hemiptera
25b. Mouthparts with palps, or without a beak 26
26a. Mouthparts vestigial, at most only the palps are distinguishable . . . . Ephemeroptera
26b. Mouthparts chewing, but frequently with a long lapping labium and/or maxillae; palps present . . . . Hymenoptera
27a. Forewing thickened, veins visible or not, texture parchmentlike or horny . 28
27b. Forewing membranous 31
28a. Forewings a pair of veinless, horny wing covers, flight wings usually folded up under the covers 29
28b. Veins of forewing stand out from parchmentlike texture 30
29a. Forewings short, exposing at least 3 abdominal segments dorsally; pincerlike cerci present . . . . Dermaptera
29b. Forewings usually covering all of abdomen dorsally, or exposing only 1 or 2 segments, if more, then without pincer-like cerci . . . . Coleoptera
30a. Hind legs jumping; femur expanded . . . . Orthoptera
30b. Hind legs not fitted for jumping; hind femur about equal in proportion to the same segments on other legs . . . . Dictyoptera
31a. Wings net-veined (with numerous longitudinal veins and many cross veins) or fore and hind wing same size and shape 32
31b. Wings with few cross veins and usually few longitudinal veins; hind wings much smaller than forewings . . . . Hymenoptera
32a. Wing veins mostly membranous and faint, a transverse line of weakness near the base of wing; fore and hind wings very similar in shape and size; tarsi 4-segmented . . . . Isoptera

32b. Wing veins strongly developed 33
33a. Tarsi 2 or 3-segmented . . . . Plecoptera
33b. Tarsi 5-segmented 34
34a. Hind wings broader at base than forewings, and with an enlarged basal area that is folded at rest . . . . Megaloptera
34b. Fore and hind wings similar in size and shape . . . . Neuroptera
35a. Abdomen with one or more pairs of fleshy legs or hooklike appendages . 36
35b. No abdominal legs 39
36a. Aquatic 37
36b. Terrestrial 38
37a. A single pair of terminal abdominal legs or hooklike appendages; frequently the larva is within a silken case (caddis) constructed of plant debris or sand particles . . . . Trichoptera
37b. Without hooklike terminal appendages . . . . Megaloptera
38a. Five or fewer pairs of fleshy abdominal prolegs, tipped with numerous crochets (hardened hooks), 1–8 pairs of simple eyes on head . . Lepidoptera
38b. Six or more pairs of abdominal prolegs, without crochets, only one pair of simple eyes on head . . . . Hymenoptera
39a. Scimitar-shaped opposable mouthparts directed anteriorly (predators of soft-bodied insects); terrestrial forms . . . . Neuroptera
39b. Mouthparts either chewing, brushlike, very weakly developed, or a pair of ventrally directed mouth hooks; if scimitar-shaped and opposable, the larva is aquatic 40
40a. Mouthparts are opposable jaws weakly developed for chewing; larvae within a colony or cell constructed to contain it and a food supply; legless . . . . Hymenoptera
40b. Mouthparts chewing, strongly developed, at least mandibles heavily sclerotized at tips, or mouthparts brushlike, or consisting of a pair of ventrally directed mouth hooks 41
41a. Without segmented thoracic legs; distinct head capsule, brushlike or chewing mouthparts and aquatic, or without distinct head capsule, body tapered anteriorly and mouthparts a pair of ventrally directed hooks . . . . Diptera
41b. Thoracic legs present; mouthparts chewing . . . . Coleoptera

SUMMARY

Collecting a wide variety of insect types requires various methods other than use of an insect net. Insects may be extracted from litter by a Berlese funnel or collected at baits set out previously. Aquatics are best collected by nets and may be separated from debris in a white pan. Some orders, e.g., lice and fleas, are only encountered in their specialized habitats.

Proper preservation of insects allows later examination without injury to the specimen. Most hard-bodied insects are preserved on insect pins because they become very brittle as they dry out. Some, like Lepidoptera, should be spread while still fresh. Small insects are glued to paper triangles on pins. Dried insects may be mounted if they are "relaxed" earlier. Soft-bodied insects are preserved in liquid because they shrivel up if dried.

All specimens should carry a label indicating location and date of collection in addition to the name of the collector. Additional information, e.g., method of collection, host, or habitat, is optional.

Carpet beetles and other insects that feed on dry animal remains can ruin an unprotected insect collection. Therefore, collections are stored in tightly closed boxes with moth crystals.

Correct identification of insects employs the use of a taxonomic key. Keys offer a series of paired choices to be made by the identifier, based on characteristics that can be seen on the insect or determined by the information on the label. A key to the orders of insects commonly collected is presented. Both immature and adult forms are included.

SUGGESTED READINGS

Borror, D. J., DeLong, D. M., and Triplehorn, C. A. 1981. "An Introduction to the Study of Insects." W. B. Saunders, Philadelphia, Pennsylvania.

Borror, D. J., and White, R. E. 1970. "A Field Guide to the Insects of America North of Mexico." Houghton Mifflin, Boston, Massachusetts.

Martin, J. E. H. 1977. The Insects and Arachnids of Canada. Part I. Collecting, Preparing, and Preserving Insects, Mites, and Spiders. Agriculture Canada, Ottawa, Ontario, Canada.

Peterson, A. 1959. "Larvae of Insects." Vols. 1 and 2. J. W. Edwards, Ann Arbor, Michigan.

Stehr, F. W. 1987. "Immature Insects." Kendall/Hunt, Dubuque, Iowa.

Glossary

Accessory glands—glands that supply seminal fluid in males. In females, the secretions glue the egg to the substrate or provide a protective covering. They are part of the reproductive system in both male and female.

Accessory pulsatile organs—organs located at the base of appendages functioning as hearts to pump blood into the appendage.

Action threshold—the density at which control measures should be applied to prevent an increasing population from reaching the economic threshold.

Aedeagus—penis.

Afferent neurons—nerve cells that carry signals away from sense organs to the central nervous system.

Aggregation pheromone—a pheromone causing aggregation in individuals.

Aggressive mimicry—behavior where a predator mimics its prey, thereby enabling the predator to approach without alarming its prey.

Allometric growth—growth of different body parts at different rates.

Allomones—chemicals released by an individual of one species inducing a response by an individual of another species that favors the emitter of the chemical.

Altruistic behavior—self-destructive behavior performed to help others of the same kind.

Ametabolous development—development without metamorphosis.

Analog—an intentionally modified molecule retaining the desired activity of its predecessor molecule, but with other characteristics changed.

Angiosperms—flowering plants.

Anterior—toward the head.

Antibiosis—against life.

Antigens—foreign substances in the blood.

Apolysis—the process the old cuticle undergoes to separate from the epidermal cells.

Assimilate—incorporate materials into living tissue.

Association neurons—nerve cells that relay signals from afferent neurons to efferent neurons.

Automimicry—behavior by which one member of a species derives protection by looking like protected members of its own species.

Autotrophs—organisms that manufacture their own organic foods.

Bioassay—comparison of the effect of a test condition on an organism with a standard condition.
Biological control—regulation of pest numbers by natural enemies.
Biological rhythms—cyclic processes in living organisms.
Biotic communities—groups of plant species and their associated animal forms.
Bivouac—mass of ants surrounding the queen, as in army ants.
Callow—new, soft, unsclerotized.
Carnivorous plants—plants that utilize the protein from the bodies of animals.
Carrying capacity—the ability to support a particular weight of organisms per unit of environment.
Caste—a set of individuals in a given colony that are both morphologically distinct and have their own specialized behaviors.
Cell theory—theory that postulates that cells are the basic units of life and the smallest components of an organism that perform all of the functions of life.
Chelicerae—a pair of feeding appendages located just in front of the mouth. Each chelicera has a jointed side piece that can be moved. Present only in Subphylum Chelicerata.
Chemosterilants—chemicals that interfere with normal reproduction.
Chorion—the shell of an insect egg.
Chromosome mapping—determination of the sequence of genes on a chromosome.
Cilia—hairlike parts of a cell that can be moved by the cell.
Classification—the process of categorizing groups of related organisms.
Coevolution—a series of reciprocal changes by each species involved in an interaction.
Colony—a group of individuals, other than a single mated pair, that constructs nests or rears offspring in a cooperative manner.
Competitive exclusion—tendency for ecological separation of similar species.
Consumers—organisms that extract energy from foods. Those organisms are incapable of converting solar energy into chemical energy.
Convergent evolution—development of similar characteristics by two or more independent lineages caused by adaptation to similar ecological status.
Crossing over—exchange of chromosomal material between a pair of homologous chromosomes.
Cryptic coloration—concealment through a matching of the insect's color and texture with its surroundings.
Cultivars—synthetic varieties produced by plant breeders.
Density dependent—the condition in which the severity of an environmental factor's action is linked to the density of the animal population.
Density independent—the condition in which the severity of action by an environmental factor is not linked to the density of the animal population.
Detritivores—organisms that feed on nonliving or decaying organic matter.
Diapause—a period of arrested development.
Diffusion—the spread of molecules from regions of high concentration to those of lower concentration, tending to equilibrate over time.
Digestion—the breakdown of complex food substances into smaller components.

Digestive enzymes—proteins that break down food.
Diluent—inert carrier by which active ingredients are diluted.
Diploid—two sets of chromosomes.
Dispersal—spread.
Dorsal—the top or back side.
Drones—sexually mature male honey bees.
Ecdysis—the shedding of the old cuticle.
Economic threshold—the lowest population density that will cause economic damage.
Ecosystem—the totality of the living and nonliving components of a given area.
Ectoparasites—external parasites.
Efferent neurons—nerve cells that effect a change, as in the contraction of a muscle.
Endocrine glands—glands that secrete hormones into the blood.
Endopterygota—insects with internally developing wings.
Entomophilic—insect liking.
Epicuticle—a proteinacious coating with a layer of waxes on its surface.
Epizootics—epidemics among animals.
Equilibrium population levels—the long-term average density levels of a population without artificial manipulation.
Eradication—extinction of a species in a defined geographic area.
Ethologist—one who makes a scientific study of behavior.
Eusocial—truly social.
Evolution—the principle of progressive change.
Exocuticle—the zone of cuticle where protein plasticizing makes the cuticle practically invulnerable to attack by chemicals as well as making it very hard.
Exopterygota—insects with external wing development.
Facultative—optional.
Fallowing—land left barren or unused for a season.
Fertilization—fusion of haploid male and female nuclei.
Filter chamber—a loop in the intestinal tract that removes water from the ingested food, thereby concentrating the food.
Flagella—*see* Cilia.
Formulation—the processing of a pesticide into a final physical form that will improve its safety, storability, applicability, and effectiveness.
Galleries—tunnels.
Ganglia—nerve masses.
Genome—the complete set of hereditary characteristics of the species.
Germ plasm—the pool of genetic variation possessed by a species.
Haploid—one set of chromosomes.
Hemimetabolous development—simple metamorphosis.
Heterosis—hybrid vigor.
Heterotrophs—organisms that acquire organic chemicals by eating other organisms or their remains.
Histogenesis—building of new structures.
Histolysis—breakdown of tissues.

Holometabolous development—complete metamorphosis.
Hormones—chemical messengers secreted into the blood by the endocrine glands.
Host resistance—that quality in a host that enables it to avoid, tolerate, or recover from the attack of pests that would cause greater damage to other hosts of the same species under the same conditions.
Host specificity—the range of species suitable for development of the organism and acceptable to it.
Hypertrophied—enlarged.
Indirect flight muscles—the muscles that provide much of the power for the wing strokes. They are not attached directly to the wing base, but to various regions of the thorax.
Innate behavior—an inherited disposition for a particular behavior.
Insecticides—pesticides that are toxic to insects.
Insect repellents—chemicals that protect animals or plants from insects by making them unattractive, unpalatable, or offensive.
Integrated pest management—the use of multiple tactics in a compatible manner to maintain pest damage below the economic threshold.
Integument—the body wall of insects.
Inversions, chromosomal—aberrant events in which pieces of chromosomes have broken out, turned around, and become fused again.
Kairomones—a chemical (or mixture) released by an individual of one species inducing a response by an individual of another species that favors the recipient.
Kinesis—an undirected response to a stimulus.
Learned behavior—a behavioral response that is altered as a result of experience.
Limiting factor—a condition, either minimum or maximum, that approaches the extremes that a species can tolerate.
Linkage—a tendency to be inherited together.
Macromolecules—large and complex molecules.
Mesal—toward the midline.
Metagenesis—the alternation of generations.
Microfilaria—offspring produced by mature female filaria worms in the host's body.
Micropyles—minute holes in the chorion of an insect egg.
Microvillae—fingerlike projections lining active locations on the plasma membrane of a cell.
Mitochondria—sites in a cell at which energy is extracted from foodstuffs, thus becoming available for energy-demanding activities.
Mode of action—the way an insecticide causes mortality.
Monophyletic—evolved from a single ancestral line.
Mycoses—fungal infections.
Nasute soldier—a termite soldier with a snout used to eject a sticky or poisonous fluid at intruders.
Natural control—pest suppression without human intervention.
Nectar—aqueous, sugar-containing secretion of those plant glandular structures called nectaries, often associated with flowers.

Nits—eggs glued on hairs or feathers, as in lice.

Nomadic phase—the period when an army ant colony moves to a new site each day.

Nuclear envelope—a variably porous double-layered membrane that separates the nucleus from the other cellular components.

Nucleus—the control center of the cell, which contains the information-bearing chromosomes.

Obligatory—required.

Oncogenic—causing tumors.

Ovaries—the female reproductive organs, equivalent to the male's testes.

Oviparity—reproduction in which the female produces eggs that develop outside her body.

Ovoviviparity—reproduction in which the fertilized egg is retained within the female's reproductive system, where it completes its embryonic development.

Paranotal lobes—lobes at the sides of the thorax, thought to have evolved into wings.

Parasite—an organism that obtains its nutritional requirements from the body material of another organism (the host) and lives in or on the body of the host, which receives no benefits from the association, although it is not usually destroyed as a result of the association, e.g., parasites of vertebrates.

Parasitoid—a parasitic insect that kills its host.

Parthenogenesis—reproduction without mating.

Paurometabolous development—the insects with a nymph in the same habitat as the adult.

Pedogenesis—reproduction by structurally immature forms.

Pest—an organism that reduces the availability of some resource valued by humans, such as a commodity, or one's health, or aesthetic pleasure.

Pesticides—pest-killing chemicals.

Pest life system—those elements of the ecosystem that are directly involved in the subject species.

Phagocytosis—the process of flowing around an object and engulfing it.

Phagostimulant—ingestion-stimulating chemical.

Phasic response—a pattern of nerve impulses in which the impulse is generated only when the sensillum is distorted and again when it returns to its original state.

Phenotype—physical manifestation of the genetic type.

Pheromones—a chemical (or mixture) that is released by one individual and elicits a response in another individual of the same species.

Photoperiod—the environmental rhythm of daylight and darkness.

Photosynthesis—the ability to transform sunlight into high-energy chemicals.

Phylogeny—relationship by descent from a common ancestor.

Physogastric—unusual degree of abdominal swelling, caused by hypertrophy of fat bodies and ovaries.

Pinocytosis—external materials taken into the cell's interior by a pinching off of a membrane-lined inpocketing, with a consequent engulfing of the adjacent material.

Plasma membrane—the outer boundary surrounding the cell.
Pollination—the placement of appropriate pollen onto the stigma of a receptive flower.
Pollinia—masses of sticky pollen.
Polymorphism—the existence of two or more forms within the same sex.
Pore canals—minute openings that are left in the cuticle as it is deposited.
Presocial behavior—social behavior beyond sexual contact, but not expressing all three characteristics that define true sociality.
Producers—the green plants with their ability to transform sunlight into high-energy chemicals.
Quarantines—restrictions on movement of infested or potentially infested plant or animal materials for the purpose of controlling the spread of a pest.
Queen—a sexually mature, egg-producing female; refers to social insects.
Reproduction—the process by which organisms give rise to offspring. It consists fundamentally of the segregation of a portion of the parent's body, subsequent growth of the segregated portion, and differentiation into a new individual.
Residual activity—continued killing action; refers to chemicals that are used to control insects.
Resonating muscles—muscles that contract more than once for every nerve impulse.
Saprophage—an organism that feeds on dead organisms.
Sclerites—hardened cuticular plates.
Sclerotization—a tanning process of the cuticle forming a plasticized (sclerotized) toughened area.
Semiochemicals—chemicals that are used in communication.
Sensilla—sense organs.
Sericulture—the mass production of silk for human use.
Sexual dimorphism—obvious differences between males and females of the same species.
Sociality—tendency to associate with other individuals of the same species.
Sociobiologists—scientists that study all aspects of social organizations and communications.
Speciation—the establishment of intrinsic barriers to gene flow between populations by the development of reproductive isolation mechanisms.
Spermatheca—sperm storage vessel in the female's reproductive system.
Spinneret—hollow spine with the opening of the duct from the silk glands in a caterpillar's head.
Spiracles—external openings (often of a complex form, with valvular arrangements) that open into large longitudinal trunks with multiple branches, allowing atmospheric air into the tracheae.
Sporulation—formation of a spore stage.
Statary phase—the period when an army ant colony stays at one site for an extended period.
Swarming—the process whereby a honey bee colony divides.
Symbiotic—interaction between organisms of different species.

Synapse—a small but distinct gap between the terminal branches of a nerve cell's axon and the dendrites of the next nerve cell.

Synergist—a chemical that increases the activity of other chemicals.

Systemic activity—the absorption and translocation of a pesticide throughout an organism.

Tagmosis—the regionalization of body segments into recognizable units.

Taxis—a directed response in which the stimulated insect moves toward or away from the source of the stimulus.

Taxonomy—the theoretical study of classifications and the basis or bases for classification.

Teratogenic—a condition causing malformation in embryos.

Termitaria—termite nests.

Testicular follicles—male's sex cell production lines enveloped in a pair of testes.

Tillage—disturbance of the soil as a means of destroying weeds by mechanical injury and other pests by exposure and as a preparation of the soil for planting or for moisture conservation.

Tolerance—the ability to suffer the attack of pests and still survive; a type of resistance.

Tonic response—a pattern of nerve impulses in which the impulse is generated throughout the time that the sense organ is stimulated.

Tracheae—the tubes of the respiratory system that branch throughout the body, bringing air from the environment to its sites of use.

Transovarial—from parent to forming egg.

Triungulin—an active "crawler" larval form.

Trophallaxis—the exchange of gut liquids between colony members.

Trophic levels—the various steps through which the sun's energy passes through the organisms in the environment.

Univoltine—one generation per year.

Vector—transmitter of a disease-causing organism.

Venter—underside.

Virions—virus particles.

Visceral muscles—the muscles that occur in sheets or layers surrounding various organs.

Viviparity—a condition in which the embryo hatches from the egg in an early embryonic state. It completes much or all of its growth and pre-adult development within the female's reproductive system, which also supplies it with nutrients from glands located within the system.

Index